HAZARDOUS MATERIALS

MANAGING THE INCIDENT

2nd edition

GREGORY G. NOLL
MICHAEL S. HILDEBRAND
JAMES G. YVORRA

FIRE PROTECTION PUBLICATIONS
OKLAHOMA STATE UNIVERSITY
STILLWATER, OKLAHOMA 74078

Produced by:
Lightworks Photography & Design
692 Genessee Street
Annapolis, Maryland 21401

Executive Producer:
Production Coordinator/Designer: **George Dodson and Kathleen Lawyer**
Editor: **Patricia Daley**
Photographer: **George Dodson, Lightworks Photography**

Library of Congress Number: 94-61530

ISBN 0-87939-111-1

Printed in the USA

First Printing, December 1994
Second Printing, September 1995
Third Printing, November 1996
Fourth Printing, June 1998
Fifth Printing, April 1999
Sixth Printing, January 2001

NOTICE

Hazardous materials emergency response work is extremely dangerous. Many responders have died or sustained serious injury and illness while attempting to mitigate an incident. There is no possible way that this text can cover the full spectrum of problems and contingencies for dealing with every type of emergency incident. The user is warned to exercise all necessary cautions when dealing with hazardous materials. Always assume a worst case scenario and place personal safety first.

It is the intent of the authors that this text be a part of the user's formal training in the management of hazardous materials emergencies. Even though this book is based on commonly used practices, references, laws, regulations, and consensus standards, it is not meant to set a standard of operations for any emergency response organization. The user's are directed to develop their own Standard Operating Procedures which follow all system, agency, or employer guidelines for handling hazardous materials. It is the user's sole responsibility to stay up to date with procedures, regulations, and product developments which may improve personal health and safety.

CONTENTS

CHAPTER 7
HAZARD AND RISK EVALUATION

CHAPTER 8
PERSONAL PROTECTIVE CLOTHING AND EQUIPMENT

CHAPTER 11
DECONTAMINATION

ACKNOWLEDGMENTS

The first edition of **Hazardous Materials: Managing the Incident** was written during 1986-87 and published in 1988. When the book was originally released, NFPA 472 was in its early stages of development, OSHA 1910.120 was only an Interim Final Rule, our partner Chuck Thomas was still running off the end of runways in Cessna 120's, and most people thought that HAZWOPER was a Burger King sandwich. A lot has changed in six years! Today, the NFPA-472 Technical Committee is currently working on the third edition, OSHA 1910.120 is a way of doing business, hazmat response has evolved into a true emergency response profession, and Chuck is flying 727's for United Airlines.

On the personal side, we logged over a million miles flying with U.S. Air and American Airlines on our way to various communities, refineries, chemical plants, and military bases (Chuck flies for United and you can never be too careful!). We also feasted on some of the finest meals that airlines can provide (roadkill roast, rubber chickens, and plastic carrots). The travel was tough but it had some benefits—Frequent Flier miles got us to Hawaii (it rained 9 out of 14 days) and the Caribbean (Hurricane Andrew was in the neighborhood). American Airlines offered some excellent ice cream sundaes in first class, and thanks to the hotels, we haven't purchased a bar of soap since 1979!

During our many travels, we've had the opportunity to work and visit with hundreds of firefighters, police officers, medics, Hazmat Response Team, Bomb Squad, and Industrial Emergency Response Team members. Many people took the time to share their personal successes and failures with us, which helped us improve ourselves as well as this book. We've tried to work the benefits of their hard work and experience into this new edition. To each person who took the time to share an idea, draw a sketch on a cocktail napkin, or "tell us like it is," we say Thank You!

Likewise, many of our peers have complimented us for the work we did in the first edition. Quite frankly, we were fortunate to work in some good organizations and for some class people, especially with the Prince George's County, Maryland Fire Department. While we appreciate the compliments, we can't take too much credit for the good ideas that friends and peers shared with us. That comment is even more appropriate for this new edition.

As authors, we view our job as being the recording secretaries, organizers, and librarians for the "Body of Knowledge" which you have shared with us. This text is our attempt to give back what the fire service and the emergency response community has given us. If you take the time to look at the Suggested Readings and References at the end of each chapter, you will note that we have built the new text around several hundred recognized authors. Their good work serves as the basis for many of the procedures and safe operating practices in this text.

Finally, we are strong advocates of a third-party review when safety issues are concerned. We have actively applied that philosophy to this revision. Approximately forty different emergency response professionals took the time to give the text a detailed technical review. In some cases, sections were reorganized and rewritten based on their recommendations. In many respects, our new book is their new book. The reviewers are listed alphabetically below. We owe each of them a special thanks, as they are truly some of the best in the business.

Captain Donald E. Abbott
Hazardous Materials Coordinator
Warren Township Fire Department
Indianapolis, Indiana

Robert C. Andrews, Jr., P.E.
Fire Chief and Vice-President
Refinery Terminal Fire Company
Corpus Christi, Texas

Phillip B. Baker
Senior Master Sergeant, U.S. Air Force
Fire Protection (retired) and
Prince George's County Fire Department
Hazardous Materials Response Team
Landover Hills, Maryland

Battalion Chief Theodore C. Bateman
Philadelphia Fire Department
Philadelphia, Pennsylvania

Ludwig Benner, Jr.
Events Analysis, Inc. and U.S. National
Transportation Safety Board (Retired)
Oakton, Virginia

Captain Lee Bergeron, CSP
Nashville Fire Department
Nashville, Tennessee

Craig W. Black
Hazardous Materials Coordinator
Prince George's County Fire Department
Hazardous Materials Division
Landover Hills, Maryland

Dr. Bryan E. Bledsoe, D.O.
Emergency Medical Services Author and
Physician
Fort Worth, Texas

Roy J. Burton
Emergency Response Liaison
U. S. Department of Transportation
Research and Special Programs Administration
Office of Hazardous Materials Initiatives
and Training
District of Columbia

Michael Callan
Captain, Wallingford, Connecticut Fire
Department (retired) and President
Fire Training Associates
Meriden, Connecticut

Dan E. Civello
Dan E. Civello and Associates, and
Hazardous Materials Coordinator
Jefferson Parish Fire Department
Metairie, Louisiana

W.D. "Dave" Cochran
Manager, Training and Education
Willams Fire & Hazard Control, Inc.
Mauriceville, Texas

Craig DeAtley, PA-C
Associate Professor, Department of
Emergency Medicine
George Washington University
District of Columbia

Leonard Deonarine, OHST, CET
President
Exodus Training Services, Inc.
Mount Airy, Maryland

Frank D. Docimo
Frank Docimo and Associates and
Turn of River Fire Department
Hazardous Materials Response Team
Stamford, Connecticut

Michael L. Donahue
Partner
Hildebrand and Noll Associates, Inc.
Olney, Maryland

Battalion Chief Jan Dunbar
Sacramento Fire Department
Sacramento, California

Chief John M. Eversole
Hazardous Materials Coordinator
Chicago Fire Department
Chicago, Illinois

John D. Flinn
U.S. Secret Service and U.S. Marine Corps, (retired)
Baton Rouge, Louisiana

Kevin Fogarty
Emergency Response Specialist
IBM Emergency Control
Hopewell Junction, New York

Scott C. Gorton
Manager—Hazardous Materials
CSX Transportation, Inc.
Richmond, Virginia

Firefighter William T. Hand
Hazardous Materials Technician
Hazardous Materials Response Team
Houston Fire Department
Houston, Texas

Edward M. Hawthorne, CSP
Safety Director
Shell Oil Company
Deer Park, Texas

Stephen L. Hermann
Hazardous Materials Coordinator
Arizona Department of Public Safety
Phoenix, Arizona

Firefighter Steven Hughes, CET
Prince George's County Fire Department
Hazardous Materials Response Team
Landover Hills, Maryland, and
HAZMAT Training, Information, and Services, Inc. (TISI)
Columbia, Maryland

Battalion Chief Paul Mauger
Chesterfield County Fire Department
Hazardous Materials Response Team
Chesterfield, Virginia

Dr. Mary Jo McMullen, MD, FACEP
Medical Advisor
Summit County Hazardous Materials Response Team
Akron, Ohio

Assistant Chief Roger McGary
Montgomery County Department of Fire & Rescue Services
Rockville, Maryland

Assistant Chief Mary Beth Michos, RN
Chief, EMS Division
Specialty Teams Coordinator
Montgomery County Department of Fire & Rescue Services
Rockville, Maryland

Robert D. Myak, CSP, CIH
Safety and Health Specialist
BP Oil Company—Refining
Cleveland, Ohio

Jeffrey E. Pollard
Firefighter/Hazardous Materials Technician
Atlanta Fire Department
Atlanta, Georgia

Battalion Chief Dennis L. Rubin
Chesterfield County Fire Department
Training and Safety Division
Chesterfield, Virginia

Richard W. Schramm
Emergency Response Manager
Rohm and Haas, Delaware Valley, Inc.
Bristol, Pennsylvania

Captain Bruce Sellon
Lincoln Fire Department
Hazardous Materials Response Team
Lincoln, Nebraska

Michael L. Spillane, MPH
Deputy Branch Chief, Emergency Management Branch
National Institutes of Health (NIH)
Division of Safety
Bethesda, Maryland

William H. Stringfield
Lieutenant, St. Petersburg, Florida Fire Department (retired) and President
William H. Stringfield and Associates, Inc.
St. Petersburg, Florida

Jeffrey O. Stull
Personnel Protection Consultant
Chairman, NFPA Technical Subcommittee on Chemical Protective Clothing
Austin, Texas

First Officer, Charles R. Thomas
United Airlines
UAL Dangerous Goods Coordinator for the Airline Pilots Association (ALPA)
Member—ALPA Central Air Safety Committee
San Francisco, California

Dr. David J. Wilson
Professor
University of Alberta
Department of Mechanical Engineering
Edmonton, Alberta, Canada

Charles J. Wright
Manager Training, Chemical Transportation Safety
Union Pacific Railroad Company
Omaha, Nebraska

Lieutenant Wayne E. Yoder
Hazardous Materials Response Team Coordinator
Delray Beach Fire Department
Delray Beach, Florida

No amount of information and original ideas can make a quality book if it doesn't look good and all the words aren't spelled right. A special group of professionals came together at the right time, with the right chemistry and background, to make this text look great. Gene Carlson and Don Davis of Fire Protection Publications made sure that we had the opportunity to publish, print, and distribute the book to you. George Dodson and Katie Lawyer of Lightworks Photography and Design did the entire photography, design, and layout for the book. Many of the illustrations were done by combining the art of photography and the image making ability of Macintosh computers. George and Katie really pushed the outside of the graphics envelope on this one.

Our friend and attorney, Charlie Donnelly helped us through the maze of legal documents required to publish this revision. Patricia Daley provided final copy editing for the manuscript.

The cartoons were a mixture of the old and the new—the old were developed by Don and Lawan Sellers, the original team members. Don passed away in 1993, from a prolonged illness. He continued to be creative in retirement, carving many award winning water fowl. The new, were drawn by Ground Zero Grafix.

A special thanks goes to Greg Socks and the members of the Washington County, Maryland, Hazardous Materials Response Team. Washington County provided the equipment and technical expertise for most of the staged photography used throughout the text. They showed great patience and discipline while standing in the hot sun in protective clothing until we got the shot just right. Washington County is proof that you don't have to have a big budget to have a first class operation.

Finally and most importantly, we would like to thank our best friends, buddies, and wives, Debbie and JoAnne. Let's face it, the women are candidates for sainthood. They held down the fort while we flew around the country, and supported us while we spent evenings and weekends pouring over comments and revisions to the chapters. Okay, we admit that Greg's son- Sean Patrick (AKA Attila the Hun) probably did drive Debbie "The General" crazy! And there is some truth to the rumor that Mike left JoAnne home alone with no electricity or running water for three days when the "Winter from Hell" turned their house into a disaster area. But hey, who's complaining, at least they have a good sense of humor!

ABOUT THE AUTHORS

Greg Noll and Mike Hildebrand are the senior partners and co-founders of Hildebrand and Noll Associates, Inc., a consulting firm which specializes in emergency planning and response issues, conducting audits and inspections of emergency response systems, and assisting facilities in developing and improving their emergency response and management organization. In addition to this text, they are the co-authors of the text entitled **Gasoline Tank Truck Emergencies: Guidelines and Procedures**, published by Oklahoma State University, Fire Protection Publications.

Gregory C. Noll. Greg was previously the Hazardous Materials Coordinator with the Prince George's County, Maryland Fire Department, and a Fire and Safety Associate with the American Petroleum Institute in Washington, DC. He is currently the President of the Pennsylvania Association of Hazardous Materials Technicians, and is a member of the Lancaster Township, Pennsylvania Fire Department.

An experienced educator, Greg is a former faculty member with the Iowa State University Fire Service Institute and is an adjunct faculty member in the hazardous materials curriculum at the National Fire Academy. He is a Certified Safety Professional (CSP), and is a member of the International Association of Fire Chiefs (IAFC) Hazardous Materials Committee and the National Fire Protection Association (NFPA) Technical Committee on Hazardous Materials Response Personnel (NFPA 472).

Michael S. Hildebrand. Mike was the former Director of Safety and Fire Protection at the American Petroleum Institute in Washington, DC. Prior to joining API, he held positions with the International Association of Fire Chiefs (IAFC), the National Transportation Safety Board (NTSB), and was a firefighter/medic with the U.S. Air Force. He is also a former Shift Officer with the Prince George's County, Maryland Fire Department Hazardous Materials Response Team.

Mike is an Associate Safety Professional (ASP) and a Professional Member with the American Society of Safety Engineers. A former member of the NFPA Standards Council and the Technical Committee on Flammable and Combustible Liquids (NFPA 30), Mike has served on numerous ANSI, ASTM and API standards committees.

James G. Yvorra. Jim was a Deputy Chief with the Berwyn Heights, Maryland Volunteer Fire Department and a Shift Officer with the Prince George's County, Maryland Fire Department Hazardous Materials Response Team. He was also a nationally known author and editor in the fields of fire, hazardous materials, and emergency medical services. While investigating a multi-vehicle accident scene on Interstate 95 in Prince George's County in January, 1988, he was struck by a car and died in the line of duty. He was posthumously awarded the Gold Medal of Valor by the Prince George's County Fire Department.

Jim dedicated his life's work to the fire service and the emergency response community. His leadership style and personal commitment to his profession helped many young men and women begin successful careers. Jim was a serious student of management and leadership, and believed the individual could make a difference in the emergency response community. His personal accomplishments are an example of that conviction. Many of his thoughts, notes, and articles are still current and live on in this text.

In light of Jim's efforts and accomplishments, several of his friends, peers and relatives formed Yvorra Leadership Development, Inc. (YLD), a non-profit foundation designed to promote leadership development among the members of the fire and emergency medical service communities. Since its inception in 1989, YLD has awarded over $15,000 in scholarships.

To contact the authors or to receive additional information on Yvorra Leadership Development, write Hildebrand and Noll Associates, Inc., Scientists Cliffs, P.O. Box 408, 2446 Azalea Road, Port Republic, Maryland 20676 or phone (410) 586-3048-0408.

*This book is dedicated
to our partner and friend,
Jim Yvorra.*

"He simply never acted
as important as he really was."

HOW TO USE THIS BOOK

ABOUT THE BOOK

"Hazardous Materials: Managing the Incident" is designed to develop the knowledge and skills necessary for emergency response personnel to safely and effectively manage a hazardous materials emergency. It is based upon a hazardous materials operations and training system which was originally developed by the authors and the members of the Prince George's County, Maryland Fire Department—Hazardous Materials Response Team. Over the years, this system has subsequently been adopted and improved upon by a number of public safety, military and industrial response teams.

This textbook is based upon a systematic procedure for managing hazardous materials emergencies, known as the Eight Step Process©. The Eight Step Process© establishes a management structure which fits any size or level of hazmat response; it provides an incident management framework with one specific goal—to maximize safety for emergency response personnel and the general public.

The book is divided into two broad areas—*"Preparing for the Hazmat Response"* and *"Implementing the Hazmat Response."* Chapters 1 through 4 provide critical information for "Preparing for the Hazmat Response." Topics include the development of a comprehensive system for managing the hazmat problem, health and safety issues and concerns, and the development and implementation of an emergency management organization.

Chapters 5 through 12 pertain to *"Implementing the Hazmat Response,"* with each chapter covering an individual function of the Eight Step Process©. In the introduction of each of these chapters will be a small icon of a Hazmat responder in a Level A suit, illustrating the tie-in to the Eight Step Process©. The information contained within each of these chapters builds upon each other, and should be read in the order presented. For example, Step 6—Implement Response Objectives, cannot be safely performed if responders have failed to (1) control the site, (2) identify the nature of the problem and the hazardous materials involved, (3) conducted a hazard and risk evaluation process, and (4) determined the appropriate level of personal protective clothing and equipment.

USING THE BOOK

This textbook may be used on an individualized self-study basis, as part of a formal training program offered in a plant, or at a emergency services training academy, or as part of an organized college curriculum. In each case, use of the text along with the accompanying workbook and instructor's guide will provide the most thorough knowledge transfer and retention for the student.

There are several features of this text you should be familiar with:

Objectives:

At the beginning of each chapter you will find a list of educational objectives which will be covered in the following pages. They will help you to compare the information provided in the text against the training requirements cited in OSHA 1910.120 (q)—Hazardous Waste Operations and Emergency Response, and the educational competencies referenced in NFPA 472—Professional Competence of Responders to Hazardous Materials Incidents. After you have completed the reading assignment, you should have a working knowledge of each chapter objective. The student workbook provides questions and exercises to aid this learning process.

Visuals:

Charts, photographs, and cartoons are used throughout the text to reinforce important concepts. Spend enough time studying each illustration to understand its point and how it addresses the appropriate educational objectives.

Design:

The text is designed to assist the student by providing:

- Easy to use, $8^1/_2$ x 11" format.
- A "NOTES" column for personal notetaking.
- Screened information and bold printing to note important facts.
- Color coded typeface to emphasize important points.
- A glossary of important terms and definitions.

Summary:

A brief statement is located at the end of each chapter which summarizes the intent of that unit.

References and Suggested Readings:

At the end of each chapter is a list of books, articles, video tapes, courses and other related educational materials that were either referenced or can be consulted for additional information on the chapter's topics. Some of these represent the current experience and thinking on the topic while others are more than thirty years old. Their lessons are timeless.

It is our personal philosophy that training and education must provide quality information which will allow the student to perform their job safely, efficiently and effectively. Likewise, we believe that training and education, particularly when dealing with adults and emergency responders, must also be fun! We have tried to live up to this philosophy by inserting some one liners and thoughts many responders have shared with us. Based upon student and reviewer feedback, we believe that we have written a text which reflects both of these points.

PREPARING FOR THE THE HAZMAT RESPONSE

THE HAZARDOUS MATERIALS MANAGEMENT SYSTEM

OBJECTIVES

1) Describe the scope and target audience of this manual.
2) Define and explain the source of, and circumstances for using, the following terms:
   - Hazardous materials
   - Hazardous substances
   - Extremely hazardous substances
   - Hazardous chemicals
   - Hazardous wastes
   - Dangerous goods
3) List the key legislative, regulatory, and voluntary consensus standards which impact hazmat planning, prevention, emergency response, and clean-up operations.
4) Describe the concept of "standard of care."
5) List and describe the components of the Hazardous Materials Management System for managing the hazardous materials problem.

"Salus populi suprema lex"

Translation:

"The people's safety is the highest law."

NOTES

INTRODUCTION

This is a text about hazardous materials (hazmat) incident response. It is designed to provide both public and private sector emergency response personnel (ERP) with a logical, building block system for managing hazardous materials emergencies. It is not a chemistry-oriented text. In fact, it assumes that most of the first-arriving ERP will have little or no formal chemistry training.

It is designed to begin at the point where ERP recognize that they are, in fact, dealing with a hazardous materials emergency, even when the exact hazardous materials have not been identified. Otherwise, normal fire, rescue and emergency medical services (EMS) guidelines will be followed.

Our primary target audiences are Hazardous Materials Technicians, the Hazmat Branch Officer, the On-Scene Incident Commander, and members of organized hazardous materials response teams. Other special operations teams, such as Bomb Squads and Confined Space Rescue Teams, will also find specific chapters of interest (e.g., Chapter 7—Hazard and Risk Assessment).

This second edition has been expanded to include additional information to assist the reader in meeting the cognitive skill requirements of OSHA 1910.120 (q) and the NFPA 472 competencies for the Hazardous Materials Technician and the On-Scene Incident Commander.

WHAT IS A HAZARDOUS MATERIAL?

You might assume that everyone knows what a hazardous material is—or at least knows one when they see one. But if one were to check the various state and federal regulations which govern the manufacture, transportation, storage, use, and clean-up of chemicals in the United States, the number of terms, definitions, and lists would be overwhelming. Key definitions are shown in Figure 1-1.

HAZARDOUS MATERIALS DEFINITIONS

- **Hazardous Materials**—Any substance or material in any form or quantity which poses an unreasonable risk to safety and health and property when transported in commerce (U.S. Department of Transportation, 49 CFR 171).
- **Hazardous Substances**—Any substance designated under the Clean Water Act and the Comprehensive Environmental Response, Compensation and Liability Act (CERCLA) as posing a threat to waterways and the environment when released (U.S. Environmental Protection Agency, 40 CFR 302).

- **NOTE**: "Hazardous Substances" as used within OSHA 1910.120 refers to every chemical regulated by EPA as a hazardous substance and by DOT as a hazardous material.
- **Extremely Hazardous Substances (EHS)**—Chemicals determined by the Environmental Protection Agency to be extremely hazardous to a community during an emergency spill or release as a result of their toxicities and physical/chemical properties (U.S. Environmental Protection Agency, 40 CFR 355).
- **Hazardous Chemicals**—Any chemical that would be a risk to employees if exposed in the workplace (U.S. Occupational Safety and Health Administration, 29 CFR 1910).
- **Hazardous Wastes**—Discarded materials regulated by the Environmental Protection Agency because of public health and safety concerns. Regulatory authority is granted under the Resource Conservation and Recovery Act (RCRA). (U.S. Environmental Protection Agency, 40 CFR 260–281).
- **Dangerous Goods**—In Canadian transportation, hazardous materials are referred to as "dangerous goods."

FIGURE 1-1: HAZARDOUS MATERIALS DEFINITIONS

NOTES

Each term in Figure 1-1 has its applications and limitations. In reality, we must recognize that hazmats can be found virtually anywhere—in industry, in transportation, in the workplace, and even in the home—so a broad definition is necessary to cover all the bases.

Hazmat emergency response primarily focuses on the interaction of the hazmat and its container. Therefore, for the purposes of this text, we will use the definition of a hazardous material developed by Ludwig Benner, Jr., a former hazardous materials specialist with the National Transportation Safety Board (NTSB) in Washington, DC.

Hazardous Material—Any substance which jumps out of its container when something goes wrong and hurts or harms the things it touches.

Benner's definition can be applied to all hazmats in all circumstances and recognizes that emergency response is as much a container or behavioral problem as it is a chemical problem.

A hazardous materials incident can then be defined as the release, or potential release, of a hazardous material from its container into the environment.

NOTES

DON SELLERS

FIGURE 1-2: HAZARDOUS MATERIAL—ANY SUBSTANCE WHICH JUMPS OUT OF ITS CONTAINER WHEN SOMETHING GOES WRONG AND HURTS OR HARMS THE THINGS IT TOUCHES

HAZMAT LAWS, REGULATIONS, AND STANDARDS

Hazmat manufacturers, transporters, users, and emergency response personnel are impacted by a large number of laws, regulations, and voluntary consensus standards. These rules touch on virtually every facet of the hazmat industry. Because of their importance to emergency planning and response operations, senior hazmat management personnel must have a working knowledge of how the regulatory system works. First, what is the difference between a law, regulation, and standard? These three terms are sometimes used interchangeably, but they do have distinctly different meanings.

Laws are primarily created through an act of Congress or by individual state legislatures. Laws typically provide broad goals and objectives, mandatory dates for compliance, and established penalties for noncompliance. Federal and state laws enacted by legislative bodies usually delegate the details for implementation to a specific federal or state agency. For example, the U.S. Occupational Safety and Health Act enacted by Congress delegates rulemaking and enforcement authority on worker health and safety issues to the Occupational Safety and Health Administration (OSHA).

Regulations, sometimes called **rules**, are created by federal or state agencies as a method of providing guidelines for complying with a law which was enacted through legislative action. A regulation permits individual governmental agencies to enforce the law through inspections, which may be conducted by federal and state officials.

NOTES

Voluntary consensus standards are normally developed through professional organizations or trade associations as a method of improving the individual quality of a product or system. Within the emergency response community, the National Fire Protection Association (NFPA) is recognized for its role in developing consensus standards and recommend practices which impact fire safety and hazmat operations. In the United States, standards are developed primarily through a democratic process whereby a committee of subject specialists representing varied interests writes the first draft of the standard. The document is then submitted to either a larger body of specialists or the general public, who then may amend, vote on, and approve the standard for publication. This procedure is known as the Consensus Standards Process.

When a consensus standard is completed, it may be voluntarily adopted by government agencies, individual corporations, or organizations. Many hazmat consensus standards are also adopted as a reference in a regulation. In effect, when a federal, state, or municipal government adopts a consensus standard by reference, the document becomes a regulation. An example of this process is the adoption of *NFPA 30—The Flammable and Combustible Liquids Code*, and *NFPA 58—The Liquefied Gas Code.*

FEDERAL HAZMAT LAWS

Hazmat laws have been enacted by Congress to regulate everything from finished products to hazardous waste. Because of their lengthy official titles, many simply use abbreviations or acronyms when referring to these laws. The following summaries outline some of the more important laws impacting hazmat emergency planning and response.

- ❑ **CERCLA—The Comprehensive Environmental Response Compensation and Liability Act.** Known as "Superfund," this law addresses hazardous substance releases into the environment and clean-up of inactive hazardous waste disposal sites. It also requires those individuals responsible for the release of the hazardous materials (commonly referred to as the responsible party) above a specified "reportable quantity" to notify the National Response Center.
- ❑ **RCRA—The Resource Conservation and Recovery Act.** This law establishes a framework for the proper management and disposal of all hazardous wastes, including treatment, storage, and disposal facilities. It also establishes installation, leak prevention, and notification requirements for underground storage tanks.
- ❑ **CAA—The Clean Air Act.** This law establishes requirements for airborne emissions and the protection of the environment. The Clean Air Act Amendments of 1990 addressed emergency response and planning issues at certain facilities with processes using highly hazardous chemicals. This included the establishment of a national Chemical Safety and Hazard Investigation Board, EPA's promulgation of *40 CFR Part 68—Risk Management Programs for Chemical Accidental Release Prevention*, and *OSHA's promulgation of 29 CFR 1910.119 Process Safety Management of Highly Hazardous Chemicals, Explosives*

NOTES

and Blasting Agents. In addition, certain facilities are required to make information available to the general public regarding the manner in which chemical risks are handled within a facility.

- **SARA—Superfund Amendments and Reauthorization Act of 1986.** SARA has had the greatest impact upon hazmat emergency planning and response operations. As the name implies, SARA amended and reauthorized the Comprehensive Environmental Response, Compensation, and Liability Act of 1980 (CERCLA or Superfund). While many of the amendments pertained to hazardous waste site clean-up, SARA's requirements also established a national baseline with regard to hazmat planning, preparedness, training, and response.

  Title I of this law required OSHA to develop health and safety standards covering numerous worker groups who handle or respond to chemical emergencies and led to the development of *OSHA 1910.120, Hazardous Waste Operations and Emergency Response (HAZWOPER).*

  Most familiar to the emergency response community is SARA, Title III. Also known as the Emergency Planning and Community Right-to-Know Act (EPCRA), SARA Title III led to the establishment of the State Emergency Response Commissions (SERC) and the Local Emergency Planning Committees (LEPC).

- **OPA—Oil Pollution Act of 1990.** Commonly referred to as OPA, this law amended the Federal Water Pollution Control Act. Its scope covers both facilities and carriers of oil and related liquid products, including deepwater marine terminals, marine vessels, pipelines, and railcars. Requirements include the development of emergency response plans, regular training and exercise sessions, and verification of spill resources and contractor capabilities. The law also requires the establishment of Area Committees and the development of Area Contingency Plans (ACP) to address oil and hazardous substance spill response in coastal zone areas.

HAZMAT REGULATIONS

Laws delegate certain details of implementation and enforcement to federal or state agencies who are then responsible for writing the actual regulations which enforce the legislative intent of the law. Regulations will either (1) define the broad performance required to meet the letter of the law (i.e., performance-oriented standards); or (2) provide very specific and detailed guidance on satisfying the regulation (i.e., specification standards).

FEDERAL REGULATIONS

The following summary includes several of the more significant federal regulations which affect hazmat emergency planning and response.

Hazardous Waste Operations and Emergency Response (29 CFR 1910.120).

NOTES

Also known as HAZWOPER, this federal regulation was issued under the authority of SARA, Title I. The regulation was written and is enforced by the Occupational Safety and Health Administration in those 23 states and 2 territories with their own OSHA-approved occupational safety and health plans. In the remaining 27 "non-OSHA" states, public sector personnel will be covered by a similar regulation enacted by the Environmental Protection Agency (40 CFR Part 311).

The regulation establishes important requirements for both industry and public safety organizations who respond to hazmat or hazardous waste emergencies. This includes firefighters, law enforcement and EMS personnel, hazmat responders, and industrial Emergency Response Team (ERT) members. Requirements cover the following areas:

- Hazmat Emergency Response Plan.
- Emergency Response Procedures, including the establishment of an Incident Management System, the use of a buddy system with back-up personnel, and the establishment of a Safety Officer.
- Specific training requirements covering instructors and both initial and refresher training.
- Medical Surveillance Programs.
- Post-emergency termination procedures.

Of particular interest to hazmat managers and responders are the specific levels of competency and associated training requirements identified within OSHA 1910.120 (q)(6). See Figure 1-3.

FIGURE 1-3

OSHA 1910.120 LEVELS OF EMERGENCY RESPONDERS

First Responder at the Awareness Level. These are individuals who are likely to witness or discover a hazardous substance release and who have been trained to initiate an emergency response notification process. The primary focus of their hazmat responsibilities is to secure the incident site, recognize and identify the materials involved, and to make the appropriate notifications. These individuals would take no further action to control or mitigate the release. First Responder–Awareness personnel shall have sufficient training or experience to objectively demonstrate the following competencies:

a) An understanding of what hazardous materials are, and the risks associated with them in an incident.

b) An understanding of the potential outcomes associated with a hazardous materials emergency.

c) The ability to recognize the presence of hazardous materials in an emergency and, if possible, identify the materials involved.

NOTES

d) An understanding of the role of the First Responder–Awareness individual within the local Emergency Operations Plan. This would include site safety, security and control, and the use of the *DOT Emergency Response Guidebook.*

e) The ability to realize the need for additional resources and to make the appropriate notifications to the communication center.

The most common examples of First Responder–Awareness personnel include law enforcement and plant security personnel, as well as some public works employees. There is no minimum hourly training requirement for this level; the employee would have to have sufficient training to objectively demonstrate the required competencies.

First Responder at the Operations Level. Most fire department suppression personnel fall into this category. These are individuals who respond to releases or potential releases of hazardous substances as part of the initial response for the purpose of protecting nearby persons, property, or the environment from the effects of the release. They are trained to respond in a defensive fashion without actually trying to stop the release. Their primary function is to contain the release from a safe distance, keep it from spreading, and protect exposures.

First Responder–Operations personnel shall have sufficient training or experience to objectively demonstrate the following competencies:

a) Knowledge of basic hazard and risk assessment techniques.

b) Knowledge of how to select and use proper personal protective clothing and equipment available to the operations-level responder.

c) An understanding of basic hazardous materials terms.

d) Knowledge of how to perform basic control, containment, and/or confinement operations within the capabilities of the resources and personal protective equipment available.

e) Knowledge of how to implement basic decontamination measures.

f) An understanding of the relevant standard operating procedures and termination procedures.

First responders at the operations level shall have received at least 8 hours of training or have had sufficient experience to objectively demonstrate competency in the previously mentioned areas, as well as the established skill and knowledge levels for the First Responder–Awareness level.

NOTES

Hazardous Materials Technician. These are individuals who respond to releases or potential releases for the purposes of stopping the release. Unlike the operations level, they generally assume a more aggressive role in that they are often able to approach the point of a release in order to plug, patch, or otherwise stop the release of a hazardous substance.

Hazardous materials technicians are required to have received at least 24 hours of training equal to the First Responder–Operations level and have competency in the established skill and knowledge levels outlined below:

a) Capable of implementing the local Emergency Operations Plan.

b) Able to classify, identify, and verify known and unknown materials by using field survey instruments and equipment (direct reading instruments).

c) Able to function within an assigned role in the Incident Management System.

d) Able to select and use the proper specialized chemical personal protective clothing and equipment provided to the Hazardous Materials Technician.

e) Able to understand hazard and risk assessment techniques.

f) Able to perform advanced control, containment, and/or confinement operations within the capabilities of the resources and equipment available to the Hazardous Materials Technician.

g) Able to understand and implement decontamination procedures.

h) Able to understand basic chemical and toxicological terminology and behavior.

Many communities and facilities have personnel trained as Emergency Medical Technicians (EMT), yet do not have the primary responsibility for providing basic or advanced life support medical care. Similarly, Hazardous Materials Technicians may not necessarily be part of a hazardous materials response team. However, if they are part of a designated team as defined by OSHA, they must also meet the medical surveillance requirements within OSHA 1910.120.

Hazardous Materials Specialists. These are individuals who respond with and provide support to Hazardous Materials Technicians. While their duties parallel those of the Technician, they require a more detailed or specific knowledge of the various substances they may be called upon to contain. This individual would also act as the site liaison with federal, state, local, and other governmental authorities in regard to site activities.

NOTES

Similar to the technician level, Hazardous Materials Specialists shall have received at least 24 hours of training equal to the technician level and have competency in the following established skill and knowledge levels:

a) Capable of implementing the local Emergency Operations Plan.
b) Able to classify, identify, and verify known and unknown materials by using advanced field survey instruments and equipment (direct reading instruments).
c) Knowledge of the state emergency response plan.
d) Able to select and use the proper specialized chemical personal protective clothing and equipment provided to the Hazardous Materials Specialist.
e) Able to understand in-depth hazard and risk assessment techniques.
f) Able to perform advanced control, containment, and/or confinement operations within the capabilities of the resources and equipment available to the Hazardous Materials Specialist.
g) Able to determine and implement decontamination procedures.
h) Able to develop a site safety and control plan.
i) Able to understand basic chemical, radiological, and toxicological terminology and behavior.

Whereas the Hazardous Materials Technician possesses an intermediate level of expertise and is often viewed as a "utility person" within the hazmat response community, the Hazardous Materials Specialist possesses an advanced level of expertise. Within the fire service, the Specialist will often assume the role of the Safety Officer or Hazmat Sector Officer, while an industrial Hazardous Materials Specialist may be "product specific." Finally, the Specialist must meet the medical surveillance requirements outlined within OSHA 1910.120.

On-Scene Incident Commander. Incident Commanders, who will assume control of the incident scene beyond the First Responder–Awareness level, shall receive at least 24 hours of training equal to the First Responder–Operations level. In addition, the employer must certify that the incident commander has competency in the following areas:

1) Know and be able to implement the local Incident Management System.
2) Know how to implement the local Emergency Operations Plan.

NOTES

3) Understand the hazards and risks associated with working in chemical protective clothing.
4) Know of the state emergency response plan and of the Federal Regional Response Team.
5) Know and understand the importance of decontamination procedures.

Skilled Support Personnel. These are personnel who are skilled in the operation of certain equipment, such as cranes and hoisting equipment, and who are needed temporarily to perform immediate emergency support work that cannot reasonably be performed in a timely fashion by emergency response personnel. It is assumed that these individuals will be exposed to the hazards of the emergency response scene.

Although these individuals are not subject to the HAZWOPER training requirements, they shall be given an initial briefing at the site prior to their participation in any emergency response effort. This briefing shall include elements such as instructions in using personal protective clothing and equipment, the chemical hazards involved, and the tasks to be performed. All other health and safety precautions provided to emergency responders and on-scene workers shall be used to assure the health and safety of these support personnel.

Specialist Employees. These are employees who, in the course of their regular job duties, work with and are trained in the hazards of specific hazardous substances, and who will be called upon to provide technical advice or assistance to the Incident Commander at a hazmat incident. This would include industry responders, chemists, and related professional or operations employees. These individuals shall receive training or demonstrate competency in the area of their specialization annually.

FIGURE 1-3: OSHA 1910.120 LEVELS OF EMERGENCY RESPONDERS

Community Emergency Planning Regulations (40 CFR 301-303).

This regulation is the result of SARA, Title III and mandates the establishment of both state and local planning groups to review or develop hazardous materials response plans. The state planning groups are referred to as the State Emergency Response Commission (SERC). The SERC is responsible for developing and maintaining the state's emergency response plan. This includes ensuring that planning and training are taking place throughout the state, as well as providing assistance to local governments, as appropriate. States generally provide an important source of technical specialists, information, and coordination. However, they typically provide only limited operational support to local government in the form of equipment, materials, and personnel during an emergency.

NOTES

The coordinating point for both planning and training activities at the local level is the Local Emergency Planning Committee (LEPC). Among the LEPC membership are representatives from the following groups:

Elected state and local officials
Fire Department
Law Enforcement
Emergency Management
Public Health officials
Hospital
Industry personnel, including facilities and carriers
Media
Community organizations

The LEPC is specifically responsible for developing and/or coordinating the local emergency response system and capabilities. A primary concern is the identification, coordination, and effective management of local resources. Among the primary responsibilities of the LEPC are:

- Develop, regularly test, and exercise the Hazmat Emergency Operations Plan.
- Conduct a hazards analysis of hazmat facilities and transportation corridors within the community.
- Receive and manage hazmat facility reporting information. This includes chemical inventories, Tier II reporting forms required under SARA, Title III, material safety data sheets (MSDS) or chemical lists, and points of contact.
- Coordinate the Community Right-to-Know aspects of SARA, Title III.

Risk Management Programs for Chemical Accidental Release Prevention (40 CFR Part 68).

Promulgated under amendments to the Clean Air Act, this was a proposed rule at the time of publication. The regulation would require that facilities that manufacture, process, use, store, or otherwise handle certain regulated substances above established threshold values develop and implement risk management programs (RMPP). The regulation is similar in scope to the OSHA Process Safety Management standard, with the primary focus being community safety as compared to employee safety.

Under the proposed rulemaking, risk management programs would consist of three elements:

- **Hazard assessment** of the facility, including the worst-case accidental release and an analysis of potential off-site consequences.
- **Prevention program,** which addresses safety precautions, maintenance, monitoring, and employee training. EPA believes that the prevention program should adopt and build upon the OSHA Process Safety Management standard.
- **Emergency response** considerations, including facility emergency response plans, informing public and local agencies, emergency medical care, and employee training.

NOTES

Hazard Communication Regulation (29 CFR 1910.1200).

HAZCOM is a federal regulation which requires hazardous materials manufacturers and handlers to develop written Material Safety Data Sheets (MSDS) on specific types of hazardous chemicals. These MSDS's must be made available to employees who request information about a chemical in the workplace. Examples of information on MSDS include known health hazards, the physical and chemical properties of the material, first aid, firefighting and spill control recommendations, protective clothing and equipment requirements, and emergency telephone contact numbers.

Under the HAZCOM requirements, hazmat health exposure information should be provided to emergency responders during the termination phase, and all exposures should be documented.

Hazardous Materials Transportation Regulations (49 CFR 100-199).

This series of regulations is issued and enforced by the U.S. Department of Transportation. The regulations govern container design, chemical compatibility, packaging and labeling requirements, shipping papers, transportation routes and restrictions, and so forth. The regulations are comprehensive and strictly govern how all hazardous materials are transported by highway, railroad, pipeline, aircraft, and by water.

National Contingency Plan or NCP (40 CFR 300, Subchapters A through J).

This outlines the policies and procedures of the federal agency members of the National Oil and Hazardous Materials Response Team (also known as the National Response Team or the NRT). The regulation provides guidance for emergency responses, remedial actions, enforcement, and funding mechanisms for federal government response to hazmat incidents. The NRT is chaired by EPA, while the vice-chairperson represents the U.S. Coast Guard (USCG).

Each of the ten federal regions also has a Regional Response Team (RRT) which mirrors the make-up of the NRT. RRT's may also include representatives from state and local government and Indian tribal governments.

When the NRT or RRT is activated for a federal response to an oil spill or hazmat incident, a federal On-Scene Coordinator will be designated to coordinate the overall response. The On-Scene Coordinator (OSC) will represent either EPA or the USCG, based upon the location of the incident. If the release or threatened release occurs in coastal areas or near major navigable waterways, the USCG will usually assume primary OSC responsibility. If the situation occurs inland and away from navigable or major waterways, the EPA will serve as the OSC. Local emergency responders should contact EPA and USCG personnel within their region to determine which agency has primary responsibility and will act as the federal OSC for their respective area.

NOTES

State Regulations

Each of the 50 states and the U.S. territories maintains an enforcement agency which has responsibility for hazardous materials. The three key players in each state usually consist of the State Fire Marshal, the State Occupational Safety and Health Administration, and the State Department of the Environment (sometimes known as Natural Resources or Environmental Quality). While there are many variations, the fire marshal is typically responsible for the regulation of flammable liquids and gases due to the close relationship between the flammability hazard and the fire prevention code, while the state environmental agency would be responsible for the development and enforcement of environmental safety regulations.

While known by various titles, most states have a government equivalent of the Federal OSHA. Approximately 25 states and territories have adopted the Federal OSHA regulations as state law. This method of adoption has increased the level of enforcement of hazardous materials regulations such as the Hazardous Waste and Emergency Response regulation previously described. State governments also maintain an environmental enforcement agency which usually enforces the Federal RCRA, CERCLA, and CAA laws at the local level. State involvement in hazardous waste regulatory enforcement has significantly increased the number of hazmat incidents reported.

VOLUNTARY CONSENSUS STANDARDS

Standards developed through the voluntary consensus process play an important role in increasing both workplace and public safety. Historically, a voluntary standard improves over time as each revision reflects recent field experience and adds more detailed requirements. As users of the standard adopt it as a way of doing business, the level of safety gradually improves over time.

Consensus standards are also updated more regularly than governmental regulations. For example, since OSHA 1910.120 was released in March, 1989, *NFPA 472—Standard for Professional Competence of Responders to Hazardous Material Incidents* (described below) has been revised twice.

In many respects, a voluntary consensus standard provides a way for individual organizations and corporations to self-regulate their business or profession. All of the national fire codes in the United States are developed through the voluntary consensus standards process, with the two key players being the NFPA and Western Fire Chiefs Association. Standards developed by these two organizations address many hazmat issues, including hazmat storage and handling, personal protective clothing and equipment, and hazmat professional competencies.

Three consensus standards used within the hazmat emergency response community are the following:

NFPA 471—Recommended Practice for Responding to Hazardous Material Incidents.

The document covers planning procedures, policies, and application of procedures for incident levels, personal protective clothing and

NOTES

equipment, decontamination, safety, and communications. The purpose of NFPA 471 is to outline the minimum requirements that should be considered when dealing with responses to hazmat incidents, and to specify operating guidelines.

NFPA 472—Standard for Professional Competence of Responders to Hazardous Material Incidents.

The purpose of NFPA 472 is to specify minimum competencies for those who will respond to hazardous material incidents. The overall objective is to reduce the number of accidents, injuries, and illnesses during response to hazmat incidents, and to prevent exposure to hazmats to reduce the possibility of fatalities, illnesses, and disabilities affecting emergency responders.

It is important to recognize that NFPA 472 covers all hazmat emergency responders from both the public and private sector.

NFPA 472 provides competencies for the following levels of hazmat responders. These levels parallel those listed within OSHA 1910.120, with the exception that the Hazardous Materials Specialist has been deleted and the Specialist Employee has been expanded and clarified upon:

- ***First Responder at the Awareness Level.*** These are individuals who, in the course of their normal duties, may be the first on scene of an emergency involving hazmats. They are expected to recognize hazmat presence, protect themselves, call for trained personnel, and secure the area.
- ***First Responder at the Operations Level.*** These are individuals who respond to releases or potential releases of hazmats as part of the initial response to the incident for the purpose of protecting nearby persons, the environment, or property from the effects of the release. They shall be trained to respond in a defensive fashion to control the release from a safe distance and keep it from spreading.
- ***Hazardous Materials Technician.*** These are individuals who respond to releases or potential releases of hazmats for the purpose of controlling the release. Hazmat technicians are expected to use specialized chemical protective clothing and specialized control equipment.
- ***Incident Commander.*** The person who is responsible for directing and coordinating all aspects of a hazmat incident.
- ***Off-Site Specialist Employee.*** These are individuals who, in the course of their regular job duties, work with or are trained in the hazards of specific materials and/or containers. In response to incidents involving chemicals, they may be called upon to provide technical advice or assistance to the incident commander relative to their area of specialization. There are three levels of off-site specialist employee:

  Level C are those persons who may respond to incidents involving chemicals and/or containers within their organization's area of specialization. They may be called upon to gather and record information, provide technical advice, and/or arrange for technical assistance consistent with their organization's emergency response

NOTES

plan and standard operating procedures. The individual is not expected to enter the hot/warm zone at an incident.

Level B are those persons who, in the course of their regular job duties, work with or are trained in the hazards of specific chemicals or containers within their organization's area of specialization. Because of their education, training, or work experience, they may be called upon to respond to incidents involving chemicals. The Level B employee may be used to gather and record information, provide technical advice, and provide technical assistance (including working within the hot zone) at the incident consistent with his or her organization's emergency response plan and standard operating procedures, and the local emergency operations plan.

Level A are those persons who are specifically trained to handle incidents involving chemicals and/or containers for chemicals used in their organization's area of specialization. Consistent with his or her organization's emergency response plan and standard operating procedures, the Level A employee shall be able to analyze an incident involving chemicals within his/her organization's area of specialization, plan a response to that incident, implement the planned response within the capabilities of the resources available, and evaluate the progress of the planned response.

At the time of publication, the NFPA 472 Technical Committee was working on the development of the third edition of the standard. Consideration was being given to the development of additional competencies for the Hazmat Branch Officer, the Hazmat Branch Safety Officer, and various areas of specialization (e.g., Railroad, Cargo Tank Trucks, etc.).

NFPA 473—Standard for Professional Competence of EMS Personnel Responding to Hazardous Material Incidents.

The purpose of NFPA 473 is to specify minimum requirements of competence and to enhance the safety and protection of response personnel and all components of the emergency medical services system. The overall objective is to reduce the number of EMS personnel accidents, exposures, and injuries and illnesses resulting from hazmat incidents. There are two levels of EMS/HM responders:

- **EMS/HM Level I.** Persons who, in the course of their normal duties, may be called on to perform patient care activities in the cold zone at a hazmat incident. EMS/HM Level I responders provide care to only those individuals who no longer pose a significant risk of secondary contamination. Level I includes different competency requirements for Basic (BLS) and Advanced Life Support (ALS) personnel.
- **EMS/HM Level II.** Persons who, in the course of their normal duties, may be called on to perform patient care activities in the warm zone at a hazmat incident. EMS/HM Level II responders may provide care to those individuals who still pose a significant risk of secondary contamination. In addition, personnel at this level shall be able to coordinate EMS activities at a hazmat

incident and provide medical support for hazmat response personnel. Level II includes different competency requirements for Basic (BLS) and Advanced Life Support (ALS) personnel.

There are many other important standards-writing bodies, including the American National Standards Institute (ANSI), the American Society of Mechanical Engineers (ASME), the American Society for Testing and Materials (ASTM), and the American Petroleum Institute (API). Each of these organizations approves or creates standards ranging from hazardous materials container design to personal protective clothing and equipment.

NOTES

STANDARD OF CARE

The concept of "standard of care" has been common in the EMS field for years, but has only recently been recognized in the hazmat emergency response field. "Standard of care" can be defined as the minimum accepted level of hazmat service to be provided as may be set forth by law, current regulations, consensus standards, local protocols and practice, and what has been accepted in the past (precedent). Standard of care may also be influenced by legal findings and case law precedent. This standard of care allows your actions to be judged based upon what is expected of someone with your level of training and experience acting in the same or similar situation.

The standard of care is a dynamic element and historically has improved over time. Looking forward, consider that a number of today's accepted hazmat response practices will no longer be recognized in five years. For example, the "washdowns" of the 1970's are viewed today as a poor operating practice.

As an emergency response professional, one must recognize that (1) a standard of care exists, and (2) that the "highbar" is constantly moving upwards. Training and continuing education are among the best ways to ensure that one will be able to provide the standard of care mandated by both society and the hazmat profession over time.

THE HAZARDOUS MATERIALS MANAGEMENT SYSTEM

The fire problem in the United States has traditionally been managed by fire suppression operations at the expense of prevention activities. Fortunately, since the release of *America Burning* in 1973, there has been an increasing emphasis on managing the fire problem from a systems perspective. Master planning, public education, residential sprinklers, improved fire code enforcement, and fire protection engineering are some examples of this change.

A similar situation exists with the hazmat problem within the community and industry. The hazmat issue must be approached from a larger, coordinated perspective if it is to be effectively managed and controlled. There are four key elements in a hazmat management system: planning and preparedness, prevention, response, and clean-up and recovery.

NOTES

If the community or a facility is performing its responsibilities within the planning and prevention functions, one will hopefully see a reduction in the number and severity of response and clean-up activities.

PLANNING AND PREPAREDNESS

Planning is the first and most critical element of the system. The ability to develop and implement an effective hazmat management plan depends upon two elements: hazards analysis and the development of a hazmat emergency operations plan.

Hazards Analysis—analysis of the hazmats present, including their location, quantity, specific physical and chemical hazards, previous history, and risk of release.

Contingency (Emergency) Planning—a comprehensive and coordinated response to the hazmat problem. This response builds upon the hazards analysis and recognizes that no single public or private sector agency is capable of managing the hazmat problem by itself.

The data and information generated by these activities will allow emergency managers to assess the potential risk to the community or facility, and to develop and allocate resources as necessary.

HAZARD ANALYSIS

Hazard analysis is the foundation of the planning process. It should be conducted for every hazmat location designated as having a moderate or high probability for an incident. In addition to risk evaluation, vulnerability—what is susceptible to damage should a release occur—must also be examined.

A hazard analysis provides the following benefits:

- It lets ERP know what to expect.
- It provides planning for less frequent incidents.
- It creates an awareness of new hazards.
- It may indicate a need for preventive actions, such as monitoring systems and facility modifications.
- It increases the chance of successful emergency operations.

An evaluation team familiar with the facility or response area can facilitate the hazard analysis process. For example, within a facility this team may include safety, environmental, and industrial engineering professionals. Similarly, fire officers and members from each battalion or district, as well as representatives from prevention and the hazmat section, would be appropriate for a fire department. The primary concern here is geographic, as most firefighters are very familiar with their "first due" area.

There are four components of a hazard analysis program:

1) ***Hazards identification***—provides specific information on situations that have the potential for causing injury to life or damage to property and the environment due to a hazardous materials spill or release. Hazards identification will initially be based upon a review of the history of incidents within the facility or industry, or evaluating those

facilities which have submitted chemical lists and reporting forms under SARA, Title III and related state and local right-to-know legislation. Information should include:

- Chemical identification
- Location of facilities that manufacture, store, use, or move hazardous materials.
- The type(s) and design of chemical container or vessel.
- The quantity of material that could be involved in a release.
- The nature of the hazard associated with the hazmat release (e.g., fire, explosion, toxicity, etc.).
- The presence of any fixed suppression or detection systems.

2) ***Vulnerability analysis***—identifies areas that may be affected or exposed, and what facilities, property, or environment may be susceptible to damage should a hazmat release occur. A comprehensive vulnerability analysis provides information on:
   - The size/extent of vulnerable zones. Specifically, what size area may be significantly affected as a result of a spill or release of a known quantity of a specific hazmat under defined conditions.
   - The population, in terms of numbers, density, and types. For example, facility employees, residents, hospitals, nursing homes, etc.
   - Private and public property that may be damaged, including essential support systems (e.g., water supply, power, communications) and transportation corridors and facilities.
   - The environment that may be affected, and the impact of a release on sensitive natural areas and wildlife.

3) ***Risk analysis***—assesses (1) the probability or likelihood of an accidental release, and (2) the actual consequences that might occur. The risk analysis is a judgment of incident probability and severity based upon the previous incident history, local experience, and the best available hazard and technological information.

4) ***Emergency response resources evaluation***—based upon the potential risks, considers emergency response resource requirements. These would include personnel, equipment, and supplies necessary for hazmat control and mitigation, EMS, protective actions, traffic control, etc. Inventories available equipment and supplies along with their ability to function. For example, are firefighting foam supplies adequate to control and suppress vapors from a gasoline tank truck rollover?

Time and resources will dictate the depth and extent to which the hazards analysis can be completed. The focus is on the hazards created by the most common and most hazardous substances. Even the simplest plan will be better than no plan at all.

A completed hazard analysis should allow emergency managers and planners to determine what level of response to emphasize, what resources will be required to achieve that response, and what type and quantity of mutual aid and other support services will be required.

NOTES

PROCESS SAFETY MANAGEMENT HAZARDS ANALYSIS TECHNIQUES

Hazards analysis is also an integral element of the Process Safety Management (PSM) process required by *OSHA 1910.119—Process Safety Management of Highly Hazardous Chemicals, Explosives and Blasting Agents* and *EPA Part 68—Risk Management Programs for Chemical Accidental Release Prevention*. Both regulations impact industries which manufacture, store, and use highly hazardous chemicals and explosives, including refineries and chemical and petrochemical manufacturing plants.

Hazards analysis methods commonly used by safety professionals within industry include:

- ***What If Analysis.*** This method asks a series of questions, such as, "What if Pump X stops running?" or "What if an operator opens the wrong valve?" to explore possible hazard scenarios and consequences. This method is often used to examine proposed changes to a facility.
- ***HAZOP Study.*** This is the most popular method of hazard analysis used within the petroleum and chemical industries. The hazard and operability (HAZOP) study brings together a multi-disciplinary team, usually of five to seven people, to brainstorm and identify the consequences of deviations from design intent for various operations. Specific guide words ("No," "More," "Less," "Reverse," etc.) are applied to parameters such as product flows and pressures in a systematic manner. The study requires the involvement of a number of people, working with an experienced team leader.
- ***Failure Modes, Effects, and Criticality Analysis (FMECA).*** This method tabulates each system or unit of equipment, along with its failure modes, the effect of each failure on the system or unit, and how critical each failure is to the integrity of the system. Then the failure modes can be ranked according to criticality to determine which are the most likely to cause a serious accident.
- ***Fault Tree Analysis.*** A formalized deductive technique that works backwards from a defined accident to identify and graphically display the combination of equipment failures and operational errors that led up to the accident. It can be used to estimate the likelihood of events.
- ***Event Tree Analysis.*** A formalized deductive technique that works forward from specific events or sequences of events that could lead to an incident. It graphically displays the events that could result in hazards and can be used to calculate the likelihood of an accident sequence occurring. It is the reverse of fault tree analysis.

FIGURE 1-4: PROCESS SAFETY MANAGEMENT HAZARDS ANALYSIS TECHNIQUES

CONTINGENCY AND EMERGENCY PLANNING

NOTES

Hazardous materials management is a multi-disciplined issue that goes beyond the resources and capabilities of any single agency or organization. As there will be a variety of "players" responding to a major hazmat emergency, the emergency operations plan and related procedures will establish the framework for how the emergency response effort will operate. To effectively manage the overall hazmat problem within the community, a comprehensive planning process must be initiated. This effort is usually referred to as "contingency planning" or "emergency planning."

There are many federal, state, and local requirements that apply to emergency planning. The one that most directly affects ERP is Title III of the Superfund Amendments and Reauthorization Act of 1986 (SARA). Title III requires the establishment of state emergency response commissions (SERC) and local emergency planning committees (LEPC). Title III also outlines specific requirements covering factors such as extremely hazardous substances (EHS), threshold planning quantities, make-up of LEPC's, dissemination of the planning, chemical lists and MSDS information to the community and the general public, facility inventories, and toxic chemical release reporting.

Figure 1-5 provides an overview of the hazmat emergency planning process, including:

1) **Organizing the planning team**—Planning requires community involvement throughout the process. Experience has shown that plans prepared by only one person or agency are doomed to failure. Emergency response requires trust, coordination, and cooperation. Remember, there is no single agency (public or private) which can effectively manage a major hazmat emergency alone.
2) **Defining and implementing the major tasks of the planning team**—These include reviewing any existing plans, identifying hazards, and analyzing and assessing current prevention and response capabilities.
3) **Writing the plan**—There are two approaches to this step: (a) develop or revise a hazmat appendix or a section of a multi-hazard emergency operations plan, or (2) develop or revise a single-hazard plan specifically for hazardous materials. Once the plan is written, it must be approved by all the respective planning groups involved.
4) **Revising, testing, and maintaining the plan**—Every emergency plan must be evaluated and kept up-to-date through the review of actual responses, simulation exercises, and the regular collection of new data and information.

While emergency planning is essential, the completion of a plan does not guarantee that the facility or community is actually prepared for a hazmat incident. Planning is only one element of the total hazmat management system.

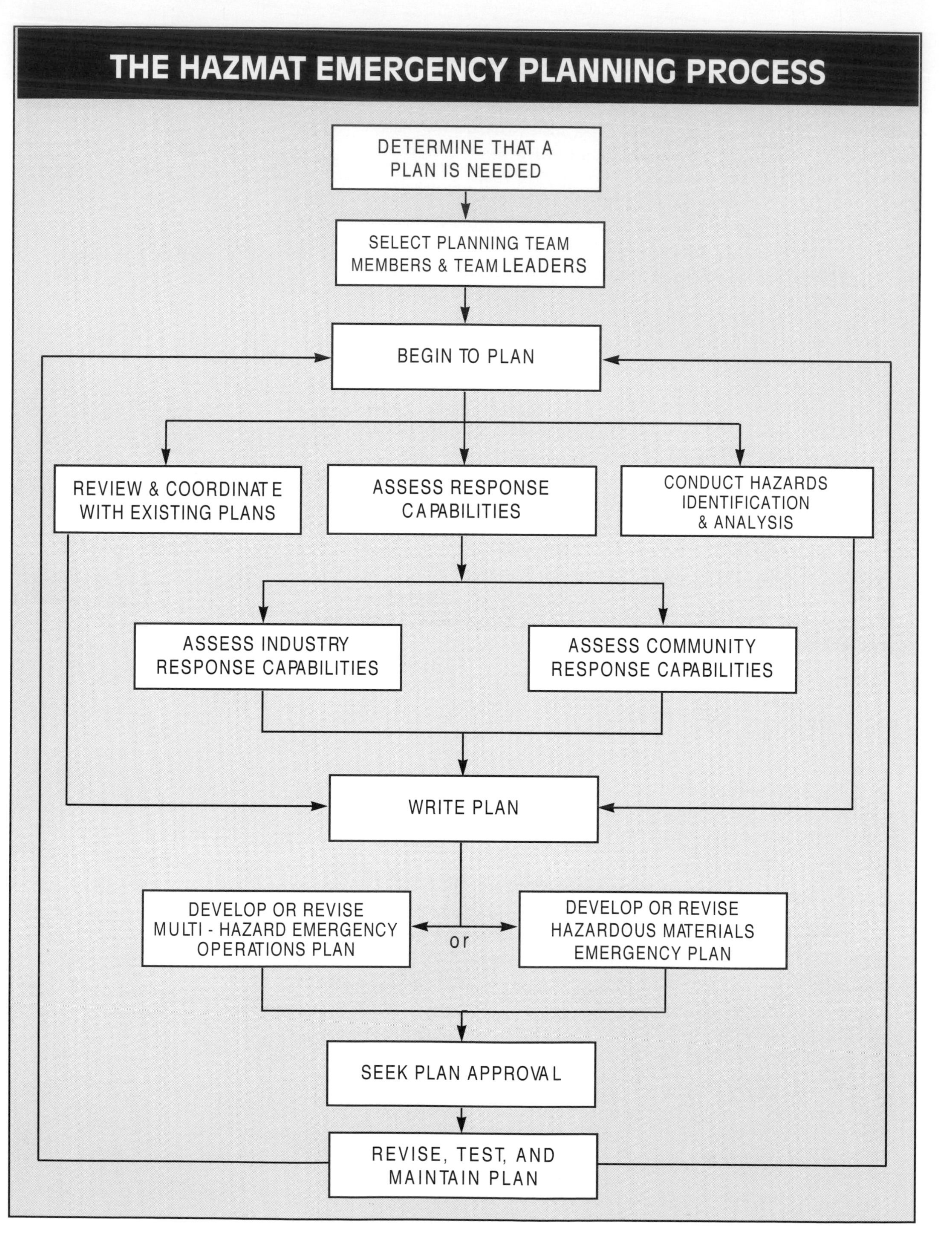

FIGURE 1.5: THE HAZMAT EMERGENCY PLANNING PROCESS

PREVENTION

NOTES

The responsibility for the prevention of hazmat releases is shared between the public and private sectors. Because of their regulatory and enforcement capabilities, however, public sector agencies generally receive the greatest attention and often "carry the biggest stick."

Prevention activities often include the following:

Hazmat Process, Container Design, and Construction Standards.

Almost all hazmat facilities, containers, and processes are designed and constructed to some standard. This "standard of care" may be based upon voluntary consensus standards, such as those developed by the National Fire Protection Association (NFPA) and the American Society of Testing and Materials (ASTM), or on government regulations. Many major petrochemical, hazmat companies, and industry trade associations have also developed their own respective engineering standards and guidelines.

All containers used for the transportation of hazmats are designed and constructed to both specification and performance regulations established by the U.S. Department of Transportation (DOT). These regulations are referenced in Title 49 of the Code of Federal Regulations (CFR). In certain situations, hazmats may be shipped in non-DOT specification containers which have received a DOT exemption.

Inspection and Enforcement.

Fixed facilities, transportation vehicles, and transportation containers are subject to some form of hazmat inspection. Fixed facilities will commonly be inspected by state and federal OSHA and EPA inspectors, in addition to state fire marshals and local fire departments. It should be recognized that many of these inspections will focus upon fire safety and life safety issues, and may not adequately address either environmental or process safety issues.

Transportation vehicle inspection is generally based upon criteria established within Title 49 CFR. The enforcing agency is often the state police, but this will vary according to the individual state, the hazmat being transported, and the mode of transportation. There are some local fire departments, such as Aurora, Colorado, which routinely perform inspections of hazmat cargo tank trucks.

Among the U.S. DOT agencies with hazmat regulatory responsibilities are the following:

- *Office of Hazardous Materials Transportation (OHMT)* of the Research and Special Programs Administration (RSPA). Responsible for all hazmat transportation regulations except bulk shipment by ship or barge. Includes designating and classifying hazmats, container safety standards, label and placarding requirements, and handling, stowing, and other in-transit requirements.
- *Office of Motor Carrier Safety (OMCS)* of the Federal Highway Administration (FHA). Responsible for inspection and

NOTES

enforcement activities relating to hazmat highway transportation and depot trans-shipment points.

- ***Federal Railroad Administration (FRA).*** Responsible for enforcement of regulations relating to hazmats carried by rail or held in depots and freight yards.
- ***Federal Aviation Administration (FAA).*** Responsible for the enforcement of regulations relating to hazmat shipments on domestic and foreign carriers operating at U.S. airports and in cargo handling areas.
- ***U.S. Coast Guard (USCG).*** Responsible for the inspection and enforcement of regulations relating to hazmats in port areas and on domestic and foreign ships and barges operating in the navigable waters of the United States.

Public Education.

Hazmats are a concern not only for industry but also for the community. The average homeowner contributes to this problem by improperly disposing of substances such as used motor oil, paints, solvents, batteries, and other chemicals used in and around the home. As a result, many communities have initiated full-time household chemical waste awareness, education, and disposal programs. In other instances, communities have established used motor oil collection stations and chemical clean-up days in an effort to reclaim and recycle these materials.

Handling, Notification, and Reporting Requirements.

These guidelines actually act as a bridge between planning and prevention functions. There are many federal, state, and local regulations which require those who manufacture, store, or transport hazmats and hazardous wastes to comply with certain handling, notification, and reporting rules. Key federal regulations include the facility reporting requirements of SARA, Title III, and the release notification requirements of CERCLA (Superfund). There are also many state regulations which are similar in scope and which often exceed the federal requirements.

RESPONSE

When the prevention and enforcement functions fail, response activities begin. Since it is impossible to eliminate all risks associated with the manufacture, storage, and use of hazmats, the need for a well-trained, effective emergency response capability will always exist.

Response activities should be based upon the information and probabilities identified during the planning process. Response activities must be based upon an evaluation of the facility or local hazmat problem. While every community should have access to a hazmat response capability, that capability does not always have to be provided by either local government or the fire service. Numerous states and regions have established both statewide and regional hazmat response team systems which ensure the delivery of a competent and effective capability in a timely manner.

Levels of Hazardous Materials Incidents—Community

RESPONSE LEVEL	DESCRIPTION	RESOURCES	EXAMPLES
I POTENTIAL EMERGENCY CONDITIONS	An incident or threat of a release which can be controlled by the first responder. It does not require evacuation, beyond the involved structure or immediate outside area. The incident is confined to a small area and poses no immediate threat to life and property.	Essentially a local level response with notification of the appropriate local, state, and federal agencies. Required resources may include: • Fire Department • Emergency Medical Services (EMS) • Law Enforcement • Public Information Officer (PIO) • Chemtrec • National Response Center	• 500-gallon fuel oil spill • Inadvertent mixture of chemicals • Natural gas leak in a building
II LIMITED EMERGENCY CONDITIONS	An incident involving a greater hazard or larger area than Level I which poses a potential threat to life and property. It may require a limited protective action of the surrounding area.	Requires resources beyond the capabilities of the initial local response personnel. May require a mutual aid response and resources from other local and state organizations. May include: • All Level I Agencies • Hazmat Response Teams • Public Works Department • Red Cross • Regional Emergency Management Staff • State Police • Public Utilities	• Minor chemical release in an industrial facility • A gasoline tank truck rollover • A chlorine leak at a water treatment facility
III FULL EMERGENCY CONDITION	An incident involving a severe hazard or a large area which poses an extreme threat to life and property and which may require a large-scale protective action.	Requires resources beyond those available in the community. May require the resources and expertise of regional, state, federal, and private organizations. May include: • All Level I and II Agencies • Mutual Aid Fire, Law Enforcement, and EMS • State Emergency Management Staff • State Department of Environmental Resources • State Department of Health • Environmental Protection Agency (EPA) • U.S. Coast Guard • Federal Emergency Management Agency (FEMA)	• Major train derailment with fire • Explosion or toxicity hazard • A migrating vapor cloud release from a petrochemical processing facility

FIGURE 1-6 (A): LEVELS OF HAZARDOUS MATERIALS INCIDENTS—COMMUNITY

Levels of Hazardous Materials Incidents—Petrochemical Industry

RESPONSE LEVEL	DESCRIPTION	RESOURCES	EXAMPLES
I INCIDENT	Minimal danger to life, property, and the environment. Problem is limited to immediate work area and public health, safety, and environment are not affected.	Handled by On-Shift Emergency Response Team (ERT) with no off-shift or mutual aid response.	• Minor spills and releases less than 55 gallons • Small pump seal fire • Minor vapor release during product transfer operations
II SERIOUS INCIDENT	Moderate danger to life, property, and the environment on the plant site. Problem is currently limited to plant property, but has the potential for involving additional exposures or migrating off-site and affecting public health, safety, and environment for a short period of time.	Requires On-Shift ERT response. Additional assistance required from off-duty ERT personnel and/or mutual aid units. Plant EOC may be activated. Corporate Crisis Emergency Plan may be activated.	• Large release of flammable, corrosive, or toxic vapors • Releases of over 100 gallons of hazardous material • Large spill fire or a seal fire on a floating roof tank
III CRISIS SITUATION	Extreme danger to life, property, and the environment. Problem goes beyond the refinery property and can impact public health, safety and the environment or a large geographic area for an indefinite period of time.	Requires multi-organizational response from plant, local fire service, industrial mutual aid units, and public safety resources. Plant EOC is activated. Corporate Crisis Emergency Plan activated.	• Process unit fire or explosion • Major release of flammable, corrosive or toxic vapors • Shipboard fire, major oil spill, or HM release which can impact major waterways

FIGURE 1-6 (B): LEVELS OF HAZARDOUS MATERIALS INCIDENTS—PETROCHEMICAL INDUSTRY

NOTES

LEVELS OF INCIDENT

Fortunately, not every incident is a major emergency. Response to a hazmat release may range from a single-engine company responding to a natural gas leak in the street, to a railroad derailment involving dozens of government and private agencies and their associated personnel.

These incidents can be categorized based upon their severity and the resources they require. Figure 1-6 (A and B) outlines these response levels.

RESPONSE GROUPS

The emergency response community consists of various agencies and individuals who respond to hazmat incidents. They can be categorized based upon their knowledge, expertise, and resources.

These responders can be compared to the levels of capability found within a typical Emergency Medical Services (EMS) system. In that system, an injury such as a fractured arm can be effectively managed by a First Responder or Emergency Medical Technician–Ambulance (EMT-A), while a cardiac emergency will require the services of an EMT-I (Intermediate) or an EMT-P (Paramedic).

In the same way, a diesel fuel spill can usually be contained by first responders, such as a fire department engine company using an absorbent. An accident involving a poison or reactive chemical will, however, require the on-scene expertise of a hazmat technician or hazmat response team. Figure 1-7 illustrates this comparison.

HAZMATS vs. EMS
COMPARING RESPONSE GROUPS

SKILL AND KNOWLEDGE LEVEL	EMERGENCY MEDICAL SERVICES	HAZARDOUS MATERIALS RESPONSE
BASIC	FIRST RESPONDER	FIRST RESPONDER–OPERATIONS
INTERMEDIATE	EMERGENCY MEDICAL TECHNICIAN	HAZARDOUS MATERIALS TECHNICIAN
ADVANCED	PARAMEDIC	HAZARDOUS MATERIALS SPECIALIST

FIGURE 1-7: COMPARING EMS AND HAZMAT EMERGENCY RESPONSE GROUPS

NOTES

HAZMAT RESPONSE TEAM (HMRT)

In order to effectively and efficiently respond to hazmat emergencies, many facilities and communities have established hazmat response teams (HMRT).

NFPA 472 defines an HMRT as an organized group of trained response personnel operating under an Emergency Operations Plan and appropriate standard operating procedures, who are expected to perform work to handle and control actual or potential leaks or spills of hazardous materials requiring close approach to the material. The HMRT members respond to releases for the purpose of control or stabilization of the incident.

In evaluating the need for an HMRT, consider the following points:

- There is no single department or agency which can effectively manage the hazmat issue by itself.
- Every community does not require a HMRT. However, every community should have access to a HMRT capability through either local, regional, or state resources.
- An HMRT will not necessarily solve the hazmat problem.
- There are numerous constraints and requirements associated with developing an effective HMRT capability. These include legal, insurance, and political issues, both initial and continuing funding sources, resource determination and acquisition, personnel and staffing, and initial and continuing training requirements.

HMRT's typically function as an Incident Management System (IMS) branch under the direct control of a Hazmat Branch Officer who reports to the Incident Commander. Based upon their assessment of the hazmat problem, the HMRT, through the Hazmat Branch Officer, provides the Incident Commander with a list of options and a recommendation for mitigation of the hazmat problem. However, the final decision is always made by the Incident Commander.

In some areas, regional or statewide hazmat response systems have been developed. Many of these have different levels of HMRT's based upon their staffing and equipment inventory. For example, the state of Virginia is divided into regions; each region is staffed by a Level III HMRT which responds to the most serious emergencies, while the more routine and less significant incidents are handled by several Level II HMRT's which are dispersed throughout each region by local public safety agencies.

Personnel on HMRT's are trained to the OSHA Hazmat Technician and Hazmat Specialist levels, and must participate in a medical surveillance program based upon the requirements of 29 CFR 1910.120. Among the specialized equipment carried by an HMRT are reference libraries, computers and communications equipment, personal protective clothing and equipment, direct-reading monitoring and detection equipment, control and mitigation supplies and equipment, and decontamination supplies and equipment.

NOTES

CLEAN-UP AND RECOVERY

Clean-up and recovery operations are designed to (1) clean-up or remove the hazmat spill or release, and (2) restore the facility and/or community back to normal as soon as possible. In many instances, chemicals involved in a hazmat release will be eventually classified as hazardous wastes.

Clean-up operations fall under the guidelines of both HAZWOPER, CERCLA (Superfund), and RCRA. Clean-up activities can be classified as:

- **Short-term**—those actions immediately following a hazmat release that are primarily directed towards the removal of any immediate hazards and restoring vital support services (i.e., reopening transportation systems, drinking water systems, etc.) to minimum operating standards. Short-term activities may last up to several weeks.

  These activities are normally the responsibility of the "responsible party"—usually the facility operator or transportation carrier (e.g., railroad, truck company). In situations where the responsible party has not been identified or does not have sufficient financial resources, they may be assumed by state or federal environmental agencies.

- **Long-term**—those remedial actions which return vital support systems back to normal or improved operating levels. Examples would include groundwater treatment operations, the mitigation of both aboveground and underground spills, and the monitoring of flammable and toxic contaminants. These activities may not be directly related to a specific hazmat incident, but are often the result of abandoned industrial or hazardous waste sites. These operations may extend over months or years.

Recovery operations focus upon restoring the facility, the community, and/or emergency response organization to normal operating conditions. Tasks would include restocking all supplies and equipment, compilation and documentation of resources purchased and/or used during the emergency response, and financial restitution, where appropriate.

ROLE OF ERP DURING CLEAN-UP OPERATIONS

Within industry, Emergency Response Team (ERT) personnel may also be responsible for the clean-up of minor spills and releases so that facility operations may continue. In contrast, public safety response personnel are usually not responsible for the clean-up of hazmat releases. However, they will often continue to be responsible for site safety until the emergency phase is stabilized and the incident is completely terminated.

At short-term operations immediately following an incident, the Incident Commander should ensure that the work area is closely controlled, that the general public is denied entry, and that the safety of emergency responders and the public is maintained during clean-up and recovery operations. When interfacing with both industry responders and contractors, the Incident Commander should ensure that they are trained to meet the requirements of OSHA 1910.120 and/or NFPA 472.

NOTES

Long-term clean-up and recovery operations do not normally require the continuous presence of ERP. Depending upon the size and scope of the clean-up, the environmental clean-up contractor or a government official (On-Scene Coordinator—"OSC" or the Remedial Project Manager—"RPM") will be the central contact point. Response personnel should be familiar with the clean-up operation, including its organizational structure, the OSC/RPM, work plan, time schedule, and site safety plan.

Clean-up operations should conform to the general health and safety requirements of both state and federal EPA and OSHA standards.

Although ERP generally do not have the authority to conduct inspections or issue citations at clean-up operations, they can bring specific concerns to the attention of the state or federal regulatory agency having jurisdiction.

SUMMARY

Hazardous materials are a multi-disciplined problem requiring an organized facility-level and community-level approach. Although this textbook is primarily oriented towards managing and implementing emergency response operations, one should recognize that response accounts for only a small portion of an effective, comprehensive hazmat management program.

REFERENCES AND SUGGESTED READINGS

Federal Emergency Management Agency, et. al., LIABILITY ISSUES IN EMERGENCY MANAGEMENT, Emmitsburg, MD: National Emergency Training Center (1992).

Fire, Frank L., Nancy K. Grant and David H. Hoover, SARA TITLE III—INTENT AND IMPLEMENTATION OF HAZARDOUS MATERIALS REGULATIONS, Tulsa, OK: Fire Engineering Books and Videos (1990).

National Fire Protection Association, HAZARDOUS MATERIALS RESPONSE HANDBOOK (2nd Edition), Boston, MA: National Fire Protection Association (1993).

National Response Team, HAZMAT EMERGENCY PLANNING GUIDE (NRT-1), Washington, DC: National Response Team (1987).

Stringfield, William H., A FIRE DEPARTMENT'S GUIDE TO IMPLEMENTING SARA, TITLE III AND THE OSHA HAZARDOUS MATERIALS STANDARDS, Ashland, MA: International Society of Fire Service Instructors (1987).

U.S. Environmental Protection Agency, et. al., HANDBOOK OF CHEMICAL HAZARD ANALYSIS PROCEDURES, Washington, DC: EPA, FEMA, DOT (1989).

U.S. Environmental Protection Agency, HAZMAT TEAM PLANNING GUIDANCE, Washington, DC: EPA (1990).

U.S. Environmental Protection Agency, et. al., TECHNICAL GUIDANCE FOR HAZARDS ANALYSIS—EMERGENCY PLANNING FOR EXTREMELY HAZARDOUS SUBSTANCES, Washington, DC: EPA, FEMA, DOT (1987).

U.S. Occupational Safety and Health Administration, HAZWOPER INTERPRETATIVE QUIPS, Washington, DC: OSHA Office of Health Compliance Assistance (March, 1993).

NOTES

Calvin and Hobbes
by Bill Watterson
DOWNTOWN TOKYO!
© 1986 Universal Press Syndicate
AARRGHHGH
GODZILLA.
WATTERSON
3-8

CHAPTER 2

HEALTH AND SAFETY

> *"Learn to avoid situations that are so exciting you don't survive."*
>
> Alan V. Brunacini, Fire Chief, Phoenix, Arizona

OBJECTIVES

1) Describe the following basic toxicological principles:
   - Exposure
   - Toxicity
   - Acute and chronic exposures
   - Acute and chronic effects
   - Routes of exposure to hazardous materials
   - Dose/response relationship
   - Local and systemic effects
   - Target organs
2) Identify the seven types of harm created by exposure to hazardous materials and their effects upon the human body.
3) Define the following toxicological terms and describe their significance in predicting the extent of health hazards at a hazmat incident:
   - Parts per Million (ppm) and Parts per Billion (ppb)
   - Lethal Dose (LD_{50})
   - Lethal Concentration (LC_{50})
4) Define the following exposure guidelines and describe their significance in predicting the extent of health hazards at a hazmat incident:
   - Threshold Limit Value—Time Weighted Average (TLV/TWA)
   - Threshold Limit Value—Ceiling (TLV/C)

NOTES

- Threshold Limit Value—Short-Term Exposure Limit (TLV/STEL)
- Permissible Exposure Limit (PEL)
- Recommended Exposure Limit (REL)
- Immediately Dangerous to Life and Health (IDLH)
- Emergency Response Planning Guideline (ERPG)

5) Define the following terms associated with radiological materials and describe their significance in predicting the extent of health hazards at a hazmat incident:
   - Alpha Radiation
   - Beta Radiation
   - Gamma Radiation
   - Half-Life
   - Time, Distance, and Shielding
6) Identify the signs and symptoms and emergency care procedures for handling heat stress emergencies.
7) Identify the procedures for reducing the effects of heat stress upon responders at a hazmat incident.
8) Identify the relative advantages and disadvantages of the following cooling devices:
   - Air Cooled Jackets and Suits
   - Water Cooled Jackets and Vests
   - Ice Vests
9) Identify the signs and symptoms of cold temperature exposures and procedures for reducing the effect upon responders at a hazmat incident.
10) Identify procedures for protecting responders against excessive noise levels at a hazmat incident.
11) Identify the components of a medical surveillance program for hazmat responders as outlined in OSHA 1910.120 (q).
12) Identify the components of a personal protective equipment (PPE) program.
13) Describe the components of a site safety plan for operations at a hazmat incident.
14) Identify the elements that should be covered in a safety briefing prior to making entry into the hot zone of a hazmat incident.
15) Identify the key safety practices and guidelines for conducting entry operations within the hot zone of a hazmat incident.
16) Describe the procedures and components for conducting pre- and post-entry medical monitoring for response personnel operating at a hazmat incident.
17) Describe the procedures for establishing and operating a Rehabilitation Sector at a hazmat incident.

NOTES

INTRODUCTION

Emergency response personnel health and safety has become an important issue for both management and labor. Regardless of its size or nature, every hazmat incident presents ERP with a potentially hostile environment. While preventing exposures to hazardous materials is always a primary concern, command personnel must also evaluate the physical working conditions, work intervals, and the stress of working in personal protective clothing and equipment.

Hazmat incidents are characterized by work environment hazards which may pose an immediate danger to life and health (IDLH), but which may not be immediately obvious or identifiable. They may also vary according to the tasks being performed and the ERP's location on the incident site and may change as response activities progress.

Protection of both emergency response and support personnel, as well as the general public, should always be the Incident Commander's (IC) primary concern. In this chapter, we will examine these health and safety concerns in detail.

TOXICOLOGY

Toxicology is the study of chemical or physical agents that produce adverse responses in the biologic systems with which they interact. Chemical agents include gases, vapors, fumes, dusts, etc., while physical agents include radiation, hot and cold environments, noise, and so forth.

Toxicity is defined as the ability of a substance to cause injury to a biologic tissue. In humans this generally refers to unwanted effects that are produced when a chemical has reached a sufficient concentration at a particular location within the body. To properly protect those personnel operating at a hazmat incident, both Command and Hazmat Branch personnel must be able to understand the basic concepts of toxicology, as well as toxicological and exposure terms, and interpret toxicity and exposure data.

There are a number of factors which determine the toxicity of a chemical and the potential harm which may result. A simple way of understanding this concept and its potential for harm is the health hazard equation of:

EXPOSURE + TOXICITY = HEALTH HAZARD

Exposure means that you have had contact with the chemical through either inhalation, ingestion, skin absorption, or direct contact. **Toxicity** refers to the ability of the chemical to harm your body once contact has occurred. The level of toxicity will depend upon the nature of the chemical, its ability to do harm to the body, and its concentration within the body. Different people may react differently to the same level of exposure.

NOTES

EXPOSURE CONCERNS

EXPOSURES AND HEALTH EFFECTS

Chemical exposures and their health effects are commonly described as acute or chronic. **Acute exposures** describe an immediate exposure, such as a single dose that might occur during an emergency response. **Chronic exposures** are low exposures repeated over time, such as responding to a number of hazmat emergencies while serving as a member of the facility Emergency Response Team or the fire department.

An **acute health effect** results from a single dose or exposure to a material, such as a single exposure to a highly toxic material or a large dose of a less toxic material. Signs and symptoms from the exposure may be immediate or may not be evident for 24 to 72 hours after the exposure. Some chemicals may produce both immediate and delayed effects. It is also possible that signs and symptoms may be "masked" by illnesses such as the common cold, nausea, and smoke inhalation. These potential delayed or "masked" effects of exposure illustrate the importance of medical debriefings as part of the incident termination phase.

Chronic health effects result from a single exposure or from repeated doses or exposures over a relatively long period of time. Although chronic exposures are usually correlated with long-term worker exposures in an industrial environment, they can also result from chemical exposures at long-term remedial clean-up operations.

Exposures to various types and doses of hazmats over a period of years, similar to those encountered by firefighters and Hazardous Materials Response Team (HMRT) members, are also associated with chronic health effects. To monitor these exposures, ERP should be provided with baseline medical profiles, participate in a medical surveillance program, and document all hazmat exposures.

ROUTES OF EXPOSURE

ERP can be exposed to hazardous materials by:

- ❑ **Inhalation**—the introduction of a chemical or toxic products of combustion into the body by way of the respiratory system. Inhalation is the most common exposure route and often the most damaging. Remember that a material does not have to be a gas in order to be inhaled—solid materials may generate fumes or dusts in a dry powdered form, while liquid chemicals will generate vapors, mists, or aerosols which can be inhaled.

  Once within the body, toxins may be absorbed into the bloodstream and carried to other internal organs, may paralyze the respiratory system, reduce the ability of the blood to transport oxygen, or affect the upper and/or lower respiratory tract. Inhalation exposures are much more likely to involve a large number of civilian exposures.

- ☐ **Skin absorption**—the introduction of a chemical or agent into the body through the skin. Skin absorption can occur with no sensation to the skin itself. Do not rely on pain or irritation as a warning sign of absorption. Some poisons are so concentrated that a few drops placed on the skin may result in death (e.g., phenol).

  Skin absorption is enhanced by abrasions, cuts, heat and moisture. This can create critical problems when working at incidents involving biological and etiological agents. Anyone with large open cuts, rashes, or abrasions should be prohibited from working in areas where they may be exposed. Smaller cuts or abrasions should be covered with a nonporous dressing.

  The rate of skin absorption can vary depending upon the body part that's exposed. For example, assuming the area of skin exposed and the duration of exposure are equal, the rate of skin absorption through the scalp or genitals area will be considerably faster than through the forearm.

  Absorption through the eyes is one of the fastest means of exposure since the eye has a high absorbency rate. This type of exposure may occur when a chemical is splashed directly into the eye, when a chemical is "carried" on toxic smoke particles into the eyes from a fire, or when gases or vapors are absorbed through the eyes. Likewise, this may be seen as an early warning signal for either PPE or SCBA failures.

- ☐ **Ingestion**—the introduction of a chemical into the body through the mouth. Inhaled chemicals may be trapped in saliva and swallowed. Exposed personnel should be prohibited from smoking, eating, or drinking except in designated rest and rehab areas after being decontaminated.

- ☐ **Direct contact**—direct skin contact with some chemicals, such as corrosives, will immediately damage skin or body tissue upon contact.

  Acids have a strong affinity for moisture and can create significant skin and respiratory tract burns. However, the injury process also creates a clot-like barrier which blocks deep skin penetration. In contrast, caustic or alkaline materials dissolve the fats and lipids which make up skin tissue and change the solid tissue into a soapy liquid (think of how drain cleaners using caustic chemicals dissolve grease and other materials in sinks and drains!). As a result, caustic burns are often much deeper and more destructive than acid burns.

  Other chemicals may be injected directly through the skin and into the bloodstream. Mechanisms of injury include needle stick cuts at EMS emergencies and the injection of high pressure gases and liquids into the body similar to the manner in which flu shots are injected with pneumatic guns.

NOTES

DOSE/RESPONSE RELATIONSHIP

A fundamental relationship exists between the chemical dose and the response produced by the human body. Given the broad range of toxicities any substance might cause, the wisdom of Paracelsus (1493–1541) becomes clear: "All substances are poisons; there is none which is not a poison. The right dose differentiates a poison and a remedy." Translation—dose makes the poison.

The dose/response concept is based upon the following assumptions:

- The magnitude of the response is dependent upon the concentration of the chemical at the biological site of action (i.e., target organ).
- The concentration of the chemical at the biological site of action is a function of the dose administered.
- Dose and response are essentially a cause/effect relationship.

Dose is the concentration or amount of material to which the body is exposed. The human body's **response** to a dose is either toxic or nontoxic. Human response may also be influenced by the age, state of health, and nutrition state of the individual. Typically, as the size of the dose increases, the potential for a toxic response also increases. For example, one gram of aspirin is an accepted dose for medicinal purposes, yet quantities in excess of one bottle have caused death.

With many chemicals, increasing moderate doses will cause no apparent biological damage. However, this increased dose rate will eventually reach a point defined as the threshold level. At that concentration and above, chemical exposure can result in biological harm. These effects may range from slight irritation to death, depending on the dose and properties of the chemical. Figure 2-1 illustrates a typical dose/response curve, and Figure 2-2 illustrates the dose/response relationship.

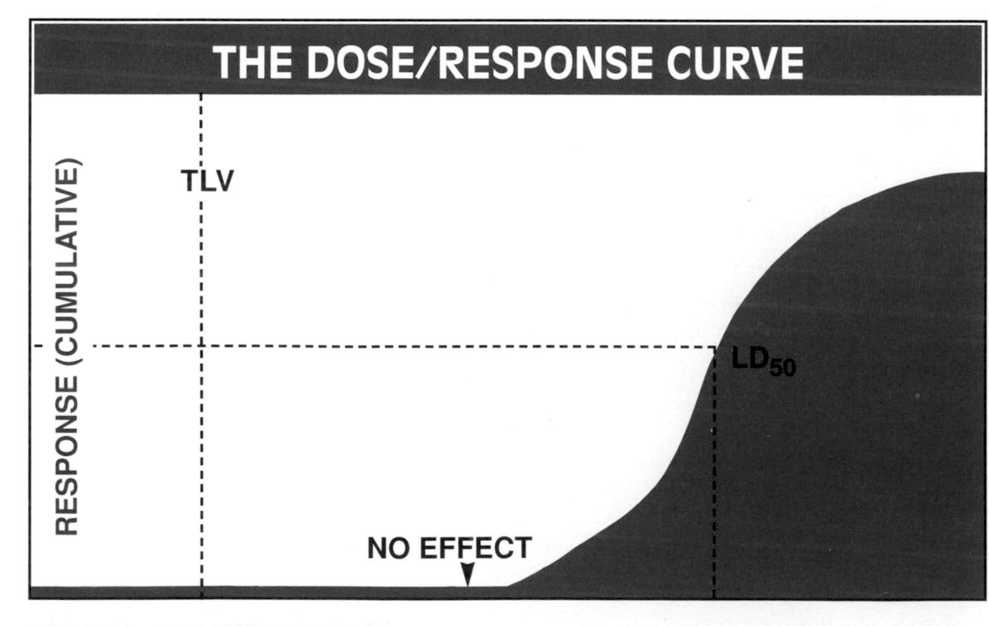

FIGURE 2-1: DOSE/RESPONSE CURVE

NOTES

DOSE/RESPONSE RELATIONSHIP		
Dose	**Acute Effect**	**Chronic Effect**
1 oz. of Bourbon Consumed in 60 Min.	MINIMAL	NONE
1 qt. of Bourbon Consumed in 60 Min.	ILLNESS OR DEATH	MINIMAL
1 oz. of Bourbon Consumed Every 60 Mins. for 12 hrs. Each Day, 365 Days a Year	MINIMAL	BRAIN/LIVER DAMAGE
1 qt. of Bourbon Consumed Over a Year	NONE	NONE

FIGURE 2-2: DOSE/RESPONSE RELATIONSHIP USING ALCOHOL CONSUMPTION AS AN EXAMPLE

EFFECTS OF HAZARDOUS MATERIALS EXPOSURES

Health effects of a hazmat exposure can be described in terms of how a hazmat attacks the body. **Local** implies an effect at the point of contact—for example, a corrosive burn. If the chemical enters the bloodstream and attacks target organs and internal areas of the human body, it is described as a **systemic effect.** Multiple systemic effects are also a distinct possibility.

Systemic effects often show up at target organs. **Target organs** are organs/tissues where a toxin exerts its effects; it is not necessarily the organ/tissue where the toxin is most highly concentrated. For example, over 90% of the lead in the adult human body is in the skeleton, but lead exerts its effects on the kidneys, the central nervous system, and the blood system.

Examples of target organs that may be affected by a systemic poison include the following:

- Liver—known as hepatotoxins, examples include carbon tetrachloride, vinyl chloride monomer, and nitroamines.
- Kidneys—known as nephrotoxins, examples include halogenated hydrocarbons and mercury.
- Central Nervous System—known as neurotoxins, examples include lead, toluene and organophosphate pesticides.
- Lungs and Pulmonary System—known as respiratory toxins, examples include asbestos and hydrogen sulfide.
- Blood System—known as hematotoxins, examples include benzene, chlordane, and DDT.
- Skeletal System—examples include hydrofluoric acid and selenium.

NOTES

- Skin—known as dermatotoxins, which may act as irritants, ulcers, chloracne, and/or cause skin pigmentation disturbances. Examples include halogenated hydrocarbons, coal tar compounds, and high levels of ultraviolet light.
- Fetus—known as teratogens, examples include lead and ethylene oxide.
- Genetic damage to cells or organisms—known as mutagens, examples include radiation, lead, and ethylene dibromide.

The human body can be harmed by seven different types of "harm events":

- **Thermal**—those events related to temperature extremes. High temperatures are common at fire-related emergencies involving flammable liquids, gases, and solids. Thermal harm resulting in frostbite can be found as a result of exposures to the extremely low temperatures associated with liquefied gases and cryogenic materials.
- **Mechanical**—those events resulting from direct contact with fragments scattered because of a container failure, explosion, or shock wave.
- **Poisonous**—those events related to exposure to toxins. Some chemicals effect the body by causing damage to specific internal organs or body systems. Examples include nephrotoxins and hepatotoxins. Other chemicals, such as benzene and phenol, are considered blood toxins because of their effects on the circulatory system. Similarly, neuorotoxins—such as organophosphate and carbamate pesticides—affect the function of the central nervous system.
- **Corrosive**—those events related to chemical burns and/or tissue damage from exposure to corrosive chemicals.

  Corrosives are divided into two chemical groups: acids and bases. Acids—such as strong inorganic acids like nitric, sulfuric, and hydrochloric—can cause severe tissue burns to the skin and permanent eye damage. Bases, or caustic or alkaline materials, break down fatty skin tissue and penetrate deeply. Sodium and potassium hydroxide are examples. Acids generally cause greater surface tissue damage while bases produce deeper, slower healing burns.

  Inhaled corrosive gases and vapors can also cause acute swelling of the upper respiratory tract and pulmonary edema. High water soluble materials, such as anhydrous ammonia, will affect the upper respiratory tract, while low water soluble materials, such as nitrogen dioxide and phosgene, will affect the lower respiratory tract. Lower respiratory tract injuries can lead to pulmonary edema and may be delayed for 24 to 72 hours after the exposure.
- **Asphyxiation**—those events related to oxygen deprivation within the body. Asphyxiants can be categorized as simple or chemical. Simple asphyxiants act on the body by displacing or reducing the oxygen in the air for normal breathing. Examples include carbon dioxide, nitrogen, and natural gas.

NOTES

Chemical asphyxiants disturb the normal body chemistry processes which control respiration. They can range from chemicals which inhibit oxygen transfer from the lungs to the cells (carbon monoxide) or prevent respiration at the cellular level (hydrogen cyanide) to those which totally paralyze the respiratory system (hydrogen sulfide in large concentrations).

- ❑ **Radiation**—those events related to the emission of radiation energy. Radiation energy is defined as the waves or particles of energy that are emitted from radioactive sources. It may be in the form of alpha or beta particles or in the form of gamma waves, depending upon the intensity of the source material. Examples include plutonium and uranium hexafluoride.

  Nonionizing radiation, such as microwaves and lasers, may also create potential harm in certain emergency situations.
- ❑ **Etiologic**—those events created by uncontrolled exposures to living microorganisms. Etiologic/biological harm is normally associated with diseases such as typhoid, tuberculosis, and bloodborne pathogens such as the hepatitis B virus and the human immunodeficiency virus (HIV), which is associated with AIDS. It is often difficult to detect when and where the physical exposure to the biological agent occurred and the route(s) of exposure.

TOXICITY CONCERNS

All of the discussion up to this point has focused upon the hazmat exposure side of the "health hazard equation." Now let's turn our attention to the toxicity side of the equation.

Specifically, toxicity depends upon the chemical's ability to enter the cells of the body and the manner in which it reacts chemically with the cells' structure. Toxicologists list four categories of factors that influence toxicity:

- ❑ **Concentration or dose.** Usually, but not always, the speed and magnitude of the hazmat's action is in direct proportion to the dose. Remember—dose makes the poison!
- ❑ **Rate of absorption.** The faster the chemical enters the body and into the bloodstream, the faster its effects will be manifested. There is often a direct relationship between the rate of absorption and the route of exposure. In order of decreasing rate of absorption, recognized routes of exposure are intravenous, inhalation, intraperitoneal (into the abdominal cavity), intramuscular, subcutaneous (beneath the skin), oral, and cutaneous (onto the skin). While this order applies to many chemicals, recognize that there are also some exceptions.
- ❑ **Rate of detoxification.** The human body has natural defenses which can break down the toxic components of a chemical. This process is possible only as long as the rate of absorption into the body is less than the rate at which the body cells can destroy or neutralize and eliminate the toxin.

NOTES

- **Rate of excretion.** Rate at which the chemical is removed or excreted from the cells and the body. Remember that there are some chemicals that cannot be excreted and will start to accumulate within the body. These include PCB's (body fat) and hydrogen fluoride (bone).

  There are also miscellaneous factors that influence the toxicity of a chemical. These include age, weight, and sex, previous health problems, the individual's dietary and smoking habits, and prescription (or other) drugs within the bloodstream at the time of exposure.

MEASURING TOXICITY

Toxicity is measured in terms of the ability of a material to injure living tissue. Since only limited data is available from accidental human exposure, toxicologists rely upon data from animal experiments to establish toxicological values for humans.

Two units of measurement are commonly cited for determining the relative toxicity of a chemical substance or compound. They are:

- **Lethal Dose, 50% kill (LD_{50})**—The concentration of an ingested, absorbed, or injected substance which results in the death of 50% of the test population. LD_{50} is an oral or dermal exposure expressed in terms of weight—mg/kg. One kilogram (kg) equals 2.2 lbs. (The lower the dose, the more toxic the substance.)

  Other lethal dose percentages may also be found in reference guidebooks, including LD_{Lo}, LD_1, LD_{10}, LD_{30}, and LD_{99}. LD_{Lo} is the lowest dose of a substance reported to have caused death in humans or animals.

- **Lethal Concentration, 50% kill (LC_{50})**—The concentration of an inhaled substance which results in the death of 50% of the test population in a specific time period (usually 1 hour). LC_{50} is an inhalation exposure expressed in terms of parts per million (PPM) for gases and vapors, and milligrams per cubic meter (mg/meter^3) or micrograms per liter (µg/liter) for dusts and mists, as well as gases and vapors. (The lower the concentration, the more toxic the substance.)

  The lethal concentration low percentage (LC_{Lo}) may also be found in reference guidebooks. LC_{Lo} is the lowest concentration of a substance in air reported to have caused death in humans or animals.

CONVERSION FORMULAS

Mg/meter^3 can be converted to parts per million by volume by using the following formula:

$$\text{Parts per million} = \frac{\text{Mg/meter}^3 \times 24.45}{\text{Molecular Weight}}$$

Parts per million can be converted to mg/meter^3 by the following formula:

$$\text{Mg/meter}^3 = \frac{\text{Parts per million} \times \text{Molecular Weight}}{24.45}$$

NOTES

PARTS PER MILLION, BILLION, TRILLON			
Abbrev-iation	Numerical Equivalent	Common Application	Example
PPM (Parts per Million)	.000,001 (10^{-6})	Harm to Humans	1 ppm = 1 oz. of gin in a tank car containing 10,000 gallons of vermouth. 1% concentration in air = 10,000 ppm 10% concentration in air = 100,000 ppm
PPB (Parts per Billion)	.000,000,001 (10^{-9})	Harm to Humans or Environment	1 ppb– 1 pinch of salt for every 10 tons of potato chips
PPT (Parts per Trillion)	.000,000,000,001 (10^{-12})	Harm to Environment	1ppt = 1 pinch of salt for every 10,000 tons of potato chips

FIGURE 2-3: THE RELATIONSHIP BETWEEN PPM, PPB, AND PPT

Various terms are used to describe the toxicity of materials on material safety data sheets (MSDS) and other reference sources. These terms were originally published by H. C. Hodge and J. H. Sterner in the *"American Industrial Hygiene Association Quarterly"* in 1949 and have become commonly accepted in describing toxicity. The Hodge–Sterner Table is shown in Figure 2-4.

HODGE–STERNER TABLE		
Experimental LD_{50} (mg/kg of body wt)	Degree of Toxicity	Probable LD_{50} for a 70 kg man (150 lb)
<1.0 mg	Dangerously Toxic	A Taste
1–50 mg	Seriously Toxic	A Teaspoonful
50–500 mg	Highly Toxic	An Ounce
0.5–5 gm	Moderately Toxic	A Pint
5–15 gm	Slightly Toxic	A Quart
>15 gm	Extremely Low Toxicity	More Than a Quart

FIGURE 2-4: THE HODGE–STERNER TABLE INDICATES THE RELATIVE DEGREE OF TOXICITY OF CHEMICALS BASED UPON THE LETHAL DOSE, 50% KILL VALUE

NOTES

EXPOSURE VALUES AND GUIDELINES

The obvious question for emergency responders is how do we limit our exposure to the hazmats at the scene of an incident? In addition, what data is available to provide guidance in determining such factors as isolation distances, the size and location of control zones, and protective action recommendations?

Numerous exposure values have been developed which can provide guidance to ERP. Exposure values are useful in understanding the degree of hazard and are only guidelines, not absolute boundaries between safe and dangerous conditions. At most hazmat emergencies, ERP will not be able to measure concentrations of specific chemicals unless monitoring instruments are available. Common exposure values and guidelines are:

❑ **Threshold Limit Value/Time Weighted Average (TLV/TWA)**—The maximum airborne concentration of a material to which an average healthy person may be exposed repeatedly for eight hours each day, forty hours per week without suffering adverse effects. They are based upon current available data and are adjusted on an annual basis by the American Conference of Governmental Industrial Hygienists (ACGIH). Because TLV's involve an eight hour exposure, they are difficult to adapt for ERP use. TLV/TWA's are expressed in ppm and mg/$meter_3$. (The lower the TLV/TWA, the more toxic the substance.)

❑ **Threshold Limit Value/Short-Term Exposure Limit (TLV/STEL)**—The 15-minute, time weighted average exposure which should not be exceeded at any time, nor repeated more than four times daily with a 60-minute rest period required between each STEL exposure. These short-term exposures can be tolerated without suffering from irritation, chronic or irreversible tissue damage, or narcosis of a sufficient degree to increase the likelihood of accidental injury, impairing self-rescue, or reducing efficiency. TLV/STEL's are expressed in ppm and mg/$meter_3$. (The lower the TLV/STEL, the more toxic the substance.)

For some substances, ACGIH could not find sufficient toxicological data to develop the TLV/STEL for chemicals which had already been assigned a TLV/TWA. In these instances, ACGIH recommends that short-term exposures not exceed three times the TLV/TWA for more than a total of 30 minutes during the day.

❑ **Threshold Limit Value/Ceiling (TLV/C)**—The maximum concentration that should not be exceeded, even instantaneously. (The lower the TLV/C, the more toxic the substance.)

For some substances, ACGIH could not find sufficient toxicological data to develop the TLV/C for chemicals which had already been assigned a TLV/TWA. In these instances, ACGIH recommends that five times the TLV/TWA be used in place of the TLV/C.

❑ **Threshold Limit Value—Skin**—Indicates possible and significant exposure to a material by way of absorption through the skin, mucous membranes, and eyes by direct or airborne contact. This attention-calling designation is intended to suggest appropriate

NOTES

measures to minimize skin absorption so that the TLV/TWA is not exceeded.

- **Permissible Exposure Limit (PEL) and Recommended Exposure Levels (REL)**—The maximum time weighted concentration at which 95% of exposed, healthy adults suffer no adverse effects over a 40-hour work week; these are comparable to ACGIH's TLV/TWA. PEL's are used by OSHA and are based on an eight-hour, time weighted average concentration. REL's are used by NIOSH and are based upon a 10-hour, time weighted average concentration. Unless otherwise noted, both are expressed in either ppm or mg/meter^3.

  PEL's were originally based upon ACGIH TLV/TWA values which were in place in 1971; however, many have not been revised since that time and do not reflect the revisions made to the ACGIH's exposure guidelines as new toxicological and exposure data has become available during the time period. PEL's are used by OSHA in evaluating workplace exposures. (The lower the PEL, the more toxic the substance.)

- **Immediately Dangerous to Life and Health (IDLH)**—An atmospheric concentration of any toxic, corrosive, or asphyxiant substance that poses an immediate threat to life, or would cause irreversible or delayed adverse health effects, or would interfere with an individual's ability to escape from a dangerous atmosphere. IDLH's are expressed in ppm and mg/meter^3. (The lower the IDLH, the more toxic the substance.)

  There are three general IDLH atmospheres: toxic, flammable, and oxygen deficient. In the absence of an IDLH value for toxic atmospheres, ERP may consider using an estimated IDLH of ten times the TLV/TWA. IDLH values for flammable atmospheres are 10 to 20% of the lower explosive limit (depending upon the situation—confined space vs. open air), while an IDLH oxygen deficient atmosphere is 19.5% oxygen or lower.

  IDLH was not originally designed as an exposure level for evaluating protective actions. However, EPA has determined that using one tenth (10%) of the IDLH value is an acceptable level of concern for evaluating hazmat release concentrations and public protective options.

- **Emergency Response Planning Guidelines (ERPG-2)**—The maximum airborne concentration below which it is believed that nearly all individuals could be exposed for up to one hour without experiencing or developing irreversible or other serious health effects or symptoms which could impair an individual's ability to take protective action.

  The ERPG exposure guidelines have been developed by the American Industrial Hygiene Association (AIHA) as a level of concern for evaluating public protective action options. AIHA has identified 100 chemicals for which it will develop ERPG's, most of which are extremely hazardous substances (EHS's) with airborne hazards (e.g., chlorine, anhydrous ammonia, hydrogen sulfide).

NOTES

These exposure values should be regarded only as guidelines, not absolute boundaries between safe and dangerous conditions. In addition, they should not be used when combinations of materials are involved. To be most effective, exposure values must be combined with monitoring instrument readings and interpreted by response personnel who are familiar with their proper application and limitations.

When evaluating the establishment of control zones at hazmat emergencies, the various TLV and the IDLH values are generally the most informative. When evaluating public protective action options, the ERPG or 1/10th of the IDLH value are useful. **REMEMBER—THE LOWER THE REPORTED CONCENTRATION, THE MORE TOXIC THE MATERIAL.**

CONTROLLING PERSONNEL EXPOSURES

The primary objective of using these various exposure guidelines is to minimize the potential for both ERP and public exposures. One method for understanding these exposure guidelines and applying them at a hazmat emergency is the concept of safe, unsafe, and dangerous. Originally developed by Mike Callan, there are three basic atmospheres at a hazmat emergency:

- Safe atmosphere—no harmful hazmat effects exist, which allows personnel to handle routine emergencies without specialized personal protective equipment (PPE).
- Unsafe atmosphere—once a hazmat is released from its container, an unsafe condition or atmosphere exists. If one is exposed to the material long enough, some form of either acute or chronic injury will often occur.
- Dangerous atmosphere—these are environments where serious irreversible injury or death may occur.

Understanding this system will assist ERP in establishing control zones, selecting respiratory and chemical protection levels, evaluating public protective action options, and determining the end of the emergency phase of the incident. Of course, it is virtually impossible to quantify an atmosphere and determine the type of atmosphere present without the use of monitoring instruments (see Figure 2-5).

Safe. Go back to our previous discussion on TLV's—TLV/TWA, TLV/STEL, and TLV/C. All of these exposure guidelines have one thing in common—remain below these values and the exposure is considered safe to the average healthy adult by all information which is known by today's health and safety professionals. However, recognize that these are dynamic values and have a tendency to be lowered over time as additional toxicological research and studies are completed (e.g., benzene). In addition, exposure to multiple chemicals may have additive or synergistic effects.

If the chemical concentration exceeds any of these three TLV exposure values, you should assume that you are now in an unsafe condition. Stay exposed long enough and some form of harmful effects may occur.

Unsafe. A general rule for ERP should be that if the material has been released from its container, assume that an unsafe atmosphere may exist

NOTES

and some form of PPE is required. Prolonged exposures at high concentrations can lead to injury; however, acute injuries may not be lethal (e.g., headaches, nausea, irritation to the eyes, nose, or throat). Unsafe atmospheres do not become seriously dangerous unless the exposure continues or the concentration of contaminants rises.

The unsafe atmosphere is an area where some ERP may ignore the signs and symptoms of overexposure. The fire service was often guilty of this during the days when "eating smoke" was fashionable. When dealing with a potential inhalation hazard, response personnel must use positive pressure self-contained breathing apparatus (SCBA) until the Incident Commander determines through the use of air monitoring that a decreased level of respiratory protection will not result in a hazardous exposure.

Dangerous. When concentrations continue to increase above unsafe levels, there is a high potential for life-threatening injuries or death to occur. This concentration level is the IDLH.

There are three general IDLH atmospheres: toxic, flammable, and oxygen deficient. As we previously noted, in the absence of an IDLH value for toxic atmospheres, ERP may consider using an estimated IDLH of ten times the TLV/TWA. IDLH values for flammable atmospheres are 10 to 20% of the lower explosive limit (depending upon the situation—confined space vs. open air), while an IDLH oxygen-deficient atmosphere is 19.5% oxygen or lower. An oxygen-enriched atmosphere contains 23.5% oxygen or higher. While these atmospheres can cause harm in various ways, the degree of harm is similar.

Responders may not always have immediate access to monitoring instruments. Therefore, what are some of the physical indicators of likely IDLH conditions?

Outside or Open Air Environment

- Visible vapor cloud—Large vapor clouds obviously indicate large concentrations of contaminants. Avoid entering a vapor cloud at all costs.
- Release from a bulk container or pressure vessel—Bulk containers can release large quantities of liquid and gas products. Bulk pressurized containers, such as horizontal tanks and spheres containing liquefied gases and cryogenic liquids, will pose the greatest hazards because of their tremendous liquid to vapor expansion ratios.
- Large liquid leaks—All liquids give off vapors when released. High vapor pressure liquids, such as many solvents and fuming corrosives, and pooled liquefied gases, such as chlorine and anhydrous ammonia, are particularly dangerous.

Inside or Limited Air Environment

- Below grade rescues or releases—Small amounts of heavier than air vapors can accumulate in low lying areas. Such areas should be avoided unless personnel are properly protected.
- Confined spaces—Any enclosed area where there is poor ventilation can result in either an oxygen-deficient, toxic, or flammable

NOTES

atmosphere, depending upon the hazmats involved. **Sixty percent of all fatalities at confined space emergencies are personnel who are acting in a rescuer capacity.**

- Artificial or Natural Barriers—Any time vapors can be trapped, they will accumulate and potentially increase in concentration. Tank dikes, highway sound barriers, and high vertical walls are potential areas where heavier than air toxic or flammable vapors can accumulate.

Biological Indicators (or using your common sense!)

- Dead birds, discolored foliage, sick animals or even humans are a pretty good indicator that a chemical release may have occurred.
- Hazmats with a potential for quick and rapid harm, such as poisonous gases, explosives, some oxidizers, and materials with very low IDLH values, should dictate the use of extreme caution.
- Physical senses and "street smarts"—Be aware of strong odors and other sensory warnings. Likewise, you don't have to be a chemist to look at a situation and determine that something is wrong with the picture. Don't underestimate your "sixth sense"—if the situation doesn't look or feel right, it probably isn't.

CARCINOGENS

Carcinogens are physical or chemical agents that cause abnormal cell growth and spread. Eventually, this action can lead to the development of malignant tumors. Carcinogens, mutagens, and teratogens are similar in that each causes some form of cell mutation. Studies show that up to 90% of all mutagens are carcinogens.

There are as many as 2,000 substances that various scientific and regulatory groups have labeled as "suspect," "probable," or "definite" human carcinogens. These groups include OSHA, the International Agency for Research on Cancer (IARC), and the National Toxicology Program (NTP). As of January, 1993, only 23 substances, including aflatoxin, asbestos, vinyl chloride, and benzene, have been proven through human epidemiological studies to increase cancer rates. The remainder received their carcinogenic classification based on animal studies.

While some carcinogens may have a TLV value, many do not. It is assumed that there is no threshold value below which these materials are "safe." In addition, carcinogens will not have an IDLH value.

RADIOACTIVE MATERIALS

Radiation is the emitting of energy from an atom in the form of either particles or electromagnetic waves. Radiation can be classified into two types:

- Nonionizing radiation—waves of energy such as radiant heat, radio waves and visible light. Examples include infrared waves, microwaves, and lasers. The amount of energy in these waves is small as compared to ionizing radiation.

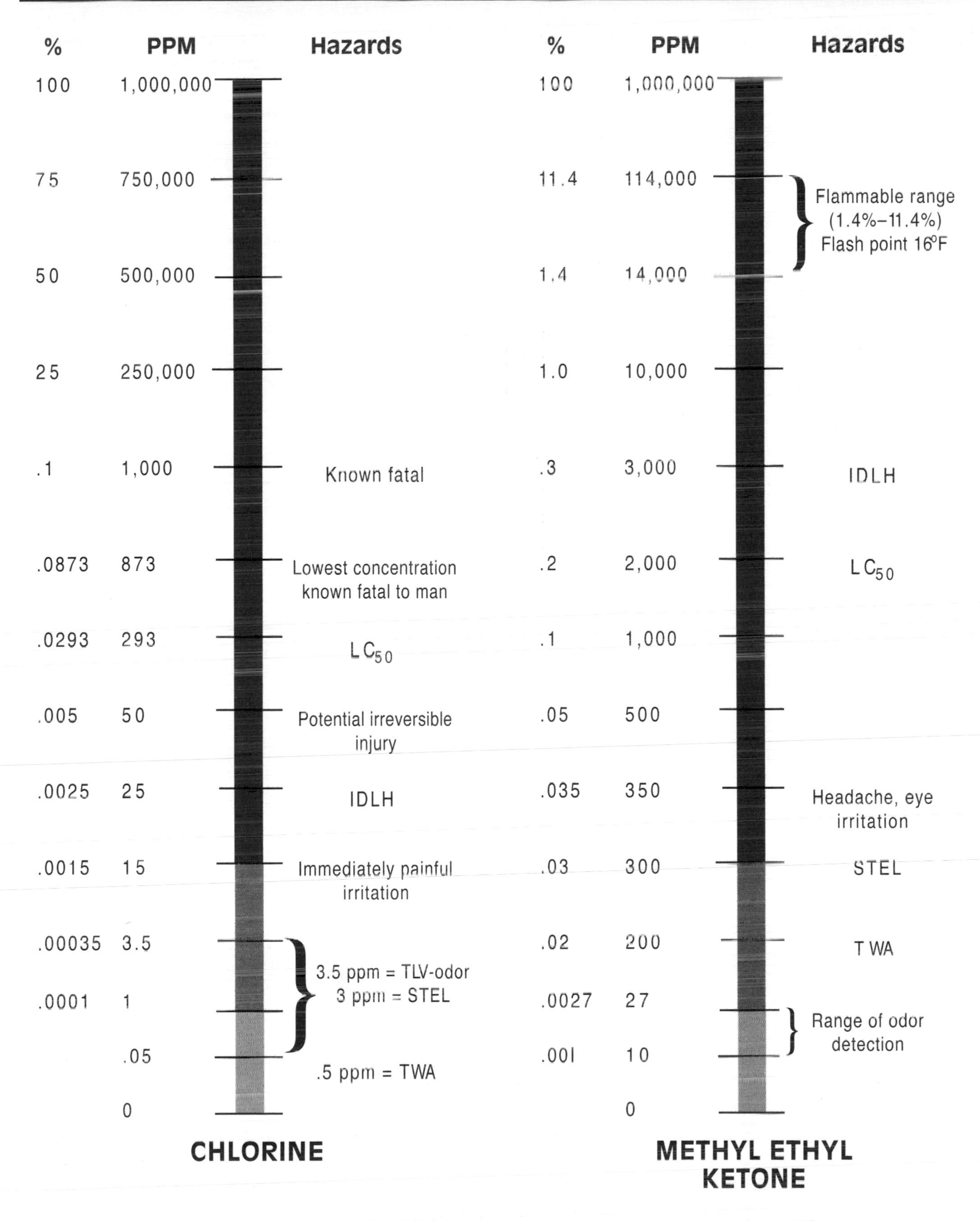

FIGURE 2-5: THE CONCEPT OF SAFE, UNSAFE, AND DANGEROUS IS AN EFFECTIVE TOOL FOR UNDERSTANDING AND INTERPRETING TOXICITY AND EXPOSURE VALUES

NOTES

- Ionizing radiation—characterized by its ability to create charged particles, or ions, in anything which it strikes. Exposure to low levels of ionizing radiation can produce short- or long-term cellular changes with potentially harmful effects, such as cancer and leukemia. X-rays are a familiar form of ionizing radiation.

There are three types of ionizing radiation:

- **Alpha particles**—largest of the common radioactive particles, it travels only 3–4 inches in air and is stopped by a sheet of paper. Greatest health hazard exists when alpha particles enter the body.
- **Beta particles**—particle which is the same size as an electron, and can penetrate materials much further than large alpha particles. Can penetrate the skin, but cannot penetrate internal organs. Can travel several yards in air, and can be stopped by a thin piece of metal or an inch of wood.
- **Gamma rays**—most dangerous form of common radiation because of the speed at which it moves, its ability to pass through human tissue, and the great distances it can cover. The range of gamma waves depends upon the energy of the source material.

Half-life is the time it takes for the activity of a radioactive material to decrease to one half of its initial value through radioactive decay. The half-life of known materials can range from a fraction of a second to millions of years. Although not normally a critical factor in incident mitigation, ERP should still note the half-life. Some extremely active radioactives used for radionucleocides and other medicinal purposes have relatively short half-lives (e.g., thallium).

In dealing with a radioactive material emergency, ERP must understand the difference between exposure and contamination. **Exposure** means that the human body has been subjected to radiation emitted from a radioactive source. **Contamination** means that the actual radioactive material has come in direct contact with one's body or clothing. As long as the material remains in contact with one's body or clothing, they are contaminated until decontamination measures are performed.

Exposure guidelines for radioactive materials are the old adage—"time, distance, and shielding." Radiation that doesn't hit anybody doesn't hurt anybody! This basic site safety concept can also apply to any hazmat release:

- **Time**—the shorter the exposure time, the less the exposure. Remember that radiation exposures are additive in their effects upon the body or any other subject—site safety and control procedures to monitor all entry operations are critical.
- **Distance**—the closer you are, the greater the exposure. The energy emitted from a radioactive source declines as one moves further away from the source. The Inverse Square Law is a simple tool for applying this safety concept. Simply stated, if you double the distance from a point radiation source, the radiation intensity is lowered by one fourth. If you increase the distance ten times, the radiation intensity is one hundredth of the original value.

NOTES

- **Shielding**—while personal protective clothing can offer protection against alpha particles, it will provide limited protection against beta particles and no protection against gamma radiation. Therefore, dense materials must be kept between ERP and the gamma source. Common shielding materials include lead, cement, and even water. Any shielding material is better than none!

Figure 2-6 outlines the Environmental Protection Agency's guidance on radioactive dose limits for personnel performing emergency duties. These limits apply to doses incurred over the duration of an emergency and should be treated as a once-in-a-lifetime exposure.

RADIOACTIVE MATERIALS DOSE LIMITS FOR WORKERS PERFORMING EMERGENCY SERVICES		
DOSE LIMIT (rem)	ACTIVITY	CONDITION
5	All Emergency Response Duties	
10	Protecting Valuable Property	Lower dose of 5 rem not practicable
25	Life Saving or Protection of Large Populations	Lower dose of 5 rem not practicable
>25	Life Saving or Protection of Large Populations	Only on a voluntary basis to persons fully aware of the risks involved

NOTES:
rem = Roentgen Equivalent Man
Source: U.S. Environmental Protection Agency

FIGURE 2-6

EXPOSURE TO ENVIRONMENTAL CONDITIONS

The physical working environment of emergency response personnel must constantly be evaluated. The human body tolerates a limited range of thermal environments. Exposure to either hot or cold weather conditions over a sustained period of time can adversely affect both the physiological and psychological conditions of response personnel. Continued exposures may result in physical discomfort, loss of efficiency, and a higher susceptibility to accidents and injuries.

NOTES

Factors that influence an individual's susceptibility to environmental conditions include:

- Lack of physical fitness
- Lack of acclimatization to the elements
- Age
- Dehydration
- Obesity
- Alcohol and drug use (including prescription drugs)
- Infection
- Allergies
- Chronic disease.

HEAT STRESS

In many respects, the human body is similar to a machine. Like cars, the body must take fuel and mix it with oxygen to produce work. A natural byproduct of this work is heat given off to the environment. Also like a car, the body has an excess heat transfer system. While a car uses a radiator with water, air, and coolant lines, the human body uses skin with sweat, air, and veins to dissipate heat buildup. However, when wearing impermeable, "nonbreathing" clothing, the body cannot effectively transfer this additional heat. The result is increased heat levels within the deep inner body core.

Heat stress is a critical concern to personnel wearing any type of impermeable, "nonbreathing" protective clothing. When wearing normal clothing, excess body heat can normally escape to the atmosphere. While chemical protective clothing (CPC) is designed to protect the user from a hostile environment and prevent the passage of harmful substances into the "protective envelope," CPC also reduces the body's ability to discard excess heat and perform natural body ventilation.

The key indicator of body heat levels is the "body core temperature." If the body heat cannot be eliminated, it will accumulate and elevate the core temperature. In addition to CPC, conditions which promote heat stress include high temperatures, high humidity, high radiant heat, and strenuous physical activities. Even after cooling is initiated, the body core temperature can take several hours to return to "normal" resting levels.

Physical reactions include:

❑ **Heat Rash**—an inflammation of the skin resulting from prolonged exposure to heat and humid air and often aggravated by chafing clothing. Heat rash is uncomfortable and decreases the ability of the body to tolerate heat.

❑ **Heat Cramps**—a cramp in the extremities or abdomen caused by the depletion of water and salt in the body. However, there is no observed increase in the body's core temperature. Usually occurs after physical exertion in an extremely hot environment or under conditions that cause profuse sweating and depletion of body fluids and electrolytes.

NOTES

- **Heat Exhaustion**—a mild form of shock caused when the circulatory system begins to fail as a result of the body's inadequate effort to give off excessive heat. Although not an immediate life-threatening condition, the individual should be immediately removed from the source of heat, rehydrated with electrolyte solutions, and the body kept cool. If not properly treated, heat exhaustion may evolve into heatstroke.
- **Heat Stroke**—a severe and sometimes fatal condition resulting from the failure of the temperature regulating capacity of the body. It is caused by exposure to the sun or high temperatures. Reduction or cessation of sweating is an early symptom. Body temperature of 105°F or higher, rapid pulse, hot skin, headache, confusion, unconsciousness, and convulsions may occur.

  Heatstroke is a TRUE MEDICAL EMERGENCY requiring immediate transport to a medical facility. Any means available should be used to cool the victim; the body temperature should be taken at five-minute intervals and should not go below 101°F. Normal intravenous saline solution may be required to rehydrate and balance electrolyte levels. However, the patient is susceptible to pulmonary edema, and careful monitoring is essential.

The signs and symptoms of heat stress and the related emergency care procedures are outlined in Figures 2-7 and 2-8.

HEAT STRESS EMERGENCIES SIGNS AND SYMPTOMS

SIGNS AND SYMPTOMS	CONDITION		
	HEAT CRAMPS	HEAT EXHAUSTION	HEAT STROKE
MUSCLE CRAMPS	YES	NO	NO
BREATHING	VARIES	RAPID SHALLOW	DEEP THEN SHALLOW
PULSE	VARIES	WEAK	FULL RAPID
WEAKNESS	YES	YES	YES
SKIN	MOIST-WARM NO CHANGE	COLD CLAMMY	DRY-HOT
PERSPIRATION	HEAVY	HEAVY	LITTLE OR NONE
LOSS OF CONSCIOUSNESS	SELDOM	SOMETIMES	OFTEN

FIGURE 2-7: SIGNS AND SYMPTOMS OF HEAT STRESS

NOTES

HEAT STRESS EMERGENCIES EMERGENCY CARE PROCEDURES	
HEAT CRAMPS	• Move patient to a nearby cool place. • Give patient salted water to drink or half-strength commercial electrolyte fluids. • Massage the cramped muscles to help ease the patient's discomfort: massaging with pressure will be more effective than light rubbing actions (optional in some EMS systems). • Apply moist towels to the patient's forehead and over cramped muscles for added relief. • If cramps persist, or if more serious signs and symptoms develop, ready the patient and transport.
HEAT EXHAUSTION	• Move the patient to a nearby cool place. • Keep the patient at rest. • Remove enough clothing to cool the patient without chilling him or her (watch for shivering). • Give the patient salted water or half-strength commercial electrolyte fluids. Do not try to administer fluids to an unconscious patient. • Treat for shock, but do not cover to the point of overheating the patient. • Provide oxygen, if needed. • If unconscious, fails to recover rapidly, has other injuries, or has a history of medical problems, transport as soon as possible.
HEAT STROKE	• Cool the patient—in any manner—rapidly, move the patient out of the sun or away from the heat source. Remove the patient's clothing and wrap in wet towels and sheets. Pour cool water over these wrappings. Body heat must be lowered rapidly or brain cells will die! • Treat for shock and administer a high concentration of oxygen. • If cold packs or ice bags are available, wrap them and place one bag or pack under each of the patient's armpits, one behind each knee, one in the groin, one on each wrist and ankle, and one on each side of the patient's neck. • Transport as soon as possible. • Should transport be delayed, find a tub or container—immerse patient up to the face in cooled water. Constantly monitor to prevent drowning. • Monitor vital signs throughout process.

BRYAN E. BLEDSOE, ROBERT S. PORTER AND BRUCE R. SHADE, PARAMEDIC EMERGENCY CARE, ENGLEWOOD CLIFFS, NJ: PRENTICE HALL, INC. (1991).

FIGURE 2-8: EMERGENCY CARE PROCEDURES FOR HEAT STREE EMERGENCIES

NOTES

To minimize the effects of heat stress, EMS personnel should be on-site to monitor and screen response personnel. Heat stress should be managed through a series of both administrative controls and through the use of personal protective equipment (PPE). Administrative controls, including the need for acclimatization or conditioning the body to working in hot environments, work/rest scheduling, rehab, and fluid replacement are outlined in Figure 2-9. PPE options are outlined below:

- **Air Cooled Jackets and Suits**—consist of small airlines attached to either vests, jackets, or CPC to provide convective cooling of the user by blowing cool air over the body inside of the suit. Cooling may be enhanced by the use of a vortex cooler or by refrigeration coils and a heat exchanger. Although sometimes found at remediation operations, they are not well-suited for emergency response applications.

  These units require an airline, large quantities of breathing air (10 to 25 cubic feet per minute), and are not as effective as the active and passive cooling units in controlling body core temperatures.

- **Ice Vests**—these consist of frozen ice packs which are inserted into a vest. This passive cooling system operates on the principle of conductive heat cooling. Although not as effective as the full-body cooling suit in controlling the body core temperature, studies have shown that ice vests are better than both the air cooled units and water cooled jackets. In addition to the physiological advantages, there are also psychological benefits (people feel a lot better in the suit!).

  Ice vests are relatively inexpensive, lightweight, improve worker comfort, decrease lens fogging, and are "user friendly" (that means firefighter-proof!). On the negative side, the vests require frozen coolant packs or an ice source at the scene of the emergency (unfortunately, the frozen coolant packs often leave the hazmat unit for someone's lunch box or cooler), and may add bulk underneath the CPC.

  These vests can also be used with heat packs for operations in extremely cold working environments.

- **Water Cooled Vests and Suits**—these units consist of a heat transfer garment (vest or full-body suit) and a cooling unit. The cooling unit normally consists of a battery or power source, a pump, and an ice/water or cooling agent container. Essentially, the cooling agent is circulated throughout the garment and operates on the principle of conductive heat transfer.

  These units often add both weight and bulk to the CPC. As a result, they tend to be more prevalent in the hazmat clean-up and remediation industry, where longer work times are required. Studies have shown that the full-body suits are substantially more effective than the water cooled vests and, overall, are the most effective method for controlling body core temperatures.

  Consideration is also being given to using this technology to develop a full-body warming system for use under cold work conditions.

NOTES

REDUCING HEAT STRESS

- Physical conditioning and acclimatization of responders is critical. The degree to which the body physiologically adjusts to work in hot environments affects its ability to actually do the work.
- Provide plenty of liquids, including prehydration with 8 to 16 ounces of fluids. Replace body fluids (water and electrolytes), use a 0.1% salt solution or commercial electrolyte mixes.
- Pre- and post-entry medical monitoring procedures are critical to ensure that the physical status of entry personnel is properly monitored and evaluated.
- Body cooling devices (e.g., cooling vests) can be used to aid natural body ventilation. These devices add weight and must be carefully balanced against worker inefficiency.
- During exceptionally hot and/or humid weather conditions, consider the installation of mobile showers and portable hose-down facilities to reduce internal CPC temperatures and to cool protective clothing.
- Rotate personnel on a shift basis. Many response organizations recommend that chemical protective clothing by worn for a maximum work time of 20 to 30 minutes, based upon the ambient temperature.
- Establish a Rehab Area and provide shelter or shaded areas to protect personnel during rest periods. Entry personnel should be placed in cool areas (e.g., air conditioned ambulances) before and after entry operations.
- When dealing with heat stress emergencies, apply wet towels or ice packs around the neck and under the armpits for maximum body cooling effectiveness.
- At remedial operations, conduct nonemergency response activities in the early morning or evening when ambient temperatures are cooler.

FIGURE 2-9: GUIDELINES FOR REDUCING HEAT STRESS

COLD TEMPERATURE EXPOSURES

Hazardous materials such as liquefied gases and cryogenic liquids expose personnel to the same hazards as those created by cold weather environments. Exposure to severe cold even for a short period of time may cause severe injury to body surfaces. Those body parts most susceptible are the ears, nose, hands, and feet.

Two factors particularly influence the development of cold injuries—the ambient temperature and wind velocity. Since still air is a poor heat conductor, ERP working in areas of low temperature and little wind can

NOTES

endure these conditions for long periods, assuming their clothing is dry. However, when a low ambient temperature is combined with an active airflow, a dangerous condition described as "wind chill" is created. For example, an ambient temperature of 20°F. with a wind of 15 miles per hour (mph) is equivalent in its chilling effect to still air at minus 5°F. Generally, the greatest increase in this effect occurs when a wind of only 5 mph increases to 10 mph.

Response personnel should also understand the term "water chill." Water chill is body heat lost through conduction. Wet clothing extracts heat from the body up to 240 times faster than dry clothing. It may also lead to hypothermia, in which the body temperature falls below 95°F. Hypothermia is a TRUE MEDICAL EMERGENCY.

Regardless of the ambient temperature, personnel will perspire heavily in impermeable chemical protective clothing. When this clothing is removed in the decontamination process, particularly in cold environments, the body can cool rapidly.

Usually, injuries from cold exposures will be local or confined to a small area of the body. Frostbite, hypothermia, and impaired ability to work are dangers at low temperatures, when the wind chill factor is low, or when working in wet clothing. Ensure that all personnel are wearing appropriate clothing (preferably layered) and have warm shelters or vehicles such as buses available.

It is essential that the layer next to the skin, especially socks, be dry. Trenchfoot can result when wet socks are worn at long-term emergencies in cool (not cold) environments. Carefully schedule work and rest periods and monitor physical working conditions. It is especially important to have warm shelters available for protective clothing donning and doffing activities.

NOISE

Hazardous materials releases often involve excessive noise levels. Examples include actuation of pressure relief devices and the use of heavy equipment on the scene. The effects of excessive noise levels can include:

- Personnel being annoyed, startled, or distracted.
- Physical damage to the ears, pain, and temporary and/or permanent hearing loss.
- Interference with communications, which may limit the ability of ERP to warn of danger or enforce proper safety precautions (verbal and radio).

Standardized hand signals should be developed for situations where excessive noise levels make verbal or radio communications impossible. Effective noise and ear protection should always be provided. This usually consists of ear plugs or ear muffs. OSHA regulations require that hearing protection be provided whenever noise levels exceed 85 dBA (decibels on the A-weighted scale).

Up to this point, we have discussed the health and safety problems created by both the physical and chemical environment at a hazmat

NOTES

incident. The following section will review the components of a health and safety management program.

HEALTH AND SAFETY MANAGEMENT PROGRAM

Personnel involved in hazmat emergency response operations can be exposed to high levels of both physiological and psychological stress. Routine activities may expose them to both chemical and physical hazards. They may develop heat stress while wearing protective clothing or while working under temperature extremes, not to mention the possibility of facing life-threatening emergencies such as fires and explosions.

A health and safety management program should be an integral element of any emergency response organization. The components of a health and safety management system for hazmat responders are outlined in *OSHA 1910.120, Hazardous Waste Site Operations and Emergency Response (HAZWOPER)*. Key areas within the regulation include medical surveillance, personal protective equipment, and site safety practices and procedures.

MEDICAL SURVEILLANCE

A medical surveillance program is the cornerstone of an effective employee health and safety management system. The primary objectives of a medical surveillance program are (1) to determine that an individual can perform his or her assigned duties, including the use of personal protective clothing and equipment; and (2) to detect any changes in body system functions caused by physical and/or chemical exposures.

A medical surveillance program should provide surveillance (pre-employment screening, periodic and follow-up medical examinations where appropriate and a termination examination), on-scene evaluation, treatment (emergency and nonemergency), recordkeeping, and program review. A sample medical surveillance program is summarized in Figure 2-10.

The success of any medical program depends on management support and employee involvement. Occupational health physicians and specialists, emergency medicine physicians, safety professionals, local or regional Poison Control Center specialists, as well as advanced EMS personnel should be consulted for their expertise. Many hazmat units have a physician with a background or interest in hazardous materials who serves as the Medical Director for their unit.

Confidentiality of all medical information is paramount. Prospective employees and new hazmat response team members must be provided a complete, detailed occupational and medical history so a baseline profile can be established. Responders should be encouraged to document any suspected exposures, regardless of the degree, along with any unusual physical or physiological conditions. Training programs must emphasize that even minor complaints (e.g., headaches, skin irritations) may be important.

NOTES

MEDICAL SURVEILLANCE PROGRAM	
COMPONENT	**RECOMMENDED**
Pre-Employment Screening	• Medical History • Occupational History • Physical Examination • Determination of fitness to work wearing protective equipment • Baseline monitoring for specific exposures
Periodic Medical Examinations	• Yearly update of medical and occupational history; yearly physical examination; testing based on 1) examination results, 2) exposures, and 3) job class and task • More frequent testing based on specific exposures • Exams may be bi-annual based upon physician's recommendation
Emergency Treatment	• Provide emergency care on site • Develop liaison with local hospital and medical specialists • Arrange for decontamination of victims • Arrange in advance for transport of victims • Transfer medical records; give details of incident and medical history to next care provider
Nonemergency Treatment	• Develop mechanism for nonemergency health care
Recordkeeping and Review	• Maintain and provide access to medical records in accordance with OSHA and state regulations • Report and record occupational injuries and illnesses • Review program periodically. Focus on current site hazards, exposures, and industrial hygiene standards

FIGURE 2-10. MEDICAL SURVEILLANCE IS AN ESSENTIAL PART OF AN EMPLOYEE HEALTH AND SAFETY PROGRAM

FIRE AND HAZARDOUS MATERIALS PERSONAL EXPOSURE REPORT

GENERAL INFORMATION

Name ______________________________

Personal ID Number ______________________ Station ______________

Incident Date ______________________ Incident Number ______________

Incident Location ______________________________

NATURE OF INCIDENT

❑ Residential Fire ❑ Industrial Fire ❑ Commerical Fire

❑ Vehicle Fire ❑ Trash / Dumpster ❑ Explosion

❑ Brush Fire ❑ Hazmat Spill ❑ Hazmat Vapor Release

❑ Other ______________________________

LEVEL OF PERSONAL PROTECTION

❑ Structural F/F Gear ❑ Chemical Liquid Splash CPC ❑ Chemical Vapor CPC

❑ Air Purifying Resp. ❑ SCBA ❑ Airline Hose (SAR)

❑ Other ______________________________

EMERGENCY RESPONSE ACTIVITY

❑ Fire Attack ❑ Exposure Protection ❑ Overhaul

❑ Spill Control ❑ Entry Operations ❑ Decon Operations

❑ Clean-up and Recovery

❑ Other ______________________________

EXPOSURE DATA

Method of Exposure

❑ Inhalation ❑ Ingestion ❑ Absorption ❑ Direct Contact

Substance(s) Exposed To (identify chemical products of combustion, etc.)

____

____

____

____

Duration of Exposure

❑ < 15 minutes ❑ 15 - 30 minutes ❑ 30 - 45 minutes

❑ 45 - 60 minutes ❑ 1 - 2 hours ❑ > 2 hours

FIGURE 2-11: EXAMPLE OF HEALTH EXPOSURE FORM

MEDICAL TREATMENT PROVIDED

Signs and Symptoms (as appropriate)

On-Scene Medical Treatment (as appropriate)

Medical Facility Treatment (as appropriate)

Follow-up Action Required (as appropriate)

MEDICAL TREATMENT PROVIDED

Individual's Signature ______________________ Date ______________

Officer's Signature ______________________ Date ______________

FIGURE 2-11: EXAMPLE OF HEALTH EXPOSURE FORM

NOTES

PRE-EMPLOYMENT SCREENING

The objectives of pre-employment screening are to determine an individual's fitness for duty, including SCBA and protective clothing use, and to provide baseline data for future medical comparisons. The screening should focus on the following areas:

- **Occupational and medical history**—This questionnaire should be completed with attention toward prior exposures to chemical and physical hazards. Also note previous illnesses, chronic diseases, hypersensitivity to specific substances, ability to use personal protective equipment (PPE), family history, and general lifestyle habits such as smoking and drug use.
- **Physical examination**—Complete a comprehensive physical examination focusing on the pulmonary, cardiovascular, and musculoskeletal systems. Additional tests which can help gauge the capacity to perform emergency response duties while wearing protective clothing include pulmonary function, electrocardiograms (EKG's), hearing, and physical "stress tests."
- **Baseline laboratory profile**—This verifies the effectiveness of protective measures and determines whether the responder is adversely affected by previous exposures. The profile may include medical screening and biological monitoring tests based on potential exposures, such as liver, renal, and blood forming functions. Pre-employment blood and serum specimens may be frozen for later testing and comparison.

PERIODIC MEDICAL EXAMINATIONS

Periodic exams must be used in conjunction with pre-employment screening. Their comparison with the baseline physical may detect trends and early warning signs of adverse health effects. Under the OSHA 1910.120 requirements, such exams shall be administered annually, and no longer than every 2 years if the attending physician believes a longer interval is appropriate. In addition, more frequent intervals may be required depending upon the nature of potential or actual exposures, type of chemicals involved, and the individual's medical and physical profile.

If an individual develops signs or symptoms indicating possible overexposure to hazardous substances or health hazards, or has been injured or exposed to substances above accepted exposure values in an emergency, medical examinations and consultations shall be provided as necessary.

Periodic screening exams can include medical history reviews which focus upon health changes, illness and exposure-related symptoms, physical examinations, and specific tests such as pulmonary function, audiometric, blood, and urine.

To ensure the completion of a comprehensive medical profile, a medical exam is required to be given to all personnel when they are removed from active duty as a hazmat responder and at the termination of their employment or membership.

NOTES

EMERGENCY TREATMENT

EMS personnel and units must be present at each hazmat incident. Their primary objectives are to implement pre- and post-entry physical monitoring activities for entry and back-up personnel, provide technical assistance for all EMS-related activities, and provide emergency medical treatment and transportation of injured, ill, or chemically contaminated civilians or response personnel. An EMS responder with a background in hazmat operations should be in charge of the EMS operations and coordinate closely with the Hazmat Branch Officer.

Some regions and industrial facilities provide either advanced life support (ALS) units or individuals specially trained and equipped for hazardous materials emergencies. They often carry drugs and antidotes for chemicals commonly found within the plant or community, along with medical information and baseline profiles of all hazmat response team personnel. Hazmat training competencies for EMS personnel can be found in *NFPA 473, Competencies for EMS Personnel Responding to Hazardous Materials Incidents.*

Standard Operating Procedures (SOP's) for the clinical management and transportation of chemically contaminated patients must be developed as part of the planning process. In addition, the specific roles, responsibilities, and capabilities of hazmat personnel, EMS personnel, and local medical facilities need to be determined.

The handling of chemically contaminated patients will be addressed in Chapter 11—Decontamination. However, remember these basic principles:

- Is it a rescue operation or a body recovery operation?
- Always ensure that EMS personnel are properly protected.
- Certain situations may exist where decontamination may aggravate patient care or further delay priority treatment. However, as a rule of thumb, all patients should receive gross decontamination.
- The ABC's can be administered to a contaminated victim if rescuers and EMS personnel are protected. It's a much better option than having a fully decontaminated but dead patient.

NONEMERGENCY TREATMENT

The signs and symptoms of certain chemical exposures may not be present for 24 to 72 hours after exposure. This is particularly true when dealing with chemicals which have delayed effects. In many instances, those exposed may already be off-duty and out of contact. All personnel operating at an incident should be medically evaluated before being released. In addition, the termination procedure should provide for a briefing for all ERP on the signs and symptoms of exposure, documentation and completion of health exposure logs or forms, and how to get immediate treatment if necessary.

NOTES

RECORDKEEPING AND PROGRAM REVIEW

Recordkeeping is an important element of the medical surveillance program. Time intervals between initial exposures and the appearance of possible chronic effects may take years to develop. Individual records should be kept for all personnel. OSHA requires that exposure and medical records be maintained for at least thirty years after the employee retires. These records must be made available to all affected employees and their representatives upon request. Procedures should also be developed for emergency access to individual records if personnel are hospitalized and their medical surveillance records are requested.

Individual medical records should include all medical exams completed, their purpose (e.g., baseline), the examining physician's observations and recommendations, and if they were a result of a specific exposure. A copy of the incident report should also be maintained in the file. Any injuries sustained during line-of-duty operations should be noted, and follow-up treatment and personal exposure logs should be maintained.

Regular evaluation of the overall medical surveillance program is important to ensure its continued effectiveness. Review the following elements on an annual basis:

- Ensure that each accident/illness is promptly investigated to determine its root cause and update health and safety procedures as necessary.
- Evaluate the effectiveness of medical testing in light of potential and confirmed exposures.
- Add or delete specific medical tests as recommended by the Medical Director and by current industrial hygiene and environmental health data.
- Review all emergency care protocols.

CRITICAL INCIDENT STRESS

Although not an element of the medical surveillance program, critical incident stress should be recognized as an issue which can impact the health and welfare of response personnel. Hazmat emergencies and the potential health risks from exposures can create high levels of psychological stress for both ERP and their families. Unknown or potential exposures can generate as much emotional stress as a documented high level exposure. The Incident Commander must recognize that these personal stressors exist and must be managed.

Medical debriefings as part of the incident termination phase are essential elements in reducing the level of stress. The debriefing should review the hazmats involved in the incident, the signs and symptoms of exposure, exposure documentation procedures, and the procedures to follow in the event an individual starts to show these signs and symptoms over the next several days. Employee Assistance Programs (EAP's) and Critical Incident Stress (CIS) debriefing teams can be an effective post-incident resource and should be used as necessary.

NOTES

PERSONAL PROTECTIVE EQUIPMENT PROGRAM

The objectives of a personal protective equipment (PPE) program are to protect personnel from both chemical and physical safety and health hazards, and to prevent injury to the user from the incorrect use and/or malfunction of protective clothing or equipment.

A comprehensive PPE program should include:

- ❑ **Hazard Assessment**—The selection and purchase of PPE should be based upon a hazard assessment of those chemicals stored, transported, or used within the facility or the community.
- ❑ **Medical Monitoring of Personnel**—Medical surveillance and medical monitoring results should be one of the criteria for the initial selection and continuing certification of "fitness for duty" of response personnel. Other issues include the effects of heat stress and environmental temperature extremes.
- ❑ **Equipment Selection and Use**—Recognize the relationship between the environment being encountered (e.g., flammable vs. toxic), the response objectives (defensive vs. offensive), the PPE user, and the PPE ensemble used.
- ❑ **Training Program**—An effective PPE program cannot exist without a comprehensive training program.
- ❑ **Inspection, Maintenance, and Storage Program**—These are key elements of a PPE program, and the absence of any one of these elements may lead to injury. This would include inspection procedures prior to, during, and after use, methods of storage, maintenance guidelines and capabilities, etc.

A written PPE program outlining these elements is required under OSHA 1910.120 (g)(5). It should include a policy statement with the guidelines and procedures listed above. Copies should be made available to all employees. Technical data concerning equipment, maintenance manuals, relevant regulations, and other essential information should also be made available as necessary.

PPE training is also required under OSHA 1910.120. It should allow the user to become familiar with the equipment in a nonhazardous environment, which builds user confidence. Understanding the capabilities and limitations of this equipment also improves the safety and efficiency of emergency operations. Finally, a well-rounded PPE training program often reduces associated maintenance expenses.

An effective PPE training program consisting of both classroom exercises and practical, hands-on field exercises should contain:

- The proper selection and use of various types of PPE, including capabilities and limitations.
- The nature of the hazards and the consequences of not using PPE.
- Human factors influencing PPE performance.
- Instructions on inspecting, donning, doffing, checking, fitting, and using PPE.

NOTES

- Practical use of PPE in a "clean" environment for an extended period of time, followed by an evaluation in a test environment.
- The buddy system and entry procedures.
- Emergency procedures and self-rescue for PPE failure.
- The user's responsibility for decontamination, cleaning, maintenance, and repair of PPE.

PPE PERSONAL SAFETY ISSUES

- FACIAL HAIR (BEARDS) AND LONG HAIR—Interferes with the use of respiratory equipment. No facial hair should pass between the face and the facepiece surface. Testing has documented that even a few days' growth of facial hair may allow contaminants to penetrate the facepiece.
- EYEGLASSES WITH CONVENTIONAL TEMPLE PIECES—Interferes with facepiece seal. A spectacle kit should be installed into the facepiece of those who require corrected vision.
- CONTACT LENSES—OSHA (20 CFR 1910.134 e, 5, ii) currently prohibits the use of contact lenses with respiratory protection devices. This prohibition was originally developed in 1971 when the use of hard contact lenses was prevalent. Contaminants and/or particulates could become trapped between the lens and the eye, creating irritation, damage, absorption, and an urge to remove the facepiece.

  In recent years, gas permeable and soft contact lenses have been routinely used by fire service personnel with self-contained breathing apparatus (SCBA). As a result of both testing and fire service experience, OSHA has indicated that it will lift the contact lens restriction during the revision of the respiratory protection standard (1910.134). In the interim, OSHA has indicated that violations of OSHA 1910.134 involving the use of gas permeable and soft contact lenses shall be recorded as de minimis violations and citations shall not be issued.
- GUM AND TOBACCO CHEWING—May cause ingestion of contaminants and compromise facepeice fit.
- PRESCRIPTION DRUG USE—Synergistic or combined effects may occur when exposed to certain substances. Unless specifically approved by a physician, prescription drug use should be avoided whenever there is a possibility of contact with toxic substances.

FIGURE 2-12: CERTAIN PERSONAL FEATURES CAN JEOPARDIZE PERSONNEL SAFETY DURING OPERATIONS. FIGURE 2-12 LISTS A NUMBER OF PERSONAL SAFETY CONSIDERATIONS PERTAINING TO THE USE OF PPE.

NOTES

SITE SAFETY PRACTICES AND PROCEDURES

SAFETY ISSUES

Safety is not simply an organizational rule or a government regulation. Safety is both an attitude and a behavior. Safety MUST be an inherent part of all operations from the development of SOP's to the selection and purchase of PPE. The operating philosophy of every emergency response organization should be, "If we cannot do this safely, then we will not do it at all."

There are two phases of an incident where the potential for ERP injury and harm is greatest—first, when we initially arrive at the scene, and second, when the incident shifts gears from the emergency phase to the clean-up and recovery phase. You cannot manage a hazmat incident if you do not have control of the scene—in fact, you will quickly find out that the incident is actually managing you!

Gaining and maintaining control of the incident scene is one of the most difficult tasks faced by the Incident Commander. A continuous problem will be everyone—including facility managers, plant personnel, the media, the general public, and even ERP—wanting to get as close as possible to the action. There will also be situations where site safety becomes lax or even nonexistent. This is especially true at "campaign" incidents extending over hours or days. It is quite easy for ERP to become bored and careless in their attitudes and actions.

Considering the validity of these observations, recognize the importance of the following safety truths:

- "What occurs during the initial 10 minutes will dictate what will occur for the next hour, and what occurs during the first hour will dictate what will occur for the next eight hours."

  TRANSLATION: If the initial-arriving ERP screws up, it will shift initial strategical goals and probably take hours to undo their mistakes and get the system back on track. If you don't know what to do, isolate the area, deny entry, and call for help. It may take a while, but eventually someone will show up who knows what to do!

- "There is nothing wrong with taking a risk. However, always remember that there are good risks and bad risks—if there is much to be gained, then perhaps much can be risked. Of course, if there is little to be gained, then little should be risked."

  TRANSLATION: Life safety is always our Number 1 priority—including the life safety of all emergency responders. There is a significant difference between a rescue and body recovery operation when hazmats are involved. Considering the average response and operational set-up time for most HMRT's, rescue is seldom a strategical priority. Remember—nowhere in your job description does it say that suicide is part of the deal!

NOTES

- "Safety must be more than a policy or procedure...it is both an attitude and a responsibility that must be shared by all responders."

  TRANSLATION: Safety must be an integral element of all hazmat operations. Yes, the Safety Officer does play a critical role in ensuring that all operations are conducted both safely and effectively. Ultimately, however, safety is the responsibility of every responder.

- "Protective clothing is not your first line of defense, but is your last line of defense."

  TRANSLATION: The selection of strategies and tactics to minimize any direct exposure to the materials involved should always be your first line of defense. Remember—evaluate defensive strategies first, then offensive. If you minimize the potential for exposure, you reduce the potential of relying solely upon your protective clothing.

- "Final accountability always rest with the Incident Commander."

  TRANSLATION: Although emergency responders are normally not responsible for product transfer and removal (except in some industrial facilities), site safety is still the Incident Commander's responsibility. Do not become lax during this phase of the emergency—site safety procedures must be continuously enforced throughout the clean-up and recovery phase.

SITE SAFETY PLAN

A site safety plan is required under HAZWOPER, paragraph b, as a mechanism for assuring the health and safety of personnel operating at clean-up and hazardous waste site operations. Although a site safety plan is not required under HAZWOPER, paragraph q, site safety must be an integral element of on-scene response operations. Standard Operating Procedures (SOP's) and checklists should be used to both verify and document that safety elements are addressed during the course of the emergency. These checklists are usually divided by positions within the Incident Management System and the Hazmat Branch (e.g., Hazmat Safety Officer, Decontamination Officer, Information Officer, etc.).

Some of the advantages of using operational checklists to meet the site safety requirements are ensuring that specific organizational guidelines and SOP's are followed, the ability to track activities and performance and document the plan of action and decision-making process.

Under HAZWOPER, paragraph b, omponents of a site safety plan should include site map or sketch, hazard and risk analysis of the identified hazmats, site monitoring, establishment of control zones, site safety practices and procedures, communications, implementation of IMS and the location of the command post, decontamination practices, EMS support, and other relevant topics. Standard site safety practices and procedures are outlined in Figure 2-13.

NOTES

STANDARD SITE SAFETY PRACTICES

- Minimize the number of personnel operating in the contaminated area.
- Avoid contact with all contaminants, contaminated surfaces, or suspected contaminated surfaces. Avoid walking through any suspected releases or placing equipment on contaminated surfaces.
- Advise all entry personnel of all site control policies including entry points, decon layout, procedures, and working times.
- Always have an escape route. Ensure that everyone knows the emergency evacuation signals.
- Ensure that all tasks and responsibilities are identified before attempting entry. If necessary, practice unfamiliar operations prior to entry.
- Use the buddy system for all entry operations. Always ensure that properly equipped back-up crews are in place.
- Maintain radio communications between entry, back-up crews, and the Safety Officer (whenever possible).
- Prohibit drinking, smoking, and any other practices which increase the possibility of hand-to-mouth transfer in all contaminated areas.
- Follow decontamination and personal cleanliness practices before eating, drinking, or smoking after leaving the contaminated area.

FIGURE 2-13: STANDARD SAFETY PRACTICES FOR OPERATIONS WITHIN A CONTAMINATED AREA

SAFETY OFFICER AND SAFETY RESPONSIBILITIES

Every hazmat incident is required to have a Safety Officer. Often, such as incidents where an HMRT is operating, safety responsibilities will be divided into two areas—first, the safety of all units operating on the incident scene and under the control of the *Incident Safety Officer*; and second, the safety of those operating within the IMS Hazmat Branch and under the control of the *Hazmat Branch Safety Officer*.

Although the Hazmat Safety Officer is subordinate to the Incident Safety Officer, he/she has certain responsibilities within the Hazmat Branch that may circumvent the normal chain of command. In either case, both Safety Officers must have the authority to stop any operations that are deemed unsafe.

Among the primary responsibilities of the Safety Officers are.

Overall Site Safety

- Ensure that the Safety Officer is identified to all personnel. The Incident Commander should advise all operating personnel, as appropriate. The use of command vests for identification is also recommended.

NOTES

- Ensure that all personnel and equipment are positioned in a safe location. Remember the basics—upwind, uphill, and always have an escape route and pre-designated withdrawal signal. Consider having vehicles and apparatus back into the incident scene for a quick exit.
- Ensure that hazard zones are established, identified, and constantly monitored and that their locations are communicated to all personnel. Consider the location of the Command Post, Hazmat Sector, and Staging Area in relation to the control zones and the potential worsening of the emergency.
- Designate a security officer to maintain overall site security, when necessary. Delegate to plant security or law enforcement whenever possible.
- Ensure that all personnel in controlled areas are in the proper level of personal protective clothing.

Entry Operations

- Coordinate with the Hazmat Medical Officer to ensure that pre-entry medical monitoring has been conducted.
- Hold a pre-entry safety briefing prior to entry. This may be provided by either the Hazmat Safety Officer or other Hazmat Branch personnel (e.g., Entry). All entry and back-up personnel must be familiar with the objectives, tasks, and procedures to be followed. Topics should include objectives of the entry operation, a review of all assignments, verification of radio procedures (designated channels) and emergency signals (both hand signals and audible), emergency escape plans and procedures, protective clothing requirements, and the location and layout of the decon area.
- Coordinate entry operations with back-up crews and the Decon Group. The Entry Team should be permitted to enter the Hot Zone only when back-up crews are in place and the decon area is prepared.
- Monitor entry operations and advise entry personnel and the Incident Commander of any unsafe practices or conditions.
- During the termination phase, advise all personnel of the possible signs and symptoms of exposure and ensure that health exposure forms are documented.

PRE- AND POST-ENTRY MEDICAL MONITORING

Medical monitoring may be defined as an ongoing, systematic evaluation of individuals at risk of suffering adverse effects of exposure to heat, stress, or hazardous materials as a result of working at a hazardous materials emergency. The objectives of medical monitoring are (1) to obtain baseline vital signs; (2) to identify and preclude from participation individuals who are at risk to sustain either injury or illness; and (3) to facilitate the early recognition and treatment of personnel with adverse physiological and/or emotional responses.

NOTES

Pre- and post-entry medical monitoring should be required at virtually every hazmat incident. This will be the responsibility of the Hazmat Medical Group. Medical monitoring provides baseline vital signs of all entry personnel, and identifies, evaluates, and eliminates those individuals who are at risk from the effects of heat stress or hazmat exposure.

Components of the pre-entry exam should include the following:

- Vital signs, including blood pressure, pulse, respiratory rate, temperature, eye movement, and body weight. If available, a 10-second EKG rhythm strip may also be taken.
- Skin evaluation, with an emphasis on rashes, lesions, and open sores or wounds.
- Lung sounds, including wheezing, unequal breath sounds, etc.
- Mental status (alert and oriented to time, location, and person).
- Recent medical history, including medications, alcohol consumption, any new medical treatment or diagnosis within the last 2 weeks, and symptoms of fever, nausea, diarrhea, vomiting, or coughing within the past 72 hours.
- Prehydration with 8 to 16 ounces of fluids (water and electrolytes).

Criteria should be established for evaluating ERP prior to entry operations. These criteria should be reviewed by an occupational health physician or specialist who is familiar with the duties and tasks of hazmat responders. The following exclusion criteria are used by a number of Hazmat Medical Groups; however, they should not supersede any existing criteria established by the local medical control.

Entry shall be denied if the following criteria are not satisfied:

- Blood Pressure—BP exceeds 100 mm Hg diastolic.
- Pulse—greater then 70% maximum heart rate (> 115) or irregular rhythm not previously known.
- Respirations—respiratory rate is greater than 24 per minute.
- Temperature—oral temperature less than 97°F or exceeds 99.5°F. Core temperature less than 98°F or greater than 100.5°F.
- Body Weight—no pre-entry exclusion.
- EKG—dysrhythmias not previously detected must be cleared by medical control.
- Mental Status—altered mental status (e.g., slurred speech, clumsiness, weakness).
- Other criteria, including:

  Skin—Open sores, large skin rashes, or significant sunburn.

  Lungs—Wheezing or congested lung sounds.

  Medical History—Recent onset of heart or lung problems, hypertension, diabetes, etc. Experienced nausea and vomiting, diarrhea, fever, or heat exhaustion within the last 72 hours. Use of prescription medication and over-the-counter medicines (e.g., decongestants, antihistamines, etc.) must be cleared through local

NOTES

medical control. Heavy alcohol consumption within the previous 24 hours or any alcohol within the past 2 hours.

Post-entry medical monitoring is performed following decontamination to determine if the responder has suffered any immediate effects from heat stress or a chemical exposure, and to determine the individual's health status for future assignment during or after the incident. Components of the post-entry exam should include the following:

- Any signs or symptoms of chemical exposures, heat stress, or cardiovascular collapse.
- Vital signs, including blood pressure, pulse, respiratory rate, temperature, and eye movement. If available, a 10-second EKG rhythm strip may also be taken.
- Skin evaluation, with an emphasis on rashes, lesions, and open sores or wounds.
- Lung sounds, including wheezing, unequal breath sounds, etc.
- Mental status (alert and oriented to time, location, and person).
- Recent medical history, including medications, alcohol consumption, any new medical treatment or diagnosis within the last 2 weeks, and symptoms of fever, nausea, diarrhea, vomiting, or coughing within the past 72 hours.
- Body weight (stripped of clothing).
- Hydration—provide plenty of liquids. Replace body fluids (water and electrolytes), use a 0.1% salt solution or commercial electrolyte mixes.

Vital signs should be monitored every 5 to 10 minutes, with the person resting, until they return to approximately 10% of the baseline. If vital signs do not return to normal, it may be necessary to transport the individual to a medical facility. Medical control should be consulted for direction and recommendations, as necessary.

EMERGENCY INCIDENT REHABILITATION

The Incident Commander should consider the circumstances of the incident and make adequate provisions early in the incident for the rest and rehabilitation of all personnel operating at the scene. This is particularly critical for "campaign" emergencies which extend over a period of hours. The Incident Commander should establish a Rehabilitation Sector or Group to coordinate rest and rehab activities. At most hazmat incidents, rehabilitation will be the responsibility of the Hazmat Medical Group.

In addition to coordinating for EMS support, treatment, and monitoring, the rehab sector is responsible for providing food and fluid replenishment, mental rest, and relief from the extreme environmental conditions associated with the incident. The Rehabilitation Area should meet the following parameters:

- Be in a location which provides physical rest by allowing the body to recuperate from the hazards and demands of the emergency. It should also be located to allow for prompt reentry back into the emergency operation upon complete rehabilitation.

NOTES

- Be located in a safe location within the Cold Zone (see Chapter 5) so that personnel can remove their protective clothing and be afforded mental rest from the stress and pressure of the emergency operation. It should also be easily accessible by EMS units.
- Provide suitable protection from the prevailing environmental conditions. During hot weather, it should be located in a cool, shaded area. In cold weather, it should be in a warm, dry area. In addition, it should be free of vehicle exhaust fumes.
- Be large enough to accommodate multiple crews, based upon the size of the incident.

An obvious safety question is how long should ERP remain in the Rehab Area before being released. Unfortunately, there is a lack of standards for heat stress monitoring and rehab periods while wearing chemical protective clothing at a hazmat emergency. Most guidelines, such as those from the EPA Standard Operating Safety Guides, are directed primarily towards hazardous waste remedial operations.

The National Fire Academy Hazardous Materials Program has developed guidelines which are primarily based upon field experience. Specifically, ERP who are in aerobically fit condition, performing under normal working conditions for a period of 20 minutes, should require the following rest and recuperation times:

Ambient Temperature	R & R Period Required
< 70°F	30 Minutes
70 to 85°F	45 Minutes
> 85°F	60 Minutes

SUMMARY

Personnel protection is the number one priority at any hazmat incident. This chapter has discussed the health and safety concerns of ERP, including exposures to hazardous materials and the physical environment, toxicity and health exposure guidelines and interpretations, the components of a hazmat health and safety management system, and site safety practices and procedures.

REFERENCES AND SUGGESTED READINGS

Agency for Toxic Substances and Disease Registry, MANAGING HAZARDOUS MATERIALS INCIDENTS—EMERGENCY MEDICAL SERVICES, Atlanta, GA: ATSDR (1992).

Agency for Toxic Substances and Disease Registry, MANAGING HAZARDOUS MATERIALS INCIDENTS—HOSPITAL EMERGENCY DEPARTMENTS, Atlanta, GA: ATSDR (1992).

American National Standards Institute, ANSI Z88.2—PRACTICES FOR RESPIRATORY PROTECTION, New York, NY: ANSI (1992).

Berger, M., W. Byrd, C. M. West, and R. C. Ricks, TRANSPORT OF RADIOACTIVE MATERIALS: Q & A ABOUT INCIDENT RESPONSE, Oak Ridge, TN: Oak Ridge Associated Universities (1992).

Bledsoe, Bryan E., Robert S. Porter, and Bruce R. Shade, PARAMEDIC EMERGENCY CARE, Englewood Cliffs, NJ: Prentice Hall, Inc. (1991).

Borak, Jonathan, M.D., Michael Callan, and William Abbott, HAZARDOUS MATERIALS EXPOSURE: EMERGENCY RESPONSE AND PATIENT CARE, Englewood Cliffs, NJ: Prentice Hall, Inc. (1991).

Bowen, John E., "Understanding Chemical Toxicity." FIRE ENGINEERING (August, 1987), pages 19 - 28.

Bronstein, Alvin C. and Philip L. Currance, EMERGENCY CARE FOR HAZARDOUS MATERIALS EXPOSURE, St. Louis, MO: C. V. Mosby Company (1988).

Callan, Michael, "Safe, Unsafe and Dangerous: The Three Worlds of Hazardous Materials." INDUSTRIAL FIRE CHIEF (September/October, 1992), pages 35–39.

Chemical Manufacturers Association, "Poisons: Identification, Toxicity and Safe Handling" (videotape). Washington, DC: Chemical Manufacturers Association (1991).

DaRoza, Robert A., "Is It Safe to Wear Contact Lenses?" Lawrence Livermore National Laboratory—UCRL-53653 (August 15, 1985).

Federal Emergency Management Agency—National Fire Academy, Firefighter Health & Safety: Program Implementation and Management Course–Student Manual, Emmitsburg, MD: FEMA.

Federal Emergency Management Agency—National Fire Academy, Firefighter Health & Safety: The Company Officer's Responsibility Course–Student Manual, Emmitsburg, MD: FEMA.

Federal Emergency Management Agency—National Fire Academy, Hazardous Materials Operating Site Practices Course, Student Manual, Emmitsburg, MD: FEMA.

Federal Emergency Management Agency—U.S. Fire Administration, EMERGENCY INCIDENT REHABILITATION, Washington, DC: FEMA (1992).

Fire, Frank L., THE COMMON SENSE APPROACH TO HAZARDOUS MATERIALS, Tulsa, OK: Fire Engineering Books and Videos (1986).

Grant, Harvey D., Robert H. Murray, Jr., and David Bergeron, EMERGENCY CARE (5th Edition), Englewood Cliffs, NJ: Prentice Hall, Inc. (1990).

International Association of Fire Fighters, Training Course for Hazardous Materials Team Members, Student Text, Washington, DC: IAFF (1991).

National Fire Protection Association, HAZARDOUS MATERIALS RESPONSE HANDBOOK (2nd Edition), Boston, MA: National Fire Protection Association (1993).

National Fire Protection Association, NFPA 1500 HANDBOOK, Boston, MA: National Fire Protection Association (1993).

National Institute for Occupational Safety and Health (NIOSH), OCCUPATIONAL SAFETY AND HEALTH GUIDANCE MANUAL FOR HAZARDOUS WASTE SITE ACTIVITIES, Washington, DC: NIOSH, OSHA, USCG, EPA (1985).

Olson, Kent R., M.D., POISONING AND DRUG OVERDOSE (San Francisco Bay Area Regional Poison Control Center), Norwalk, CT and San Mateo, CA: Appleton and Lange (1990).

Sheean, Dennis M., "Braving the Elements of Radiological Response." RESCUE (March/April, 1993), pages 51–55.

Stutz, Douglas R. and Stanley J. Janusz, HAZARDOUS MATERIALS INJURIES: A HANDBOOK FOR PRE-HOSPITAL CARE (2nd Edition), Beltsville, MD: Bradford Communications Corp. (1988).

U.S. Environmental Protection Agency, MANUAL OF PROTECTIVE ACTION GUIDES AND PROTECTIVE ACTIONS FOR NUCLEAR INCIDENTS, Washington, DC: EPA (1992).

U.S. Environmental Protection Agency, et. al., TECHNICAL GUIDANCE FOR HAZARDS ANALYSIS—EMERGENCY PLANNING FOR EXTREMELY HAZARDOUS SUBSTANCES, Washington, DC: EPA, FEMA, DOT (1987).

Veghte, James H., "Physiologic Field Evaluation of Hazardous Materials Protective Ensembles." U.S. Fire Administration Report FA-109 (September, 1991).

Williams, Philip L. and James L. Burson, INDUSTRIAL TOXICOLOGY, New York, NY: Van Nostrand Reinhold (1985).

Wray, Thomas K., "Mutagens, Teratogens and Carcinogens." HAZMAT WORLD (March, 1991), pages 88–89.

Wray, Thomas K., "Risk Assessment." HAZMAT WORLD (January, 1993), pages 45–47.

Calvin and Hobbes **by Bill Watterson**

THE INCIDENT MANAGEMENT SYSTEM

OBJECTIVES

1) List the categories of participants at a hazmat incident.
2) Identify the purpose, need, benefits, and elements of the Incident Management System.
3) Identify the key elements of the Incident Management System necessary to coordinate response activities at a hazmat incident.
4) Identify the elements and procedures for establishing command during a hazmat incident.
5) Identify the elements and procedures for transferring command during a hazmat incident.
6) Identify the considerations for determining the location of the command post for a hazmat incident.
7) Identify the duties and functions of the following Hazmat Branch functions within the Incident Management System:
   - Safety
   - Information/Research
   - Entry/Reconnaissance
   - Decontamination
   - Medical
   - Resources

"Leadership is the ability to decide what is to be done, and then to get others to want to do it."

General Dwight D. Eisenhower, U.S. Army

NOTES

8) Identify the duties and responsibilities of the Hazmat Branch Officer and describe how to coordinate the activities of the branch.
9) Identify the following incident management terms and their relationship to each other:
   - Incident priorities
   - Strategic goals
   - Tactical objectives
   - Tactical response modes (offensive, defensive, non-intervention)
   - Tasks
10) Identify the process for determining the effectiveness of the Incident Action Plan.

INTRODUCTION

Direct, effective command operations are necessary at every type of emergency. However, hazardous materials incidents place a special burden on the command system since they often involve communications between separate agencies and the coordination of many different functions and personnel assignments—from public protective actions to the use of specialized personal protective equipment.

This chapter will review the fundamental concepts of incident management and its application at hazmat emergencies. Primary topics will include the various players who characteristically appear at a hazmat incident, the elements of the Incident Management System (IMS), the functions and responsibilities of the Hazmat Branch, and commanding the hazmat response.

THE PLAYERS

A hazmat incident often attracts an interesting collection of participants. Some can be found at the scene of any major fire or medical emergency, while others (e.g., clean-up contractors) are unique to hazmat incidents.

Regardless of who they are and how they materialize on the scene, the Incident Commander must be able to quickly identify and categorize them. These participants include:

The Incident Commander (Command or IC)—the individual responsible for establishing and managing the overall operational plan. This process includes developing an effective organizational structure, developing an incident strategy and action plan, allocating resources, making appropriate assignments, managing information, and continually attempting to achieve the basic command goals.

Section Chiefs and Branch/Sector Officers—those individuals assigned by the Incident Commander to manage specific geographic areas or specific functions. Key players include the Operations Section Chief, the Hazmat Branch Officer, and the Hazmat Safety Officer.

Command Staff—those individuals appointed by and directly reporting to the Incident Commander. These include the Safety Officer, the Liaison Officer, and the Public Information Officer (PIO).

Fire/Rescue/EMS Companies—those individuals and teams assigned to assist the Incident Commander and/or command and general staff by gathering and managing information, implementing assigned tasks, and coordinating communications. They may be firefighters, EMS personnel, company officers, or any other knowledgeable individuals on the scene.

Emergency Response Team (ERT)—crews of specially trained personnel used within industrial facilities for the control and mitigation of emergency situations. They consist of both shift personnel with ERT responsibilities as part of their job assignment (e.g., plant operators), or volunteer members. ERT's may be responsible for both fire, hazmat, medical, and technical rescue emergencies, depending upon the size and operation of their facility.

Hazardous Materials Response Teams (HMRT)—crews of specially trained and medically evaluated individuals responsible for directly managing and controlling hazmat problems. They may include people from the emergency services, private industry, governmental agencies, or any combination. They generally perform more complex and technical functions than the fire, rescue, and EMS companies. Depending upon the jurisdiction, HMRT's can function as an advisory group or provide "hands-on" operational capabilities.

Communications Personnel—the central communications function for emergency services. They receive calls for assistance and dispatch appropriate units to the correct incident locations. They provide a crucial link for those working on the scene by dispatching additional needed resources, including technical information specialists such as off-site specialty employees, poison control specialists, and chemical manufacturers.

Facility Managers—individuals who normally do not have an on-scene emergency response function, but who are key players within the plant environment. They report to the plant Emergency Operations Center and are typically responsible for providing logistical support to field emergency response units and for coordinating external issues, including community liaison, media relations, and governmental notifications.

Support Personnel—individuals who provide important support services at the incident. Water and utility company employees, heavy equipment operators, and food service personnel are some examples.

Technical Information Specialists—individuals who provide specific expertise to the Incident Commander either in person, by telephone, or through other means. They may represent the manufacturer or shipper or be otherwise familiar with the chemicals or problems involved. Within a plant, examples would include Industrial Hygiene and Environmental Affairs representatives. CHEMTREC™ and the Local Emergency Planning Committee often provide an excellent way to reach information resources in any given area of expertise.

NOTES

NOTES

Environmental Clean-up Contractors—individuals who may provide both mitigation and support services at the incident. Capabilities may include spill control, performing product transfer operations, site clean-up and recovery, and remediation operations. Personnel should be trained to meet the training requirements of OSHA 1910.120.

Government Officials—individuals who normally do not have an emergency response function, but who can bring a lot of political clout to the incident. Failure to professionally address their questions and concerns within the IMS organization can have significant political and/or organizational impacts both during and after the emergency.

News Media—aggressive, usually intelligent individuals who work to inform the public of major happenings within their community or region. Because of the unusual, and often frightening nature of hazardous materials incidents, it is very important that the public be accurately informed quickly and regularly of the incident. Television and radio are excellent methods with which to coordinate and manage large-scale public protective actions activities. On-scene media may also be willing to provide quality videotape equipment and helicopter shots for emergency service use in exchange for coverage.

Investigators—individuals who are responsible for determining the origin and cause of the hazmat release. A hazmat incident is not really concluded until this investigation is complete. Future legal proceedings, possible regulatory citations or criminal charges, and financial reimbursement for the time, equipment, and supplies of emergency services may well depend upon their efforts. Certain types of incidents require interaction between investigators on the federal, state, and local levels, as well as in the private sector.

Victims—individuals who may be exposed, contaminated, injured, killed, or displaced as a result of the hazardous materials incident. Once emergency responders begin to provide treatment and care, they become patients. Special care should be given to their welfare and to informing them of potential short- and long-term signs and symptoms of exposure.

Spectators—curious, usually well-meaning members of the facility and/or general public who arrive at the scene to assist or watch the adventure. Since they are often difficult to control—especially during campaign incidents—spectators need to be monitored and managed constantly to ensure their safety.

The Hazardous Material—a potentially harmful substance or material that has escaped or threatens to escape from its container. It should be considered an active, mobile opponent which must be carefully monitored at all times. Whenever its container is stressed or it has already escaped, the hazardous material should be considered a threat to the other players.

THE INCIDENT MANAGEMENT SYSTEM

Why so much interest in incident management? One primary reason is the HAZWOPER requirement that both public safety and industrial emergency response organizations use a "nationally recognized Incident

Command System for emergencies involving hazardous materials." But beyond regulatory requirements, experience has shown that the normal, day-to-day bureaucracy is not well-suited to meeting demands created by emergency situations.

The IMS and Hazmat Branch information presented in this chapter is based upon the National Interagency Incident Management System (NIIMS). NIIMS is a "baseline"incident management organization which is utilized by both federal, state, and local governments, as well as many private sector organizations throughout the United States.

The Incident Management System (also recognized as the Incident Command System) is an organized system of roles, responsibilities, and procedures for the command and control of emergency operations. It is a procedure-driven system based upon the same business and organizational management principles which govern organizations on a daily basis. As is the case with the day-to-day management of any organization, IMS has both technical and "political" aspects which must be understood by the key players in the emergency response system.

In the past, these players were concerned primarily with the technical or operational aspects of an emergency. Today, however, the playing field has changed significantly. Large-scale incidents have very real political, legal, and financial effects on how both the public and the corporate shareholders view emergency response professionals. Recent incidents have taught us that an emergency can have a favorable technical or operational outcome and still be a political disaster. These political issues will be reviewed in Chapter Four.

INCIDENT MANAGEMENT vs. CRISIS MANAGEMENT

Crisis management programs are becoming more commonplace in many corporate and industrial organizations. Experience has shown that there is a direct relationship between incident management and crisis management concerns, as well as the organizational structure for managing each event.

What is the difference between an incident and a crisis? In the broadest sense, an *incident* can be defined as an occurrence or event, either natural or man-made, that requires action by emergency response personnel to prevent or minimize loss of life or damage to property and/or natural resources. Essentially, an incident interrupts normal procedures, has limited and definable characteristics, and has the potential to precipitate a crisis. Examples of incidents include fires, hazmat, medical, and rescue emergencies. In short, an incident does not necessarily mean that an organization or a community has a crisis.

The definition of a *crisis* will vary significantly depending upon the type or organization (e.g., public vs. private sector) and situations anticipated. A good starting point is the following definition: A crisis is an unplanned event that can exceed the level of available resources, and has the potential to significantly impact an organization's operability and credibility, or pose a significant environmental, economic, or legal liability. From a corporate perspective, a crisis often goes beyond the fenceline and impacts the community, requires community and mutual

NOTES

Incident – occurrence or event.

Crisis – Unplanned event.

NOTES

IMS - Incident Management System

aid resources, is longer in duration, and requires both facility and corporate resources for its successful management.

Many have problems in both identifying and implementing the transition from incident management to crisis management. The scale of an incident is not the sole factor which determines its potential to develop into a crisis. A broad range of additional factors may come into play, including the political environment in which the incident occurs and the occurrence of similar or recent incidents in either the region or the industry.

Crisis management builds upon the philosophy of incident management. **If IMS is not utilized on a regular basis for the "more routine" incidents, it will be very difficult to successfully implement it during a major incident or crisis situation.**

IMS LESSONS LEARNED

IMS must be an inherent element in any successful hazmat response program. Response experience has provided us with the following lessons learned:

1) ***The management and control of routine, day-to-day emergencies establishes the framework for how the larger, more significant situations will be managed.***

IMS should be the basic operating system for all emergencies. The incident management structure should logically expand as the nature and complexity of the emergency expands, resulting in a smooth transition with minimum organizational changes or adjustments.

The corollary of this activity is quite simple—if IMS is not utilized for all "routine" emergencies, one cannot expect the organizational structure to function and adapt effectively, efficiently, and safely when a major emergency occurs. The routine establishes the basics, upon which the nonroutine must build. The more routine decisions that are made prior to the emergency, the more time the Incident Commander will have to make critical decisions during the emergency.

2) ***There is no single agency which can effectively manage a major emergency alone.***

An emergency within a community or facility is going to require the resources and expertise of various organizations and agencies. All organizations with an emergency response mission bring their own agenda to the emergency scene. Each of these agendas represents real, valid, and significant concerns. Problems are often created, however, when there is no communication prior to the emergency (i.e., you don't know the key players and their emergency response mission) and everyone feels that their specific agenda or interest is the most important. Remember, the ability to mount a safe and effective response builds upon what is accomplished during planning and preparedness activities.

In the absence of planning, organizational relationships are often based upon perceptions, and perceptions are often based upon our experiences with one individual or one incident. If that experience was positive, we tend to view the respective organization in a positive light until proven otherwise. Similarly, if that experience was negative...you know the rest.

There is no single organization which can effectively manage a major hazmat emergency alone. Organizations which attempt to maintain the normal bureaucracy in managing a major incident will have inherent problems in implementing a timely and effective emergency response. There is no excuse for not knowing who the key public safety and mutual aid players are within your area, and with whom you are going to interface on a regular basis or at major emergencies.

3) ***A variety of different "players" will respond to a "working" hazmat incident. Therefore, what occurs during planning and preparedness will establish the framework for how the emergency response effort will operate.***

Both SARA, Title III, and HAZWOPER have focused much attention on the need for hazmat planning and the implementation of an incident management system. Communities and facilities which previously may have done very little emergency planning are now required to develop planning documents to meet pre-established regulatory criteria. Unfortunately, the result has often been emergency response plans (ERP) which look good on paper, but don't really work on the street. In the rush to develop ERP's which meet the letter of the law, many have lost sight of the importance of the operational utility of their ERP.

In some cases, more emphasis has been placed on "Do we meet the requirements of the law?" as compared to "Do we meet the requirements of the law, AND can our personnel perform the duties that we expect?" In fact, a subjective assessment of emergency planning programs could be summarized as follows: The majority of industrial emergency response programs are "compliance-oriented," where the letter of the law is satisfied, but the performance of facility emergency responders is often untested and not verified through either exercises or actual experience. In contrast, many public safety response programs are "operationally oriented," where personnel are able to perform the expected emergency response tasks, but often lack the required regulatory documentation.

Planning and preparedness establish the framework for how the emergency response effort will function. While ERP's must satisfy minimum regulatory requirements, the ERP must be both operationally oriented and representative of actual personnel and resource capabilities. ERP's which are not user friendly do not get used.

4) ***Many hazmat response programs tend to be "people-dependent" programs.***

Emergency response programs can be categorized as being either "people-dependent" or "system-dependent." Specialized response teams, such as hazmat, bomb squads, and technical rescue teams, are

NOTES

NOTES

often very "people-dependent." These organizations rely almost totally on the experience of a few key individuals and can result in failed emergency response efforts if these key individuals are not present at an incident.

In contrast, a "system-dependent" organization has clearly defined objectives, specific duties and responsibilities spelled out in standard operating procedures (SOP's) and operational checklists, and available resources listed. A system-dependent response allows individuals to assume different roles in an emergency regardless of their daily activities. Written procedures, operational checklists, and an effective training and critique program ensure that less experienced personnel can get the job done with an acceptable level of safety and efficiency.

In essence, a "system-dependent" organization delivers a consistent level of quality and service, regardless of personnel or location. For example, when you order a Big Mac© from any McDonald's throughout the world, there will be little difference in either quality or taste. How do they do it several billion times in a row? Simply by emphasizing procedures and personnel training. In short, Mickey D's is the epitome of a "system-dependent" organization.

An organizational philosophy and management goal of emergency response programs should be the development of operational procedures which will bring consistency to emergency operations. The components of this system are:

1) The development of standard operating procedures (SOP's);
2) Training all personnel in the scope, application, and implementation of the SOP's;
3) The execution of the SOP's on the emergency scene;
4) Post-incident review and critique of their operational effectiveness; and
5) Revision and updating of the SOP's on a regular basis.

This standard management cycle helps build an organization with the ability to "self-improve" over time. This is critical, as the accepted "standard of care" keeps rising over time. What was considered an adequate emergency response program five years ago may be viewed as inadequate by today's standards.

Of course, having procedures alone is not sufficient, as the procedures must reflect the ability to handle not only the major emergency, but also the "day-to-day" operations. Many organizations prepare for the "big one" but still can't handle the everyday occurrence. If you don't have your act together on the "day-to-day" operations, you are not going to pull it out of your hat for a major emergency.

5) ***In those cases where IMS has not resulted in the operational improvements expected, the problems are typically associated with planning, training, and the organization "buying into" the IMS program, as compared to the IMS system itself.***

IMS is not a panacea, but is an organizational process and a manage-

NOTES

ment tool. As with any new effort, it will be necessary to establish and communicate a policy regarding the application and use of IMS for incident management and crisis management purposes. This policy must be established and supported by the highest levels of management and communicated throughout the organization. If the boss doesn't "buy in" to the program, neither will the troops!

IMS ELEMENTS

IMS has a number of common characteristics which permit different organizations to work together safely and effectively in order to bring about a favorable outcome to the emergency. It is predicated upon several basic management concepts, including the following:

- **Division of labor**—work is assigned based on the functions to be performed, the equipment available, and the training and capabilities of those performing the tasks.
- Clearly defined **lines of authority**.
- **Delegation** of authority and responsibility, as appropriate. However, ultimate responsibility always rests with the Incident Commander.
- **Unity of command**—each person reports to only one supervisor.
- Optimum **span of control** of five individuals (range of 3 to 7), depending upon the tasks to be performed, the associated danger or difficulty, and the level of delegated authority.
- Establishment of both **line and staff functions** within the organization. Line functions are directly associated with the implementation of incident operations, while staff functions support incident operations.

The key elements of IMS are:

- Common Terminology
- Modular Organization
- Predesignated Incident Facilities
- Integrated Communications
- Unified Command Structure
- Consolidated Plan of Action
- Comprehensive Resource Management

COMMON TERMINOLOGY

A hazmat response program must be built around an IMS which uses standardized terminology for organizational functions, resource elements, and incident facilities. This is particularly critical at multi-agency and multi-jurisdiction incidents. Basic IMS organizational terms are:

1) **Incident Commander**: The individual responsible for the management of on-scene emergency incident operations. The Incident Commander must be thoroughly trained to assume these responsibilities and is not automatically authorized to perform these activities by virtue of his/her position within the organization. At most routine

INCIDENT MANAGEMENT STRUCTURE

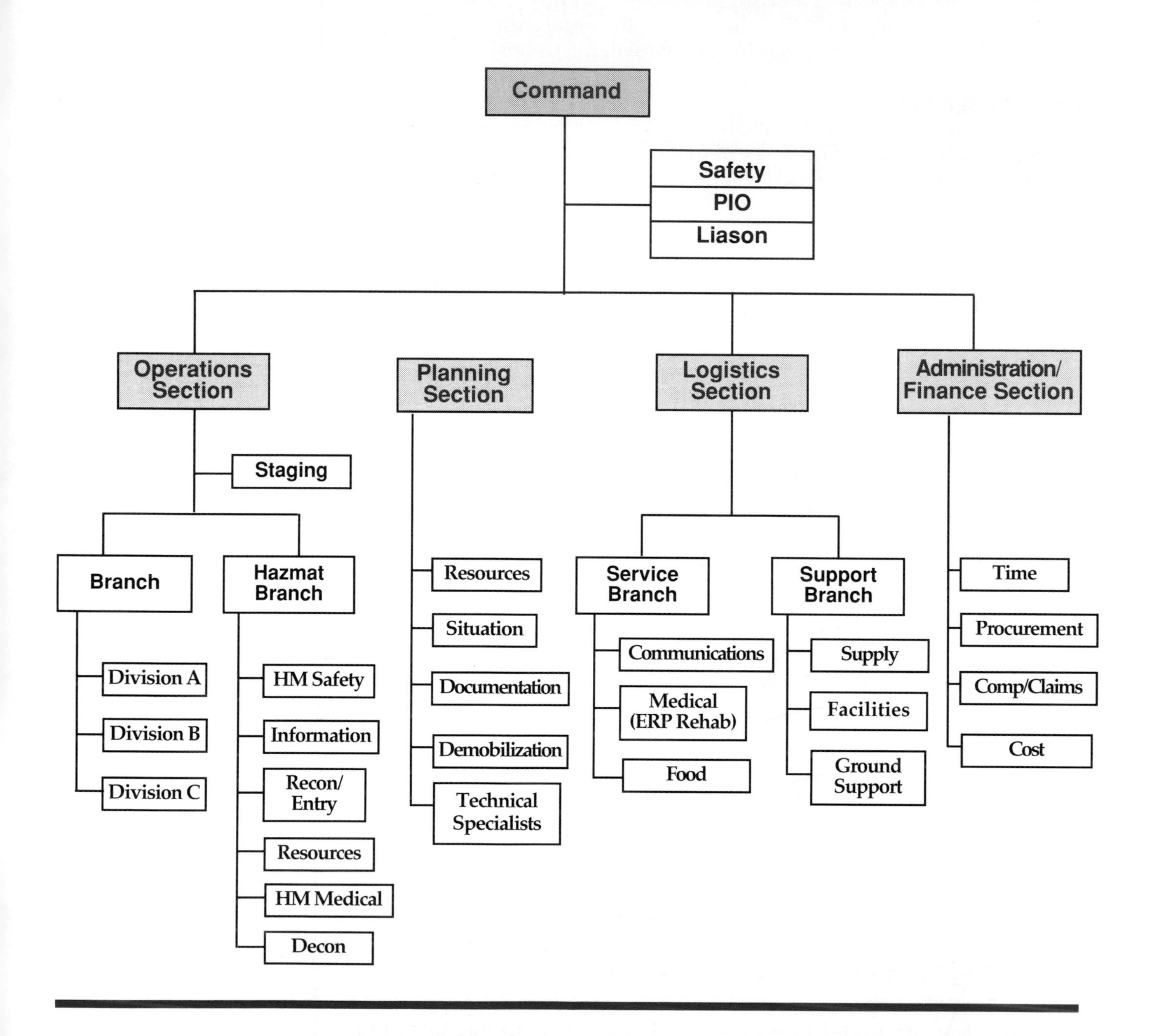

FIGURE 3-1: THE INCIDENT MANAGEMENT SYSTEM CONSISTS OF THE INCIDENT COMMANDER, COMMAND STAFF OFFICERS, AND FOUR FUNCTIONAL SECTIONS—OPERATIONS, PLANNING, LOGISTICS AND ADMINISTRATION/FINANCE

NOTES

incidents, the Incident Commander will be located at the emergency scene and will operate from a designated "command post" location. However, as the scope of the incident escalates and senior officers and managers are activated, overall Command may be transferred from the emergency scene to the Emergency Operations Center (EOC).

A single command structure is used when one response agency has total responsibility for the overall incident. However, some hazmat emergencies will require that command be unified or shared between several organizations. Unified command is described in greater detail later in this chapter.

2) **Sections**: The organizational level which has functional responsibility for primary segments of emergency incident operations. Sections and their respective unit-level positions are only activated when the corresponding functions are required by the incident.

   Within the incident management system, sections represent broad functional areas, such as Operations, Planning, Logistics, or Administration/Finance. Section Chiefs would report directly to the Incident Commander.

   The primary responsibilities of each of the sections are as follows:

   - Operations Section—Responsible for the management, coordination, and control of all on-scene tactical operations. This would encompass both fire, hazmat, oil spill, technical rescue, and emergency medical operations.

     (NOTE: This IMS terminology should not be confused with an industrial facility's Operations Department, and does not specifically refer to process/operations personnel or activities.)

   - Planning Section—Typically regarded as the "information" arm of the IMS process. Responsibilities include (1) collecting, evaluating, and disseminating incident information; and (2) maintaining information on both the current and forecasted situation, the status of resources assigned to the incident, and the preparation and documentation of action plans. Failure to establish a Planning Section for a moderate to large incident is similar to a General attempting to manage the battleground without any reconnaissance or military intelligence support. Planning units include situation status, resource status, documentation and demobilization, and technical specialists.

     The Planning Section plays a critical role in providing specialized expertise based upon the needs of the incident. Common technical areas include health and safety, industrial hygiene, environmental, and process engineering.

   - Logistics Section—Once staged resources are exhausted, the response effort becomes extremely dependent upon the availability of the resupply and logistics effort. Logistics is responsible for providing all incident support needs, including facilities, services, and materials. This would encompass communications,

NOTES

rehabilitation and food unit concerns within the Service Branch, and supply, facilities, and ground support unit concerns within the Support Branch.

The Logistics Section should not be viewed as an "order taker" to support the response effort. To be effective, the Logistics Section must be proactive and anticipate future needs, be knowledgeable of both the emergency and project procurement process, and be able to facilitate the transition from the emergency response to project phase of a long-term emergency.

- Administration/Finance Section—Responsible for all costs and financial actions of the incident. Their primary responsibility is to get funds where they are needed, ensure that adequate, yet simple, financial controls are in-place, and keep track of funds. Primary finance units would include time, procurement, cost, and compensation/claims. In some corporate IMS structures, the legal unit may also be assigned to the Finance Section.

3) **Branch:** The organizational level having functional or geographic responsibility for major segments of incident operations. The branch level is organizationally between the section and division/group/sector levels, and is supervised by a Branch Director. Branches are often established when the number of divisions/groups or sectors exceeds the recommended span of control.

   A Hazmat Branch will normally be established within the Operations Section to coordinate all hazmat response team activities at Level II and Level III hazmat incidents.

4) **Division/Groups/Sectors**: Divisions are the organizational level having responsibility for operations within a defined geographic area. A building floor, plant location, or process area may be designated as a division, such as the Division 4 (i.e., 4th floor area) or the Alky Division (i.e., Alkylation Process Unit). Divisions are under the direction of a Supervisor.

   Groups are the organizational level responsible for a specified functional assignment at an incident. Hazmat units may operate as a Hazmat Group at Level I and smaller Level II hazmat incidents. Groups are under the direction of a Supervisor.

   Sectors are the organizational level having responsibility for operations within a defined geographic area or with a specific functional assignment. At major incidents, sectors are essentially sub-units within a section or branch, and are under the supervision of a Sector Officer. Examples of functional sectors within the Operations Section would be the Safety, Rescue, or Medical Sectors. A location may also be designated as a geographic sector, such as the North Sector or Marine Dock Sector.

5) **Director/Supervisor/Leader/Officer**: The individual(s) responsible for implementing the Incident Commander's and/or Section Chief's objectives and controlling all people and equipment assigned to their respective branch (Director), division (Supervisor), group (Leader), or

NOTES

sector (Officer). These positions are usually assigned to the most knowledgeable person from the sector's area of discipline.

As with the On-Scene Incident Commander, these responsibilities may be transferred to other personnel during the course of an emergency as senior officers arrive on scene and personnel assignments are changed, or when relief is needed due to the duration of the incident.

*6) **Command and Staff Officers:** These officers are appointed by and report directly to the Incident Commander. These include the Safety Officer, the Liaison Officer, and the Public Information Officer (PIO).
   - The **Safety Officer** is responsible for the safety of all personnel, including monitoring and assessing safety hazards, unsafe situations, and developing measures for ensuring personnel safety. Assistant safety officers, such as the Hazmat Safety Officer, may be designated as necessary. Safety Officers have the authority to stop any activity that poses an imminent danger.
   - The **Liaison Officer** serves as a coordination point between the Incident Commander and any assisting or coordinating agencies not involved in the unified command structure who have responded to the emergency.
   - The **Public Information Officer** will have responsibility for coordinating all media contacts and activities during an emergency.

MODULAR ORGANIZATION

The IMS organizational structure develops in a modular fashion based upon the size and nature of the incident. The system builds from the top down, with initial responsibility and performance placed upon the On-Scene Incident Commander. At the very least, an Incident Commander must be identified on all incidents, regardless of their size. As the need exists, separate sections can be developed, each with several divisions/groups or sectors that may be established.

The specific organizational structure established will be based upon the management needs of the incident. For example, at a simple incident, such as a 55-gallon spill of sulfuric acid on a loading dock, personnel are not required to staff each major functional IMS area. The operational demands and the number of resources do not require delegation of management functions. However, a complex incident, such as a major train derailment or a large toxic vapor cloud release, may require staffing all sections to manage each major functional area and delegate management functions.

PRE-DESIGNATED INCIDENT FACILITIES

Emergencies require a central point for communications and coordination. Depending on the incident size, several types of pre-designated facilities may be established to meet IMS requirements. These are:

1) **Command Post**: The "on-scene" location where the Incident Commander develops goals and objectives, communicates with

NOTES

subordinates, and coordinates activities between various agencies and organizations. The command post is the "field office" for on-scene emergency operations, and requires access to communications, information, and both technical and administrative support. The Incident Commander should remain at the command post so that he/she is readily accessible to all personnel. Where safely possible, the command post should be located so that the incident site can be viewed.

FIGURE 3-2: COMMAND POST OPERATIONS

A number of emergency response organizations have specialized command vehicles which function as the command post. However, the necessary equipment for a field command post can be pre-packaged into a "command post kit." Minimum equipment includes:

- Radio capability to communicate with both ERP, mutual aid units, and facility maintenance/operations personnel. This should include mutual aid radios, programmable scanner radios for monitoring emergency radio frequencies, and access to the NOAA weather frequencies, where appropriate.
- Cellular telephone capability.
- Copies of appropriate emergency response guidebooks and other reference sources. These should include the Emergency Response Plan and a response folder/booklet containing copies of all IMS checklists and worksheets.
- IMS Command Vests.
- Tactical control chart.
- Pair of binoculars.

NOTES

FIGURE 3-3: EOC OPERATIONS

2) **Emergency Operations Center (EOC):** The command post is the nerve center of the emergency operation and is usually located near the scene of the emergency. However, if the scope of the incident increases, the plant or community EOC would then be activated. In this situation, overall command for the incident would be transferred to the EOC. All communications with the media and outside agencies would be coordinated through the EOC.

Based upon physical needs and safety requirements, the EOC is normally remote from the emergency scene. The EOC should provide both phone and radio communications, information resources, and the ability for a large number of personnel to work in a comfortable and secure area. These elements become essential as the number of "players" increases and the incident stretches into days as opposed to hours.

It is important to recognize the differences between the command post and the EOC when both are operating simultaneously at a major emergency. The command post is primarily oriented towards tactical control issues, while the EOC deals with both strategical and external issues and coordinates all logistical and resource support for on-scene operations.

Emergencies have occurred where the EOC was impacted by the incident and could not be used. As a result, many facilities and some communities have identified both a primary and alternate EOC location. Both EOC's should be similarly equipped and provided with comparable information and resource capabilities. In evaluating potential EOC locations, consideration should be given to the impact of potential fire, hazardous materials spills, or vapor releases upon

NOTES

the site. It should be noted that control rooms are typically poor options for EOC's within high risk industrial facilities.

An EOC should be equipped with the following:

- Radio and telephone communications. This should include mutual aid radios, programmable scanner radios for monitoring emergency radio frequencies, and access to the NOAA weather frequencies, where appropriate. Sufficient telephone lines should be available for both incoming and outgoing calls.
- Detailed copies of both area and facility maps, site plot plans, emergency pre-plans, hazard analysis documentation and other related information.
- Copies of appropriate emergency response guidebooks and other reference sources. These should include the Emergency Response Plan, the LEPC Plan, material safety data sheets (MSDS), and other pertinent plans and procedures.
- General administrative support, including writing boards, incident status and documentation boards, computer capabilities, telefax and copying machines.
- Television sets and AM/FM radios to monitor local news coverage.
- Back-up emergency power capability to support EOC lighting, telephone, and radio base stations.

3) **Staging Area:** The designated location where emergency response equipment and personnel are assigned on an immediately available basis until they are needed. Staging is effective when the Incident Commander anticipates that additional resources may be required and orders them to respond to a pre-designated area approximately three minutes from the scene.

   The Staging Area should be clearly identified through the use of signs, color-coded flags or lights, or other suitable means. The exact location of the Staging Area will be based upon prevailing wind conditions and the nature of the emergency, and are assigned to the emergency scene from the Staging Area as needed. Staging ensures that resources are close by but not in the way.

   Staging becomes a sector within the Operations Section. The Staging Officer is responsible to account for all incoming emergency response units, to dispatch resources to the emergency scene at the request of the Incident Commander, and to request additional emergency resources, as necessary.

INTEGRATED COMMUNICATIONS

Communications are critical to the safe and efficient management and control of the emergency. Ideally, the Incident Commander should be able to communicate directly with all on-scene units and support personnel. However, the more players at the incident, the less likely that they will share the same radio frequency. The more people using the same frequency, the more crowded it becomes.

NOTES

Communications are managed most effectively through the use of a common communications center. Emergency communications work best when one radio from each responding department or organization is represented at the Command Post. When there is a large amount of radio traffic, it may be necessary to have an aide assigned to each radio frequency used in the Command Post.

Where common or mutual aid radio frequencies are unavailable, the Incident Commander should request that a designated individual report to the Command Post with a radio from his or her organization. Using these individual radios, the Incident Commander can ensure that communications flow horizontally and vertically within the command structure. Experience has shown that it is more effective to have all companies operating on a different radio frequency to work together as a sector, rather than dividing their resources between several sectors.

Whenever a situation is encountered which could immediately cause or has caused injuries to emergency response personnel, the term "EMERGENCY TRAFFIC" should precede the radio transmission. This will be given priority over all other radio traffic. If an "emergency traffic" message is issued by the Communications Center, it may be preceded by a radio alert tone.

The status of an incident will obviously change as response countermeasures are implemented and progress is made in returning to normal operations. The following radio designations can be used to indicate the status of an emergency:

- ❑ *Active*. An incident is active from the time of discovery and notification until it no longer impacts the facility or community. While an emergency is active, communications should be dedicated to the support of the emergency. Routine activities within the area where the emergency occurs may be suspended.
- ❑ *All Clear*. As used within firefighting operations, the report indicates that the primary search of the area has been completed.
- ❑ *Under Control*. Once the emergency is controlled to a level where routine operations can resume without impacting recovery operations, the emergency can be designated as "under control." The "under control" designation allows communication procedures to return to normal and moves the incident to the restoration and recovery phase.
- ❑ *Termination*. The incident is terminated when the Incident Commander declares that field response activities are now completed and that all response units can be released from the emergency.

While difficult, it is critical that the Incident Commander be provided with regular and timely progress reports throughout the course of the emergency. Information is power; when the Operations Section becomes a "black hole" for all incident information, it breaks down the ability of the overall IMS structure to support the operational response effort and address the myriad of external issues in a timely and effective manner. Communications of a sensitive nature should not be given over nonsecure cellular telephones or radios which can be monitored.

NOTES

UNIFIED COMMAND STRUCTURE

Hazmat emergencies often involve situations where more than one organization shares management responsibility, or where the incident is multi-jurisdictional in nature. A unified command structure simply means that all organizations/jurisdictions who have an emergency response mission jointly contribute to the process of:

- Determining overall incident objectives and strategies.
- Establishing incident priorities.
- Ensuring that objectives are carried out safely.
- Making maximum use of all available resources.

The number of players in the unified command structure depends on the type and nature of the emergency, the location of the incident (which political jurisdiction the incident falls within), and the number of internal and external agencies involved.

As in a single command structure, the Operations Section Chief will have responsibility for implementation of the incident action plan. The determination of which agency or organization the Operations Section Chief represents must be made by mutual agreement of the unified command. This may be done on the basis of greatest jurisdictional involvement, number of resources involved, existing statutory authority, or by mutual knowledge of the individual's qualifications.

CONSOLIDATED PLAN OF ACTION

Every emergency incident needs some form of an incident action plan. The incident action plan consists of strategic goals, tactical objectives, incident priorities, and resource requirements. For Level I hazmat incidents, the plan is usually simple and can be communicated directly to the individual(s) carrying out the Incident Commander's instructions. However, on large-scale incidents this may not be practical.

Emergencies involving multi-organizations or multi-jurisdictions working within a unified command structure require consolidated action planning. As more organizations arrive at the emergency scene, they bring with them individual agendas and objectives. These may be driven by:

- Facility responsibilities.
- Legally mandated requirements.
- Financial interests.
- Contractual responsibilities.
- Specific mission goals and charters.

For example, local law enforcement agencies arriving at an incident may be primarily interested in traffic safety, while the U.S. Environmental Protection Agency (EPA) and the State Department of Environmental Quality (DEQ) are both interested in water pollution and airborne releases. All three agencies have a legal right to be involved in the emergency, but neither group has the same objective.

NOTES

A consolidated action plan is used to ensure that:

- Everyone works together toward a common emergency response goal; that is, protecting life safety, the environment, and property.
- Individual response agendas are coordinated so that personnel and equipment are used effectively and in a spirit of cooperation and mutual respect.
- Everyone works safely at the scene of the emergency.

The most effective way to ensure that a consolidated plan of action is implemented is to have the senior representative of each "major player" at the incident present at the command post and/or EOC at all times.

When multiple organizations are involved in an emergency, the Incident Commander functions as the "Chairman of the Board." Command runs the incident as one would run a meeting, making sure that every organization has its say and that the entire group works toward a resolution of the emergency as quickly and as safely as possible. The Incident Commander should remain focused on realistic objectives and ensure that each entity or special interest has input into the plan.

COMPREHENSIVE RESOURCE MANAGEMENT

The Incident Commander must analyze overall incident resource requirements and deploy available resources in a well-coordinated manner. In the case of a major hazmat emergency, resources will include both personnel, equipment, and supplies. The bottom line is simple—the proper type and level of resources must be available to support emergency operations in a timely manner.

Among the resource management lessons learned as a result of previous emergencies and exercises are the following:

- Get it done rather than argue about whose problem it is.
- It is easier to "gear down" operations than it is to play "catch up." If you think you will need it, call for it!
- Overreact until the emergency situation is fully understood. React to the potential, not the existing situation.
- Accept help from others.

COMMAND OPERATIONS

STANDARD OPERATING PROCEDURES

SOP's are required for any organization to function effectively. They become even more critical when facing hazmat emergencies. A sloppy or unorganized response to a hazmat incident may well place personnel and emergency units at an unacceptable position of risk, which could mean major injuries, unnecessary contamination, and even death.

A hazmat emergency response program can be compared to a computer system. Both consist of two elements—hardware and software. While computer hardware can cost thousands of dollars, it is relatively useless

NOTES

unless guided and directed by an effective computer software program. And like computer hardware, emergency response hardware (i.e., personnel, vehicles, supplies, and equipment) are useless unless they can be operated in a safe, effective, and coordinated manner. SOP's are the "software program" which drive and direct the actions of ERP and their equipment at a hazmat emergency.

SOP's offer the following benefits and advantages:

- Reflect organizational policy on incident management and hazmat response.
- Provide a standardized approach to train for and manage a hazmat incident.
- Provide predictable approaches to incident management, and help to make the IC's operations more effective.
- Provide a training tool for ERP reference.
- Provide a baseline for critiques and the review of emergency operations.

ESTABLISHING COMMAND

Like any other type of emergency, a hazmat incident requires strong, central command. Without it, the scene will usually degenerate into an unsafe, disorganized "free-for-all."

A strong, central command will:

- Fix command responsibility on one particular individual through a standard identification system.
- Ensure that strong, direct, and visible command is established as soon as possible.
- Establish a management framework that clearly outlines the objectives and functions of the operations.

As part of the command process, the Incident Commander is responsible for establishing incident priorities. These will include:

1) Life safety, including the removal of any endangered personnel and the treatment of the injured.
2) Safety, accountability, and welfare of all responders throughout the course of the incident.
3) Incident stabilization or activities required to minimize the potential of additional danger or harm.
4) Environmental and property conservation.

Much of the subsequent success or failure of the command system to manage a specific incident will depend on the manner in which the first-arriving officer established command. Regardless of rank, this individual should always initiate the following functions:

- Correctly assume command. Is he/she the highest ranking or designated officer present?
- Confirm command. Have the Communication Center and all personnel on the scene and enroute been notified of the command structure?

NOTES

- Select the proper command mode. Has the initial Incident Commander (1) assumed a stationary, command presence or (2) stayed with the first-arriving units while using a portable or mobile radio to continue command?
- Establish a Staging Area. Has a safe, easily accessible location been designated for incoming personnel and apparatus?
- Request necessary assistance. Is a Level I, II, or III hazmat assignment needed?
- Relinquish command as required. Has a higher ranking or better qualified individual arrived on scene who is prepared to assume command?

TRANSFERRING COMMAND

As the incident progresses, additional senior officers and managers may arrive on the scene. Command is transferred to improve the quality of the response organization.

The Incident Commander has the overall responsibility and authority for managing an incident. The arrival of a ranking officer on the incident scene does not mean that command has been transferred to that officer. Command is only transferred when the outlined transfer-of-command procedure has been completed. Likewise, if a higher ranking officer wants to effect a change in the management of an incident, he or she must first be on the scene of the incident, then utilize the transfer-of-command procedure.

Within the chain of command, the transfer of command should be governed by the following procedures:

- Command should never be transferred or passed to an officer who is not on the scene.
- The officer assuming command will communicate with the person being relieved face-to-face or by radio. Face-to-face is the preferred method.
- The person being relieved will brief the officer assuming command of the following:
  1) Current incident conditions.
  2) Incident action plan.
  3) Progress towards the completion of the tactical objectives.
  4) Safety considerations.
  5) Deployment and assignment of operating units and personnel.
  6) Appraisals of the need for additional resources.
- Review the tactical worksheet with the officer assuming command, as it should outline the location and status of personnel and resources in a standardized form that should be recognized throughout the organization.
- The person being relieved will then be assigned to the best advantage by the officer assuming command.

NOTES

DELEGATING RESPONSIBILITY

The Incident Commander may delegate both responsibility and authority by activating section and unit positions within the IMS. This transition from the initial response to a major incident organization will be evolutionary and positions will only be filled as corresponding tasks and activities are required.

Section and unit officers can be any emergency service individual who is familiar with the assigned tasks. Detailed checklists can make their job much easier. These officers should be provided with the same protective gear as their subordinates wear and should wear a command vest or other identification which makes them conspicuous. They are responsible for:

- Direct (face-to-face) supervision of their subordinates' efforts.
- Monitoring of personnel safety and welfare.
- Requesting additional resources, as required.
- Coordinating actions with other sections and units, as required.
- Advising the IC of situation status, changing conditions, and progress and exception reports.

THE COMMAND POST

The command post is the "field office" of the Incident Commander. This may be in the front seat of a security truck at an industrial site or inside a large, specially designed command vehicle with computers, fax machines, radio operators, and coffee machines on an interstate highway.

The command post should be positioned in a predictable and conspicuous location which affords the IC a good view of the incident without interfering with operations. Obviously, there will be times when the size or location of the incident prohibits a direct view.

In addition to the communications and information resources previously listed under "Pre-designated Incident Facilities," a command post should provide the Incident Commander with the following:

- A place safe from the hazardous material(s).
- A quiet (relatively) place where one can think, discuss, and decide.
- A vantage point from which to see (when possible).
- Good communications capabilities (e.g., radio, phone, fax, etc.).
- Inside lighting.
- A place to write and record.
- Protection from the weather.
- Protection from the media.
- Staff space.
- Computers.

Some incidents may require that two or three vehicles and/or tables, copy machines, and other supplies be incorporated into the command post. Such "office equipment" should be located during pre-planning to allow for its immediate transport and use when needed.

NOTES

HAZMAT BRANCH OPERATIONS

Hazmat operations are usually the responsibility of a Hazardous Materials Response Team (HMRT), but may be assigned to a specific unit or company when a HMRT is not immediately available. Most technical hazmat operations are performed by personnel trained to the Hazardous Materials Technician level. Organizations which do not possess this capability must limit their response activities to the training and resource levels of their personnel.

Hazmat personnel will function as an IMS Group for Level I and some Level II incidents and as an IMS Branch for most Level II and all Level III incidents. The Hazmat Officer will usually report to the Operations Section Chief, but would report directly to the Incident Commander if the Operations Section is not established. To simplify our discussions, we will use the term "Hazmat Branch" to denote both group or branch operations for the remainder of the text.

The Hazmat Branch is responsible for all hazmat operations which occur at an incident. Primary Hazmat Branch functions include safety, site control, research, entry, and decontamination, while secondary functions include medical and resource concerns. Specific tasks associated with each of these functions include:

- **Safety Function**—primarily the responsibility of the Incident Safety Officer and the Hazmat Safety Officer. Responsible for ensuring that safe and accepted practices and procedures are followed throughout the course of the incident. Possess both the authority and responsibility to stop any unsafe actions and correct unsafe practices.
- **Site Control Function**—establish control zones, establish and monitor access routes at the incident site, and ensure that contaminants are not being spread.
- **Information Function**—responsible for gathering, compiling, coordinating and disseminating all data and information relative to the incident. This data and information will be used within the Hazmat Branch for assessing hazard and evaluating risks, evaluating public protective options, selecting PPE, and developing the incident action plan.
- **Entry Function**—responsible for all entry and back-up operations within the Hot Zone, including reconnaissance, monitoring, sampling, and mitigation.
- **Decontamination Function**—responsible for the research and development of the decon plan, set-up, and operation of an effective decontamination area capable of handling all potential exposures, including entry personnel, contaminated patients, and equipment.
- **Medical Function**—responsible for pre- and post-entry medical monitoring and evaluation of all entry personnel, and provides technical medical guidance to the Hazmat Branch as requested.

- **Resource Function**—responsible for control and tracking of all supplies and equipment used by the Hazmat Branch during the course of an emergency, including documenting the use of all expendable supplies and materials. Coordinates, as necessary, with the Logistics Section Chief.

HAZMAT BRANCH STAFFING

Figure 3-4 illustrates the organizational structure and staffing for the Hazmat Branch. The respective positions and responsibilities of the Hazmat Branch are outlined below.

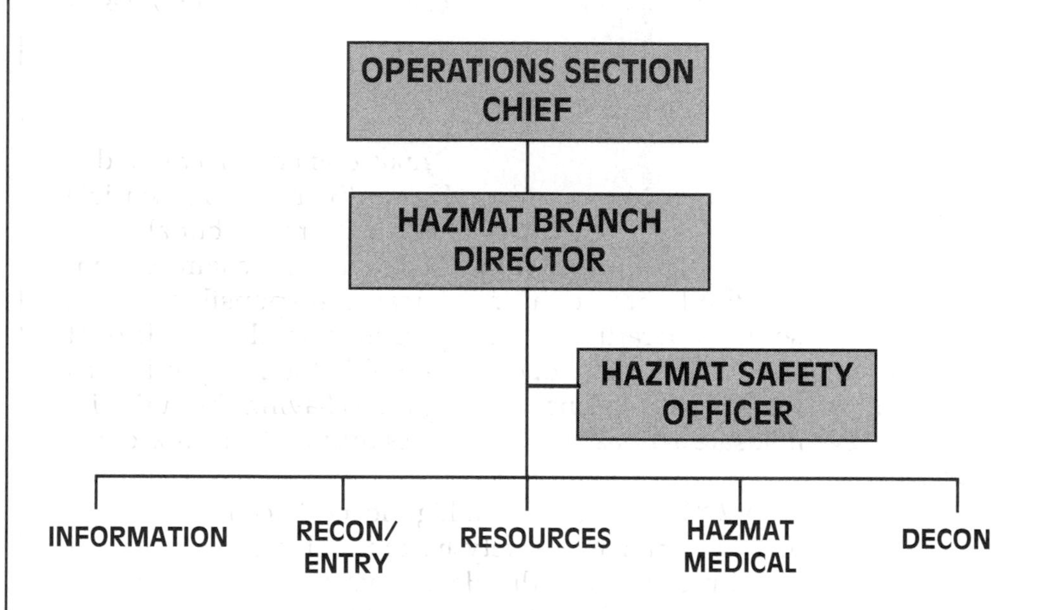

FIGURE 3-4: THE HAZMAT BRANCH IS RESPONSIBLE FOR ALL HAZMAT OPERATIONS WHICH OCCUR AT AN INCIDENT

HAZMAT BRANCH DIRECTOR

Responsible for the management and coordination of all functional responsibilities assigned to the Hazmat Branch, including safety, site control, research, entry, and decontamination. The **Hazmat Branch Director** must have a high level of technical knowledge and be knowledgeable of both the strategical and tactical aspects of hazmat response. The Hazmat Branch Director will typically be trained to the Hazardous Materials Technician level and will often be an officer on the HMRT. Depending upon the scope and nature of the incident, the Hazmat Branch Director will report to either the Operations Section Chief or the Incident Commander.

Based upon the IC's strategic goals, the Hazmat Branch Director develops the tactical options to fulfill the hazmat portion of the Incident Action Plan. Working at the command post, the Hazmat Branch Director is responsible for ensuring that the following tasks are completed:

- Control zones are established and monitored.

- Site monitoring is conducted to determine the presence and concentration of contaminants.
- Site safety plan is developed and implemented.
- Tactical objectives are established for the Hazmat Entry Group within the limits of the team's training and equipment limitations.
- All Hot Zone operations are coordinated with the Operations Section Chief or Incident Commander to ensure that tactical goals are being met.

The Hazmat Branch Director will also have a staff consisting of the Hazmat Safety Officer and the Hazmat Liaison Officer. The radio designation is "HAZMAT."

HAZMAT SAFETY OFFICER

The Hazmat Safety Officer reports to the Hazmat Branch Director and is subordinate to the Incident Safety Officer. This individual is responsible for coordinating safety activities within the Hazmat Branch but also has certain responsibilities that may circumvent the normal chain of command. Remember—the Incident Safety Officer is responsible for the safety of all personnel operating at the emergency, while the Hazmat Safety Officer is responsible for all operations within the Hazmat Branch and within the Hot and Warm Zones. This includes having the authority to stop or prevent unsafe actions and procedures during the course of the incident.

The Hazmat Safety Officer must have a high level of technical knowledge to anticipate a wide range of safety hazards. This should include being hazmat trained, preferably to the Hazardous Materials Technician level, and being knowledgeable of both the strategical and tactical aspects of hazmat response. While it is not the Hazmat Safety Officer's job to make tactical decisions or to set goals and objectives, it is his or her responsibility to ensure that operations are implemented in a safe manner.

Among the specific functions and responsibilities of the Hazmat Safety Officer are:

- Participate in the development and implementation of the Site Safety Plan.
- Advise the Hazmat Branch Director of all aspects of health and safety, including work/rest cycles for the Entry Team.
- Possess the authority to alter, suspend, or terminate any activity that may be judged to be unsafe.
- Ensure the protection of all Hazmat Branch personnel from physical, chemical, and/or environmental hazards and exposures.
- Identify and monitor personnel operating within the Hot Zone, including documenting and confirming both "stay times" (i.e., time using air supply) and "work times" (i.e., time within the hot or warm zone performing work) for all entry and decon personnel.
- Ensure that EMS personnel and/or units are provided, and coordinate with the Hazmat Medical Officer.

- Ensure that health exposure logs and records are maintained for all Hazmat Branch personnel, as necessary.

The radio designation is "HAZMAT SAFETY."

HAZMAT LIAISON

HAZMAT LIAISON OFFICER

At working incidents requiring continuous coordination and communications between the Hazmat Branch and individuals/organizations outside of the Branch, a **Hazmat Liaison Officer** may be appointed. This individual is a member of the Hazmat Branch Director's staff who is responsible for the following:

- Inform, update, and coordinate the activities of the Hazmat Branch within the rest of the IMS organization. Depending upon the level of the incident, the Hazmat Liaison Officer may regularly provide updates to the Operations Section Chief or the Incident Commander.
- Serve as a point of contact between the Hazmat Branch and other emergency service units, private industry representatives, environmental agencies, etc.
- Distribute written updates and explanations from the Hazmat Branch to other IMS members.

The radio designation is "HAZMAT LIAISON."

HAZMAT INFORMATION/RESEARCH GROUP/TEAM

The Hazmat Information/Research Group is managed by the **Information Officer** (may also be known as Research or Science). Depending upon the level of the incident and the number of hazardous materials involved, the Information Group may consist of several persons or teams. The Hazmat Information Group and the Information Officer are responsible for the following:

- Provide technical support to the Hazmat Branch.
- Research, gather and compile technical information and assistance from both public and private agencies.
- Provide and interpret environmental monitoring information, including the analysis of hazmat samples and the classification and/or identification of unknown substances.
- Provide recommendations for the selection and use of protective clothing clothing and equipment.
- Project the potential environmental impacts of the hazmat release.

The Information Officer's radio designation is "INFORMATION."

ENTRY GROUP/TEAM

The Entry Group is managed by the **Entry Officer**. This individual is responsible for all entry operations within the Hot Zone and should be in constant communication with the Entry Team.

The Entry Group and the Entry Officer are responsible for the following:

- Recommend actions to the Hazmat Branch Officer to control the emergency situation within the Hot Zone.
- Implement all offensive and defensive actions, as directed by the Hazmat Branch Officer, to control and mitigate the actual or potential hazmat release.
- Direct rescue operations within the Hot Zone, as necessary.
- Coordinate all entry operations with the Decon, Information, and Hazmat Medical Groups.

The Entry Group Officer's radio designation is "ENTRY."

ENTRY

Personnel assigned to the Entry Group will include the entry and back-up teams, and personnel assigned for entry support. The **Entry Team** consists of all personnel who will enter and operate within the Hot Zone to accomplish the tactical objectives specified within the Incident Action Plan. Entry Teams will always operate using a buddy system.

The **Back-Up Team** is the safety team which will extract the entry team in the event of an emergency. The Back-Up Team must be in-place and ready whenever entry personnel are operating within the Hot Zone.

Entry support personnel (may also be known as the **Dressing Team**) are responsible for the proper donning and outfitting of both the Entry and Back-Up Teams. The Dressing Team will report to the Dressing Officer, who is responsible for determining the number of entry personnel, having sufficient entry support personnel available, and ensuring that the appropriate levels of PPE are available for donning.

DECONTAMINATION GROUP/TEAM

DECON

The Decontamination Group is managed by the **Decon Officer**. This individual is responsible for the Decon Team.

The Decontamination Group and the Decon Officer are responsible for the following:

- Determine the appropriate level of decontamination to be provided.
- Ensure that proper decon procedures are used by the Decon Team, including decon area set-up, decon methods and procedures, staffing, and protective clothing requirements.
- Coordinate decon operations with the Entry Officer and other personnel within the Hazmat Branch.
- Coordinate the transfer of decontaminated patients requiring medical treatment and transportation with the Hazmat Medical Group.
- Ensure that the Decon Area is established before any entry personnel are allowed to enter the Hot Zone.
- Monitor the effectiveness of decon operations.
- Control all personnel entering and operating within the decon area.

The Decontamination Officer's radio designation is "DECON."

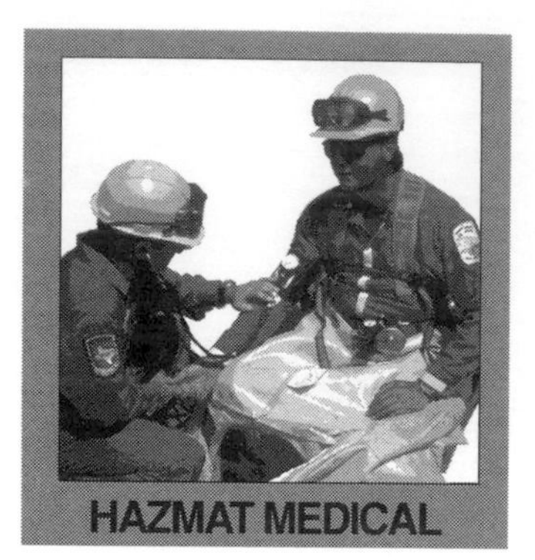
HAZMAT MEDICAL

HAZMAT MEDICAL GROUP/TEAM

The Hazmat Medical Group is led by the **Hazmat Medical Officer** Personnel assigned to this group will be located in the Entry Team dressing area and in the Rehabilitation Area. The Hazmat Medical Group and Hazmat Medical Officer are responsible for the following:

- Provide pre-entry and post-entry medical monitoring of all entry and back-up personnel.
- Provide technical assistance for all EMS-related activities during the course of the incident.
- Provide emergency medical treatment and recommendations for ill, injured, or chemically contaminated civilians or emergency response personnel.
- Provide EMS support for the Rehab Area.

The Hazmat Medical Officer's radio designation is "HAZMAT MEDIC."

Although subordinate to the Decon Officer, the **Hazmat Rehabilitation Officer** must also work closely with the Hazmat Medical Group. The Rehab Officer is responsible for overseeing all operations within the Rehab Area. Guidelines for the placement and location of the Rehab Area were previously covered in Chapter 2.

The Hazmat Medical Group will conduct post-entry medical monitoring, cooling, and rehydration of entry and back-up personnel in the Rehab Area. All operating personnel should not be given anything to eat or drink unless approved by Hazmat Medical personnel. Medical findings and personal exposure forms should be forwarded to the Hazmat Safety Officer and/or the Entry Group Officer.

HAZMAT RESOURCE

HAZMAT RESOURCES GROUP/TEAM

In some organizations, a hazmat resource function is established. The Hazmat Resources Group is led by the **Hazmat Resource Officer** Personnel assigned to this group will be located in the Cold Zone and will be responsible for acquiring all supplies and equipment required for Hazmat Branch operations, including protective clothing, monitoring instruments, leak control kits, etc. In addition, the Hazmat Resources Officer will also be responsible for documenting all supplies and equipment expended as part of the emergency response effort. The Hazmat Resources Group and Hazmat Resources Officer must work closely with the Logistics Section Chief.

The Hazmat Medical Officer's radio designation is "HAZMAT RESOURCES."

INCIDENT ANALYSIS

SITUATION EVALUATION

The initial phase of a situation evaluation is the size-up process. Size-up is the rapid, yet deliberate, consideration of all critical scene factors. Its

NOTES

result is the development of a rational incident action plan. A rapid and initial size-up must be completed on every hazmat alarm before personnel or apparatus are committed to potentially hazardous operations.

Size-up begins with an effective pre-incident planning program. Once an alarm is received, the Communications Center can provide the nature of the emergency, occupancy, location, and units responding. While enroute, the IC should note the weather conditions, time of day, and any additional information such as reports of persons collapsed.

The IC can complete this "mental picture" as he/she approaches the scene by answering the questions discussed listed below within the Operations Review.

INFORMATION EVALUATION

The Incident Commander should demonstrate certain characteristics to effectively do the job. These include:

- Distinguish between assumptions and facts. The IC is undeniably responsible for the entire operation and the welfare of everyone involved. Command must be completely aware of all major decisions and operations made under his/her jurisdiction. If the IC is unsure or uncomfortable with any part of the plan or any of the information received, the strategy and tactics should be put on hold until the IC is satisfied. The IC is a risk evaluator and a resource allocator.
- Maintain a flexible approach to decision-making. The overall hazmat plan must be constantly updated as more and better information is received from operational sections and units and outside sources.
- Develop a standardized response to reported conditions. There are some basic facts and observations that must accompany any assumptions used in formulating initial operations. For example, personnel should not begin entry operations inside a hazard area until sufficient chemical and hazard information has been accessed and reviewed.
- Shift to a management role as soon as possible. The IC must begin delegation of tactical responsibilities as soon as possible. Otherwise the IC will quickly become overwhelmed with both people and information.
- Solicit opinions and ideas. The ultimate decision rests with the IC, but it should be based upon input from the entire incident staff. Allow subordinates (through their section and unit officers) to voice their opinions. Be careful to avoid "group-think," as voiced by those "yes-men" who surround every Incident Commander.

From the time of arrival on scene, the IC must prioritize problems and develop solutions by collecting information. The effective IC will:

- Seek out data that is current, accurate, and specific.
- Delegate information retrieval.

NOTES

- Know how to find reference data and how to use it.
- Collect the right information in the right order.
- Use a wide variety of sources.

STRATEGICAL AND TACTICAL EVALUATION

The basic configuration of command consists of three levels—strategic, tactical, and task. The strategic level is the broadest level and provides overall direction of the incident. Based upon these goals, the tactical level assigns operational objectives. Finally, the task level assigns specific tasks to response units or personnel in order to meet the tactical objectives. The Incident Action Plan should cover all strategic responsibilities, all tactical objectives, and all support activities needed during the incident. This includes defining when and where all resources will be assigned to the incident to control the situation.

The combination of the IC's evaluation of the current conditions and forecast of future conditions leads to the development of the overall operational strategy. **Strategic goals** are the broad game plan which is developed to meet the incident priorities (life safety, incident stabilization, environmental and property conservation). Examples of strategic goals would include fire control, spill and leak control, and public protection.

Strategic goals, in turn, are achieved by the completion of the **tactical objectives**. Whereas strategic goals are broad and general, tactical objectives are specific and measurable. Examples of tactical objectives for a public protection strategic goal would include isolation, evacuation, or protection in-place, while a fire control strategy would consider fire control, fire extinguishment, or controlled burning as alternative tactical objectives.

Tactical response objectives to control and mitigate the hazmat problem are implemented in either an offensive, defensive, or nonintervention mode. Criteria for evaluating these options include the level of available resources (e.g., personnel and equipment), the level of training and capabilities of emergency responders, and the potential harm created by the hazmat release.

A hazmat operation is functioning in the **offensive** mode when it allows hazmat personnel, with appropriate protection and training, to operate in the Hot Zone to attempt to control the hazard at its source. Emphasis is often towards hazmat containment or fire extinguishment. Applying a Chlorine B kit on a one-ton chlorine cylinder and applying dome clamps on an MC-306/DOT-406 gasoline tank truck are examples of offensive tactical options.

In comparison, **defensive** hazmat operations are actions which ERP can safely take without coming in direct contact with the hazmats involved. Personal exposures are minimized, with the emphasis towards hazmat confinement or keeping the hazmat release to a specific area. Diking, damming, and retention of liquid spills, exposure protection during a fire, and vapor dispersion are examples of defensive tactical options.

Nonintervention is an option where no ERP actions are taken to control or mitigate the hazmat release. Tactical objectives are directed towards site management and control, evacuation, protecting exposures, and allowing the hazmat incident to run its natural course. Sometimes no response is the best response. Monitoring a leaking flammable or toxic compressed gas cylinder from a safe distance until it is empty or withdrawing all units from a well-involved building fire involving reactive chemicals are good examples.

Most firefighting operations are conducted in the offensive mode. A quick interior attack is usually the best attack. The opposite is true for hazmat operations. Most operations will begin from a defensive point of view. The most important question the Incident Commander should ask is, "What happens if I do nothing?"

There will be times when an operation will be in a marginal mode. In other words, initial information indicates that it is relatively safe to attempt an offensive tactical objective, yet it is very possible that things may turn for the worse during that process.

This is the time at which ERP are most at risk. The IC, the Hazmat Branch Officer, and the Safety Officer must constantly monitor the operation and its results—or lack of them. If the operation cannot be completed as expected, if any safety problem arises, or if there is a change in the operational plan, they must immediately "blow the whistle" and stop the activities. The entire scene immediately reverts, then, to a defensive game plan.

It is very easy for the IC to be lulled into a false sense of security by personnel actually operating in the Hot Zone. "Just a few more minutes, and we'll have the leak sealed" is not an acceptable response to a "clear the Hot Zone" decision. Don't overlook gut reactions. The IC has a unique point of view that the other players do not. One doesn't need a degree in chemistry to know that something just doesn't look right.

For some situations, the IC should consider establishing specific time limits for the implementation and completion of tactical response objectives. Experience shows that responders can easily lose track of time during a working incident and not recognize some of the early warning signs of things going bad. To minimize this problem, get into the habit of providing regular and timely (e.g., every 15 minutes) tactical updates.

In summary, the IC should pursue strategic goals and tactical objectives which will (1) provide the most favorable outcome in terms of harm to people, property, the environment, and systems disruption; (2) satisfy incident priorities; and (3) present the least risk to emergency responders.

Tactical Worksheet

The IC needs to keep track of the various assignments made, along with the progress of section and subordinate officers in completing those tasks. This operational note-taking becomes particularly critical when Command is faced with an extended operation.

The tactical worksheet provides a standard approach that everyone can understand. It is designed to be used by the IC and the command

NOTES

N NOTES

TECHNICAL SPECIALISTS AND INFORMATION SOURCES

When gathering information, ERP are often the true nonbelievers. They have been lied to so many times that they automatically question most information when it is initially presented to them. This is not necessarily a bad trait, as long one has a structured procedure to guide one through the information gathering process. In many respects, the role of responders during the information gathering process is similar to that of a detective.

A likely source of hazard information will be individuals who either work with chemical(s) or process units or who have some specialized knowledge, such as container design, toxicology, or chemistry. When evaluating these sources and the information they provide, consider these observations:

- Many individuals who are specialists in a narrow, specific technical area may not have an understanding of the broad, multi-disciplined nature of emergency response. For example, information sources may provide extensive data on process engineering and design, yet may be unfamiliar with basic site safety and personal protective clothing practices.
- Some technical specialists have knowledge which is based upon dealing with a chemical or process in a structured and controlled environment. When faced with the same chemical or process in an uncontrolled or emergency response situation, they may provide only limited information.
- There are no experts, but only information sources! Each individual source will have its own advantages and limitations. A colleague who responded to the Persian Gulf oil spill during Operation Desert Storm in 1991 provided this response after being referred to as an expert: "We aren't experts, but we do have good judgment. Recognize, however, that good judgment is based upon experience, and experience is often the product of bad judgment." Translation—we screwed up enough to know what will probably work and what won't.
- Sometimes ERP interact with individuals with whom they have had no previous contact. Before relying upon their recommendations, ascertain their level of expertise and job classification by asking specific questions. Remember these two points: (1) Technical smarts is not equivalent to street smarts! Having an alphabet behind one's name does not automatically mean that an individual necessarily understands the world of emergency response and operations in a field setting; and (2) Twenty years of experience may actually be one year of experience repeated twenty times.
- Questioning information sources is an art and requires the skills of both a detective and a diplomat. While this is certainly not an interrogation process, you must be confident of the source's authority and expertise. Always conduct the interview with respect to the person's rank or position within the field or

NOTES

organization. One method is to ask questions for which you already know the answer in order to evaluate that person's competency and knowledge level. Remember, final accountability rests with the Incident Commander.

Finally, remember that everybody brings their own agenda and scorecard to a hazmat incident. Don't assume that your concerns (1) are the most important, and (2) are always going to be the same as "their" concerns.

HAZARD AND RISK ASSESSMENT CONCERNS

Hazard and risk assessment is the cornerstone of decision-making and is directly tied to developing strategic goals and tactical response objectives. Unfortunately, while many IMS training programs stress the *gathering* of hazard information, they fail to adequately address the concept of *assessing* the level of risk. ERP have been injured because they did not understand or underestimated the level of risk. In their efforts to do something, responders often became part of the problem, rather than being part of the solution.

Historically, many responders have often blindly adopted the philosophy of "acceptance of risk." Simply stated, in their classic role of protecting lives and property, responders accept totally unreasonable risks where the potential of injury and harm is overwhelming, as compared to the probability of making conditions better. The net result of this "acceptance of risk" are situations where injuries and loss of life by responders exceeds the cost of damage by the incident. A classic example are confined space rescue situations, where National Institute For Occupational Safety and health (NIOSH) statistics have shown that over 60% of all fatalities are personnel functioning as a rescuer.

The IC must be able to distinguish between assumptions and facts. Command is undeniably responsible for the entire operation and the welfare of everyone involved. If the IC is unsure or uncomfortable with any part of the Incident Action Plan or information received, operations should be put "on-hold" until satisfied. In short, the IC should become a risk evaluator, not a risk taker.

DECISION-MAKING ISSUES AND CONCERNS

The management style and the decision-making structure should vary based upon the timeline and requirements of the incident. Three forms of decision-making are commonly used during incident management:

- **Autocratic**—commonly used on the fireground. Essentially, senior officers give commands which are implemented with little question or feedback. This method gets the best results in fire and rescue situations where time constraints are critical.
- **Bureaucratic**—common to incident clean-up and recovery operations, as well as remediation and removal activities. In this case, bureaucratic, legal, and administrative influences kick in and do not allow for quick or rapid decisions. Some have referred to this as "decision paralysis." Remember, you cannot micromanage a major emergency!

NOTES

- **Participative**—common to special operations, such as hazmat, bomb squad, and technical rescue. The incident timeline usually allows for more discussions at both the strategical and tactical level. This would be similar to consolidated action planning, one of the primary IMS elements.

Stress

From a human factors perspective, we often fail to recognize or appreciate the effect of stress upon decision-making, particularly when operating at long-term incidents. Stress is more often a product of frustration, particularly when ERP perceive there are organizational or political roadblocks in managing the situation. Experience has shown that the normal bureaucracy will not facilitate rapid decision-making.

A reluctance by the IC to delegate tasks and responsibilities will increase the level of stress, reduce the efficiency of the emergency response, and often leave major issues unaddressed. In some instances, Command tries to support all on-scene operations from within the Operations Section, rather than establish a Planning Section or Logistics Section to facilitate the delegation and implementation of supporting tasks.

The Process

Managing a major hazmat incident is no different than taking an army to war—the emergency response effort will be no better than the information, forecasting, and technical expertise available (Planning Section), the physical and personnel resources available for timely response and support (Logistics Section), and the financial and administrative support provided (Administrative/Finance Section).

While final policy and strategic decisions should always rest with Command, they should be based upon input from the entire IMS organization. Solicit opinions and ideas—they foster both individual and organizational "buy-in" into the decision-making process. Allow everyone (through their section and sector officers) to voice opinions, particularly when dealing with situations where the hazards are exceptionally high. Remember, the people being asked to take the risks should have a voice in the decision-making process. This collective input will strengthen the decision-making process, as well as present the IC with more options.

Emergency responders often seek direction on how to handle a decision-making situation where the IC or a senior officer does not agree with a subordinate's recommendation, or where a senior plant manager has no on-scene responsibility and does not belong on the emergency scene. There are no easy answers here other than these two fundamental points: (1) always be professional in this situation (i.e., don't make your boss look bad!); and (2) always try to have any conversation on a one-to-one basis away from the rest of the troops.

Lessons Learned

Some decision-making lessons learned include the following:

- Never say never, particularly when dealing with a long-term, campaign operation. History is full of incidents where tactical options which appeared to be totally unrealistic during the first hour eventually looked real good and were implemented during the twentieth hour.

- Consider the art of communications. Effective communications is one part talking and ten parts listening. Beware of individuals whose hearing is affected by management position or promotion, as well as the "yes men" who show up at major emergencies and often flock around the IC.
- When an incident goes bad or is particularly politically sensitive, anticipate being the scapegoat. In order to minimize political vulnerability, the IC must continuously (1) consult and build a consensus on the Incident Action Plan; (2) document; and (3) not assume anything.
- A time-tested rule for minimizing political vulnerability is the "rule of threes." Simply, when faced with significant or politically sensitive decisions, consult at least three independent reference sources. The more politically sensitive the incident, the greater the need for the reference sources to be respected and reputable individual(s).

  For example, using a single emergency response guidebook as the *sole* technical justification for evacuating 5,000 people is only asking for well-justified technical and political criticism of the emergency response effort, even if it was the correct decision! To minimize your political vulnerability, also seek input from CHEMTREC™, the shipper, or local technical information specialists.

NOTES

LIABILITY ISSUES

Liability concerns can stop or strongly influence the IC from making certain decisions. Accepting responsibility to perform certain tasks is not necessarily accepting liability for the emergency. Liability issues are particularly prevalent during the clean-up and recovery phase and in situations involving community and public damage claims.

While the IC must be aware of the legal implications of decision-making, this issue should be addressed during the planning phase. Well-written procedures that are reviewed and approved before an incident, reflect current standards of care, and are regularly practiced in the field help to shift the focus of liability issues from the "individual" decision-maker to the overall "organization," where they properly belong.

The bottom line—accept the basic premise that at some point you and/or your organization are likely to be sued, cited, or questioned for your decisions as an officer. In many cases, these accusations will be either politically or financially motivated and totally lack any technical foundation. The key, therefore, is to minimize the potential of a successful challenge. Remember—consult, document, and do not assume!

NOTES

"EVERYBODY HAS THE ANSWER FOR YOUR PROBLEM"

Hazmat emergencies attract a great deal of internal and external attention. They also bring a number of entrepreneurs, salespeople, managers, and "do gooders" to the attention of the IC, many of them professing to have the answer to your problem.

The reality of incident management is that while these people can be helpful, they are usually a major distraction to the Incident Commander. This is particularly true at long-term emergencies or if they have the ear of a local political or governmental official.

The Incident Commander may have to designate an IMS representative (e.g., Liaison Officer or Planning Section Officer) to address these external contacts, as they occasionally do provide worthwhile information or resources and can influence both public and political opinions. At the least, failure to seriously address these people may generate bad publicity and political backlash.

THE ETERNAL OPTIMISTS

Be aware of the eternal optimists within your own ranks. Don't allow your own people to "walk you down the yellow brick road" without exploring all viable options and alternatives. Remember, initial observations often underestimate the significance of a problem. Command should always be prepared to implement an alternative action plan if the current plan fails.

Experience has shown that there is often an inadequate flow of timely and accurate information from the incident scene to the Incident Commander. When asking subordinates for progress reports, don't continuously ask questions which can be answered with a simple "yes" or "no." If you only ask if everything is okay, don't be surprised when your people consistently say it is!

Both the IC and the Safety Officer have a unique view of the incident that many of the other players do not. Remember the difference between "street smarts" and "technical smarts"—one doesn't need a degree in chemistry to know when something just doesn't look right.

RISK COMMUNICATIONS

With the advent of SARA, Title III, Risk Management Programs, and community right-to-know legislation, the concept of risk communication has taken on greater importance. For example, as part of the hazards analysis process under *40 CFR Part 68—Risk Management Programs for Chemical Accidental Release Prevention*, facilities must identify their "worst-case scenario" and how the situation will be managed.

Effective risk communications between responders and the community must be built upon trust. Many risk communication texts cite that the general public tends to overestimate the risks associated with the catastrophic-type event (e.g., Bhopal-type scenario), and underestimate the risks associated with the more common, routine-type emergency (e.g., flammable liquid emergencies). Unfortunately, in the absence of previous communications between the community, a facility, and the general

NOTES

public, experience shows that risk communication often becomes an emotional rather than a factual issue.

The issue of organizational credibility has also received increased attention. A number of studies have asked the public who they trust as sources of information about chemical risks, with emphasis on the qualities of credibility and expertise. Among the lessons learned are that local government officials, including emergency responders, are generally perceived to be low on expertise, but are high on trustworthiness. Other "trustworthy" information sources include health professionals and the Local Emergency Planning Committee (LEPC). In contrast, industrial and other government officials are generally perceived to be high on expertise, but low on trustworthiness.

POST-INCIDENT ISSUES AND CONCERNS

A major hazmat incident is normally a long, mentally demanding event. Once the emergency is finally stabilized, most individuals want to return to the status quo. Unfortunately, there are post-incident political issues which must still be addressed.

Legislation and Regulations

The history of regulations within the United States has traditionally been in reaction to a major incident or crisis. Today, the long-term costs of implementing new governmental regulations and standards can far exceed the total costs associated with providing a timely, professional, and well-managed emergency response program.

Similarly, recent changes in federal OSHA regulations have led to an increased fine and penalty structure. Deficiencies in emergency planning and response programs have been cited in a number of instances, and six figure fines have not been uncommon.

Opportunities

Major incidents are often tragedies for ERP and the community. Yet major incidents represent opportunities, not only for resources but also for policy and management action. Some refer to this as "crisis diplomacy."

Many people and organizations are often referred to as being lucky. Perhaps, but luck is also where preparation meets opportunity. A good officer or manager will always have a "wish list" or a "pet project" position paper in the top desk drawer awaiting that right opportunity. Like it or not, emergencies represent opportunities!

Share the Lessons Learned

Post-incident debriefing sessions and critiques are an integral element in developing a "system dependent" emergency response organization (see Chapter 12). Keep in mind that while you can fool the spectators, you can't fool the players. To a large degree, critiques mirror the personality of the critique facilitator. This is an important element to consider when selecting a facilitator for a "high profile" or politically sensitive operation.

NOTES

Careful planning and consideration are required to make the critique process both effective and accepted. Experience shows that there are three types of critiques: (1) personnel lie to each other about what a great job they did; (2) personnel yell at each other for screwing up; and (3) personnel focus upon lessons learned, emphasizing changes or improvements which need to be made in SOP's, resources, and training, so that the same errors are not repeated. Obviously, the latter type is the preferred.

Finally, always be a student. Organizations and individuals often state that they cannot share their experiences with respect to a specific incident because of either legal implications or because of their attorney's advice. One certainly can understand the rationale for such statements. However, as emergency responders, one could argue that it is equally unethical and immoral to have experiences and not share the lessons learned with one's peers within the emergency response community. Sharing lessons can often be done behind the scenes via informal networking, without the spotlight of a formal presentation or published paper.

MEDIA RELATIONS

Media relations is an integral element of incident management. The media is a key player in shaping public perception of both the severity and impact of the actual emergency and the effectiveness of the response. Recognize that (1) the media is going to get a story with or without your assistance; and (2) it is better for you (as the Incident Commander) to provide that information rather than Harry Homeowner down the street. If you doubt these statements, simply think back to the 1991 Gulf War and the number of media representatives who risked personal harm to travel alone into the military theater to get a story.

THE PUBLIC INFORMATION OFFICER

The key player within the IMS responsible for coordinating the collection and release of information is the Public Information Officer (PIO). Only the IC, the PIO, or their representative should be authorized to release information to the media. All requests for information or interviews should be referred to the PIO so as to maintain consistency and accuracy.

The PIO is responsible for the following tasks:

- Assemble and prepare news information, bulletins, and press releases, and release information to the public or media as approved by the Incident Commander.
- Establish communications with all IMS players, facility representatives, and governmental agencies, and assure uniformity of all messages.
- Attend incident briefings and meetings to update public information releases.
- Arrange for meetings between the media and incident personnel as directed by the IC.

NOTES

FIGURE 4-3: A PIO IS A KEY PLAYER IN COORDINATING THE COLLECTION AND RELEASE OF INFORMATION.

The PIO should establish a media information center in a safe and central location, dependent upon the nature of the emergency. The media information center should be easily accessible by the IC and other command officers, so that interviews and press releases can be provided in both a timely and controlled manner with minimal disruption of emergency management and incident response operations.

Where possible, a media area should be designated in a safe location where media representatives can videotape and photograph the incident site, command post, and emergency response operations. However, never put the media area at the command post. Think Hollywood—perceptions are reality. Visual shots and pictures of emergency responders in action help to "sell" and "market" the capabilities of emergency response agencies.

Aircraft may be used by both the media and spectators for aerial views of the incident. These aerial shots are often useful for the incident size-up process. However, flight operations can create unsafe conditions on the ground. The rotary downwash of a helicopter may blow hazardous vapors or burning materials onto on-scene response personnel, while the shock wave and concussion from an explosion can damage aircraft or blow the aircraft out of control.

To minimize these problems, the IC should consider controlling the airspace around the incident scene. To initiate this process, Command should contact the regional office of the Federal Aviation Administration (FAA).

NOTES

PROVIDING INFORMATION TO THE MEDIA

In managing media relations and responding to risk communication concerns, both the IC and the PIO should remember these fundamental principles:

- If it's a fact, get it out. Be the best source of all facts, both good and bad. Be timely, avoid "spinning" the story, and always protect your credibility.
- From the public's perspective, the media is the messenger and perceptions are reality. If the public perceives that an incident is being well-managed, their perceptions of the response effort are usually positive.

  Of course, the reverse is also true! Consider, for example, how ERP may be perceived by the public when they initiate major traffic shutdowns or public evacuations as a result of an incident where there is no physical indicator of a problem (e.g., smoke, fire, odors, etc.).
- For public safety agencies, emergencies are an opportunity to market their skills, capabilities, and resource requirements to the public (i.e., their market).
- In his textbook *Fire Command*, Phoenix, AZ Fire Chief Alan Brunacini notes the following tactical truth: "The Incident Commander must be careful of what he/she says and how he/she acts in difficult situations. Offhanded, dumb command comments are like aluminum beer cans—they last forever in the environment."

Serving as the PIO is both a challenging and sometimes unnerving process. The following guidelines and hints may prove helpful when preparing to deal with the media:

- EXPECT TO BE NERVOUS! Think of the media interview as simply a conversation.
- Keep track of what you say.
- Don't be afraid to correct yourself if you have made an error. Likewise, know when to "shut up."
- Don't ever assume that the media or a reporter knows anything about the topic of the interview. Likewise, don't be afraid to correct a reporter when he/she is wrong or uninformed.
- Remember the "Twelve Second Rule"—A TV reporter will look for a sound bite or about 12 seconds worth of interview in response. The first statement should be a concise, positive summation of the incident and the facility or community's response to the incident.
- Talk to be understood. Always remember who your audience is. Avoid technical jargon.
- Never lie, speculate, or give personal opinions. Avoid "spinning" the story.
- Reaction to inflammatory questions should be measured and planned. PAUSE and collect your thoughts before speaking.

NOTES

Do not argue or lose control; maintain composure.

- Be public oriented. Show genuine concern for public and community welfare.
- Control rumors. If it is a fact, get it out. Do not withhold information (except death notifications).
- Don't play favorites. It is important to always be viewed as a credible source.
- Never provide comments and information "off the record."
- Stay away from "no comment." If you don't know the answer, say so, then try and find out.
- Don't guess at causes or damage estimates.
- Don't make any admissions regarding potential liability.
- Accommodate interview requests as soon as possible. Provide updates on a frequent and regular basis.
- Prepare post-incident news releases or summary information for the media.

ANTICIPATING QUESTIONS

When dealing with the media at a hazmat incident, certain types of questions should be anticipated. Many of these questions can be addressed prior to an incident and assembled into a "Hazmat Fact Sheet." The following questions were referenced from the textbook, *Chemicals, The Press & The Public*, Chapter 5—"Reporting on a Chemical Emergency."

❑ The Incident
  - What is the nature of the emergency?
  - How many injuries? Fatalities? What is the nature of the injuries and fatalities?
  - How many people were evacuated from the facility? Within the community?
  - How is the surrounding environment affected?
  - Have similar incidents occurred in the past?

❑ The Hazardous Material(s) Involved
  - What hazmat(s) are involved in the emergency?
  - Is it a solid, liquid, or gas?
  - What are the public health implications?
  - What quantity was released?
  - Are there other extremely hazardous substances (EHS's) stored, manufactured, or used within the facility?

❑ The Facility
  - Does the facility have an Emergency Response Plan?
  - Has the facility participated in the Local Emergency Planning Committee (LEPC) and the development of the community Emergency Response Plan?

NOTES

- Has the facility notified the LEPC of the hazardous materials stored and used on the site?
- Has the facility and/or community conducted a risk assessment of the potential threats posed by the facility to the community?

❑ Meteorological Conditions and Factors

- What are the current temperature, wind velocity, and humidity conditions? Are they considered favorable or unfavorable as they affect the spread of the hazmat?
- What are the immediate and short-term weather forecasts? Will the changes affect the dispersion of the hazmat?

❑ Physical Surroundings

- Will the terrain and ground contour around the incident site affect the hazmat dispersion in any manner?
- Are there nearby population centers that might be at particular risk, such as the schools, hospitals, shopping centers, etc.?
- Will nearby residents be evacuated or "sheltered-in-place"? What are the criteria for making this decision?

❑ Health Risks

- By what routes are humans exposed to the chemical (e.g., inhalation, ingestion, skin absorption, etc.)?
- What are the potential health effects? Are these effects acute or chronic?
- Are particular population groups particularly susceptible?
- Can the hazmat(s) involved react with other hazardous materials in the facility or in the area?

❑ Post-Incident Follow-up Questions

- What types of safeguards were in place?
- Did the facility have to report under any of the sections of SARA, Title III? Did it submit reports?

  ___Section 302—Presence of Extremely Hazardous Substances

  ___Section 304—Accidental Releases and Emergency Notification

  ___Section 311—Hazardous Chemicals MSDS's or Lists

  ___Section 312—Tier II Emergency and Hazardous Chemical Inventory Forms

  ___Section 313—Toxic Chemical Release Form

- What prevention measures and approaches has the facility implemented?
- What is the accident history of the facility or carrier?
- Does the facility or carrier provide training for its employees? What types of training are provided with respect to the handling of emergencies?
- What routes are used to ship and transfer hazardous materials through the community?

- Does the facility and/or community have equipment or instruments to detect and track a release?
- What types of emergency response equipment does the facility and/or community have?
- Was emergency medical care available? What level of care?
- Does the facility know of any possible substitutes which could be used for the hazardous materials released? What environmental and health issues are posed by these substitutes? What are the economic issues involved in using substitutes?

NOTES

SUMMARY

This chapter has provided an overview of the common interpersonal, organizational, and external issues associated with the management of a hazmat incident. It is based upon "lessons learned" while managing emergencies, and it supports the fundamental concepts of the Incident Management System. Many of the issues raised are not necessarily new or unique, but experience shows that they are often overlooked by many command officers.

Remember, your emergency response performance will be evaluated on two interrelated factors: (1) the implementation of a timely, well-trained and equipped emergency response effort in the field, and (2) the effective management of the interpersonal, organizational, and external impacts created by the incident. An effective response effort can be compromised or completely negated by poor management of the political and external issues.

REFERENCES AND SUGGESTED READINGS

Anderson, Richard, "Meeting the Press Doesn't Have to be a Disaster." INDUSTRIAL FIRE CHIEF (May/June, 1993), pages 32–34.

Brunancini, Alan V., FIRE COMMAND, Boston, MA: National Fire Protection Association (1985).

Chemical Manufacturers Association, RISK COMMUNICATION, RISK STATISTICS & RISK COMPARISONS, Washington, DC: Chemical Manufacturers Association (1988).

Emergency Film Group, Industrial Incident Management (video-tape), Plymouth, MA: Emergency Film Group (1994).

Hadden, Susan G. A CITIZEN'S RIGHT TO KNOW—RISK COMMUNICATION AND PUBLIC POLICY, Boulder, CO: Westview Press (1989).

McCallum, David B. and Vincent T. Covello, "Chemical Risks and the Chemical Industry: Public Knowledge, Attitudes and Behaviors in Six Communities." Briefing paper prepared for the Chemical Manufacturers Association by The Center for Risk Communication, Columbia University School of Public Health (January, 1990).

National Fire Protection Association, HAZARDOUS MATERIALS RESPONSE HANDBOOK (2nd Edition), Boston, MA: National Fire Protection Association (1992).

National Safety Council—Environmental Health Center, CHEMICALS, THE PRESS, AND THE PUBLIC. Washington, DC (1989).

Noll, Gregory G. "The Politics of Incident Management." FIRE ENGINEERING, March, 1992, pages 53-59.

IMPLEMENTING THE HAZMAT RESPONSE

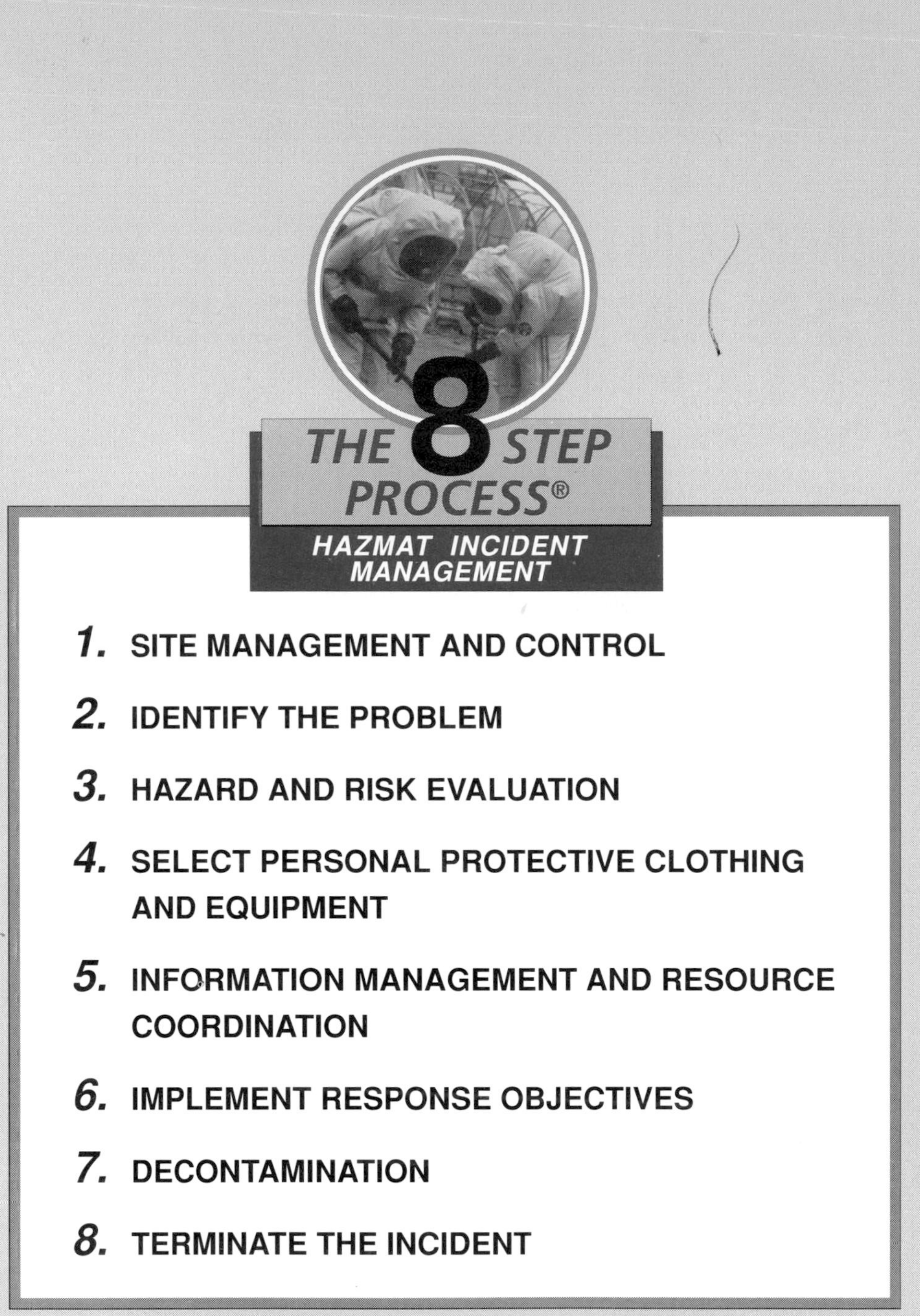
THE 8 STEP PROCESS®
HAZMAT INCIDENT MANAGEMENT
1. SITE MANAGEMENT AND CONTROL
2. IDENTIFY THE PROBLEM
3. HAZARD AND RISK EVALUATION
4. SELECT PERSONAL PROTECTIVE CLOTHING AND EQUIPMENT
5. INFORMATION MANAGEMENT AND RESOURCE COORDINATION
6. IMPLEMENT RESPONSE OBJECTIVES
7. DECONTAMINATION
8. TERMINATE THE INCIDENT

CHAPTER 5

SITE MANAGEMENT AND CONTROL

OBJECTIVES

1. Define Site Management and Control.
2. Identify the procedures for initially establishing command at a hazmat incident.
3. Identify the guidelines for the safe approach and positioning of emergency response personnel at a hazmat incident.
4. Define the following incident management terms and describe their significance in controlling emergency response resources at a hazmat incident:
   a) Staging
   b) Level 1 Staging
   c) Level 2 Staging
5. Identify the procedures required to protect the public by establishing a perimeter at a hazmat incident.
6. Describe the role of security and law enforcement personnel in establishing perimeters at a hazmat emergency.
7. Identify the procedures for establishing scene control through the use of Hazard Control Zones at a hazmat incident.
8. Define the following terms and describe their significance in establishing Hazard Control Zones:
   a) Hot Zone

"Panic is fine as long as you're the first one to panic."

Francis L. Brannigan
Fire Service Author

NOTES

b) Warm Zone
c) Cold Zone
d) Area of Refuge

9. Define the following terms and describe their significance in protecting the public at a hazmat incident:
   a) Public Protective Actions
   b) Evacuation
   c) Protection-in-Place
10. Describe three criteria for evaluating Protection-in-Place as a Public Protective Action option.
11. Describe the guidelines and procedures for implementing Protection-in-Place at a hazmat incident.
12. Describe three criteria for evaluating evacuation as a Public Protective Action option.

SITE MANAGEMENT TASKS

Site Management is the first step in the **Eight Step Incident Management Process©**. Its major focus is on establishing control of the incident scene and isolating people from the problem. The Incident Commander simply cannot begin extended operations until the hazard area has been identified and the isolation perimeter secured. People standing around the hazmat scene are potential rescues until isolation has been established and the scene cleared.

From a tactical perspective, Site Management can be divided into six major tasks. These include:

1. Assuming Command and establishing control of the incident scene.
2. Assuring safe approach and positioning of emergency response resources at the incident scene.
3. Establishing Staging as a method of controlling arriving resources.
4. Establishing a security perimeter around the incident scene.
5. Establishing Hazard Control Zones to assure a safe work area for emergency responders and supporting resources.
6. Sizing up the need for immediate rescue and implementing initial Public Protective Actions.

Life safety is the highest tactical priority of any Incident Commander. There will always be situations where initial size-up warrants that ERP move directly into rescue operations (e.g., a driver who is obviously alive and trapped in the cab of a burning gasoline tank truck). However, even under the most extreme situations, implementing initial Site Management tasks will save lives. Don't let a bad situation become worse by getting sucked into letting ERP charge into rescue situations without following safe operating procedures. Rescue is discussed in more detail in Chapter 10, Implementing Response Objectives.

NOTES

ESTABLISHING COMMAND

Like any emergency, a hazmat incident requires strong, central command. Without it, the scene will usually degenerate into an unsafe, disorganized "free-for-all."

As pointed out in Chapter Three, a strong, central command will:

- Fix command responsibility on one particular individual through a standard identification system.
- Ensure that strong, direct, and visible command is established as soon as possible.
- Establish a management framework that clearly outlines the objectives and functions of the operations.

The success or failure of emergency operations will depend on the manner in which the first-arriving officer or responder establishes command. Regardless of job title or rank, this individual should always initiate the following functions:

- Correctly assume command. The person assuming command should be the highest ranking or most experienced person present.
- Confirm command. Confirm that all personnel on the scene and enroute have been notified of the command structure.
- Select a stationary location for the command post. An experienced commander only gives up the advantage of a stationary command post when it is *absolutely necessary* for the IC to personally provide one-on-one direction to ERP operating in forward positions. In either case, the IC must maintain a command presence on the radio.
- Establish a Staging Area. Make sure that Staging is in an easily accessible location and has been announced over the radio for incoming personnel and apparatus.
- Request necessary assistance. Is the problem a Level I, II, or III hazmat incident?

Chapter Three provides an in-depth review of the Incident Management System as it relates to hazardous materials emergencies.

APPROACH AND POSITIONING

Safe approach and positioning by the initial emergency responders is critical to how the overall incident will be managed. Emergencies that start bad because of poor positioning sometimes stay bad. If initial emergency responders become "part of the problem," the IC has to change the Action Plan to deal with new circumstances. For example, if firefighters become contaminated, the Action Plan shifts from protecting the public to rescuing and decontaminating the responders.

This isn't rocket science—just a common-sense application of basic safe operating principles. Some general guidelines include:

NOTES

- When possible, approach from uphill and upwind. (We recognize that the engineers don't always build roads upwind.) If you find yourself approaching from the downwind side, then use distance to the maximum advantage or switch to SCBA.
- Look for physical clues. For example, avoid wet areas, vapor clouds, spilled material, etc. Again, use some common sense; if birds are flying in one side of the vapor cloud and not coming out the other side, you probably have a problem.

Conditions can change quickly at hazmat incidents. A drum marked "flammable" may also be found to be poisonous, or a migrating vapor cloud can envelop apparatus. Don't position too close until a proper size-up has been completed.

STAGING AREAS

Staging is the designated location where emergency response equipment and personnel (resources) are assigned until they are needed.

Staging becomes a sector within the Operations Section (see Chapter Three). The Staging Officer accounts for all incoming emergency response units, dispatches resources to the emergency scene at the request of the Incident Commander, and requests additional emergency resources as necessary.

The ideal Staging Area is close enough to the perimeter to significantly reduce response time, yet far enough away to allow units to remain highly mobile for assignment. Staging is effective when the Incident Commander anticipates that additional resources may be required and orders them to respond to a predesignated area approximately three minutes from the scene.

Large campaign-style incidents can bring extensive resources to the scene which may be needed at different times throughout the emergency. If resources will not be required for some time, the IC should consider establishing primary and secondary staging areas. Within the Incident Management System these are sometimes referred to as Level I and Level II Staging Areas.

Level I Staging is the primary location for the initial-arriving emergency response units. As first responders arrive, they go directly to the incident scene following standard operating procedures. The first unit arriving at the scene of the emergency assumes command and begins site management operations. All other responding units stage at a safe distance away from the scene, until ordered into action by the Incident Commander. Normally, Level I Staging takes place outside the facility's main gate or on a side street very close to the problem. Level I Staging is always in a safer upwind location. Obviously, you should not drive through an unsafe location to take up a position in a safe location.

Level II Staging is the secondary or base location for resources. It is reserved for large, complex, or lengthy hazmat operations. As additional units arrive near the emergency, they are staged together in a specific location under the command of a Staging Officer. The crews can be briefed on the situation by the Staging Officer and wait for assignments in more comfortable surroundings where they are protected from the weather. When resources are required on the scene, they can be moved forward to an area located near the incident perimeter and deployed by the IC.

Staging Areas should be clearly identified through the use of signs, color-coded flags or lights, or other suitable means. The exact location of the Staging Area will be based upon prevailing wind conditions and the nature of the emergency.

NOTES

ESTABLISHING AN ISOLATION PERIMETER

Good firefighting tactics dictate that, when faced with a fire on the fourth floor of a building, evacuation of the fire floor happens first. The crews then search and clear the floors above or below the fire as necessary. When dealing with a hazmat spill in the same building, however, committing all resources to the floor of origin may eventually contribute to the rescue problem as firefighters and sometimes the general public continue to enter the building.

Isolating the area and establishing a perimeter can be as simple as stretching banner tape across access roads approaching a spill or coordinating fire, police, EMS, and Emergency Preparedness personnel in a massive evacuation effort. Regardless of the complexity, this is one of the first tactical considerations.

The first objective of the isolation procedure, after rescue, is to immediately limit the number of civilian and public service personnel exposed to the hazmat. This begins by identifying and establishing an isolation perimeter. When confronted with an incident inside a structure, the best place to begin is at the points of entry such as the main entrance doors. Once the doorways are secured and the entry of unauthorized personnel (including firefighters) is denied, crews can begin to isolate above and below the hazard. Obviously, proper protective clothing and equipment must be worn.

The same concept applies for outdoor scenarios. First, secure the entry points and then establish an isolation perimeter around the hazard. Begin by controlling intersections, on/off ramps, service roads, or any other access to the scene. At this point, a Recon crew can begin a size-up. Injured civilians are still a top priority, but highways and access points can quickly choke up and totally restrict any type of access to the scene. Surrounding the problem with grid-locked, occupied vehicles will quickly compound rescue operations and generally bring the entire operation to a grinding halt.

As the situation continues, conditions can change and the hazmat can migrate into the area where vehicles are stopped, waiting for traffic to move. The occupants can become victims with no immediate means of escape.

While a large perimeter surrounding the incident is desirable, a common mistake is to seal off more real estate than can be effectively

NOTES

controlled. Just as in military operations, it takes a given amount of manpower to patrol a certain perimeter. If the patrols are too sparsely manned or too infrequent, someone is sure to penetrate the line. Given limited manpower, it is better to completely secure a smaller area and expand the perimeter outward as additional resources become available.

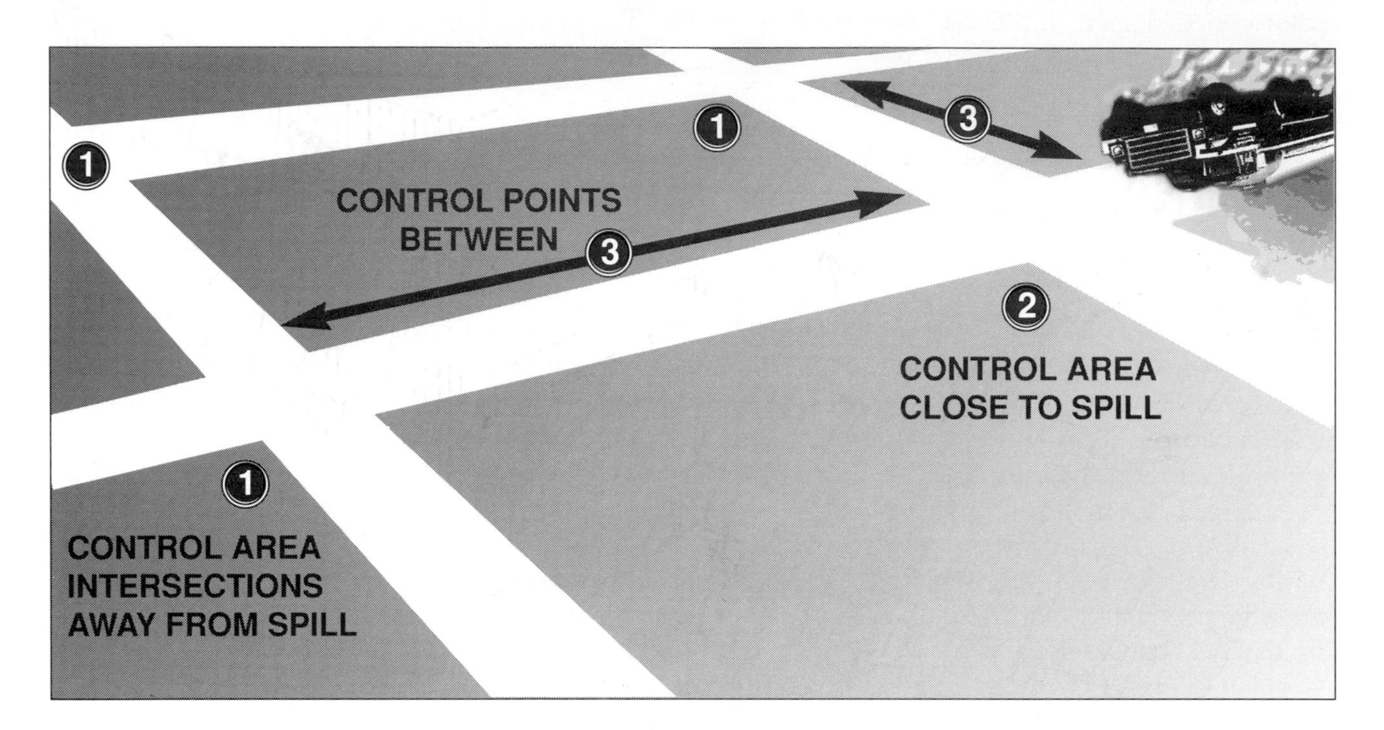

FIGURE 5-1: CONTROL THOSE POINTS AWAY FROM THE SPILL FIRST

SECURITY AND LAW ENFORCEMENT

The IC should make perimeter isolation assignments as soon as possible. This usually begins with summoning the police or security supervisor to the Command Post. This individual will become a key player who will help establish communications between agencies and determine what area(s) will be controlled first, and how it will be managed throughout the incident. He or she should be briefed with all available information.

The people who are actually involved in establishing a perimeter or in the hands-on acquisition of buildings need to know exactly what the potential hazards and risks appear to be. If there is even a remote chance that these officers may be exposed to the hazard as the isolation area expands, they must be provided with proper safety equipment along with specific directions concerning where to go if things go bad.

Law enforcement personnel are best utilized where traffic and crowd control will involve large groups of people on public property. Another important law enforcement function is patrolling the perimeter for civilians trying to sneak a closer look and for the occasional renegade photographer or camera operator trying to get the closest shot.

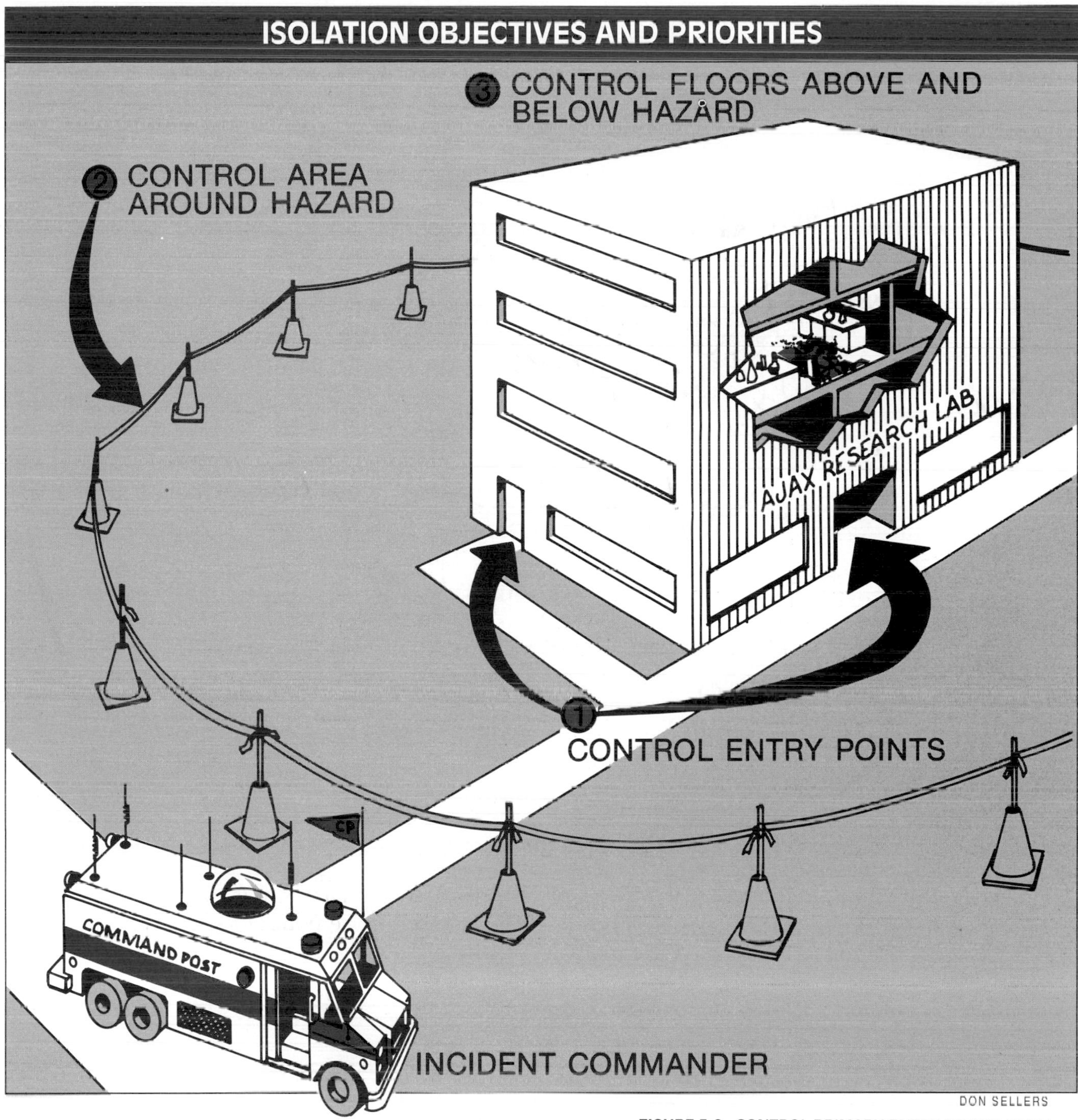

FIGURE 5-2: CONTROL PRIMARY ENTRY POINTS FIRST

Police are better trained for perimeter security than firefighters. They look the part and have the credibility to convince people to relocate to safer areas. If the situation gets ugly, they usually have the authority and equipment to relocate people against their wishes.

When operating on a private facility such as an industrial plant, the on-site security force fills the same slots in the system. Most industrial

NOTES

security officers are well trained and extremely familiar with the site and its resources. They often know the employees inside the plant by sight and can provide specifics on evacuation plans, emergency procedures, and availability of special tools. Usually, they can handle security functions within the plant while police officers control the areas outside of the fence. Working together under a Unified Command System, the law enforcement and security team can be a valuable asset to the IC.

HAZARD CONTROL ZONES

DEFINING HAZARD CONTROL ZONES

Now that the primary isolation perimeter has been secured, the IC can begin to work on his or her second isolation objective by establishing Hazard Control Zones. Essentially, the IC divides the turf he already controls into three distinctly different zones, beginning at the hazmat and working outward toward the perimeter. Hazard Control Zones are designated from most to least dangerous as Hot, Warm, and Cold Zones. See Figure 5-3.

The primary purpose of establishing three different Hazard Control Zones within the isolation perimeter is to provide the highest level of control and personnel accountability for ERP working at the emergency scene. Defined zones help ensure that workers do not inadvertently cross into a contaminated area or place themselves in locations which could be quickly endangered by explosions or migrating vapor clouds.

As a general rule, the public should always be located outside of the isolation perimeter, the field command post and support personnel should be located in the Cold Zone, emergency operations personnel supporting the HazMat Team should be positioned in the Cold and Warm Zones, and the "hands-on" entry team should be located in the Hot Zone as necessary.

An area of refuge should also be established within the Hot Zone to control personnel who may have been exposed to the hazmat. They should be held within this limited area until they can be safely handled (e.g., a Decon area has been established). The Hot Zone should be large enough to provide one or more Areas of Refuge as necessary.

IDENTIFYING HAZARD CONTROL ZONES

Hazard Control Zones should be marked physically and posted on the IC's command and control chart. The Hot Zone can be indicated with colored banner tape, color-coded traffic cones, or color-coded light sticks.

In outdoor situations, Hazard Control Zones can be designated by using key geographical reference points such as a tank dike wall, fenceline, or street name. Geographic areas should be communicated verbally by radio or in a face-to-face briefing between the IC and sector officers.

When the hazard is confined to a building, these zones can be denoted by their location within the structure. For example, a spill in room 321 may dictate that rooms 320 to 322 would be a Hot Zone, the rest of the

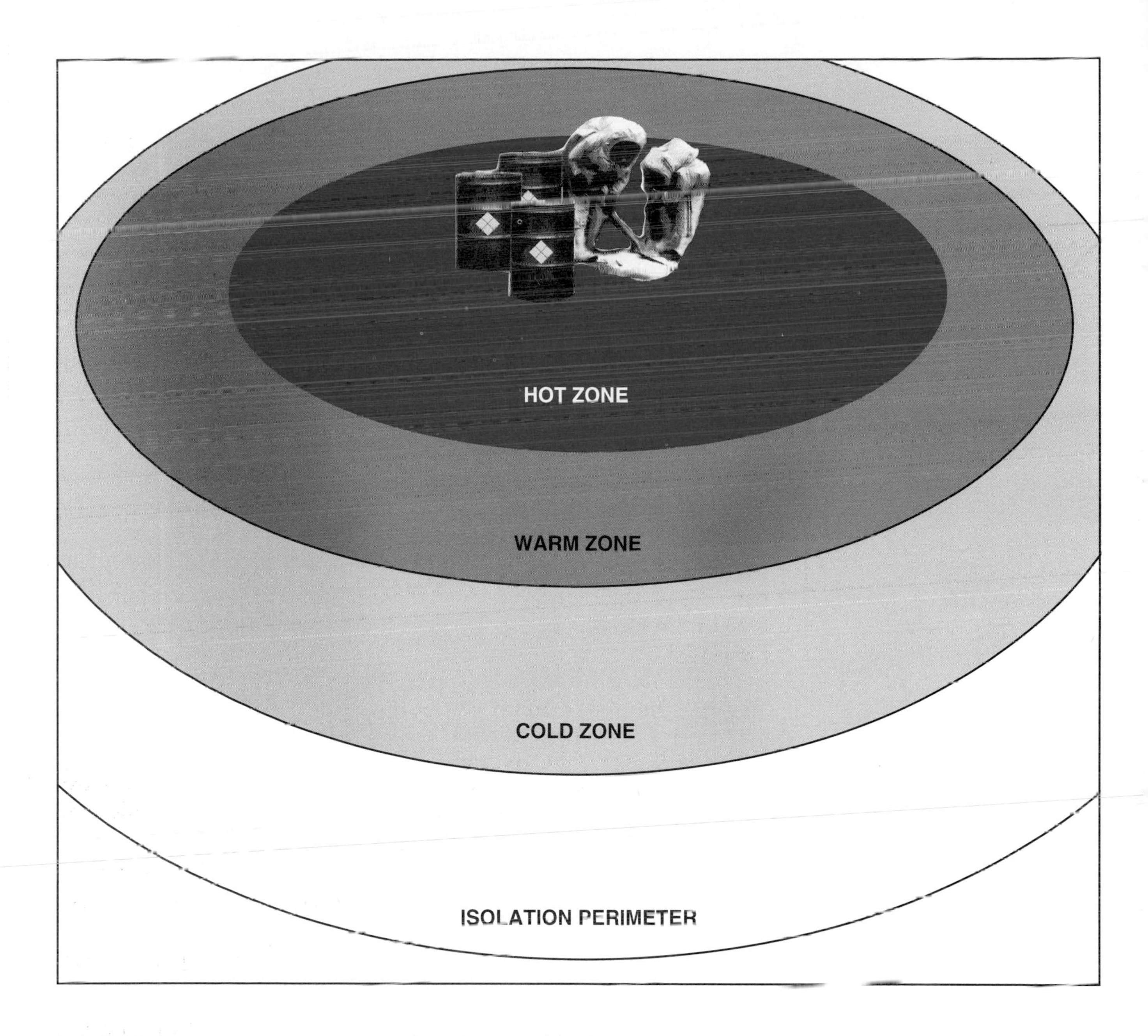

Isolation Perimeter

The designated crowd control line surrounding the Hazard Control Zone. The Isolation Perimeter is always the line between the general public and the Cold Zone.

Hazard Control Zones

The designation of areas at hazardous materials incidents based upon safety and the degree of hazard. Many terms are used to describe these control zones; however, for the purposes of this text, these zones are defined as the hot, warm, and cold zones.

Hot Zone

Area immediately surrounding a hazardous materials incident, which extends far enough to prevent adverse effects from hazardous materials releases to personnel outside the zone. This zone is also referred to as the exclusion zone or the restricted zone by some ERP. The Hot Zone is usually Immediately Dangerous to Life and Health or contains physical hazards. IDLH means any atmospheric concentration of a toxic, corrosive or asphyxiant substance that poses an immediate threat to life, or would cause irreversible or delayed adverse health effects, or would interfere with an individual's ability to escape from a dangerous atmosphere.

Warm Zone

The control zone at a hazardous materials incident site where personnel and equipment decontamination and hot zone support take place. It includes control points for the access corridor, helping to reduce the spread of contamination. This zone is also referred to as the decontamination, contamination reduction, or limited access zone by some ERP

Cold Zone

The safety control zone at a hazardous materials incident that contains the command post and other support functions required to control the incident. This zone is also referred to as the clean zone or the support zone by some ERP.

Area of Refuge

Area within the hot zone where exposed or contaminated personnel are protected from further contact and/or exposure. This is a "holding area" where personnel are controlled until they can be safely decontaminated or treated.

FIGURE 5-3: HAZARD CONTROL ZONES

NOTES

building would be the Warm Zone, and the area outside of the building itself would be designated a Cold Zone. See Figure 5-2.

While it is acceptable to estimate the size of the Hazard Control Zones early in the incident based on visible clues, the IC should move toward a more definitive assessment using monitoring instruments.

Initial monitoring efforts should concentrate on determining if IDLH concentrations are present. Decisions regarding the size of Hazard Control Zones should be based on the following:

1) **Flammability**—if dealing with a confined space or indoor release, the IDLH/action level is 10% of the lower explosive limit (LEL). If dealing with an open-air release, the initial action level is 20% of the LEL. Any areas falling within these parameters are clearly inside the Hot Zone.
2) **Oxygen**—an IDLH oxygen-deficient atmosphere is 19.5% oxygen or lower, while an oxygen-enriched atmosphere contains 23.5% oxygen or higher. In evaluating an oxygen-deficient atmosphere, consider that the level of available oxygen may be influenced by contaminants which are present. Areas containing atmospheres which are either oxygen deficient or enriched should be designated as the Hot Zone.
3) **Toxicity**—unless a published action level or similar guideline (e.g., ERPG-2) is available, the STEL or IDLH values should initially be used. If there is no published IDLH value, ERP may consider using an estimated IDLH of ten times the TLV/TWA. Hazard Control Zones can be established for toxic materials using the following guideline:
   - Hot Zone—monitoring readings above IDLH exposure values.
   - Warm Zone—monitoring readings equal to or greater than TLV/TWA or PEL exposure values.
   - Cold Zone—monitoring readings less than TLV/TWA or PEL exposure values.
4) **Radioactivity**—any positive reading above background level would confirm the existence of a radiation hazard and should be used as the basis for establishing a hot zone.

Chapter Seven includes a more detailed discussion concerning the application and interpretation of monitoring instruments as it relates to hazard and risk assessment.

Hazard Control Zones should change with time by expanding or contracting depending on the size of the incident and the territory the hazards and risks require. As the incident winds down, zones should be reduced accordingly. Retaining large Hazard Control Zones without good technical reasons will create problems with property owners and outside agencies. This is especially true at incidents involving critical highways or at manufacturing facilities. On longer-duration incidents, holding onto large chunks of real estate could generate political problems which can erode the IC's credibility. Don't commit political suicide at an incident by keeping the property owner or law enforcement personnel out of the information loop. Brief them on how and why you have established Hazard Control Zones early in the incident.

NOTES

Safe operating procedures should strictly control and limit the number of personnel working in the Hot Zone. Most hazmat operations can be accomplished with two to four personnel working for specified time periods using the Buddy System. This is a system of organizing emergency response personnel into work groups in such a manner that each person in the work group is designated to be observed by at least one other ERP. The purpose of the Buddy System is to provide rapid assistance in the event of an emergency.

INITIATING PUBLIC PROTECTIVE ACTIONS

Public Protective Actions (PPA) is the strategy used by the Incident Commander to protect the general population from the hazardous material by protecting-in-place or evacuation. This strategy is usually implemented after the IC has established an Isolation Perimeter and defined the Hazard Control Zones for emergency responders.

There are no clear benchmarks available for this decision-making process, but rather a combination of factors, including the size and nature of the release, hazards of the material(s) involved, weather conditions, type of facility, and the availability of "air tight" structures.

The selection between Protection-in-Place or evacuation is NOT an either–or choice; at some incidents, it may be most effective to evacuate one portion of the threatened facility or community while instructing others to protect-in-place. See Figure 5-4.

INITIAL PPA DECISION-MAKING

Protective action decisions are very incident-specific and require the use of the IC's judgment and experience. For example, if a release occurs over an extended period of time, or if there is a fire that cannot be controlled within a short time, evacuation is typically the preferred option for non-essential personnel. However, evacuation may not always be necessary during incidents involving the airborne release of extremely hazardous substances, such as hydrofluoric acid, chlorine, and anhydrous ammonia. Airborne materials can move downwind so rapidly that there may be no time to evacuate a large number of plant employees or the surrounding community. In other situations, evacuating people may actually expose them to greater potential risk. For short-term releases, the most prudent course of action may be to remain inside of a structure.

The IC's decision to either evacuate or seek Protection-in-Place should be based upon an initial evaluation of the following factors:

- ❑ **Hazardous material(s) involved**, including their characteristics and properties, amount, concentrations, physical state, and location of release.
- ❑ **The population at risk, including both facility personnel and the general public.** In addition, the IC must consider the resources required to implement the recommended protective action, including notification, movement/transportation, and possible relocation shelters.

NOTES

- **The time factors involved in the release.** Consideration must be given to the rate of escalation of the incident, the size and observed or projected duration of the release, the rate of movement of the hazardous material, and the estimated time required to implement protective actions.
- **The effects of both the present and projected meteorological conditions upon the control and movement of the hazardous materials release.** These would include atmospheric stability, temperature, precipitation, and wind conditions.
- **The capability to communicate with the population at risk** and emergency response personnel prior to, during, and after the emergency.
- **The capabilities of the HMRT and other personnel to implement, control, monitor, and terminate the protective action.** This should include a size-up of the structural integrity and infiltration rates of structures potentially available for Protection-in-Place throughout the area.

Prior knowledge of the hazmat or the facility through planning information or computer dispersion models acquired through the hazards analysis process can also assist the Incident Commander in this evaluation.

Regardless of the tactic used, achieving public protective action objectives translates into gaining control of a specified area beyond the isolation perimeter, securing and clearing that area, and then controlling a second downwind or adjacent area. In this manner, more and more threatened areas can be secured as more resources become available to the IC.

It is imperative that the Incident Commander use a systematic and structured approach to clearing the public away from the hazard area. Without coordination and direction from the command post, ERP crews can easily turn into a loosely organized band of free-lancers (all for one, none for all).

Establishing priorities and communicating the plan for Public Protective Action tactics are important from the beginning and should be updated on a map at the Command Post as new zones are identified and controlled.

In the early stages of an incident, the IC is often preoccupied with size-up and rescue activities and can easily overlook "people" problems in the immediate area. Be aware that if the situation deteriorates rapidly, exposures within 1,000 feet may be contaminated. In other words, everyone inside the isolation perimeter is a potential rescue.

Areas which should receive immediate attention by the IC include:

- **Locations within 1,000 feet of the incident which will be rapidly overtaken by the hazmat.** This is especially a concern when a flammable or toxic gas is drifting downwind.
- **Locations near the incident where people are already reasonably safe from the hazmat.** People near the hazmat should be alerted to keep clear of the hazard and remain indoors until given other instructions.

- **Key locations that control the flow of traffic and pedestrians into the hazard area.** For example, doorways, on-ramps, and grade crossings.
- **Special high-occupancy structures such as schools.**
- **Structures containing sick, disabled, or incarcerated persons.**

The *DOT Emergency Response Guidebook* is a good resource document to guide the Incident Commander in making quick initial judgment calls on which PPA option to implement. The Guidebook also provides some basic guidelines concerning the size of the initial isolation zone based on the type of hazardous material and size of container. The IC should be thoroughly familiar with how to use the *Emergency Response Guidebook*. The instructions to the ERG provide some useful background information on Public Protective Action decision-making.

There is a fine line between isolation objectives and evacuation. For our purposes, Isolation requires quick action to protect the public and first responders from an immediate, life-threatening situation. Isolation is a necessity; failure to act when people are outside, in exposed locations, will result in injuries. In contrast, Evacuation implies a prolonged, precautionary stay away from the affected location.

NOTES

EVACUATION AND PROTECTION-IN-PLACE ARE NOT EITHER/OR OPTIONS

Public Protective Actions

The strategy used by the Incident Commander to protect unexposed people from the hazardous material by protecting-in-place or evacuation. This strategy is usually implemented after the IC has established an Isolation Perimeter and defined the Hazard Control Zones for emergency responders.

Evacuation

The movement of fixed facility personnel and the public from a threatened area to a safer location. It is typically regarded as the controlled relocation of people from an area of known danger or unacceptable risk to a safer area, or one in which the risk is considered to be acceptable.

Protection-In-Place

Directing fixed facility personnel and the general public to go inside of a building or structure and remain indoors until the danger from a hazardous materials release has passed. It may also be referred to as in-place protection, sheltering-in-place, sheltering, taking refuge, and other terms.

FIGURE 5-4

NOTES

PROTECTION-IN-PLACE

Protection-in-Place is a concept that is a familiar part of people's routine activities. For example, it is not unusual for people to close windows to keep out dust or noise, or to keep the house cool on a hot summer day. The concept of Protection-in-Place as applied to a hazmat release is identical, but the objective is to prevent the migration of toxic vapors into the structure.

Protection-in-Place activities are based upon the concept that toxic vapors will pass over structures without moving inside them. General Protection-in-Place procedures which the IC can issue to the public include:

- ❑ Close all doors to the outside and close and lock all windows (windows seal better when locked). Seal any obvious gaps around windows, doors, vents, etc. with tape, plastic wrap, wet towels, or other materials.
- ❑ Turn off all HVAC systems and air conditioners. If applicable, place inlet vents in the closed position.
- ❑ Close fireplace dampers.
- ❑ Turn off and cover all exhaust fans.
- ❑ Close as many internal doors as possible.
- ❑ Monitor the local AM/FM radio or local television stations for further information.

While following these guidelines will increase the effectiveness of Protection-in-Place as a protective action, it does not necessarily ensure that this type of protective action will always be effective.

Experience has also shown that the public's compliance with the IC's recommendations and instructions to protect-in-place will be dependent upon the following factors:

- ❑ Receipt of a timely and an effective warning message. For example, a warning message broadcast in English in a Spanish-speaking neighborhood may be misunderstood. A warning to take shelter may be interpreted to mean evacuate.
- ❑ Clear rationale for the decision to protect-in-place, as compared to an evacuation. If the instructions sound dumb, they won't be implemented.
- ❑ An absence of visual danger clues, such as large vapor clouds, fires and explosions, etc.
- ❑ High credibility of emergency response personnel with the general public.
- ❑ Previous training and education by fixed facility personnel and the public on the application and use of Protection-in-Place.

Research and accident investigations indicate that staying indoors may provide a safe haven during toxic vapor releases. However, sustained continuous releases will eventually filter into a structure and endanger

NOTES

the occupants. In addition, Protection-in-Place may not be the best option if the vapors are flammable, such as with liquefied petroleum gas. The Incident Commander may have to make critical decisions based upon weather conditions and forecasts. High humidity and warm air can force vapors towards the ground. In addition, air ventilation and air conditioning ducts may force toxic vapors into the building before the public is warned and the order to take shelter is issued by the IC.

Incidents which may justify Protection-in-Place of the surrounding community involve:

- The hazardous material has been totally released from its container and is dissipating.
- The released material forms a "puff" or migrating plume pattern (e.g., vapor clouds that will quickly disperse and are not from a fixed, continuous point source).
- A fast-moving toxic vapor cloud will quickly overrun exposed people.
- Short-duration solid or liquid leaks are present.
- Migrating vapor clouds of known low toxicity and quantity are occurring.
- Leaks that can be rapidly controlled at their source by either engineered suppression or mitigation systems, or through ERP containment and confinement tactics.

The concept of sheltering the public inside of a safe structure has gained acceptance as a viable tactical option and is now a recognized alternative to evacuation. However, in order to make good judgment calls in the field, it is important for the IC to understand how and why Protection-in-Place works and when it may not be an appropriate tactic. (See Figures 5-5 and 5-6.)

FIGURE 5-5

CASE STUDY

I-610 AT SOUTHWEST FREEWAY, HOUSTON, TEXAS MAY 11, 1976

On a bright sunny day at 11:08 am, on May 11, 1976, a Transport Company of Texas tractor-semitrailer tank truck transporting 7,509 gallons of anhydrous ammonia struck and penetrated a bridge rail on a ramp connecting I-610 with the Southwest Freeway (U.S. 59) in Houston, Texas. The truck left the ramp, struck a bridge support column of an adjacent overpass, and fell onto the Southwest Freeway, approximately 15 feet below.

The tank truck breached immediately on impact, releasing most of its contents into the atmosphere. At the time of the accident there were about 500 persons within 1/4 mile of the release. The released ammonia fumes rapidly penetrated automobiles and buildings. When their occupants left to escape the fumes during the early minutes of the

NTSB

FIGURE 5.5: NORTH VIEW OF ACCIDENT SITE SHOWING CLOUD AT 2 MINUTES AFTER CRASH

release, many were exposed to fatal doses of ammonia.

The temperature at the time of release was in the low 80's. The released ammonia immediately vaporized and the 7 mph wind gradually decreased the vapor concentration at ground level. Witnesses reported the white ammonia vapor cloud initially reached a height of 100 feet before being carried by the 7 mph wind for approximately 1/2 mile. After five minutes, most of the liquefied ammonia had boiled off and the vapor cloud was completely dispersed.

Seventy-eight of the 178 victims, who were within 1,000 feet of the release point, were hospitalized and treated for symptoms of ammonia inhalation. Over 100 persons were treated for less severe injuries. Five of the six fatalities were due to ammonia exposure. Because all fatalities were within 200 feet of the estimated release, it is estimated that the ammonia concentration within this distance was greater than 6,500 parts per million for at least two minutes. The IDLH for ammonia is 500 ppm.

A detailed investigation of this incident conducted by the U.S. National Transportation Safety Board in 1979 revealed that there were significant differences in the degree of injury among the exposed victims who evacuated buildings and those who protected-in-place.

The Board's conclusion was that the protection offered survivors by the vehicles and buildings demonstrated that there were alternatives to simply running away from the released hazmat. A detailed investigation conducted by the Board showed that people who sheltered and remained inside buildings received no harm from the ammonia. Also, people who remained inside of their automobiles generally received less severe injuries than those who left their cars and tried to escape the ammonia.

While there have been other investigations conducted with similar conclusions, the Houston case was the first the authors are aware of concerning Protection-in-Place issues which were documented using forensic science. As a result of NTSB's report, many emergency response and safety professionals began to re-think whether evacuation was always the best tactical option.

The investigation conducted by the Board also documented that the actions taken by the Houston Fire Department saved lives. Within 10 minutes the HFD dispatched 14 emergency rescue units and 4 pieces of fire apparatus. The Incident Command System was used to coordinate EMS, fire, and police agencies. As a result, many contaminated and injured victims were located by search and rescue teams and escorted to safety, where they received medical treatment.

FIGURE 5-5

UNDERSTANDING WHY PROTECTION-IN-PLACE WORKS

The Department of Mechanical Engineering at the University of Alberta, Edmonton, Canada, has conducted extensive tests on Canadian and American homes to determine whether Protection-In-Place tactics would actually work. Test findings revealed that for an accidental toxic gas release that occurs over several minutes to half an hour, even a very leaky building contains a sufficient reservoir of fresh air to provide effective sheltering-in-place. However, for longer-duration releases of one to three-hours, the average indoor concentration may reach 80% or more of the outdoor average during a steady continuous release.

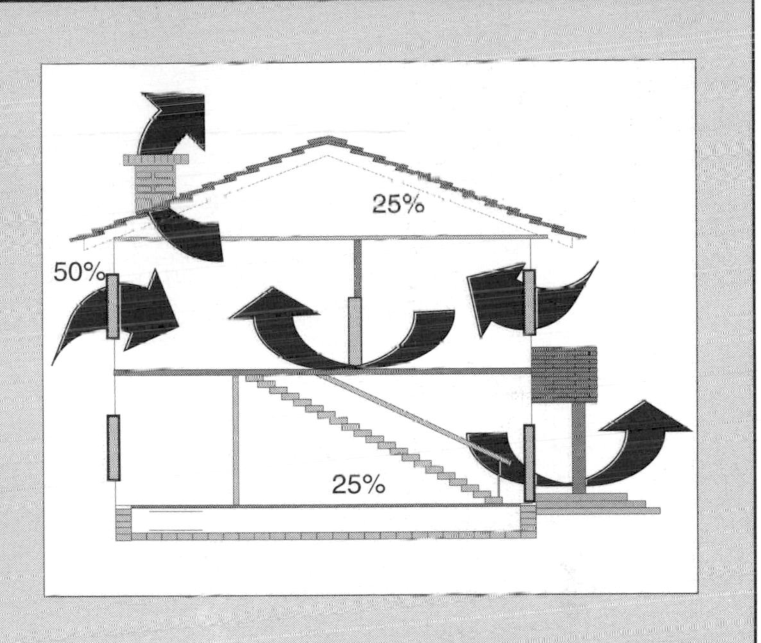

For releases that have a long duration of an hour or more, the choice between shelter or evacuation is difficult to make. Typical air exchange rates in a house are about 0.5 air changes per hour (ACH). For a three-hour release, this exchange rate causes the air in the house to be replaced 1.5 times during the event. After this air exchange, the indoor concentration is about 80% of the average outdoor value. Obviously, the more energy efficient or tight the building is, the slower the air exchange rate will be.

For more information see, "Effectiveness of Indoor Sheltering During Long Duration Toxic Gas Releases" by D. J. Wilson, Department of Mechanical Engineering, University of Alberta, Edmonton, Alberta, T6G 2G8.

FIGURE 5-6

The age and construction of a building has a lot to do with how successful Protection-in-Place will be. The Incident Commander should become familiar with the types of structures in the community. Pre-incident plans and Hazards Analysis Surveys should incorporate this type of information. As a general rule, the older a building is, the less likely it will provide a safe place of refuge for periods of longer than one hour.

Neighborhood surveys can reveal a significant amount of information concerning the types of buildings and influence organizational policy concerning whether Protection-in-Place is a viable option. For example, if an initial survey of the homes surrounding a chemical plant reveals that they are of 1940's construction and have not been retrofitted with energy efficient doors and windows, then Protection-in-Place may not be the best option to implement. On the other hand, if the same neighborhood is made up primarily of modern, energy efficient homes, the rate of air changes per hour will be substantially lower.

Some jurisdictions and petro-chemical companies have adopted a simple survey tool to evaluate the various types of structures. An example is provided in Figure 5-7.

RATING SYSTEM FOR PROTECTION-IN-PLACE

Type-I Structure

Modern, energy efficient building constructed since 1970.
Or, older building with upgraded energy efficiency.

Type-II Structure

Older construction, with limited energy efficiency. Constructed between 1950 and 1970.

Type-III Structure

Oldest type construction built between 1920 and 1950.

Type-IV Structure

Mobil home, trailer, shed, etc. without energy efficiency.

FIGURE 5-7

PROTECTION-IN-PLACE FIELD SURVEY

Please briefly describe the location of the area by cross-street, neighborhood or facility.

Complete Family Name ____________________

Street Address ____________________

Phone Number ____________________

Number of Occupants ____________________

Special Comments ____________________

____

Are there any additional buildings ☐ YES ☐ NO
How many? ____________________

1. Check the box which best describes the block, neighborhood or area that you are surveying and list the geographic location.

   ☐ Residential area consisting primarily of single family structures.

   ☐ Residential area consisting primarily of multiple family structures.

   ☐ Commercial area consisting primarily of stores, restaurants, office buildings.

   ☐ Industrial area consisting primarily of warehouses and manufacturing type structures.

   ☐ Special structures consisting of high life occupancy such as a hospital, nursing home, school, or prison. For example, structures requiring special assistance.

2. Check the box that best describes the type of building construction for the majority if the structures in the area you are surveying.

   ☐ TYPE I: Modern construction built since1970.

   ☐ TYPE II: Older construction built between 1950 and 1970.

   ☐ TYPE III: Oldest construction or historic built between 1920 and 1950.

   ☐ TYPE IV: Mobil home, trailer, etc. no energy efficiency.

FIGURE 5-7

NOTES

EVACUATION

LIMITED-SCALE EVACUATION

Evacuation of both industrial fixed facility personnel and the general public is an attempt to avoid their exposure to *any* quantity of the released hazardous material. Under good conditions, evacuation will remove these individuals from any exposure to the released hazmat for a given length of time. Fortunately, most hazmat incidents only require a limited-scale evacuation. This typically involves just one or two buildings. Some examples which might justify evacuating buildings include:

- ❑ Whenever the building is on fire or the hazmat is leaking inside the building.
- ❑ Whenever explosives or reactive materials are involved and can detonate or explode, producing flying glass or causing structural collapse.
- ❑ Whenever there are leaks involving toxics that cannot be controlled and are expected to continue leaking.
- ❑ Whenever the IC determines that the leak cannot be controlled by ERP and civilians are at risk.

SPECIAL OCCUPANCIES

Many Incident Commanders question whether special occupancies such as hospitals and nursing homes should be evacuated, even if the majority of the surrounding population is evacuating. Very little has been written on the experience of hospital evacuations during hazardous materials incidents. One study of a Mississauga, Ontario accident suggests that mass evacuations of in-patient populations can be accomplished in short time periods. The Mississaugua experience of one hospital showed that there were 300 patients discharged and 200 transferred in 4 hours without any adverse effects. It doesn't take much imagination to visualize the complications involved in evacuating two or three hospitals while dealing with the general population.

FIXED INDUSTRIAL FACILITIES

Unlike their public safety counterparts, industrial emergency response teams have the advantage of knowing exactly what types of hazardous materials are in their facilities. This prior knowledge allows evacuation decisions to be more specific.

OSHA requires fixed facilities to have written evacuation procedures. Well-written Public Protective Action procedures can provide useful guidelines to supervisors and the facility IC concerning whether a limited- or full-scale evacuation is necessary. Many facilities have developed a tiered approach to implementing Public Protective Actions. One method used is to define three levels of Public Protective Action that are associated with the Levels of Incident described in Chapter One. See Figure 5-8.

LEVELS OF PROTECTIVE ACTION

Fixed industrial facilities should provide written guidelines to employees and contractors concerning when it is appropriate to evacuate or protect-in-place. Industrial facilities are familiar with the hazards and risks of the products they manufacture and can develop specific guidelines. The following example was prepared by a gas plant handling hydrogen sulfide.

LEVEL ONE INCIDENT: PROTECTIVE ACTION

A Level One Protective Action requires all employees and contractors to evacuate the work site in an upwind direction and report to their pre-designated briefing area as defined in the site-specific emergency plan. Personnel will be accounted for and emergency work assignments will be issued by the on-site supervisor. Individual employees who are working alone at a remote location and must evacuate a work site should contact the Control Room Operator and notify him or her that they are leaving the area. Examples of incidents requiring a Level One Protective Action would include a small flammable or toxic leak from a valve or flange which produces concentrations in the 10 ppm range on company property.

The leaking gas is not likely to go beyond company property. The leak is easily repairable using safety equipment on-site. There is no risk to the public, but the fire department is notified immediately that there is a Level One Incident in progress.

LEVEL TWO INCIDENT: PROTECTIVE ACTION

A Level Two incident will require personnel to evacuate the immediate work site and meet at the pre-designated assembly area for accountability and emergency work assignment as specified in the site-specific emergency plan. The public surrounding the work site should be protected-in-place. Examples of incidents requiring a Level Two Incident Protective Action would include a moderate release of hydrogen sulfide which is producing atmospheres of up to 300 ppm on company property with atmospheres of 10 ppm immediately adjacent and downwind of company property. The leak can be rapidly repaired, and the hydrogen sulfide gas is rapidly dispersing. The fire department is alerted via 911, and the Fire Chief assumes command of the incident.

LEVEL THREE INCIDENT: PROTECTIVE ACTION

A Level Three incident is a major emergency which requires the total evacuation of all company personnel and the surrounding public to a pre-designated location(s) outside of the immediate area. Depending upon the nature of the incident and the potential for the problem to migrate off-site, there will also be public protective action instructions given to the surrounding community. Concentrations of hydrogen sulfide are rapidly exceeding 300 ppm immediately adjacent to company property, and the off-site concentrations will reach 50 to 100 ppm. The leak cannot be immediately repaired, atmospheric concentrations are unfavorable for rapid dispersing of hydrogen sulfide gas, and homes within the H_2S isopleth are not of energy efficient construction.

FIGURE 5-8

FULL-SCALE EVACUATION

Full-scale evacuations are difficult at best. They disrupt people's lives and present several serious complications that deserve the Incident Commander's attention before a decision is made to implement the evacuation option. Two of the more important considerations are:

NOTES

Life Safety: In some cases, evacuations may even endanger the lives of the people being evacuated. Traffic accidents, stress induced heart attack, and accidental exposure to the hazmat being released are real-world examples.

Expense: Full-scale evacuations are expensive. One study conducted by the Battelle Human Affairs Research Center for the Atomic Industrial Forum indicated that the cost to individual households for evacuation would be expected to be almost seven times the cost of Protection-in-Place. Costs to the public sector are approximately three times as high and fifteen times as high for the manufacturing sector. An expensive operation is very hard to justify, and there will be no shortage of critics the day after the evacuation (even if you made the right decision).

The location of the general population and the time of day should always be a factor in the IC's decision-making process. For example, FEMA studies indicate that on any given weekday, 40% of the community are in their home while 60% are at work or in school. In contrast, the nighttime figures are roughly reversed. In either case, the majority of the population is under roof and would require relocation during an evacuation. One research study indicated that it takes 2.5 to 3 hours to warn 90% of the public through door-to-door contact but only 20 to 35 minutes with sirens or the Emergency Broadcast System.

In urban areas, up to 20% of the population could be in transit from or to their homes. Rush-hour traffic and the time of day are significant factors in deciding whether or not to evacuate. Some studies conducted by the nuclear power industry show that with good planning and traffic control assistance from police agencies, many urban highways are capable of handling large traffic flows created by a full-scale evacuation. The same studies, however, also point out that high density traffic jams can be created at critical traffic arteries when large crowds attempt to evacuate locations like athletic stadiums and concert halls. Figure 5-9 provides a more detailed picture of principal locations of the populations.

POPULATION LOCATIONS

LOCATION/ACTIVITY	HOURS PER DAY	PERCENT OF TIME
• HOME	16.6	69.2
• SCHOOL OR WORK	4.7	19.6
• COMMUTING	1.2	5.0
• OUTDOORS	1.5	6.2

FEMA

FIGURE 5-9: AVERAGE 24HR POPULATION LOCATION

NOTES

The decision to commit most emergency response resources to a full-scale evacuation should be initially determined by the Incident Commander based upon the specific conditions of the emergency. However, some situations that may justify a full-scale evacuation include:

- ❑ Large leaks involving flammable and/or toxic gases from large-capacity storage containers and process units.
- ❑ Large quantities of materials which could detonate or explode, damaging additional process units, structures, and storage containers in the immediate area.
- ❑ Leaks and releases that are difficult to control and could increase in size or duration.
- ❑ Whenever the Incident Commander determines that the release cannot be controlled and facility personnel and/or the general public are at risk.

When the decision is made to commit to a full-scale public evacuation, four critical sectors must be established and managed effectively in order for the operation to succeed.

1. **Alerting.** Alert people and tell them where to go and what to do.
2. **Transportation.** Move displaced people to a safer location well outside of the area at risk.
3. **Relocation.** Keep displaced people housed, comfortable, and fed.
4. **Information.** Keep displaced people informed of your progress toward making their problem go away.

In most jurisdictions, there is no single government or private agency with the charter or expertise to manage and service all four of these areas of responsibility. Usually a public safety agency takes the lead for overall incident coordination, and other public agencies and private organizations handle specific elements of the evacuation effort. For example, working under a Unified Command structure, the Fire Department may coordinate the tactical aspects of the evacuation, while Emergency Preparedness and Police agencies handle the alerting and notification responsibilities. Working under the direction of the Emergency Management Agency, the local Public Transit Authority or School District may provide bus transportation from the affected area to relocation shelters. The School District, in cooperation with the local Red Cross, may provide temporary shelter and food for the displaced.

Each part of the country has its own variation on the theme, and it really doesn't matter who provides what service as long as it is planned in advance and coordinated at a central location using Unified Command. See Chapter Three for more detail on how Unified Command works.

Good working relationships on a one-on-one basis at the local level can help produce results fast in a crisis, but full-scale evacuations simply have too many moving parts, and there is far too much at stake to simply "wing it". Well thought out evacuation plans should be part of every industrial and public safety agency Emergency Preparedness program. The plan should lay out the big picture and clearly spell out areas of

NOTES

responsibility for each agency. These should also be backed up by Memorandums of Understanding and written agreements between different organizations who may have overlapping areas of responsibility. For example, if the evacuation plan calls for the School District to provide buses and schools for transportation and relocation of residents, who provides the drivers and opens up the school during summer vacation? Who buys and cooks the food, how will it be requisitioned in the middle of the night, and who will serve it to the people once they arrive? Don't look stupid because you didn't have a plan.

In order to make all these moving parts work, the IC should have direct and frequent communications with representatives of the lead agencies supporting the evacuation. During extended operations, these representatives should relocate from the field command post and operate from the Emergency Operations Center.

It is also necessary to have direct radio communications between the Emergency Operations Center and each Relocation Shelter. Conditions can change quickly, especially when migrating vapor clouds are involved. The Relocation Shelters may have to be moved, or displaced people may temporarily need to take shelter (protect-in-place).

ALERTING

In order for evacuation to be successful, the IC must assure that people are quickly alerted that there is an emergency in progress. The methods of notification will vary depending on the location of the emergency and the type of plan and hardware in place.

FIXED FACILITY ALERTING

At a fixed facility such as a refinery or chemical plant, the notification process normally occurs by activation of sirens or by use of an on-site public address system.

One of the most frequent problems encountered in fixed facilities is the confusion created by a single warning tone which may also be used to indicate the beginning or ending of a work shift, a fire, toxic gas release, etc. The same tone is used with one, two, or three blasts on the horn, which have different meanings. As a general rule, evacuation alarms should be unique and distinctly different from any other type of alarm in the facility.

As part of the pre-incident planning process, fixed facility personnel should have special knowledge of the following:

- Methods by which their personnel are notified of an emergency evacuation, including sound of the alarm system.
- Instructions on where personnel should report to and assemble when the evacuation alarm is sounded.
- Facility evacuation routes and corridors.
- Ability to communicate with facility personnel at evacuation assembly areas. This is critical in accounting for evacuees and initiating search and rescue operations.
- Location of both primary and alternate assembly locations.

NOTES

From a tactical perspective, the IC should be aware that an activated evacuation alarm at a fixed facility does not necessarily mean that the occupants have either protected-in-place or evacuated. Additional direction and assistance from Emergency Response Personnel may be required to complete the evacuation.

Once notified and evacuated, a head count should be taken of facility personnel, contractors, and visitors to ensure that all personnel are accounted for. Supervisors are usually responsible for coordinating all personnel accountability activities. Information regarding any missing personnel and their previously known location within the facility should be relayed to the IC so that search and rescue operations can begin. However, the IC should be aware that initial reports of people missing based on head counts are usually not correct.

PUBLIC ALERTING

Alerting and notifying the public to protect-in-place or evacuate creates additional complications which need to be factored into the IC's decision-making process. There is no single "best" way to alert the public that there is a problem. See Figure 5-10 for an example.

A variety of communication technologies exist to assist the IC with the warning process. There are many advantages and disadvantages of each of these systems, and it is important to recognize that one is not necessarily better than another. Each one is a different type of tool and must be selected based on local conditions. See Figure 5-11.

A brief summary of the different alerting methods is as follows:

- **Personal Notification:** This consists of a simple door-to-door visit to residents in the affected area. This is usually completed by police agencies. This initial contact has a multiplier effect as residents call neighbors and relatives, alerting them that there is a problem. While useful, the effectiveness of this type of notification cannot be factored into the IC's decision-making.
- **Loudspeakers/Public Address Systems:** Loudspeakers on emergency vehicle siren systems are an effective way of alerting people outside in public areas such as parks. Public Address Systems inside shopping malls and public assembly buildings may also be used.
- **Tone Alerted Radios:** Some fixed facilities such as chemical plants and oil and gas plants have provided special Tone Activated Radios for residents living near their facility. These operate on the same principle as a volunteer firefighter's radio. A radio signal is sent from the control room at the plant which sets off an alerting tone inside each home's radio. A live, real-time message can then be broadcast. Special weather radios are also available that are activated by the National Weather Service for severe storm warnings. This system can be used to issue special warnings if prior arrangements have been made with the NWS.
- **Emergency Broadcast System:** The Emergency Broadcast System or EBS is an effective method of alerting people in buildings and

NOTES

automobiles. Obviously, the system only works if the public has an AM/FM radio and it is tuned in. However, a well-trained public may be directed to turn on their radio whenever they hear community warning sirens. This method is especially effective in areas where the public is used to receiving tornado warnings by radio. See Figures 5-12 and 5-13.

- **Scanner Radios:** Scanner radios are widely used by fire and police buffs, boaters, and truck drivers. The National Oceanographic and Atmospheric Administration (NOAA) weather radio system covers a major portion of the population within the country. The station broadcasts continuously and can be used to warn people of special atmospheric emergencies such as migrating toxic vapor clouds. This is especially useful for boaters who may be downwind on lakes or rivers. The U.S. Coast Guard may also contact boaters directly by issuing special broadcasts.
- **Television:** Television Capture Systems are becoming more popular as alerting tools in communities where cable television systems are used. Emergency services can "capture" the cable station and transmit a message which scrolls across the bottom of the viewer's screen. The media may also break into normal programming with a special broadcast.
- **Sirens and Alarms:** These may include the community Emergency Preparedness Agency sirens (Air Raid Sirens) or special sirens installed in areas around fixed facilities.
- **Aircraft:** Helicopter loudspeakers can be an effective method of alerting people in outdoor and remote areas (e.g., flying over parks, campgrounds, and hunting areas).
- **Signs:** Many urban highways and commuter routes have electronic message signs for alerting drivers of traffic conditions. These can be used to alert drivers about problems. Bus and train stations also have similar electronic bulletin boards.
- **Computerized Telephone Notification Systems:** CT/NS can reach a potentially large number of people by simultaneously telephoning blocks of people by computer. Pre-recorded messages provide instructions to residents.

The reliability of computer alerting systems has improved significantly in recent years. For example, more powerful and user friendly software has been developed to drive the computer components in alerting systems.

A good community alerting and notification system is based on a variety of systems which are described in the Community Emergency Response Plan. The plan should spell out who has the authority and responsibility to activate each system. Each system component should be tested on a regular basis. More than one war story has been told about a warning siren that failed to work or a system which accidentally activated, creating chaos.

Public Protective Action Alerting Sequence

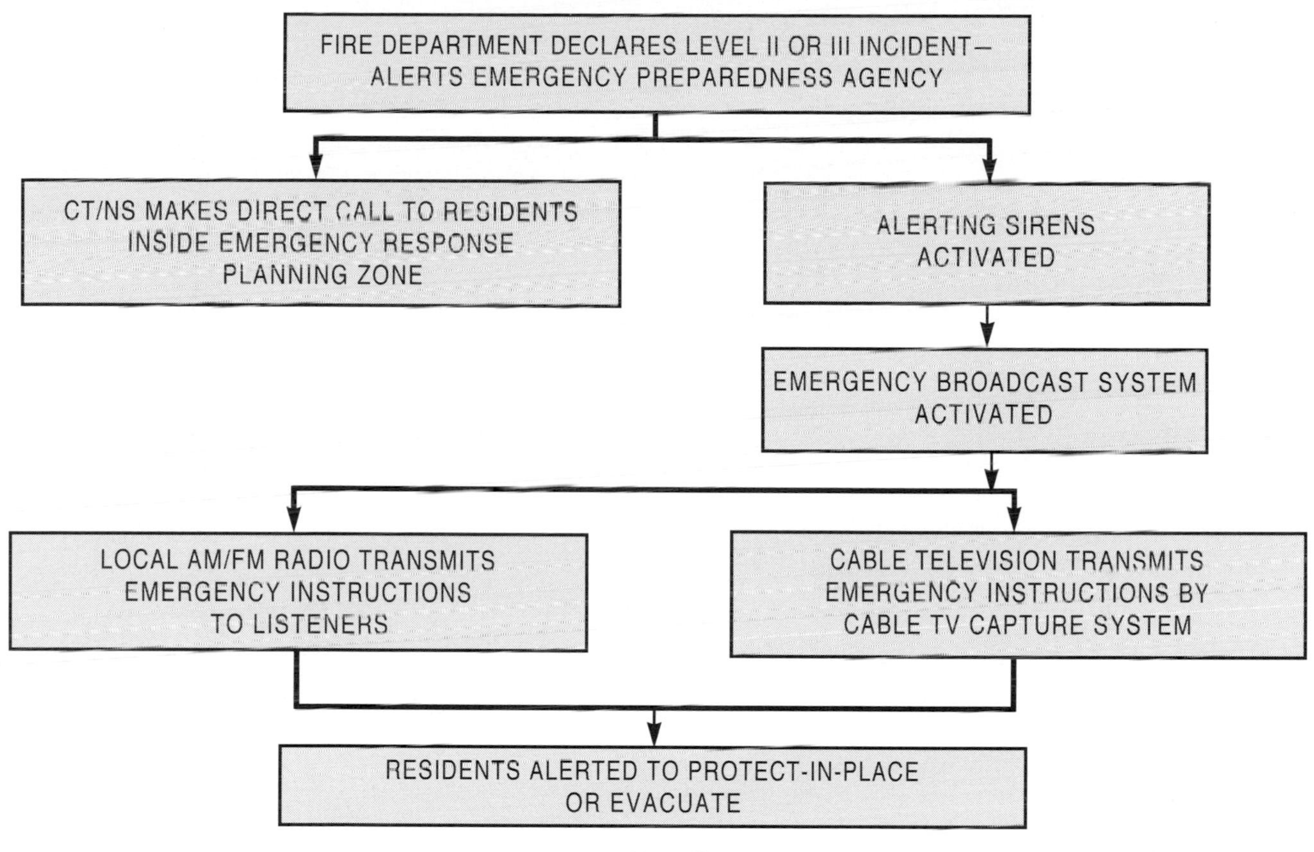

FIGURE 5-10: OVERVIEW OF HOW A COMMUNITY WARNING SYSTEM WORKS

Estimates of Warning Dissemination Times for Alternative Systems (in minutes)

Warning System	Percent of Population Warned[a]			
	25	50	75	90
MEDIA	20–30	45–60	80–120	180–240
DOOR TO DOOR	40–45	60–80	100–120	150–180
ROUTE ALERT	25–35	40–50	60–70	90–150
TONE ALERT RADIO OR AUTO TELEPHONE	2–3	4–5	7–10	10–15
SIREN/MEDIA	5–10	12–15	15–20	20–35
SIREN/FIXED RESPONSE	1–2	2–3	14–5	10–15

[a]Under good weather conditions and assuming systems that are maintained.

FIGURE 5-11: ESTIMATES OF ALERTING TIMES

TYPES OF WARNING SYSTEM MESSAGES

Emergency Broadcast System (EBS):

The national emergency notification system that uses commercial AM and FM radio stations for emergency broadcasts. The EBS is usually initiated and controlled by Emergency Preparedness agencies.

Sample EBS Messages:

Initial Warning: During the initial size-up stages of an incident, the IC often does not have enough information to make a decision whether to protect-in-place or evacuate. Alerting time is critical. An initial warning message can be transmitted to alert people that there is a problem.

Example:

"We interrupt your normally scheduled program for the following emergency broadcast. The County Fire Department has just issued a hazardous materials emergency warning for the Evanston, Wyoming area. You are instructed to take immediate shelter inside your home or any public building until the danger has passed. Close all doors and windows and turn off your heating or air conditioning system. Stay tuned to this station for more details. Additional emergency instructions for hazardous materials emergencies are provided on page 2 of the local telephone directory." See opposite page.

Protection-in-Place Warning: When the IC has decided to implement the Protection-in-Place option, the EBS message should be designed to get people indoors and to seal off sources of contaminated air that could filter into the home.

Example:

"This is the County Fire Protection District. Emergency response units have responded to a hazardous materials incident along the Interstate-80 area. The emergency is under control; however, you are advised to remain indoors with your doors and windows closed and the heating and air conditioning system off. This is only a precaution which will keep any contaminated air which may be in your area outside of your home. You are completely safe as long as you remain indoors. Emergency repairs are underway and we will advise you as soon as it is safe to go back outside. Stay tuned to this station for more information. If you have an emergency at your location, dial 911 for help."

Evacuation Warning: When the IC issues an evacuation order by EBS, the instructions must tell the public where to evacuate to. General instructions may cause residents to enter dangerous areas.

Example:

"This is the County Fire Protection District. Emergency response units have responded to a hazardous materials incident in the Interstate-80 area. Emergency response teams are on the scene at this time and are taking immediate action to control the hazardous materials release. As a precaution, we have ordered an immediate evacuation of the following areas. These include all buildings in the East Park and Quail Creek neighborhoods. If you are in one of these areas, you are instructed to evacuate immediately. Proceed to a designated emergency shelter at the Evanston High School. Representatives from the County will meet you at the relocation center. Police agencies will patrol your area until the emergency is over. If you have an emergency at your location dial 911. Please do not call 911 if you do not have an emergency. Additional information will be provided at the relocation center. Again, please leave immediately."

FIGURE 5-12

Many locations have improved their alerting systems by providing detailed Public Protective Action instructions in the local telephone directory. Alerting messages received by the Emergency Broadcast System direct residents to a specific page number in their telephone book, where detailed instructions are provided.

EVANSTON AREA HAZARDOUS MATERIAL EMERGENCY PROCEDURES

HOW YOU WILL BE NOTIFIED OF A HAZARDOUS MATERIAL EMERGENCY

If there is a hazardous material emergency in the Evanston area, the community's warning and notification system will alert you. The system consists of eight sirens strategically placed around the community and a public address network that immediately captures the Emergency Broad- cast System (EBS) radio stations and the cable TV stations. The official EBS stations are KEVA (1240 AM) and J KOTB (106.3 FM) radio.

The sirens sound for three-minute periods several times to raise the alert at the beginning of an emergency and to indicate all clear at the end.

In areas not served by warning sirens, official cars with loudspeaker systems will alert residents throughout the neighborhoods.

WHAT TO DO IF YOU HEAR THE COMMUNITY ALERT SIREN

- Stay calm.
- Go indoors, close all windows, doors, fireplace dampers, and vents, and turn off fans, heating systems, or air conditioners.
- If you are in a vehicle, close all windows and vents and stop at the nearest building for shelter.
- Turn on an EBS radio station for information and instructions. The message will be repeated on the EBS stations as necessary until conditions change.
- If you are instructed to protect your breathing, cover your nose and mouth with a handkerchief or other cloth, wet if possible.
- Stay indoors until instructed to do otherwise.
- Do not use the telephone unless you have an emergency. The lines are needed for official business, and your call could delay emergency response.
- As soon as it can be determined that the hazardous condition has passed, local authorities will announce the emergency is over.
- If the emergency involves a toxic gas release, at the "All Clear" you may be instructed to open windows and doors, ventilate the building, and go outside.

INSTRUCTIONS

All EBS messages begin with a description of the emergency, including time and location. The instructions that follow depend on the severity of the emergency. Residents will hear directions for one of four emergency procedures: Warning Alert, Shelter in Place, Prepare to Evacuate, or Evacuate.

Hazardous Material Warning Alert

The WARNING ALERT indicates there is a problem that poses no present danger to the community However, there could be a potential for escalation to a more serious situation. The WARNING ALERT informs residents to stand by.

The following are typical instructions for a WARNING ALERT:

1. Stay indoors.
2. Stay tuned to an EBS radio station for further instructions.

Shelter in Place

In certain hazardous material incidents it is safer to keep community residents indoors rather than to evacuate them.

The following are typical insructions for SHELTERING IN PLACE:

1. If you are outdoors, protect your breathing until you can reach a building.
2. Go to an inside room, preferably one with no or few windows.
3. Close all windows, doors, and vents and cover cracks with plactic wrap, tape, or wet rags.
4. Keep your pets inside.
5. Listen to an EBS radio station for further advice.

Prepare to Evacuate

You may be asked to PREPARE TO EVACUATE if a situation has the potential of escalating to the point where an evacuation is required.

During this time, authorities will act to alleviate the emergency and also will prepare for an orderly evacuation if it becomes necessary.

The following are typical instructions for PREPARING TO EVACUATE:

1. All person in (names of areas) should stay indoors and prepare to evacuate.
2. If you are in your home, gather all necessary belongings. Pack only the items you need most, such as clothing, medicine, baby supplies, portable radio, credit cards, flashlight, and checkbook.
3. Locate and review your Community Emergency Action Card. You need not evacuate at this time.
4. Stay tuned to your EBS radio station for further instructions.

Evacuate

An EVACUATION may be ordered if the community is threatened and there is time to evacuate safely. Make sure the EVACUATE order applies to your area.

The following are typical instructions for an EVACUATION:

1. All persons in (names of areas) should evacuate in an orderly manner.
2. In certain circumstances you may need special equipment for evacuation. Under those conditions you are required to remain in your home until fire department personnel arrive with the equipment and instruct you in its use.
3. Children in school will be taken to an evacuation shelter if necessary.
4. Lock your house and turn on the porch light as you leave. Your neighborhood will be guarded while you are away.
5. Drive or walk toward the main roadway in your area.

Emergency personnel stationed along these routes will direct you away from the emergency area toward an evacuation shelter.

6. Use your own car if you can. Keep all vehicle windows, doors, and vents closed. If you have room, take passengers. Please observe traffic laws.
7. Turn on your car radio for information.
8. If you can't find transportation, walk to one of the pickup points along the nearest MAIN emergency route in your area. You can get a ride there.
9. Law enforcement personnel will be in place in the evacuated area to prevent looting, vandalism, etc.
10. You may return home as soon as the emergency is declared over and it is safe to return.

SPECIAL ARRANGEMENTS FOR MOBILITY-IMPAIRED PERSONS

If you are disabled or need assistance, send the Uinta County Fire department a postcard with your name, address and telephone number, a description of the extent of your difficulty, and any other pertinent information about the state of your health, such as diet, medicines, etc.

This information will be placed on file so you will receive the necessary assistance in the event of an emergency. To help the department maintain current files, it is important for you to mail them an update card if there are changes in your status.

Uinta County Fire Department P.O.Box 6401
Evanston, WY 82931
(307) 789-3013

WHERE TO GET MEDICAL, FIRE, OR POLICE HELP DURING AN EMERGENCY

If you need medical assistance, or if you need to report a fire or violation of the law, call 911. Give your name, address (including ZIP code), the nature of the emergency, and stay on the line.

INFORMATION

For more information about hazardous material emergencies contact the Uinta County Fire Department at 789 3013.

Information for the Evanston Area Hazardous Material Emergency Pages was provided courtesy of Union Pacific Resources Company.

FIGURE 5-13: ADDITIONAL INSTRUCTIONS CAN BE PROVIDED IN THE TELEPHONE DIRECTORY

NOTES

TRANSPORTATION

Experience in large-scale evacuations indicates that the majority of the effected population will leave the hazardous area by use of their own automobile or by catching a ride with a friend or neighbor. However, a significant portion of the population cannot drive or do not own an automobile. This is especially the case in urban areas, where many people use public transportation as their only means of getting around.

There is also a large segment of the population that uses public transit to commute to and from work. A full-scale evacuation of an urban center during a normal work week would require thousands of people to relocate using trains, rapid transit systems, and buses.

The best data available concerning how long it takes to implement community evacuations is the Mississauga, Canada, train derailment. Their experience shows that 40% of the population took almost immediate action to leave, 65% evacuated after 30 minutes, and almost 90% evacuated 45 minutes after being warned. Figure 5-11 provides an overview of the time required to alert and evacuate people.

Good advanced planning is required to prepare for transporting people during an evacuation. The best sources of transportation in most areas are school buses and public transit buses. Many communities have designated specific buses for emergency response duty. These buses may be equipped with Emergency Preparedness radios, and their drivers are specially trained.

While school bus fleets are an excellent form of transportation, they may not be immediately available during the summer months. A Memorandum of Understanding should be in place between emergency response organizations and the agency providing bus services. This should outline how many busses can be committed to emergency evacuation duty, who has the authority to request and assign buses, and how the drivers will be contacted on a 24-hour basis.

Special emergency extraction may be required to transport people from areas close to the hazardous materials release. Some fire departments have addressed this problem by purchasing Emergency Escape Packs, also known as Emergency Breathing Apparatus (EBA). These devices typically have 5 to 10 minutes of breathing air in their cylinders. The facepiece is a simple plastic hood which is placed over the head to provide a fresh-air breathing supply for limited duration. This is usually adequate to move someone from inside their home and into an awaiting vehicle, where they will be transported to a safe area.

EBA's may be carried on Special Air Units operated by the fire department or be staged in strategic locations for use by Emergency Response Personnel. They may be found at or near special project sites, such as oil and gas wells where hydrogen sulfide is present or toxic waste dump remediation projects. Figure 5-14 provides an example.

FIGURE 5-14: EMERGENCY ESCAPE BREATHING APPARATUS

NOTES

RELOCATION SHELTERS

Relocation Shelters are used to temporarily house people displaced during an evacuation. One study suggests that approximately 65% of the evacuees do not stay at public Relocation Shelters. They may check into a hotel or stay with friends and relatives outside of the evacuated area.

The remaining 35% of the population will require some form of public shelter. These are typically located at schools, National Guard Armories, or at community centers.

In order for Relocation Shelters to be effective, they need the following elements in place:

- **Appropriate Building:** Relocation shelters may be in service for just several hours or for several days. The building must have a food service area, adequate restrooms and bathing facilities, and an area large enough for temporary sleeping furniture such as cots. The building must be air conditioned or heated and have adequate security to protect residents. Relocation Shelters should always be energy efficient in case the occupants need to protected-in-place.
- **Shelter Manager:** An individual trained in Shelter Management techniques should be assigned to each Relocation Shelter. The Shelter Manager organizes and supervises shelter activities and is the single point of contact between the shelter and the field command post. Many jurisdictions have a cooperative arrangement with the local Red Cross chapter or religious and civic organizations. The local Emergency Preparedness Agency coordinates the shelter program, provides the funding, and trains the Shelter Managers, while the cooperating organizations provide the personnel to staff the shelter.

NOTES

- **Shelter Support Staff:** If the Relocation Shelter will be operating for an extended period, a shelter staff should be provided to assist with its operation. Examples include receptionists to document who arrives and departs the shelter, EMT's or nurses to attend to the sick and disabled, food service personnel to prepare meals, building engineers or maintenance personnel, and counselors. It is also advisable to assign a police officer at each shelter for security. Direct radio contact with the Emergency Operations Center is recommended.

If shelters will remain in operation for an extended period, arrangements must be made for around-the-clock staffing and provisions. Relocation Shelters should be managed under the Incident Command System as a Sector reporting to the Evacuation Section.

INFORMATION

The Incident Commander should assure that displaced people are kept informed of the actions being taken by Emergency Response Personnel to mitigate the problem. Failure to keep people informed may create panic.

Relocation Shelters should receive special attention by providing regular briefings to the Shelter Manager. This can be accomplished by direct radio or telephone contact on an hourly basis or by faxing a brief status report from the Emergency Operations Center to each shelter location.

When the incident covers a period of days, the IC should consider issuing a written progress report, which can be posted twice daily in the shelter. If the displaced are not kept informed, they will quickly form a negative opinion of your operation and make their feelings known to the rest of the world through the media.

The press can be a powerful tool in confirming that the initial evacuation was well handled and is still necessary for public safety. It is important that the IC project to the media an image of professionalism and control during the evacuation. The IC should hold regularly scheduled joint press briefings with senior representatives of the media present. For example, the Emergency Preparedness, Police, and Fire agencies should conduct their briefings as a team and project unity in their decision-making.

SUMMARY

The first step toward gaining control of a hazardous materials incident is to isolate the people from the problem. This is accomplished by establishing a perimeter and Hazard Control Zones.

The Incident Commander has two distinct choices when faced with the protection of the population from potential exposure to a hazardous material; these include Protection-in-Place and evacuation. There are no clear cut benchmarks available for making this decision but, rather, a combination of factors, including the weather, time of day, type of occupancy, and other conditions. The key factor for protecting-in-place is whether there will be a higher exposure trying to evacuate outdoors than

NOTES

there will be by remaining indoors. This is independent of the type of material being released and depends only on the warning time, the duration of the release, and the infiltration rate of the building. For short-duration releases in areas where the buildings are of modern construction, sheltering is almost always the best choice, regardless of the hazmat's toxicity.

The IC must appreciate that the decision to evacuate initiates a process that is expensive and difficult to call off once it is started. Before initiating any evacuation, the IC should consider whether the population at risk can be protected just as well in place, within the confines of the structure where they are presently located.

REFERENCES AND SUGGESTED READINGS

Carter, Harry, "Downtown HazMat," PENNSYLVANIA FIREMAN, (May 1993). Pages 96-100.

Brunancini, Alan V., FIRE COMMAND, Boston, MA: National Fire Protection Association (1985).

ESTABLISHING WORK CONTROL ZONES AT UNCONTROLLED HAZARDOUS WASTE SITES, Environmental Protection Agency Publication 9285.2-06FS, Washington, DC (April, 1991).

HAZARDOUS MATERIALS WORKSHOP FOR LAW ENFORCEMENT, Federal Emergency Management Agency, Washington, DC (April, 1992).

Lindell, Michael, K., Bolton, Patricia, A., Perry, Ronald, W., et al., "Planning Concepts and Decision Criteria for Sheltering and Evacuation in a Nuclear Power Plant Emergency." A report prepared for the National Environmental Studies Project of the Atomic Industrial Forum, Inc. by the Battelle Human Affairs Research Centers, Seattle, Washington (June, 1985).

Maloney, Daniel, M., Policastro, Anthony, J., and Coke, Larry, "The Development of Initial Isolation and Protection Action Distances Table for the U.S. DOT Publication-1990 Emergency Response Guidebook." A paper presented at the 3rd Annual HazMat 90 Central Conference (March, 1990).

National Fire Service Incident Management System Consortium, MODEL PROCEDURES GUIDE FOR STRUCTURAL FIREFIGHTING, Stillwater, OK: Oklahoma State University–Fire Protection Publications (1993).

National Transportation Safety Board, "Special Investigation Report: Survival In Hazardous Materials Transportation Accidents." NTSB Report No. NTSB-HZM-79-4 (December, 1979).

Sherman, M. H., Wilson, D. J., and Kiel, D. E., "Variability in Residential Air Leakage." A technical paper presented at the ASTM Symposium on Measured Air Leakage Performance in Buildings, Philadelphia, Pennsylvania (April, 1984).

Sorenson, John, H., "Evaluation of Warning and Protective Action Implementation Times for Chemical Weapons Accidents." A report prepared for the Aberdeen Proving Ground, Maryland by Oak Ridge National Laboratory. Report No. ORNL/TM-10437 (April, 1988).

Wheeler, W. H. and Byrd, W. R., "Emergency Planning for H_2S Releases: Utilizing Shelter in Place and Interagency Drills." A report prepared for presentation at the Society of Petroleum Engineers Conference held in San Antonio, Texas. SPE Report No. 25979 (March, 1993). Pages 373-377.

Wilson, D. J., "Variation of Indoor Shelter Effectiveness Caused by Air Leakage Variability of Houses in Canada and the USA." Prepared by the Department of Mechanical Engineering, University of Alberta, Edmonton, Alberta. Presented at the U.S. EPA/FEMA Conference on the Effective Use of Sheltering-in-Place as a Potential Option to Evacuation During Chemical Release Emergencies, Emmitsburg, Maryland (December, 1988).

Wilson, D. J., "Effectiveness of Indoor Sheltering During Long Duration Toxic Gas Releases", Department of Mechanical Engineering, University of Alberta, Edmonton, Alberta (May, 1991).

Wilson, D. J., "Wind Shelter Effects on Air Infiltration for a Row of Houses," Department of Mechanical Engineering, University of Alberta, Edmonton, Alberta (September, 1991).

Wilson, D. J., "Accounting for Peak Concentrations in Atmospheric Dispersion for Worst Case Hazard Assessments," Department of Mechanical Engineering, University of Alberta, Edmonton, Alberta (May, 1991).

IDENTIFY THE PROBLEM

OBJECTIVES

1) Describe the principles of recognition, identification, classification and verification as they apply to hazardous materials emergencies.
2) List and describe the seven methods of identifying hazardous materials.
3) Identify the basic design and construction features of the following bulk packages, nonbulk packages, and storage vessels:
   a) Fixed tanks, storage tanks
   b) Tank containers (intermodal portable tanks)
   c) Piping
   d) Railroad tank cars
   e) Cargo tanks (tank trucks and trailers)
   f) Carboys
   g) Cylinders
   h) Drums
4) Identify each of the railroad tank cars by type:
   a) Nonpressure tank cars with and without expansion domes
   b) Pressure tank cars
   c) Cryogenic liquid tank cars
   d) High pressure tube cars
   e) Pneumatically unloaded hopper cars

"Appearances often are deceiving."

Aesop,
Philosopher (550 BC)

NOTES

5) Identify each of the intermodal tank containers by type:
    a) IM-101 portable tank
    b) IM-102 portable tank
    c) DOT Spec. 51 portable tank
    d) Specialized intermodal tank containers, including cryogenic intermodal tank containers and tube modules
6) Describe the types of specialized marking systems found at fixed facilities.
7) Describe the DOT specification markings for nonbulk and bulk packaging, and their significance in identifying the design and construction of the packaging and the types of hazmats likely found.
8) Given an emergency involving a ruptured underground pipeline, identify the following:
    a) How the pipeline may carry different products
    b) Ownership of the pipeline
    c) Type of product in the pipeline
    d) Procedures for shutting down the pipeline or controlling the leak
9) Identify and describe the placards, labels, markings, and shipping documents used for the transportation of hazardous materials.

INTRODUCTION

In 1971, a railroad derailment in the Houston, Texas metropolitan area caused a breach in a pressure tank car transporting propane, which ignited. The fire impinged upon an adjoining tank car of vinyl chloride. After approximately 45 minutes of exposure to fire, the vinyl chloride tank car violently ruptured. As a result, a fire department photographer was killed and a number of emergency response personnel and civilians were injured.

Because of the nature of the incident, the National Transportation Safety Board (NTSB) investigated. The investigators concluded that the following factors contributed to the severity of the accident:

- Lack of adequate training, information, and documented procedures for on-scene identification.
- Lack of adequate assessment of threats to safety.
- Reliance on firefighting recommendations which did not take into consideration the full range of hazards.

Although this tragedy occurred over 20 years ago, the lessons learned are timeless. Timely identification and verification of the hazmats involved are critical to the safe and effective management of a hazmat incident. Failure to perform basic tasks of recognition, identification, and verification are like going to war without knowing who the enemy is.

NOTES

This chapter will discuss the second step in the **Eight Step Process©—Identify the Problem**. Although this text is directed toward the Hazardous Materials Technician and the On-Scene Incident Commander, the recognition and identification process actually starts as soon as emergency responders are notified of the emergency. Problem identification cannot be safely and effectively accomplished if responders have not first controlled the emergency scene. Likewise, strategic goals and tactical response objectives cannot be formulated if the nature of the problem is not defined. Remember—a problem well-defined is half-solved!

This chapter will review the basic principles of problem identification and methods of hazmat recognition, identification, and classification. The authors would like to especially thank Charles J. Wright, Manager of Hazardous Materials Training for the Union Pacific Railroad, for his assistance in revising this chapter.

BASIC PRINCIPLES

KNOWING THE ENEMY

Managing a hazmat emergency is much like trying to manage a war. Neither effort will be very successful unless one learns as much as possible about the "enemy," where it can be found, and its general tendencies and behaviors. Among the most critical tasks in managing a hazmat emergency are surveying the incident scene to detect the presence of hazmats, identifying the nature of the problem and the materials involved, and identifying the type of hazmat container and the nature of its release. This effort is made more difficult by the large number and variety of hazardous materials found in society. Federal studies have shown that there are between 5 to 6 million chemicals and formulations.

Despite the potential enormity of the hazmat problem, experience shows that only a handful of materials are responsible for the majority of hazmat incidents. Various state and federal studies have shown that approximately 50% of all hazmat emergencies involve flammable and combustible liquids (e.g., gasoline, diesel fuel, and fuel oils) and flammable gases (e.g., natural gas, propane, butane). The next most common hazard class is corrosive materials.

A national study commissioned by EPA in 1985 analyzed approximately 7,000 hazmat incidents, the materials involved, and the source of their release. The study showed that approximately 75% of the releases occurred in facilities which produce, store, manufacture, or use chemicals; the remaining 25% occurred during transportation.

The most useful information, however, focused upon the hazmats involved. Ten chemicals accounted for 50% of all the incidents studied; these ten chemicals also accounted for approximately 35% of the total deaths and injuries (see Figure 6-1).

While ERP may potentially deal with thousands of chemicals, a rather small list of hazmats accounts for the majority of our problems. Be practical and realistic—the routine establishes the foundation upon which the

NOTES

TOP 10 CHEMICALS INVOLVED IN HAZMAT INCIDENTS

CHEMICAL NAME	% OF RELEASES	% OF DEATHS AND INJURIES
PCB (Polychlorinated Biphenyls)	23.0%	2.8%
Sulfuric Acid	6.5%	4.7%
Anhydrous Ammonia	3.7%	6.8%
Chlorine	3.5%	9.6%
Hydrochloric Acid	3.1%	5.6%
Sodium Hydroxide	2.6%	1.9%
Methyl Alcohol	1.7%	0.4%
Nitric Acid	1.7%	1.5%
Toluene	1.4%	2.4%
Methyl Chloride	1.4%	-

U.S. ENVIRONMENTAL PROTECTION AGENCY

FIGURE 6-1

nonroutine must build. If you don't "have your act together" managing high probability flammable liquid and gas emergencies, it's unlikely that you'll perform very well trying to manage an incident involving a more exotic hazardous material.

This background information is no substitute for conducting a hazard analysis and developing an Emergency Response Plan for your plant or community. However, it does support the point that with proper analysis and planning, your hazmat emergency response program can become more effective and efficient.

SURVEYING THE INCIDENT

The identification process starts with a survey of the incident site and surrounding conditions. ERP should complete the following tasks: (1) identify the hazmats involved; (2) identify the presence and condition of the containers involved; and (3) assess the conditions at the incident site, including exposures, topography, injuries, etc.

The identification process is built upon the following basic elements:

1) **Recognition**—Recognize the presence of hazardous materials. Positive recognition automatically shifts the operation into the hazmat mode. Basic recognition clues include occupancy and location, container shapes, markings and colors, placards and labels, shipping and facility documents, monitoring and detection instruments, and senses.

2) **Identification**—Identify the hazardous materials involved and the nature of the problem. Primary hazmat identification clues include markings and colors, shipping and facility documents, and monitoring and detection instruments.

NOTES

Regardless of the method of recognition or identification, always verify the information obtained. Don't take anybody's word at face value, including other responders. **Always verify—verify—verify!**

3) **Classification**—Determine the general hazard class or chemical family of the hazardous materials involved. A rule-of-thumb is, "If you can't identify, then try to classify." Basic classification clues include occupancy and location, container shapes, placards and labels, monitoring and detection instruments, and senses.

When dealing with unknown substances, HMRT's often rely upon instruments and chemical analytical kits which use a systematic process to determine the unknown's identity and hazards. The most widely used system is the HazCat® Chemical Identification System. Although responders may not always be able to identify the hazmat(s) involved, they will usually be able to determine the unknown's hazard class or chemical family.

IDENTIFICATION METHODS AND PROCEDURES

In some situations, emergency responders will initiate emergency operations without immediately realizing that hazardous materials are involved. In other cases, they may initiate aggressive, offensive operations before the material(s) are positively identified, verified, and evaluated. In either case, emergency responders have placed themselves at an unacceptable risk.

Responders rely upon seven basic clues as part of their identification process. Look for hazmats in every incident; then identify or at least classify the material. The seven clues are:

- Occupancy and location
- Container shapes
- Markings and colors
- Placards and labels
- Shipping papers and facility documents
- Monitoring and detection equipment
- Senses

The relationship between each of these clues and the identification process is illustrated in Figure 6-2.

These clues are also listed in relation to their distance from the hazmat (See Figure 6-3). The closer you are to the release site, the greater the likelihood that you can accurately identify the material(s). The catch is that you also face a substantially higher risk of exposure and/or contamination. Your goal should be to learn as much as possible from as far away as possible.

To ensure personnel safety during size-up, many emergency responders rely upon binoculars or spotting telescopes for identification. Binoculars should not be less than 10x35 power. While binoculars are commonly used by public safety responders, they can also be used by industry responders to safely verify container labels and other markings

NOTES

CLUES AND THE IDENTIFICATION PROCESS

CLUE	RECOGNIZE	IDENTIFY	CLASSIFY
Occupancy & Location	✓		✓
Container Shapes	✓		✓
Markings & Colors	✓	✓	
Labels & Placards	✓		✓
Shipping Papers & Facility Documents	✓	✓	
Monitoring & Detection Equipment	✓	✓	✓
Senses	✓		✓

FIGURE 6-2

from a distance. Although they provide a narrow field of vision, telescopes can also be a useful tool for the identification process.

OCCUPANCY AND LOCATION

Hazardous materials and chemicals surround us every day—not only in transportation and industrial facilities but in stores, hospitals, supermarkets, warehouses, garages, and even in our homes. These potential locations can be categorized into four basic areas—production, transportation, storage, and use. Occupancy and location can be used as a clue for both the recognition and classification of the hazmats involved.

The key for determining these potential sites is through the hazard analysis process. SARA Title III and state/local "Right-to-Know" legislation also requires facilities to notify the fire department, the Local Emergency Planning Committee (LEPC), and other government agencies when on-site quantities exceed established threshold values (e.g., 500 pounds, 10,000 pounds, etc.). Hazard analysis information should include a list of the hazmats on-site, their quantity and location, and hazards. In addition, material safety data sheets (MSDS) or comparable information should be available. Many local jurisdictions require fixed facilities to use Knox Boxes™ or similar devices at the front gate or main entrance for the storage of all pertinent facility information.

Experienced responders are able to associate certain hazmats with different types of occupancies. For example, what types of hazmats would you find in a metal plating operation? If you said sodium cyanide, potassium cyanide, and strong inorganic acids (e.g., nitric acid, sulfuric acid, etc.), you've got the right idea.

NOTES

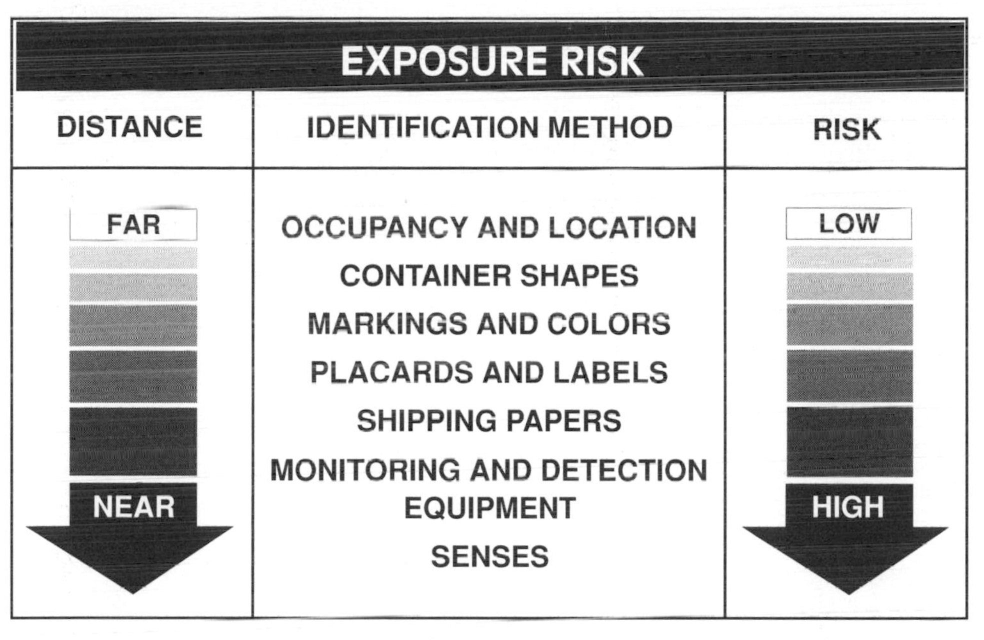

FIGURE 6-3: THE CLOSER YOU ARE TO A HAZMAT WHEN IDENTIFYING THE HAZMAT(S) INVOLVED, THE GREATER YOUR RISK OF EXPOSURE.

Responders must recognize that hazmat exposures are no longer limited to industrial or transportation emergencies. Not every emergency is a hazmat emergency; however, responders have the potential of being exposed to hazmats at ANY emergency.

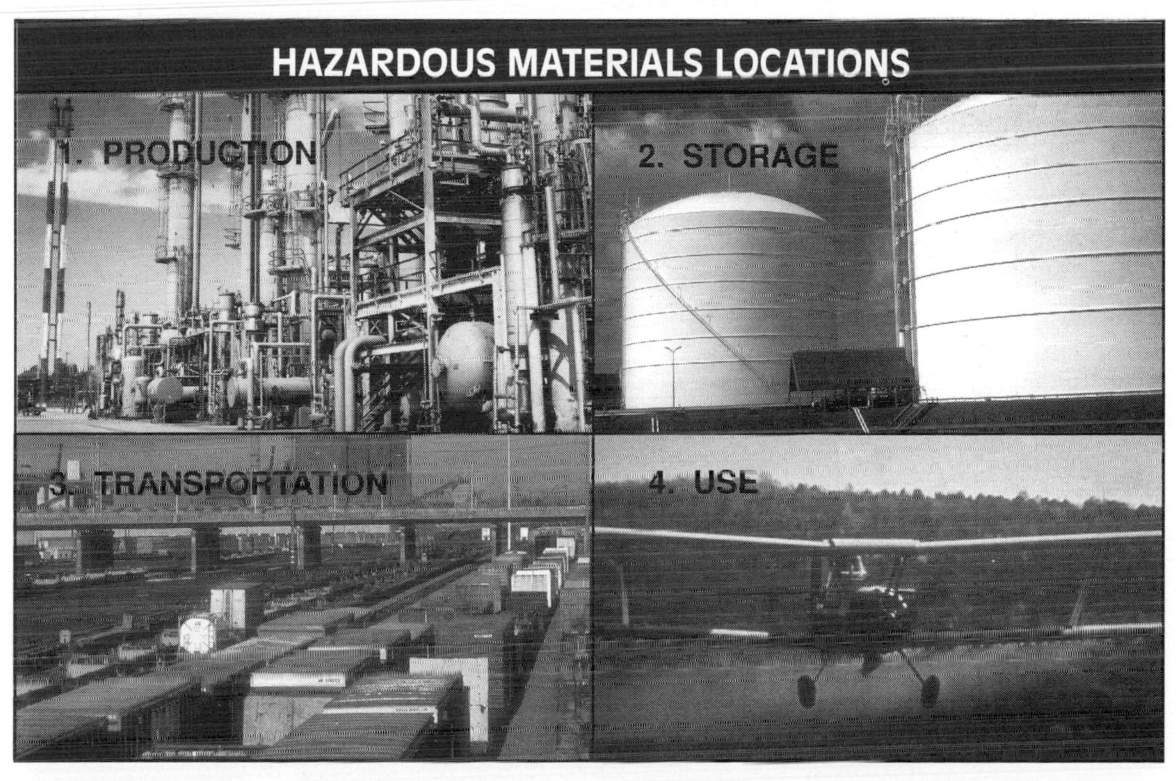

FIGURE 6-4: HAZARDS ANALYSIS IS ESSENTIAL FOR DETERMINING HAZARDOUS MATERIALS LOCATIONS WITHIN THE PLANT AND COMMUNITY.

NOTES

CONTAINER SHAPES

The size, shape, and construction features of a container/packaging are the second clue to the hazmat identification process. Container shapes can be used as a clue for both the recognition and classification of the hazmats involved. The DOT defines packaging as a receptacle, which may require an outer packaging, and any other components or materials necessary for the receptacle to perform its containment function and to ensure compliance with minimum packaging requirements.

Packaging used for transporting hazardous materials is regulated by DOT. Other types of containers are used only at fixed facilities, such as process towers, piping systems, and reactors. Packaging used for production, storage, and use is usually built according to nationally recognized consensus standards such as those provided by the American Society of Mechanical Engineers (ASME), the American Society of Testing and Materials (ASTM), and the American National Standards Institute (ANSI). For example, welded steel atmospheric pressure petroleum storage tanks are constructed according to specifications of the American Petroleum Institute (API), Standard 650. This standard is, in turn, referenced by NFPA's Technical Standard on Flammable and Combustible Liquids (NFPA 30).

Packaging is divided into three general groups: nonbulk packaging, bulk packaging, and facility containment systems.

❑ **Nonbulk Packaging**—Packaging with:

- Liquid capacity of 119 gallons (450 liters) or less.
- Solid net mass of 882 pounds (400 kg) or less for solids, or capacity of 119 gallons (450 liters) or less.
- Gas capacity of 1,001 pounds (454 kg) or less.

Nonbulk packaging may consist of single packaging (e.g., drum, carboy, cylinder) or combination packaging—one or more inner packagings inside of an outer packaging (e.g., glass bottles inside a fiberboard box, infectious disease sample containers). Nonbulk packaging may be palletized or placed in overpacks for transport in vehicles, vessels, and freight containers. Examples include bags, boxes, carboys, cylinders, and drums.

❑ **Bulk Packaging**—Packaging with:

- Liquid capacity greater than 119 gallons (450 liters).
- Solid net mass greater than 882 pounds (400 kg) for solids, or capacity greater than 119 gallons (450 liters).
- Gas capacity greater than 1,001 pounds (454 kg).

Bulk packages can be an integral part of a transport vehicle (e.g., cargo tank trucks, tank cars, and barges) or packaging placed on or in a transport vehicle (e.g., intermodal portable tanks, ton containers).

- **Facility Containment Systems**—Packaging, containers, and containment systems which are part of a fixed facility's operations. Depending upon the nature of the facility and its operations, the types of containment systems can vary greatly. Examples include pressurized and nonpressurized storage tanks, process towers, chemical and nuclear reactors, piping systems, pumps, storage bins and cabinets, dryers and degreasers, machinery, and so forth.

NOTES

Responders should become familiar with container and packaging shapes and be able to relate them to potential contents and hazard classes (i.e., container vs. contents). The military has used aircraft silhouettes for years as a training tool for recognizing both friendly and enemy aircraft. This same concept can be applied by emergency responders in recognizing and identifying hazmat containers and packaging.

NONBULK PACKAGING

Historically, nonbulk packaging was constructed to DOT specifications. When HM-181 was implemented in 1991, DOT replaced approximately 100 non-bulk specifications with 20 performance-oriented standards. Specification regulations still remain for some containers, including DOT cylinders, portable tanks (e.g., DOT 51, 56, 57, 60 and IM 101, 102), and cargo tanks (e.g., DOT 306, 307, 312 and MC-331, 338). Performance tests for nonbulk packaging include a drop test, leakproof test, hydrostatic test, stacking test, and vibration standard test.

Figure 6-5 illustrates the common types of nonbulk packaging.

FIGURE 6-5
UNION PACIFIC RAILROAD

NONBULK PACKAGING

BAGS

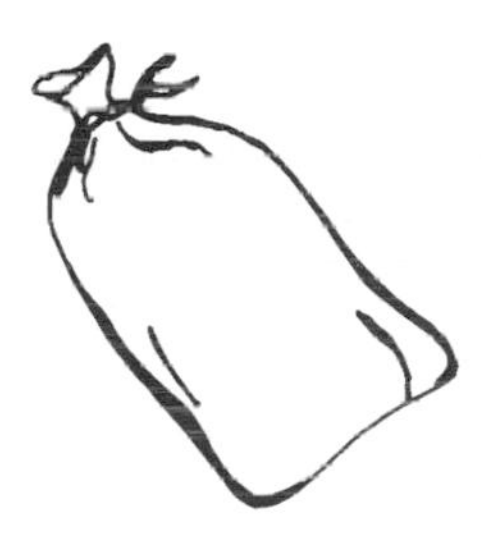

Twisted and tied bag

Folded and glued bag

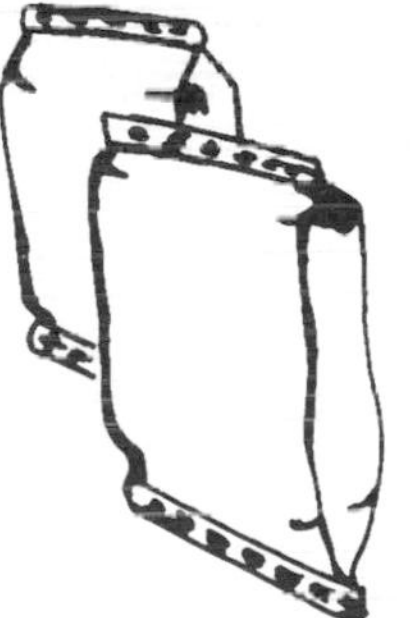

Stitched bags

DESCRIPTION	CONTENTS
• Flexible packaging constructed of cloth, burlap, kraft paper, plastic, or a combination of these materials. • Closed by folding and glueing, heat sealing, stitching, crimping with metal, or twisting and tying. • Typically contain up to 100 lbs. of material and usually palletized.	• Used for solid materials, such as fertilizers, pesticides, and caustic powders.

BOTTLES

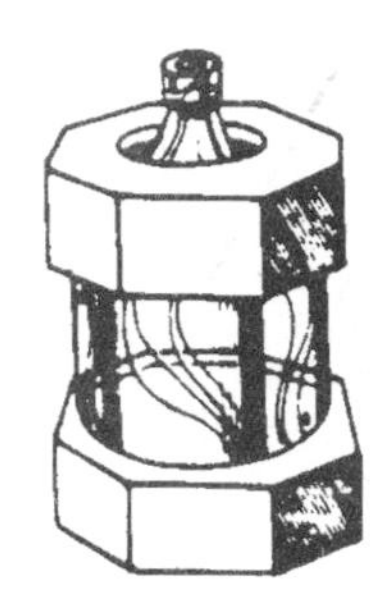

Protected bottle

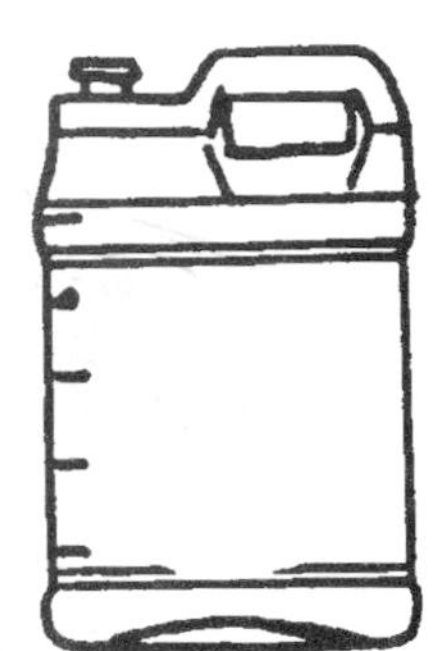

Plastic bottle

Glass bottle

- Constructed of glass and plastic, although metal and ceramic are sometimes used.
- Closed with threaded caps or stoppers.
- Range from ounces to 20 gallons or more in capacity.
- Usually placed in an outside packaging for transport.
- Sometimes referred to as jugs or jars.

- Used for liquids and solids, including laboratory reagents, corrosive liquids, and anti-freeze.
- Brown bottles are commonly used for light sensitive and reactive materials, such as organic peroxides.

BOXES

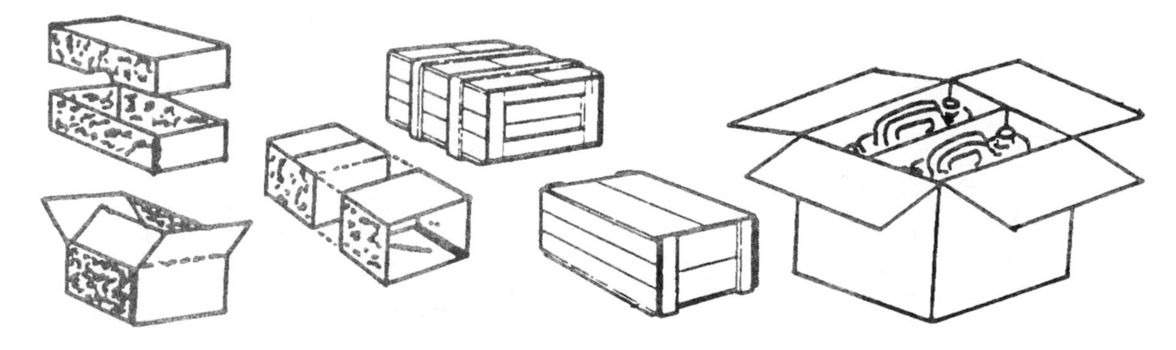

Fiberboard boxes | Wooden boxes | Divided fiberboard box

DESCRIPTION	CONTENTS

DESCRIPTION

- Rigid packaging that completely encloses the contents.
- Commonly used as the outside packaging for other nonbulk packages. Inner packaging may be surrounded with absorbent or vermiculite.
- Constructed of fiberboard, wood, metal, plywood, plastic, or other suitable materials.
- Fiberboard boxes may contain up to 65 lbs. of material; wooden boxes up to 550 lbs.

CONTENTS

- Used for liquids and solids.
- Almost any hazmat can be found inside a box (e.g., battery acid, laboratory reagents).
- Combination packaging using a box within inner package commonly used for radioactive materials and etiological agents, infectious disease samples.

MULTI-CELL PACKAGING

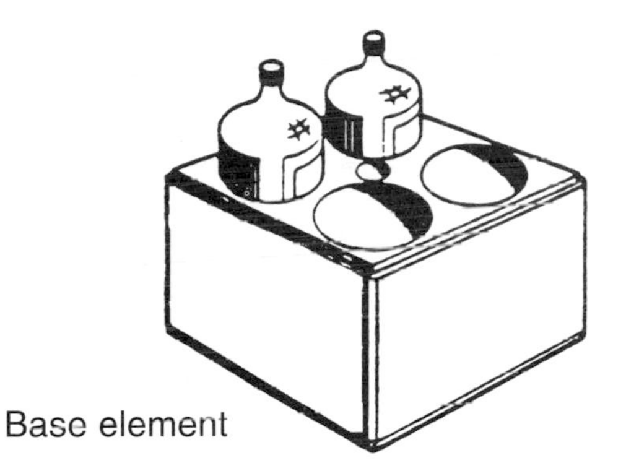

Base element

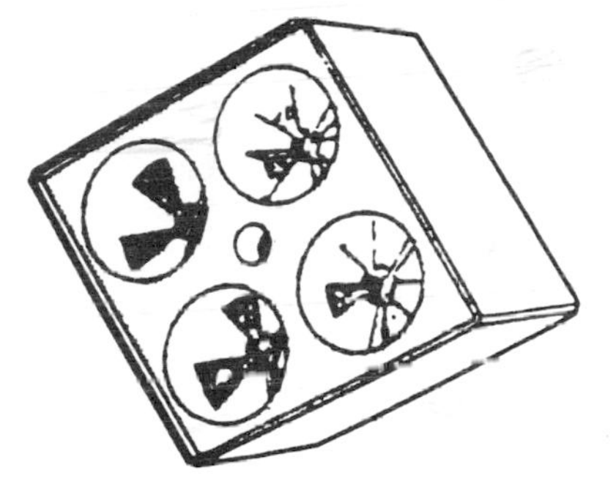

Lid element

- Consist of a form-fitting, expanded polystyrene box encasing one or more bottles.
- When transporting certain DOT hazmats, maximum bottle capacity is 4 liters (just over 1 gallon), and up to 6 bottles may be placed in one multi-cell package.

- Used for liquids, such as specialty chemicals, corrosive liquids, and various solvents.

CARBOYS

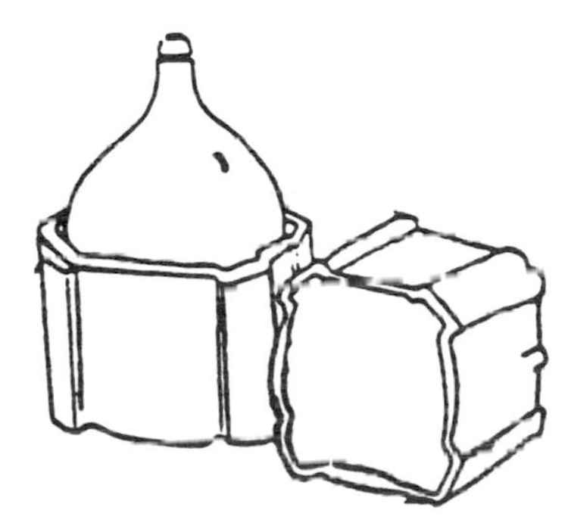

Carboy in polystyrene

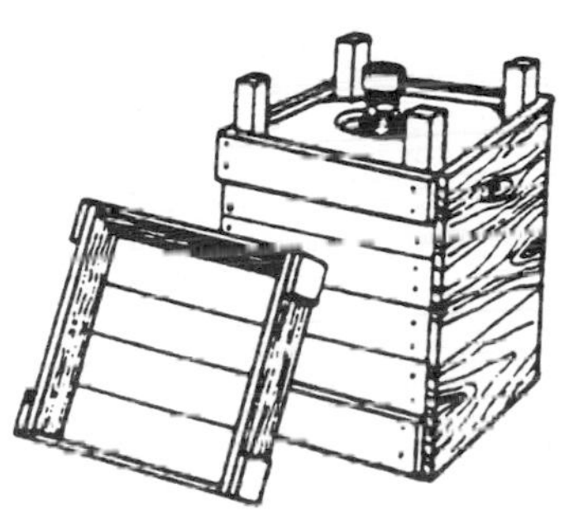

Carboy in wooden box

DESCRIPTION

- Glass or plastic "bottles" that may be encased in outer packaging (e.g., polystyrene box, wooden crate, plywood drum).
- Range in capacity to over 20 gallons.

CONTENTS

- Used for liquids, such as acids and caustics, and water.

CYLINDERS

- Three types of cylinders—aerosol containers, uninsulated cylinders, and cryogenic (insulated) cylinders.
- May have outer packaging.
- Service pressures range from a few psi to several thousand psi.
- Majority equipped with pressure relief device (e.g., relief valve, rupture disk, fusible plug).

- Used for pressurized, liquefied, and dissolved gases.

AEROSOL CONTAINERS

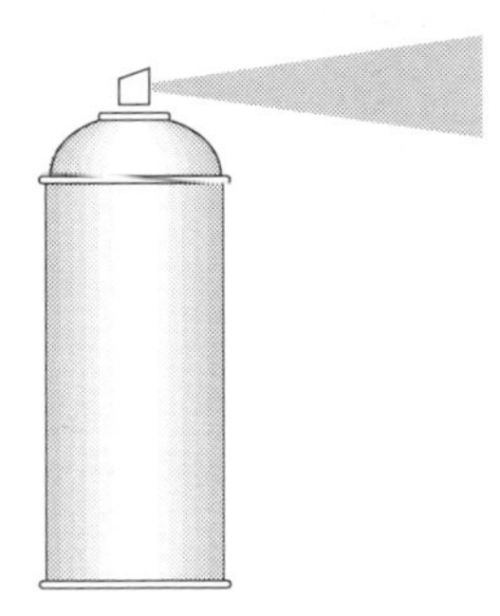

DESCRIPTION

- Small cylinders made of metal, glass, or plastic.
- Transported in boxes.
- Propellant may be flammable gas (e.g., propane).

CONTENTS

- Contain hazmats with a propellant. Examples are cleaners, lubricants, paint, toiletries, and pesticides.

UNINSULATED CONTAINERS

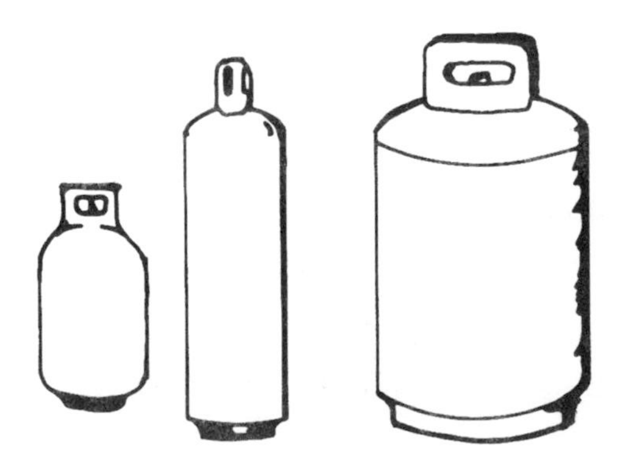

DESCRIPTION

- Typically made of steel, although aluminum or fiberglass aluminum may be found.
- Found in variety of sizes (e.g., 20 lb. propane cylinder to 1 ton chlorine cylinder).
- Do not have uniform taper on cylinder head, and thread design will vary depending upon contents.

CONTENTS

- Used for pressurized and liquefied gases, such as acetylene, LPG, chlorine, and oxygen.

CRYOGENIC (INSULATED) CYLINDERS

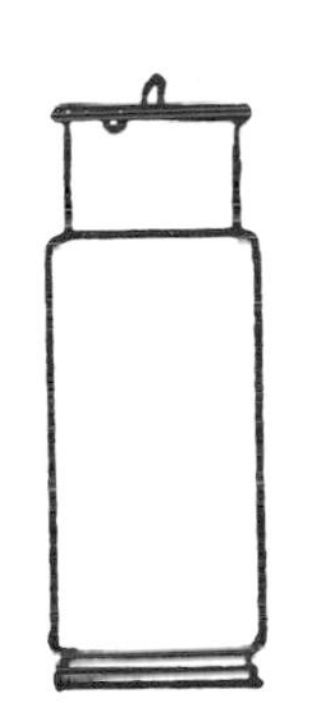

DESCRIPTION

- Consist of an insulated metal cylinder contained within an outer protective metal jacket.
- Area between cylinder and jacket normally under vacuum.
- Designed for a specific range of service pressures and temperatures.
- Found in a range of sizes.

CONTENTS

- Used for cryogenic liquids, such as liquid argon, liquid helium, liquid nitrogen, and liquid oxygen.

DRUMS

5 gallon drum
(pail, bucket, can)

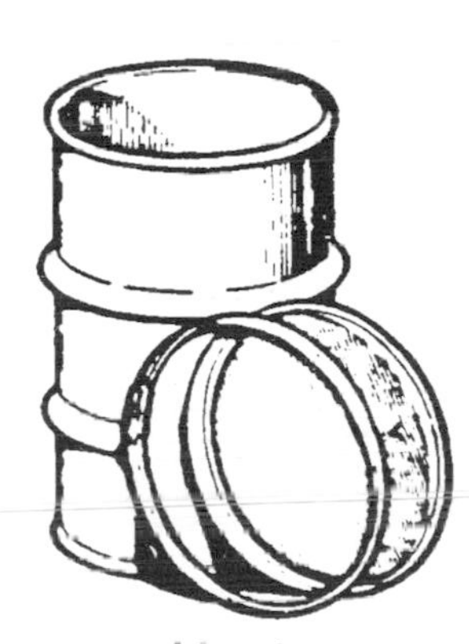

Metal open
head drum

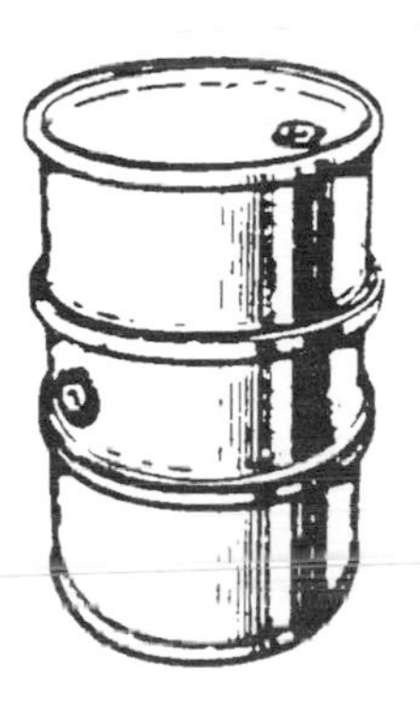

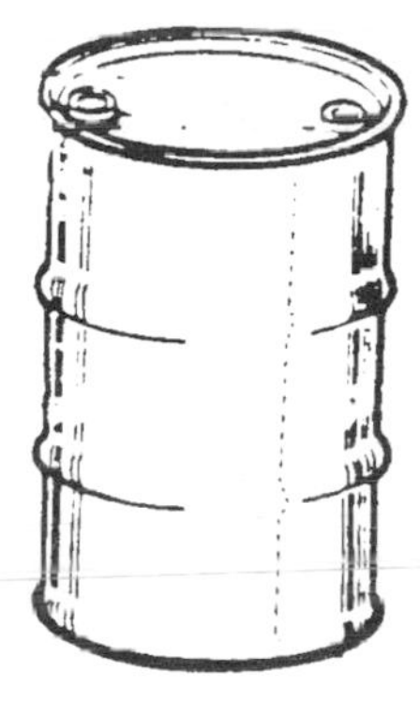

Tight or closed head
metal drums

Open head
plastic drum

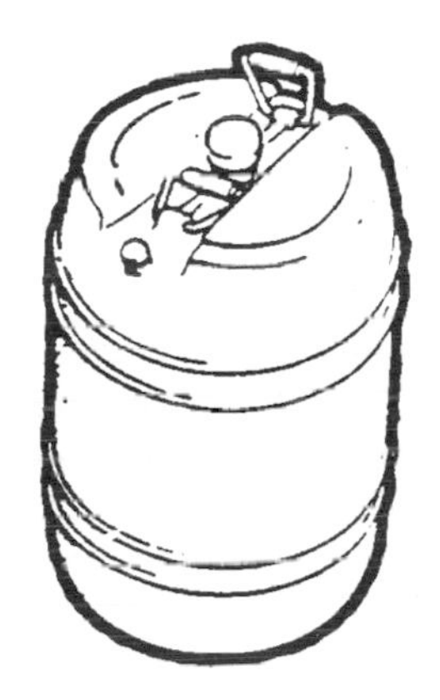

Tight or closed head
plastic drum

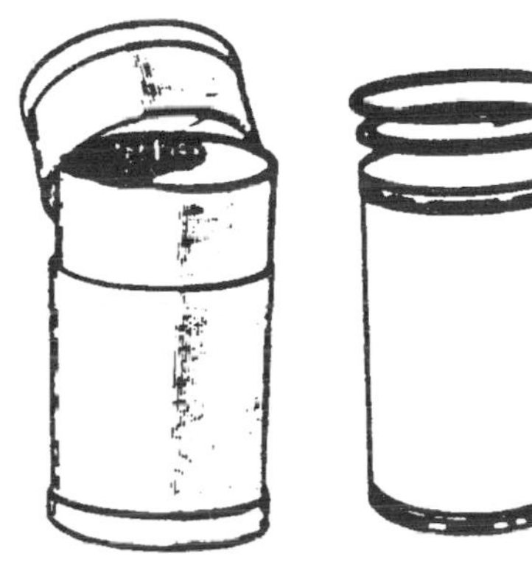

Fiber (fibre) drums

Plywood drum

DESCRIPTION

- Cylindrical packagings made of metal, plastic, fiberboard, plywood, or other suitable materials.
- Typical capacity is 55 gallons, although smaller drums can be found.
- Overpack drums used to hold damaged or leaking nonbulk packaging have 85 gallon capacity.
- Have removable ("open head") or nonremovable heads ("tight or closed head").
- Closed head drum usually contains two openings—2-inch and 3/4-inch diameter plugs called "bungs."

CONTENTS

- Used for liquids and solids. Depending upon the construction of the drum, examples include lubricating grease, caustic powders, corrosive liquids, flammable solvents, and poisons.

FIGURE 6-5

BULK PACKAGING

Bulk packaging may either be placed on or in a transport vehicle, or may be an integral part of a transport vehicle (e.g., cargo tank truck, tank car).

The following charts provide further information on shapes, descriptions and contents of both bulk transportation and storage containers:

- Bulk Packaging Placed on or in a Transport Vehicle (Figure 6-6)
- Cargo Tank Trucks (Figure 6-7)
- Railroad Tank Cars (Figure 6-9)

FACILITY CONTAINMENT SYSTEMS

Facility containment systems are packaging, containers, and/or associated systems which are part of a fixed facility's operations. Examples may include storage tanks, process towers, chemical and nuclear reactors, piping systems, pumps, storage bins and cabinets, dryers and degreasers, machinery, and so forth.

The following charts provide further information on shapes, descriptions, and contents of common facility storage tanks and vessels:

- Atmospheric and Low Pressure Liquid Storage Tanks (Figure 6-11)
- Pressurized Storage Vessels (Figure 6-12)

FIGURE 6-6
UNION PACIFIC RAILROAD

BULK PACKAGING PLACED ON/IN TRANSPORT VEHICLES

BULK BAGS

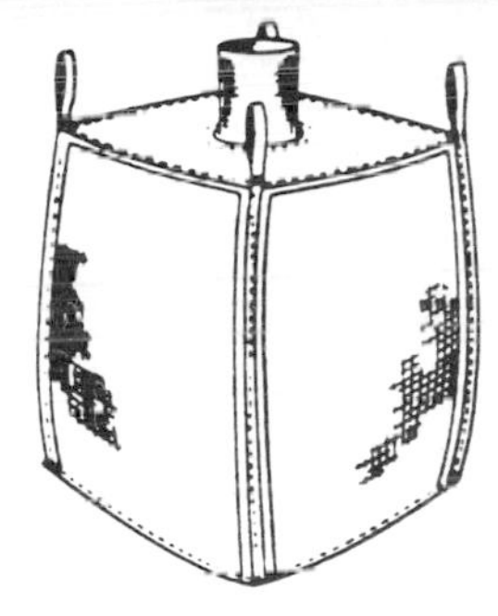

Strap design

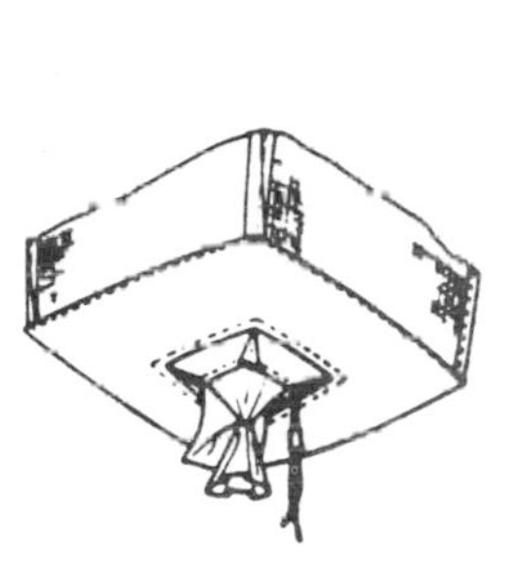

Bottom outlet

Sleeve design

DESCRIPTION

- Pre-formed packaging constructed of flexible materials (e.g., polypropylene), available plain, coated, or with liners.
- Standard sizes range from 15 to 85 cubic ft., with capacities of 500 to 5,000 lbs.
- Transported in both open and closed transport vehicles.

CONTENTS

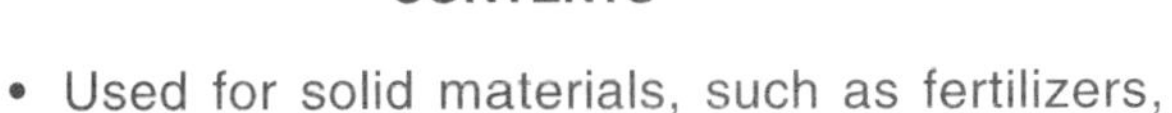

- Used for solid materials, such as fertilizers, pesticides, and water treatment chemicals.

PORTABLE BINS

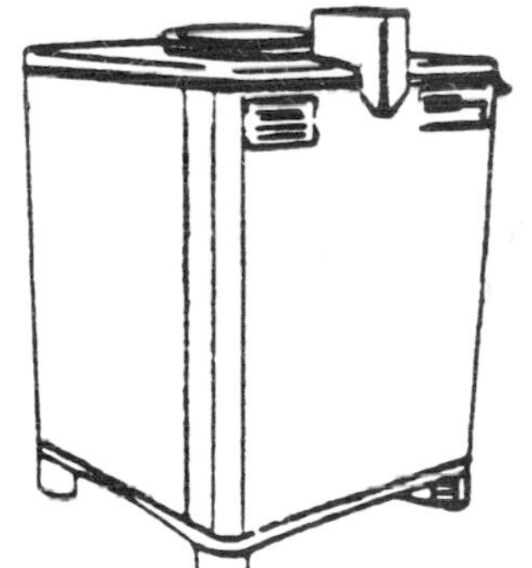

DESCRIPTION

- Approximately 4 ft. square and 6 ft. high.
- May contain up to 7,700 lbs.
- Loaded through the top and unloaded from the side or bottom.
- Usually placed on flat-bed trucks and trailers in agricultural areas.

CONTENTS

- Used for solid materials, such as ammonium nitrate fertilizer and calcium carbide.

NONPRESSURE PORTABLE TANK

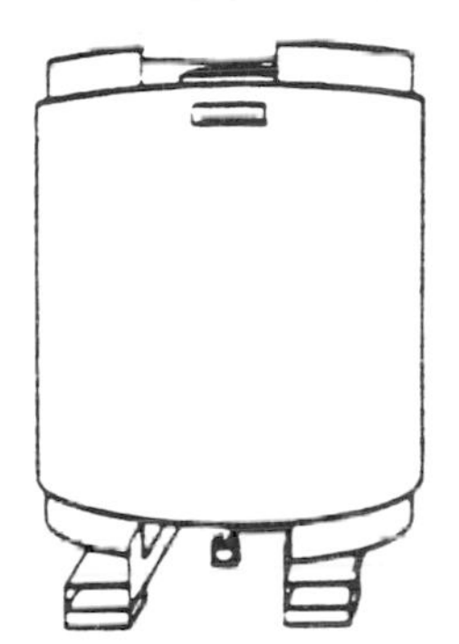

DESCRIPTION

- May have rectangular or circular cross sections, approximately 6 ft. high.
- Capacity of approximately 300 to 400 gallons.
- May contain internal pressures up to 100 psi.

CONTENTS

- Used for liquid materials, such as liquid fertilizers, water treatment chemicals, and flammable solvents.

INTERMODAL NONPRESSURE PORTABLE TANK (IM 101 AND IM 102)

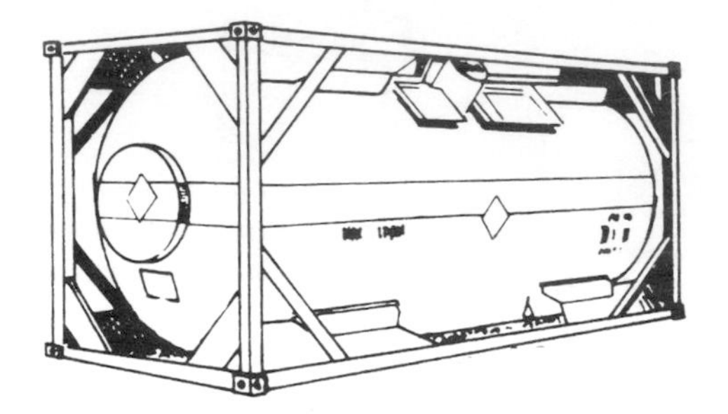

Box type frame

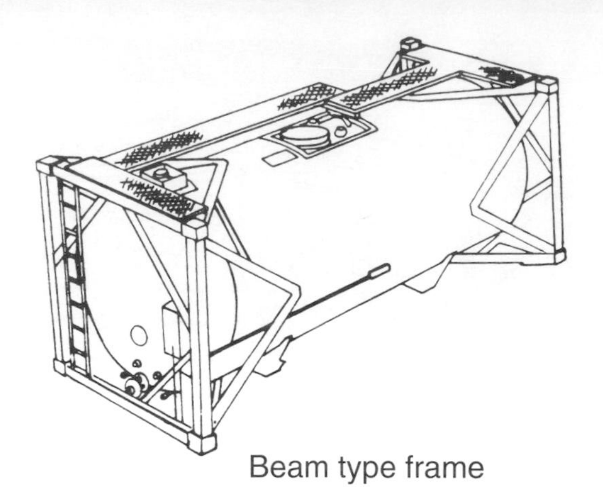

Beam type frame

DESCRIPTION

- May have rectangular, oval, or circular cross sections. Commonly 20 ft. long, 8 or 8-1/2 ft. wide, and from 8 to 9-1/2 ft. high.
- Found within a "box" or "beam" supporting frame.
- Capacities generally do not exceed 6,300 gallons.
- Maximum allowable working pressure for IM 101 (IMO Type 1) ranges from 25.4 to 100 psig.
- Maximum allowable working pressure for IM 102 (IMO Type 2) ranges from 14.5 to 25.4 psig.
- Often use a combination pressure/vacuum relief device. Rupture disks may also be found.

CONTENTS

- Used for liquid materials, such as food grade commodities, whiskey, liquid fertilizers, solvents, and corrosive liquids.

INTERMODAL PRESSURE PORTABLE TANKS (IMO TYPE 5)

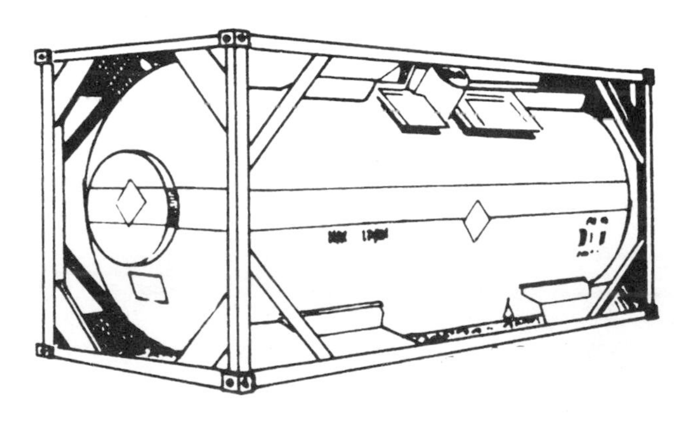

- Have a circular cross section and may be as large as 6 ft. diameter and 20 ft. long.
- Found within a "box" or "beam" supporting frame.
- Capacities range up to 5,500 gallons.
- Service pressures range from 100 to 500 psi.

- Used for liquefied gases and liquids, such as LPG, anhydrous ammonia, bromine, sodium, and aluminum alkyls.

INTERMODAL SPECIALIZED PORTABLE TANKS–CRYOGENIC

(IMO TYPE 7)

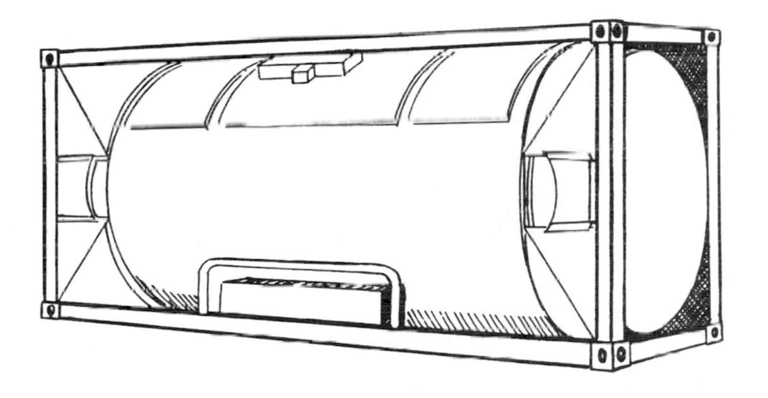

DESCRIPTION

- Approved for use by DOT special exemption.
- Consist of tank within a tank design with insulation between inner and outer tanks.
- Area between inner and outer tank normally maintained under vacuum.

CONTENTS

- Used for cryogenic liquids, such as liquid argon, liquid helium, liquid nitrogen, and liquid oxygen.

INTERMODAL SPECIALIZED PORTABLE TANKS–TUBE MODULE

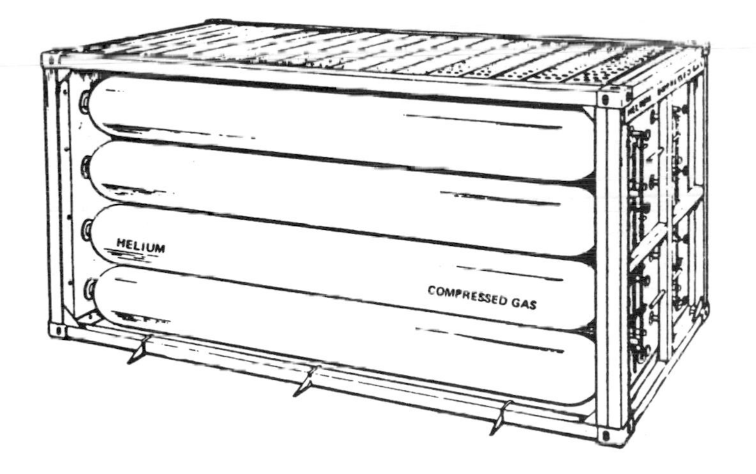

DESCRIPTION

- Consist of several seamless steel cylinders from 9 to 48 inches in diameter, permanently mounted inside an open frame.
- Box-like compartment at one end enclosing the valving.
- Service pressures range up to 2,400 psi and higher.

CONTENTS

- Used for pressurized gases, such as helium, nitrogen, and oxygen.

TON CONTAINERS

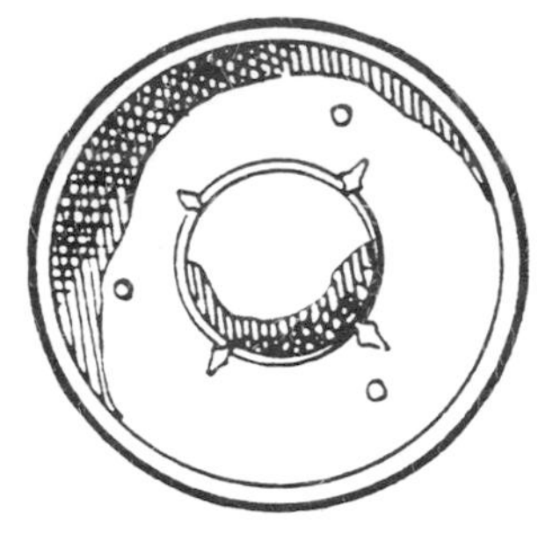

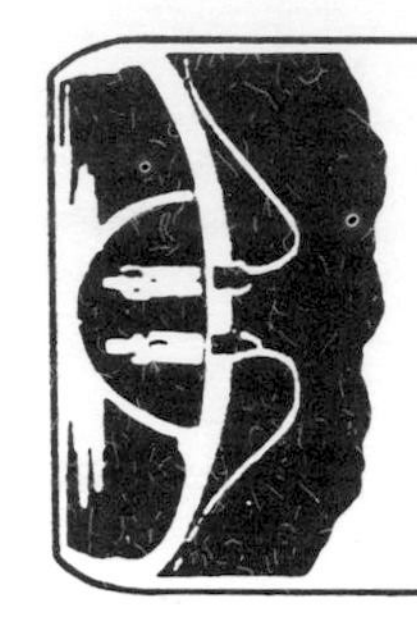

DESCRIPTION

- Cylindrical pressure tanks approximately 3 ft. in diameter and 8 ft. long with concave or convex heads.
- Transport one ton of chlorine; actual cylinder weighs approximately 1,800 lbs. empty.
- Container valves are found at one end under a protective cap.
- Chlorine and sulfur dioxide containers have fusible plugs on each head. Phosgene containers have no pressure relief device.
- Transported on both railcars and trucks.

CONTENTS

- Used for liquefied gases, such as chlorine, sulfur dioxide, and phosgene.

RADIOACTIVE PROTECTIVE OVERPACKS AND CASKS

DESCRIPTION

- Also referred to as Type A and Type B packaging.
- Type A—designed to prevent loss or dispersal of contents under normal conditions of transport.
- Type B—Meets same criteria as Type A, but designed to meet standards for performance under hypothetical accident conditions.
- Consist of rigid metal materials with a cylindrical or box-like configuration.

CONTENTS

- Used for the transportation of radioactive materials and radioactive wastes.

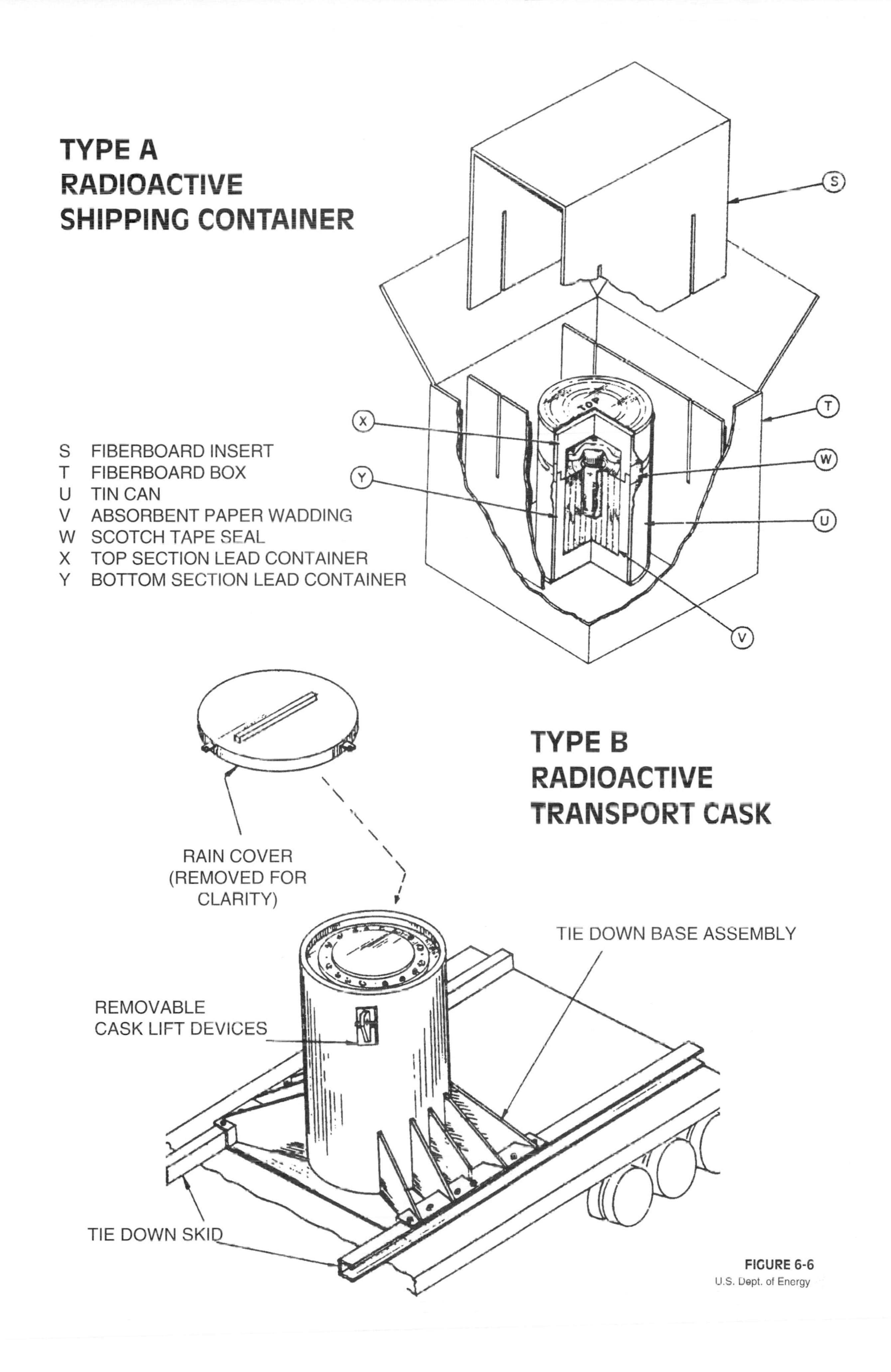

FIGURE 6-6
U.S. Dept. of Energy

CARGO TANK TRUCKS

CONTAINER SHAPE	DESCRIPTION	CONTENTS
MC-306/DOT-406 **ATMOSPHERIC PRESSURE CARGO TANK TRUCKS**	• OVAL CROSS SECTION INDICATES NONPRESSURIZED TANK (LESS THAN 3 PSI) • USUALLY SINGLE-SHELL, ALUMINUM CONSTRUCTION. OLDER STEEL CONSTRUCTED TANKS MAY BE FOUND. • GENERALLY 9,000 GALLONS MAXIMUM CAPACITY.	• TRANSPORT PETROLEUM PRODUCTS (GASOLINE, FUEL, OIL).
MC-307/DOT-407 **LOW PRESSURE CHEMICAL CARGO TANK TRUCKS**	• CIRCULAR CROSS SECTION WITH LOW PRESSURES (UP TO 40 PSI). • DOUBLE SHELL CONSTRUCTION WITH INSULATION THE MOST COMMON. • INSULATED TANKS MAY NOT APPEAR CIRCULAR IN CROSS SECTION. • ONE OR TWO COMPARTMENTS WITH OVERTURN PROTECTION. • GENERALLY 6,000 TO 7,000 GALLONS MAXIMUM CAPACITY.	• TRANSPORTS FLAMMABLE AND COMBUSTIBLE LIQUIDS, POISONS, AND CHEMICALS WITH A VAPOR PRESSURE OF 18 PSI AT 100°F OR GREATER , BUT NOT MORE THAN 40 PSI AT 170°F.
MC-312/DOT-412 **CORROSIVE CARGO TANK TRUCKS**	• CIRCULAR CROSS SECTION, SMALLER DIAMETER WITH EXTERNAL REINFORCING RIBS OFTEN VISIBLE. • MAY ALSO BE FOUND IN DOUBLE SHELL CONFIGURATION. • INSULATED TANKS MAY NOT APPEAR CIRCULAR IN CROSS SECTION. • OVERTURN AND SPLASH PROTECTION AT DOME COVER/VALVE LOCATIONS. • GENERALLY 5,000 TO 6,000 MAXIMUM CAPACITY.	• TRANSPORTS CORROSIVE LIQUIDS AND HIGH DENSITY LIQUIDS.
MC-331 **HIGH PRESSURE GAS CARGO TANK TRUCKS**	• CIRCULAR CROSS SECTION WITH ROUNDED ENDS OR HEADS. • DESIGN PRESSURES OF NOT LESS THAN 100 PSI OR MORE THAN 500 PSI. • SINGLE-SHELL, NONINSULATED TANK. • UPPER TWO-THIRDS PAINTED WHITE OR HIGHLY REFLECTIVE COLOR. • CAPACITY RANGES FROM 2,5000 ("BOBTAIL" DELIVERY TRUCK) TO 11,500 GALLONS (CARGO TANK TRUCK).	• TRANSPORTS LP GASES AND ANHYDROUS AMMONIA (PARTICULARLY IN THE SPRING).

FIGURE 6-7

CARGO TANK TRUCKS

CONTAINER SHAPE	DESCRIPTION	CONTENTS
MC-338 CRYOGENIC LIQUID TANK TRUCKS 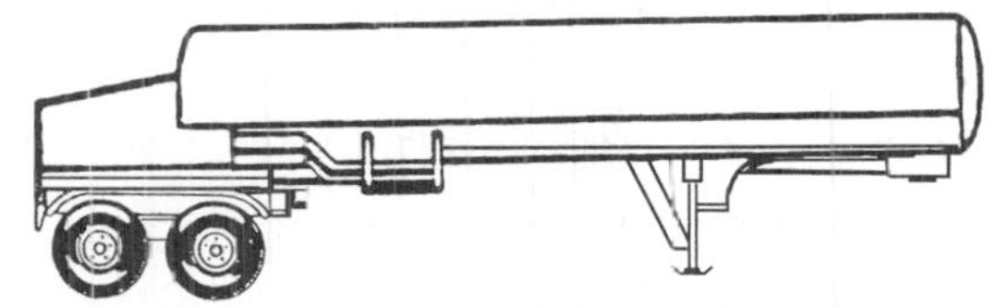	• WELL-INSULATED "THERMOS BOTTLE" DESIGN WITH FL TANK ENDS. • DOUBLE SHELL TANK WITH RELIEF PROTECTION. • OFTEN HAVE VAPORS DISCHARGING NORMALLY FROM RELIEF VALVES.	• TRANSPORT CRYOGENIC LIQUIDS (E.G., LOX, LIQUID NITROGEN, LIQUID ARGON, AND LIQUID CARBON DIOXIDE).
COMPRESSED GAS TRAILER 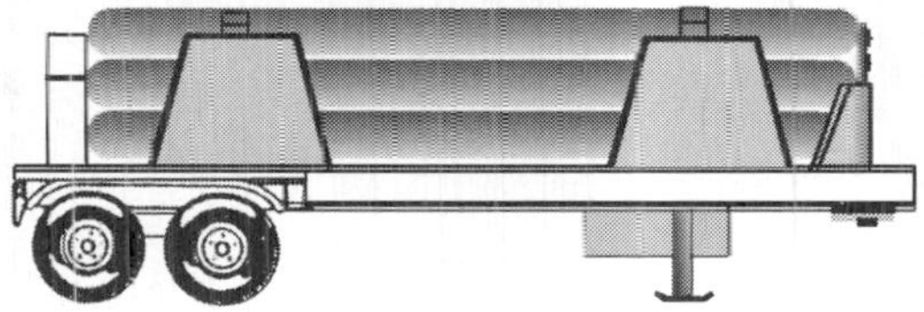	• OFTEN REFERRED TO AS A "TUBE TRAILER." • CYLINDERS ARE STACKED AND MANIFOLDED TOGETHER. • MANIFOLD AT REAR. • PRESSURES RANGE FROM 3,000 TO 5,000 PSI. • OFTEN FOUND AT CONSTRUCTION AND INDUSTRIAL SITES.	• TRANSPORT COMPRESSED GASES (E.G., OXYGEN, NITROGEN, HYDROGEN).
MOLTEN SULFUR CARGO TANK 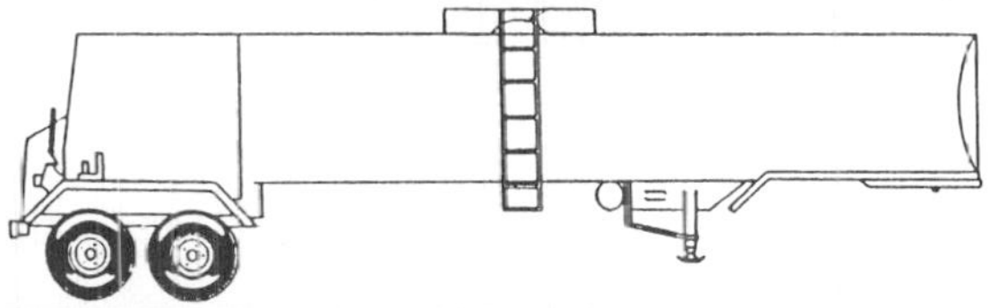	• CONSTRUCTED OF MILD STEEL OR ALUMINUM. • COVERED WITH INSULATION, WITH STEAM COALS INSIDE SHELL TO ENSURE SULFUR REMAINS MOLTEN. • REQUIRED TO BE STENCILED ON SIDES AND ENDS WITH WORDS "MOLTEN SULFUR."	• MOLTEN SULFUR.
PNEUMATICALLY OFF-LOADED HOPPER TRAILERS 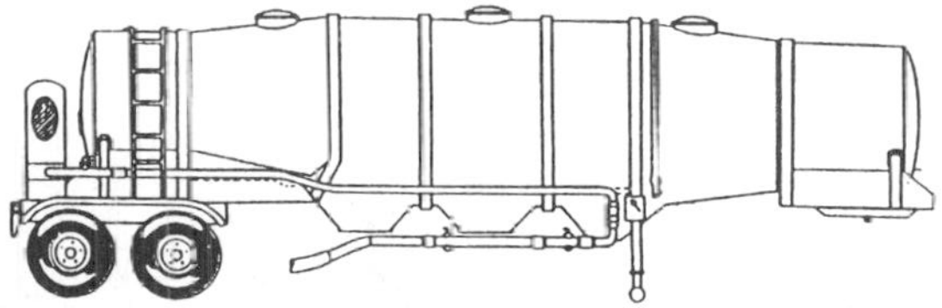	• ALSO KNOWN AS DRY BULK HOPPERS. • CAPACITY UP TO 1,500 CUBIC FT. • CARRY VERY HEAVY LOADS; CENTRIFUGAL FORCE CAUSE OF MANY ROLLOVERS. • USE AIR PRESSURE FOR PRODUCT TRANSFER. • STATIC CHARGES ARE A COMMON HAZARD.	• AMMONIUM NITRATE FERTILIZER, CEMENT, DRY CAUSTIC SODA, PLASTIC PELLETS.

NOTES

FIGURE 6-8: CARGO TANK TRUCKS UNDER PRESSURE (Eg. MC 307/DOT 407, MC 312/DOT 412, MC 331, MC 338) WILL HAVE A ROUND SHAPE, WHILE NON-PRESURIZED CARGO TANK TRUCKS (Eg. MC 306/DOT 406) WILL BE ELIPTICAL

RAILROAD TANK CARS

CONTAINER SHAPE	DESCRIPTION	CONTENTS
NONPRESSURIZED TANK CARS	• HORIZONTAL TANK WITH FLAT OR NEARLY FLAT ENDS. • FITTINGS AND VALVING VISIBLE ON TOP OF CAR. • OLDER CARS WILL HAVE AN EXPANSION DOME WITH VISIBLE FITTINGS. • TANK PRESSURES RANGE FROM 35 TO 100 PSI. • CAPACITIES OF 4,000 TO 45,000 GALLONS. • OFTEN HAS BOTTOM UNLOADING VALVES. • NONPRESSURE TANK CAR CLASSES: DOT 103 AAR 201 DOT 104 AAR 203 DOT 111 AAR 206 DOT 115 AAR 211 • SOME NON-PRESSURIZED CARS ARE NOW BEING CONSTRUCTED WITH A HOUSING AROUND ALL FITTINGS. Has one or more compartments	• TRANSPORTS WIDE VARIETY OF LIQUIDS, MOLTEN SOLIDS, AND SOME LIQUEFIED GASES.
TANK TRAIN®	• SERIES OF NONPRESSURE TANK CARS INTER-CONNECTED WITH FLEXIBLE HOSES TO ALLOW FOR LOADING/UNLOADING OF THE CARS FROM ONE END. • CONSIST OF SPRING LOADED, BUTTERFLY VALVE ARRANGEMENT ON EACH CAR WHICH IS PNEU-MATICALLY CONTROLLED FROM LOADING/UNLOADING POINT.	• TRANSPORTS FUEL OIL, AND PESTICIDES.
PRESSURIZED TANK CARS	• CYLINDRICAL TANK WITH ROUNDED ENDS. • FITTINGS AND VALVES ENCLOSED IN DOME. • OFF-WHITE PAINT INDICATES SPRAYED-ON THERMAL INSULATION. • BLACK PAINT USUALLY WILL INDICATE A JACKETED TANK CAR. • TANK PRESSURES RANGE FROM 100 TO 600 PSI. • CAPACITIES OF 4,000 TO 45,000 GALLONS.	• TRANSPORTS FLAMMABLE, NONFLAMMABLE, AND POISONOUS GASES. • PRESSURE TANK CAR CLASSES: DOT 105 DOT 114 DOT 109 DOT 120 DOT 112

FIGURE 6-9

RAILROAD TANK TRUCKS

CONTAINER SHAPE	DESCRIPTION	CONTENTS
CRYOGENIC LIQUID TANK CARS 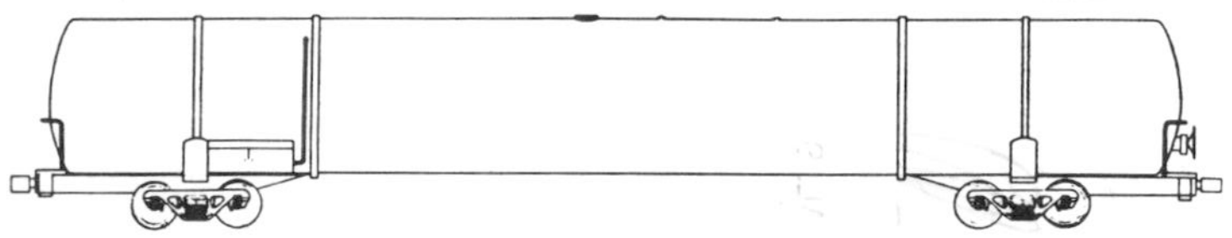	• WELL-INSULATED "THERMOS BOTTLE" DESIGN. • DOUBLE SHELL TANK SIMILAR TO FIXED STORAGE TANKS. • TRANSPORT LOW PRESSURE REFRIGERATED LIQUIDS (PRESSURES 25 PSIG OR LOWER). • ABSENCE OF ANY TOP FITTINGS. • LOADING/UNLOADING FITTINGS AND SAFETY RELIEF DEVICE OFTEN FOUND IN CABINETS AT DIAGONAL CORNERS OR ON ONE END AT GROUND LEVEL. • CRYOGENIC TANK CAR CLASSES: DOT 113 AND AAR 204W.	• TRANSPORTS CRYOGENIC LIQUID OXYGEN, LIQUID HYDROGEN, LIQUID NITROGEN.
BOXED CRYOGENIC TANK CAR 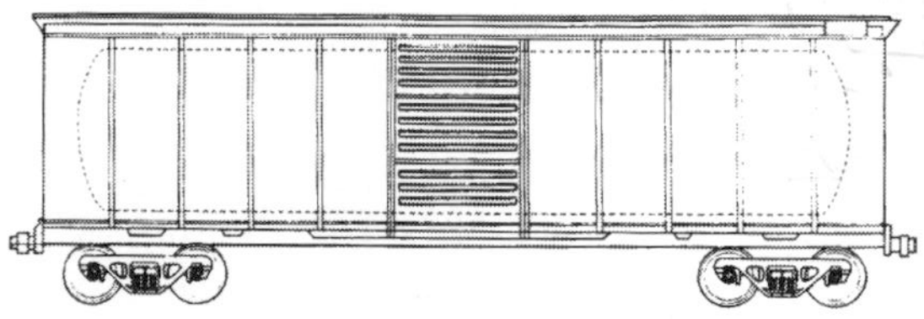	• CRYOGENIC TANK INSIDE OF A BOX CAR. • ALSO KNOWN AS "XT" BOXED TANK. • VALUES AND FITTINGS FOUND INSIDE DOORS ON BOTH SIDES. • TANK CAR CLASS AAR 204XT.	• TRANSPORTS CRYOGENIC LIQUIDS SUCH AS LIQUID OXYGEN, LIQUID HYDRO-GEN, LIQUID NITROGEN.
HIGH PRESSURE TUBE CARS 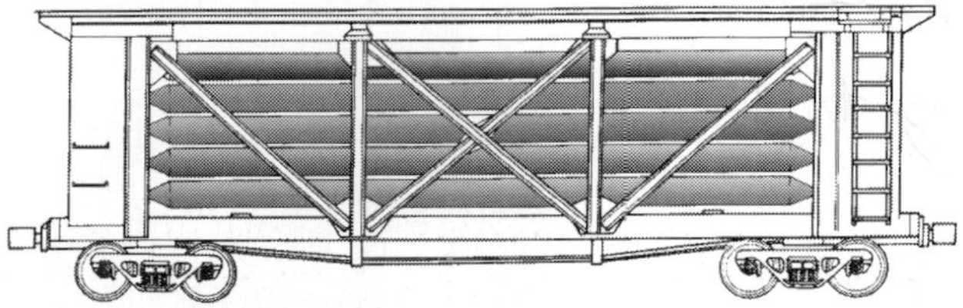	• FORTY (40) FT. BOX TYPE, OPEN FRAME CAR. • CONTAINS 30 SEAMLESS, UNINSULATED STEEL CYCLINDERS ARRANGED HORIZONTALLY AND PERMANENTLY ATTACHED TO CAR. • CYLINDER PRESSURES UP TO 5,000 PSI. • LOADING/UNLOADING FITTINGS AND SAFETY DEVICES IN CABINET AT END OF CAR. • TANK CAR CLASS DOT 107.	• TRANSPORTS HELIUM, HYDROGEN, OR OXYGEN.
PNEUMATICALLY UNLOADED COVERED HOPPER CARS 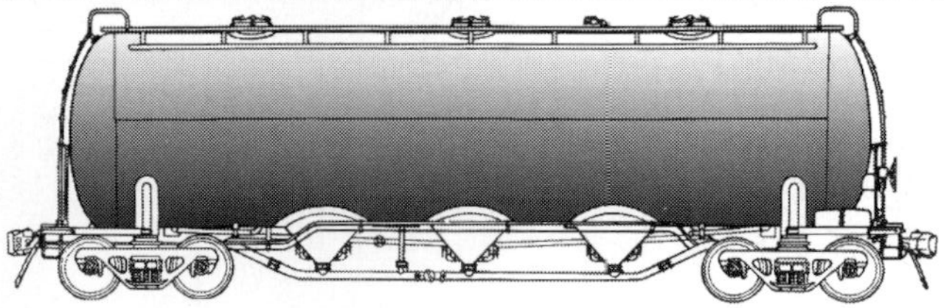	• COVERED HOPPER CAR UNLOADED BY AIR PRESSURE. • TANK TEST PRESSURES RANGE FROM 20 TO 80 PSI. • TANK CAR CLASS ARR 207.	• TRANSPORTS AMMONIUM NITRATE FERTILIZER, DRY CAUSTIC SODA, PLASTIC PELLETS.

RAILROAD TANK CAR NOMENCLATURE

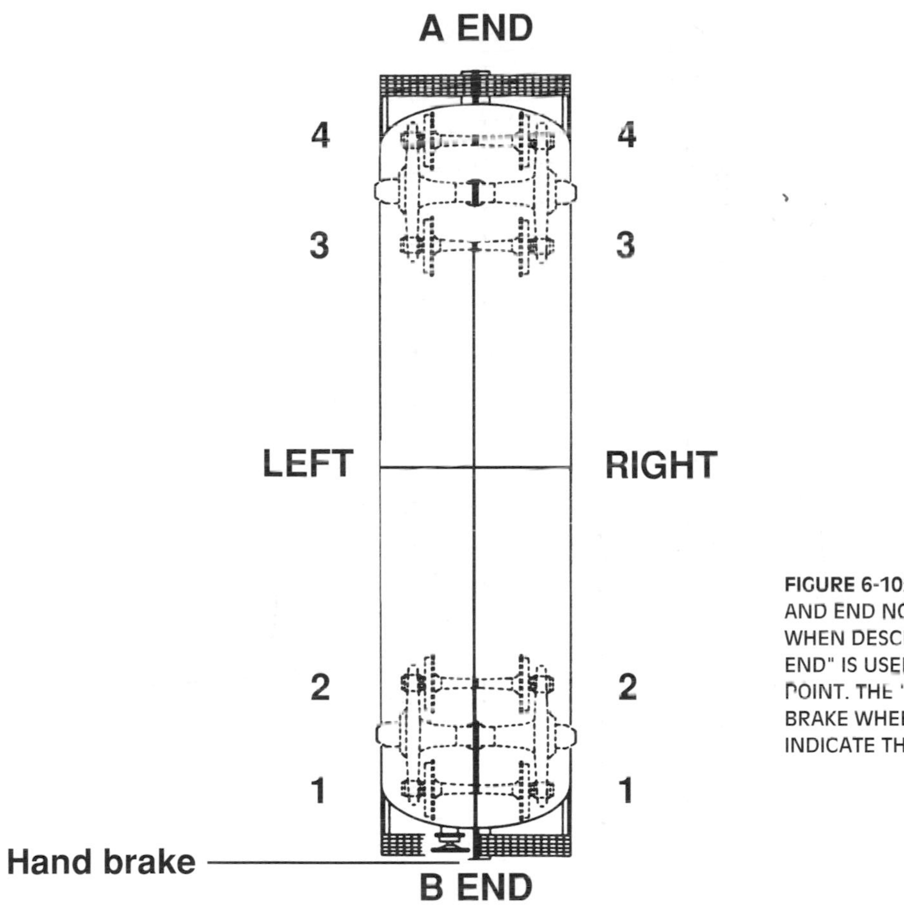

FIGURE 6-10: RAILROAD TANK CAR SIDE AND END NOMENCLATURE (TOP VIEW). WHEN DESCRIBING A TANK CAR, THE "B-END" IS USED AS THE INITIAL REFERENCE POINT. THE "B-END" IS WHERE THE HAND BRAKE WHEEL IS LOCATED; NUMBERS 1-4 INDICATE THE WHEELS.

ATMOSPHERIC AND LOW PRESSURE LIQUID STORAGE TANKS

CONTAINER SHAPE	DESCRIPTION	CONTENTS
CONE ROOF TANK	• TANK WITH VERTICAL CYLINDRICAL WALLS SUPPORTING A FIXED, INVERTED CONE ROOF. • OPERATES AT ATMOSPHERIC PRESSURE. • MAY HAVE INSULATION, PARTICULARLY FOR HEAVY FUEL OIL AND ASPHALT SERVICE. • IF CONSTRUCTED TO API 650 SPECIFICATIONS, ROOF-TO-SHELL SEAM DESIGNED TO FAIL IN CASE OF FIRE OR EXPLOSION.	• STORES FLAMMABLE, COMBUSTIBLE, AND CORROSIVE LIQUIDS.
OPEN FLOATING ROOF TANK	• WIND GIRDER AROUND TOP OF TANK SHELL. • LADDER ON ROOF. • ROOF ACTUALLY FLOATS ON LIQUID SURFACE. • SEAL AREA BETWEEN TANK SHELL AND ROOF.	• STORES FLAMMABLE AND COMBUSTIBLE LIQUIDS.

CONTAINER SHAPE	DESCRIPTION	CONTENTS
OPEN FLOATING ROOF TANK WITH GEODESIC DOME 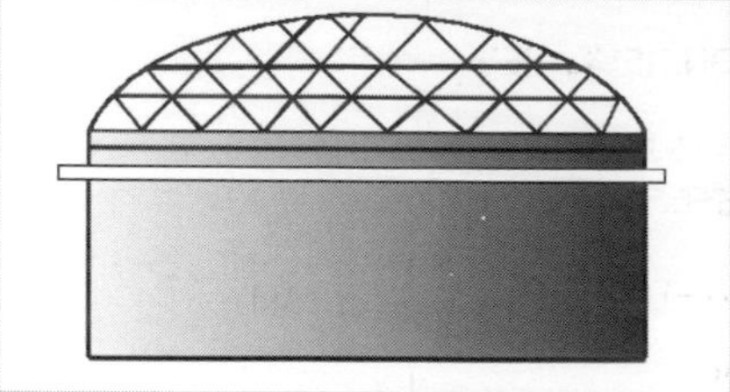	• OPEN FLOATING ROOF WITH LIGHTWEIGHT ALUMINUM GEODESIC DOME.	• STORES FLAMMABLE LIQUIDS.
COVERED FLOATING ROOF TANK 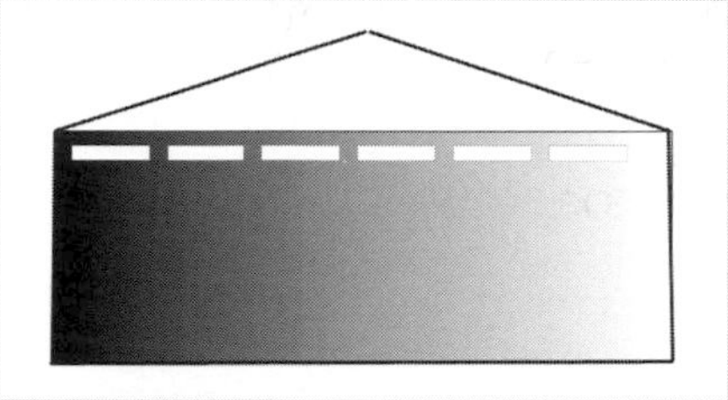	• ALSO REFERRED TO AS AN INTERNAL FLOATING ROOF. • CONE ROOF TANK WITH AN INTERNAL FLOATING ROOF. • LARGE VENTS FOUND AT THE TOP OF THE TANK SHELL. • SEAL AREA BETWEEN TANK SHELL AND ROOF.	• STORES FLAMMABLE AND COMBUSTIBLE LIQUIDS.
HORIZONTAL TANK 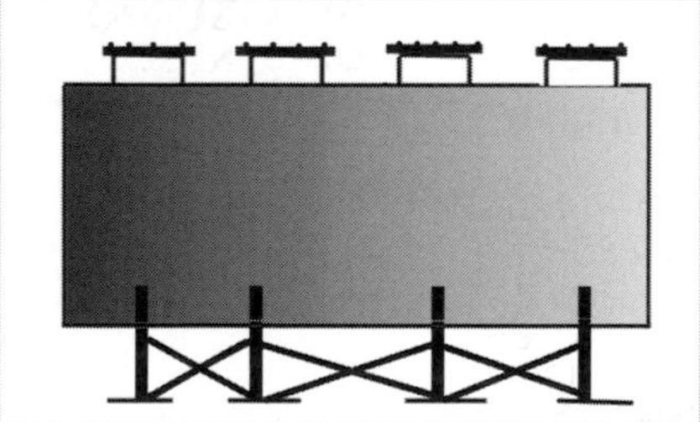	• HORIZONTAL CYLINDRICAL TANK SITTING ON LEGS, BLOCKS, ETC. • STRUCTURAL INTEGRITY OF THE SUPPORTS IS CRITICAL. • OLDER TANKS HAVE BOLTED CONSTRUCTION: TANKS SINCE 1950'S GENERALLY ARE WELDED.	• STORES FLAMMABLE AND COMBUSTIBLE LIQUIDS, CORROSIVES, POISONS, ETC.
DOME ROOF TANK 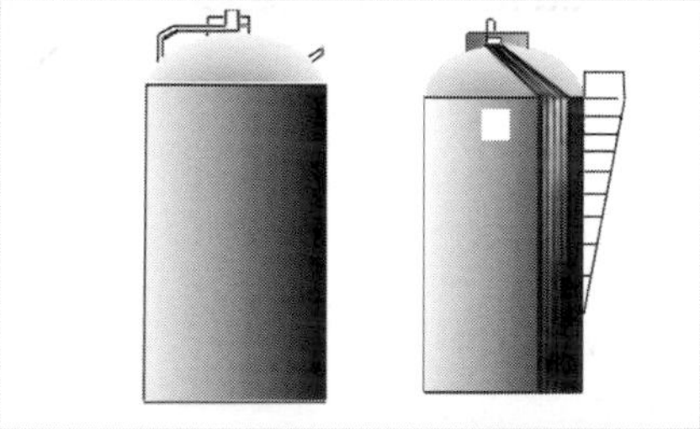	• TANK WITH VERTICAL CYLINDRICAL WALLS SUPPORTING A FIXED DOME-SHAPED ROOF. • OPERATING PRESSURE OF 2.5 TO 15 PSI. • ROOF WILL NOT ALWAYS FAIL AS DESIGNED.	• STORES FLAMMABLE AND COMBUSTIBLE LIQUIDS, FERTILIZERS, CHEMICAL SOLVENTS, ETC.
UNDERGROUND STORAGE TANK 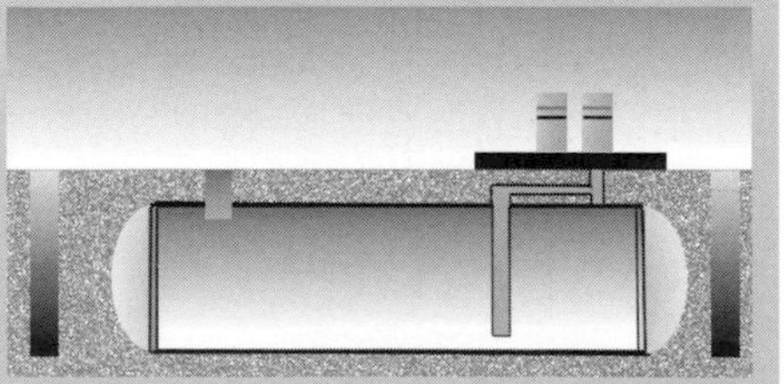	• HORIZONTAL TANK CONSTRUCTED OF STEEL, FIBERGLASS, OR STEEL WITH FIBERGLASS COATING. • ANY TANK WITH GREATER THAN 10% SURFACE AREA UNDERGROUND IS CONSIDERED AN UNDERGROUND TANK. • VISIBLE CLUES ARE VENTS, FILL POINTS, AND POTENTIAL OCCUPANCY/LOCATIONS (E.G., SERVICE STATION, FLEET MAINTENANCE).	• PRIMARILY STORES PETROLEUM PRODUCTS.

FIGURE 6-11

PRESSURIZED STORAGE VESSELS		
CONTAINER SHAPE	**DESCRIPTION**	**CONTENTS**
HIGH PRESSURE HORIZONTAL TANK 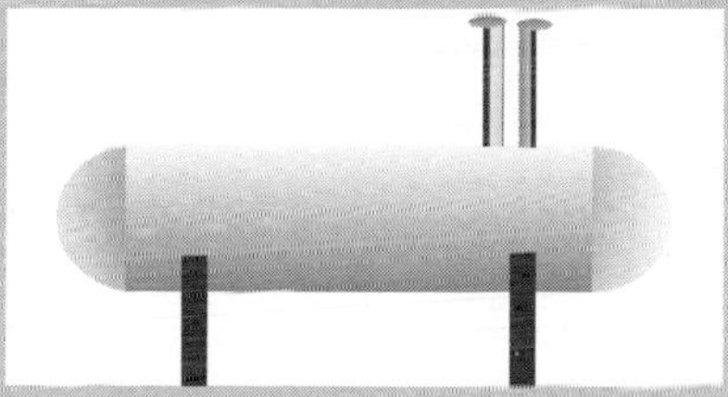	• GENERALLY SINGLE SHELL, NONINSULATED TANK. • ROUNDED ENDS INDICATE HIGH PRESSURE. • PAINTED WHITE OR HIGHLY REFLECTIVE COLOR. • SIZE VARIES WITH OCCUPANCY–1,000 TO 30,000+ GALLONS. • PRESSURES OF 100 TO 500 PSI.	• STORES LP GASES, ANYHYDROUS AMMONIA, VINYL CHLORIDE, HIGH VAPOR PRESSURE FLAMMABLE LIQUIDS.
HIGH PRESSURE SPHERICAL STORAGE TANK 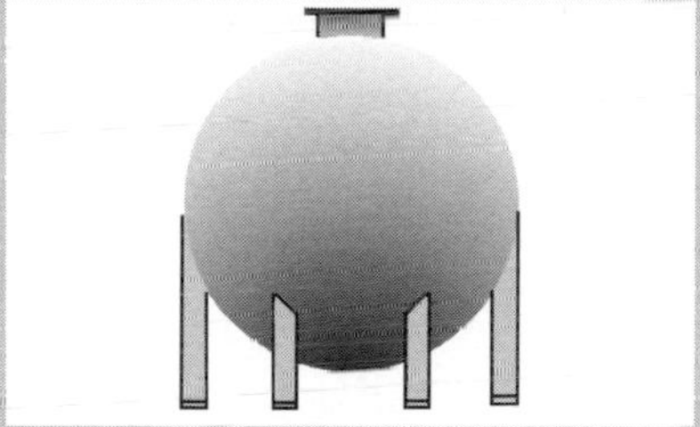	• SINGLE SHELL, NONINSULATED TANK. • PAINTED WHITE OR HIGHLY REFLECTED COLOR. • CAPACITIES TO 600,000 GALLONS. • MAY HAVE WATER SPRAY SYSTEM FOR FIRE PROTECTION. • PRESSURES OF 100 TO 500 PSI.	• STORES LP GASES, VINYL CHLORIDE.
CRYOGENIC LIQUID STORAGE TANK 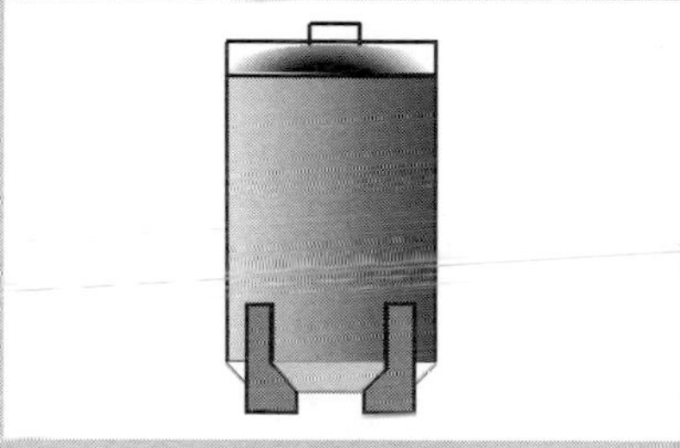	• WELL-INSULATED "THERMOS BOTTLE" DESIGN. • PRIMARILY FOUND AT HEAVY INDUSTIRAL FACILITIES, HOSPITALS, GAS PROCESSING FACILITIES, ETC. • CAPACITIES RANGE FROM SEVERAL HUNDRED TO 1,000 GALLONS AT FIXED FACILITIES. • LARGE INSULATED, REFRIGERATED, LOW PRESSURE STORAGE CONTAINERS MAY BE FOUND FOR LPG, LNG AND ANHYDROUS AMMONIA. MAY BE GREATER THAN 2,000,000 GALLONS. • DESIGN PRESSURE OF UP TO 250 PSI.	• STORES LIQUID OXYGEN (LOX), LIQUID NITROGEN, LIQUID CARBON DIOXIDE, ETC.

FIGURE 6-12

MARKINGS AND COLORS

Hazardous materials packaging and facility containment systems will often have specific markings or colors which provide some indication of their hazards or contents. This is especially true for bulk transportation containers. These clues may include color codes, container specification numbers, signal words, or even the content's name and associated hazards. At facilities, clues may include Hazard Communication markings, piping color code systems, and specific signs and/or signal words (e.g., "Hydrofluoric Acid Area").

Markings and colors can be used as a clue for hazmat recognition, identification, and classification. We will evaluate these systems based on the nature and use of the package.

FACILITY MARKINGS

NFPA 704 System. Many state and local fire codes mandate the use of the NFPA 704 marking system at all fixed facilities in their jurisdiction, including tanks and storage areas. The 704 system, which is shown in Figure 6-13, is not used on transport vehicles.

THE NFPA 704 MARKING SYSTEM

The NFPA 704 Marking System distinctively indicates the properties and potential dangers of hazardous materials. The following is an explanation of the meanings of the Quadrant Numerical Codes:

HEALTH - (Blue)

IN GENERAL, HEALTH HAZARD IN FIREFIGHTING IS THAT OF A SINGLE EXPOSURE WHICH MAY VARY FROM A FEW SECONDS UP TO AN HOUR. THE PHYSICAL EXERTION DEMANDED IN FIREFIGHTING OR OTHER EMERGENCY CONDITIONS MAY BE EXPECTED TO INTENSIFY THE EFFECTS OF ANY EXPOSURE. ONLY HAZARDS ARISING OUT OF AN INHERENT PROPERTY OF THE MATERIAL ARE CONSIDERED. THE FOLLOWING EXPLANATION IS BASED UPON PROTECTIVE EQUIPMENT NORMALLY USED BY FIREFIGHTERS:

4 MATERIALS TOO DANGEROUS TO HEALTH TO EXPOSE FIREFIGHTERS. A FEW WHIFFS OF THE VAPOR COULD CAUSE DEATH OR THE VAPOR OF LIQUID COULD BE FATAL ON PENETRATING THE FIREFIGHTER'S NORMAL FULL PROTECTIVE CLOTHING. THE NORMAL, FULL-PROTECTIVE CLOTHING AND BREATHING APPARATUS AVAILABLE TO THE AVERAGE FIRE DEPARTMENT WILL NOT PROVIDE ADEQUATE PROTECTION AGAINST INHALATION OR SKIN CONTACT WITH THESE MATERIALS.

3 MATERIALS EXTREMELY HAZARDOUS TO HEALTH, BUT AREAS MAY BE ENTERED WITH EXTREME CARE. FULL-PROTECTIVE CLOTHING INCLUDING SELF-CONTAINED BREATHING APPARATUS, COAT, PANTS, GLOVES, BOOTS AND BANDS AROUND LEGS, ARMS, AND WAIST SHOULD BE PROVIDED. NO SKIN SURFACE SHOULD BE EXPOSED.

2 MATERIALS HAZARDOUS TO HEALTH, BUT AREAS MAY BE ENTERED FREELY WITH FULL-FACE MASK AND SELF-CONTAINED BREATHING APPARATUS WHICH PROVIDES EYE PROTECTION.

1 MATERIALS ONLY SLIGHTLY HAZARDOUS TO HEALTH. IT MAY BE DESIRABLE TO WEAR SELF-CONTAINED BREATHING APPARATUS.

0 MATERIALS WHICH WOULD OFFER NO HAZARD BEYOND THAT OF ORDINARY COMBUSTIBLE MATERIAL UPON EXPOSURE UNDER FIRE CONDITIONS.

FLAMMABILITY (Red)

SUSCEPTIBILITY TO BURNING IS THE BASIS FOR ASSIGNING DEGREES WITHIN THIS CATEGORY. THE METHOD OF ATTACKING THE FIRE IS INFLUENCED BY THIS SUSCEPTIBILITY FACTOR.

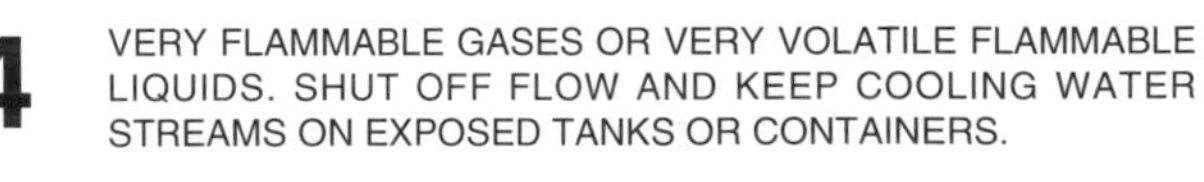

4 VERY FLAMMABLE GASES OR VERY VOLATILE FLAMMABLE LIQUIDS. SHUT OFF FLOW AND KEEP COOLING WATER STREAMS ON EXPOSED TANKS OR CONTAINERS.

3 MATERIALS WHICH CAN BE IGNITED UNDER ALMOST ALL NORMALTEMPERATURE CONDITIONS. WATER MAY BE INEFFECTIVE BECAUSE OF THE LOW FLASH POINT.

2 MATERIALS WHICH MUST BE MODERATELY HEATED BEFORE IGNITION WILL OCCUR. WATER SPRAY MUST BE USED TO EXTINGUISH THE FIRE BECAUSE THE MATERIAL CAN BE COOLED BELOW ITS FLASH POINT.

1 MATERIALS THAT MUST BE PREHEATED BEFORE IGNITION CAN OCCUR. WATER MAY CAUSE FROTHING IF IT GETS BELOW THE SURFACE OF THE LIQUID AND TURNS TO STEAM. HOWEVER, WATER FOG GENTLY APPLIED TO THE SURFACE WILL CAUSE A FROTHING WHICH WILL EXTINGUISH THE FIRE.

0 MATERIALS THAT WILL NOT BURN.

REACTIVITY (STABILITY) (Yellow)

THE ASSIGNMENT OF DEGREES IN THE REACTIVITY CATEGORY IS BASED UPON THE SUSCEPTIBILITY OF MATERIALS TO RELEASE ENERGY EITHER BY THEMSELVES OR IN COMBINATION WITH WATER. FIRE EXPOSURE WAS ONE OF THE FACTORS CONSIDERED ALONG WITH CONDITIONS OF SHOCK AND PRESSURE.

4 MATERIALS WHICH (IN THEMSELVES) ARE READILY CAPABLE OF DETONATION OR OF EXPLOSIVE DECOMPOSITION OR EXPLOSIVE REACTION AT NORMAL TEMPERATURES AND PRESSURES. INCLUDES MATERIALS WHICH ARE SENSITIVE TO MECHANICAL OR LOCALIZED THERMAL SHOCK. IF A CHEMICAL WITH THIS HAZARD RATING IS IN AN ADVANCED OR MASSIVE FIRE, THE AREA SHOULD BE EVACUATED.

3 MATERIALS WHICH (IN THEMSELVES) ARE CAPABLE OF DETONATION OR OF EXPLOSIVE DECOMPOSITION OR EXPLOSIVE REACTION BUT WHICH REQUIRE A STRONG INITIATING SOURCE OR WHICH MUST BE HEATED UNDER CONFINEMENT BEFORE INITIATION. INCLUDES MATERIALS WHICH ARE SENSITIVE TO THERMAL OR MECHANICAL SHOCK AT ELEVATED TEMPERATURES AND PRESSURES OR WHICH REACT EXPLOSIVELY WITH WATER WITHOUT REQUIRING HEAT OR CONFINEMENT. FIREFIGHTING SHOULD BE DONE FROM AN EXPLOSIVE-RESISTANT LOCATION.

2 MATERIALS WHICH (IN THEMSELVES) ARE NORMALLY UNSTABLE AND RAPIDLY UNDERGO VIOLENT CHEMICAL CHANGE BUT DO NOT DETONATE. INCLUDES MATERIALS WHICH CAN UNDERGO CHEMICAL CHANGE WITH RAPID RELEASE OF ENERGY AT NORMAL TEMPERATURES AND PRESSURES OR WHICH CAN UNDERGO VIOLENT CHEMICAL CHANGE AT ELEVATED TEMPERATURES AND PRESSURES. ALSO INCLUDES THOSE MATERIALS WHICH MAY REACT VIOLENTLY WITH WATER OR WHICH MAY FORM POTENTIALLY EXPLOSIVE MIXTURES WITH WATER. IN ADVANCE OR MASSIVE FIRES, FIREFIGHTING SHOULD BE DONE FROM A SAFE DISTANCE OR FROM A PROTECTED LOCATION.

1 MATERIALS WHICH (IN THEMSELVES) ARE NORMALLY STABLE BUT WHICH MAY BECOME UNSTABLE AT ELEVATED TEMPERATURES AND PRESSURES OR WHICH MAY REACT WITH WATER WITH SOME RELEASE OF ENERGY BUT NOT VIOLENTLY. CAUTION MUST BE USED IN APPROACHING THE FIRE AND APPLYING WATER.

0 MATERIALS WHICH (IN THEMSELVES) ARE NORMALLY STABLE EVEN UNDER FIRE EXPOSURE CONDITIONS AND WHICH ARE NOT REACTIVE WITH WATER. NORMAL FIREFIGHTING PROCEDURES MAY BE USED.

SPECIAL INFORMATION (White)

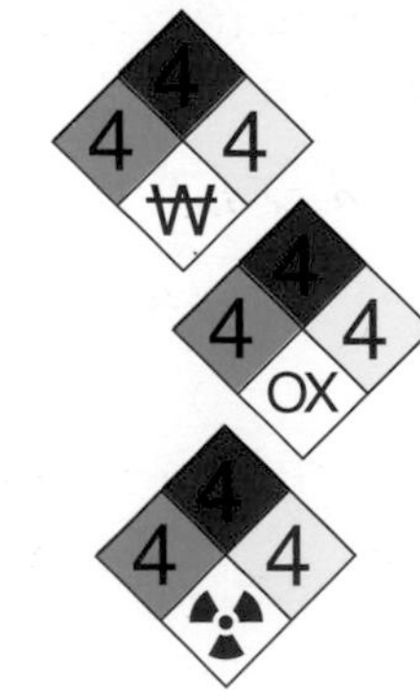

MATERIALS WHICH DEMONSTRATE UNUSUAL REACTIVITY WITH WATER SHALL BE IDENTIFIED WITH THE LETTER W WITH A HORIZONTAL LINE THROUGH THE CENTER (W).

MATERIALS WHICH POSSESS OXIDIZING PROPERTIES SHALL BE IDENTIFIED BY THE LETTERS OX.

MATERIALS POSSESSING RADIOACTIVITY HAZARDS SHALL BE IDENTIFIED BY THE STANDARD RADIOCTIVITY SYMBOL.

FIGURE 6-13
NATIONAL FIRE PROTECTION ASSOCIATION

NOTES

Hazard Communication Marking Systems. There are several hazard communication systems which are used within industry. They may be known by several titles, including the Hazardous Materials Information System (HMIS) and the Hazardous Materials Identification Guide (HMIG). Similar to NFPA 704, these systems provide a standardized hazard rating scale from 0 (minimal hazard) to 4 (extreme hazard) for health, flammability, and reactivity. In addition, there are alphabetical designations for the required level of personal protective clothing. These markings may be found at facility entrances, room entrances, and directly on containers.

Polychlorinated Biphenyls (PCB). A great deal of attention has been focused on the hazards of polychlorinated biphenyls (PCB's) in recent years. Although severely restricted, PCB-laden electrical transformers may still be found in use today. These transformers are required to be marked with black-on-yellow warning labels. Unfortunately, these labels are small (6 inches by 6 inches or less) and often cannot be seen from a safe distance (see Figure 6-14).

Only those transformers containing a concentration of 50 ppm or greater must be marked. As a general rule, regard all transformer incidents as PCB related until tests prove otherwise.

Unfortunately, there are situations in which high concentrations of PCB's can be encountered with no warning labels. Examples include fluorescent light ballasts manufactured prior to 1977, hydraulic and vacuum pump systems, some adhesives, microscope immersion oils, and the manufacture of "carbonless" carbon paper. Although the hazards of PCB's must be recognized, greater hazards are created if the PCB fluids burn and break down into toxic dioxin and furan compounds.

FIGURE 6-14: PCB-CONTAMINATED TRANSFORMERS CONTAINING MORE THAN A 50 PPM CONCENTRATION MUST BE MARKED WITH THIS LABEL.

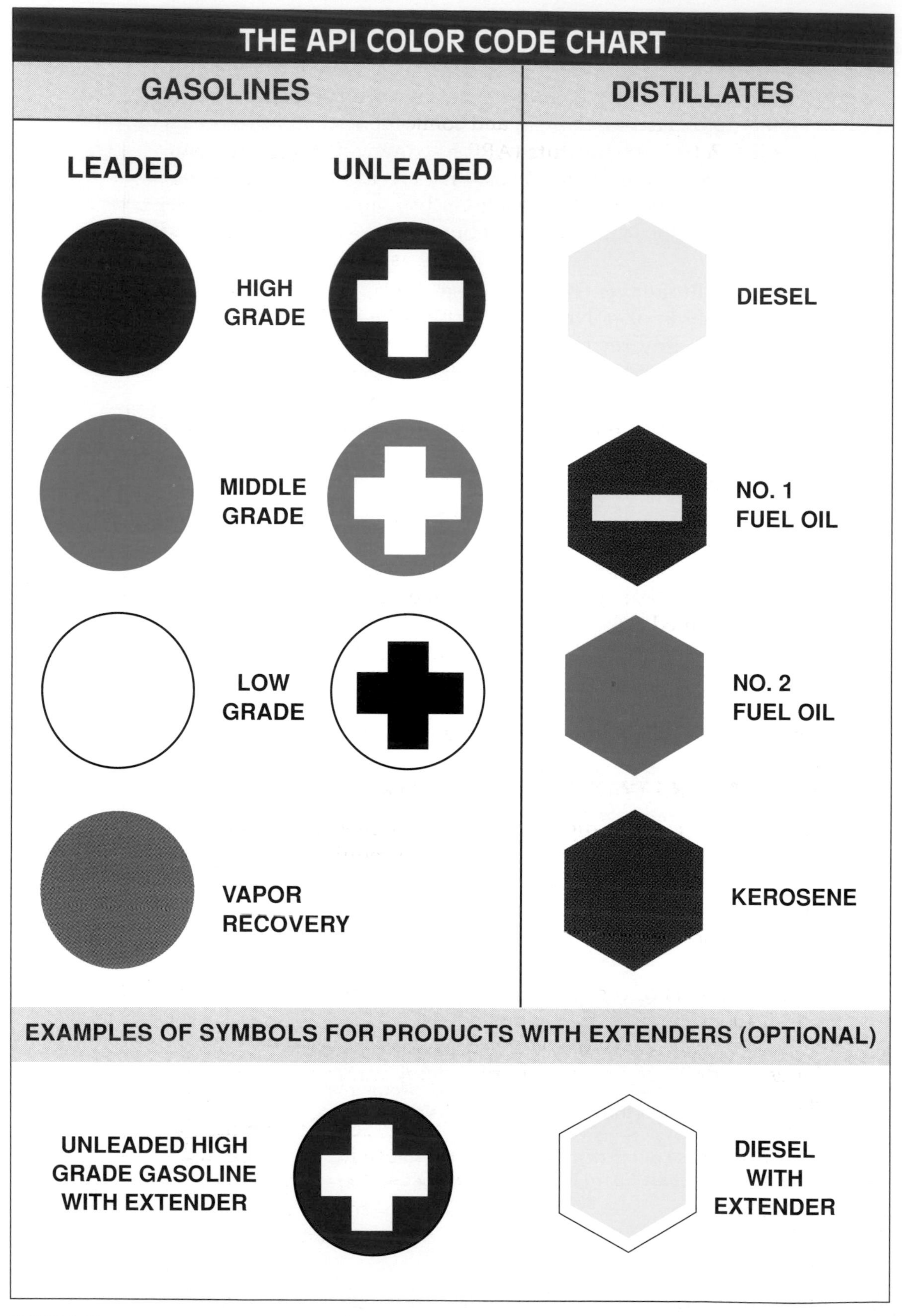

FIGURE 6-15

SOURCE: AMERICAN PETROLEUM INSTITUTE

API Petroleum Products Color Code Chart. Bulk liquid petroleum marketing and storage facilities which move more than one product typically use a color code and symbol system which denotes the different grades of gasoline and fuel oils. These markings are typically found at the loading rack and at transfer valves and connections.

The American Petroleum Institute (API) has developed a recommended uniform marking system which is shown in Figure 6-15. This system classifies the many hydrocarbon fuels and blends into leaded and unleaded gasoline (regular, premium, super), gasoline additives, and distillates and fuel oils. This marking system has also been adopted by several states for use at liquid petroleum facilities and service stations for identifying piping and connections at loading racks, as well as the fill point connections for service station tanks.

U.S. Military Marking System. This marking system will primarily be found on both structures and containers at U.S. military facilities. The system consists of both fire and chemical hazard symbols (see Figure 6-16). The four fire division symbols parallel Division 1.1 through 1.4 explosive materials.

Other Marking Systems. Color codes are also used at bulk LPG facilities. Liquid product lines will often be color-coded either dark blue or orange while vapor lines are light blue or yellow. Liquid lines will ALWAYS be larger in diameter than vapor lines. Fire protection piping will normally be painted red.

The OSHA Hazard Communication Standard and many state and local Right-to-Know regulations require the marking of hazardous materials storage vessels, piping, and process units within industry. However, there are no universal warning labels or systems required.

BULK PACKAGING AND TRANSPORTATION MARKINGS

Identification Number. Four-digit identification numbers are assigned to each proper hazmat shipping name. They are required on shipping papers and on, or near, bulk transport container placards. This includes cargo tanks, rail cars, and portable tanks transporting hazmats. They can also be displayed on other bulk containers, such as vans or railroad hopper cars.

Acceptable methods of displaying this marking are shown in Figure 6-17. Identification numbers are prohibited on DOT explosives, dangerous, radioactive, and subsidiary hazard placards. The identification number must be displayed on the supplemental orange panel for these materials.

When viewing shipping papers or MSDS's, the identification number may be found with the prefix "UN" or "NA." Those identification numbers preceded by the letters "UN (United Nations)" are appropriate for domestic and international transportation. Those preceded by the letters "NA (North America)" are not recognized for international transportation, except to and from Canada.

Inhalation Hazard Markings. The words "Inhalation Hazard" indicate that the material is considered hazardous by inhalation (e.g., anhydrous ammonia).

NOTES

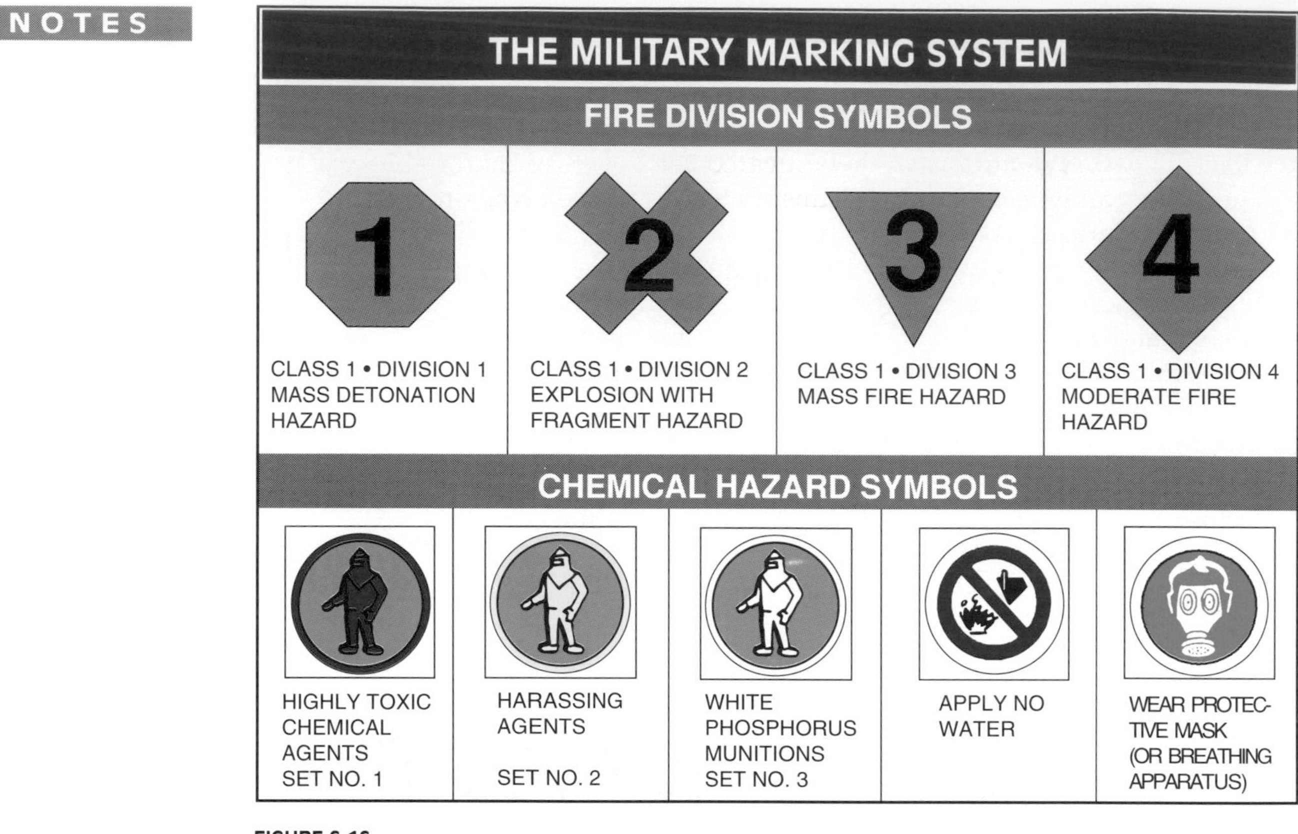

FIGURE 6-16

FIGURE 6-17: THE FOUR-DIGIT IDENTIFICATION NUMBER MAY BE DISPLAYED ON BULK CONTAINERS IN ONE OF THREE WAYS.

Marine Pollutant Mark. These markings are displayed on both sides and both ends of bulk packages of materials designated on the shipping papers as a marine pollutant, except when the container is properly placarded (e.g., Poison, Flammable Liquid, etc.).

Elevated Temperature Materials. Elevated temperature materials are materials which, when offered for transportation in a bulk container, are:

- Liquids at or above 212°F (100°C).
- Liquids with a flash point at or above 100°F (37.8°C) that are intentionally heated and are transported at or above their flash point.
- Solids at a temperature at or above 464°F (240°C).

Except for a bulk container transporting molten aluminum or molten sulfur (which must be marked "MOLTEN ALUMINUM" or "MOLTEN SULFUR," respectively), the container must be marked on each side and each end with the word "HOT" in black or white lettering on a contrasting background. The word "HOT" may be displayed on the bulk packaging itself or in black lettering on a white square-on-point configuration similar in size to the placard.

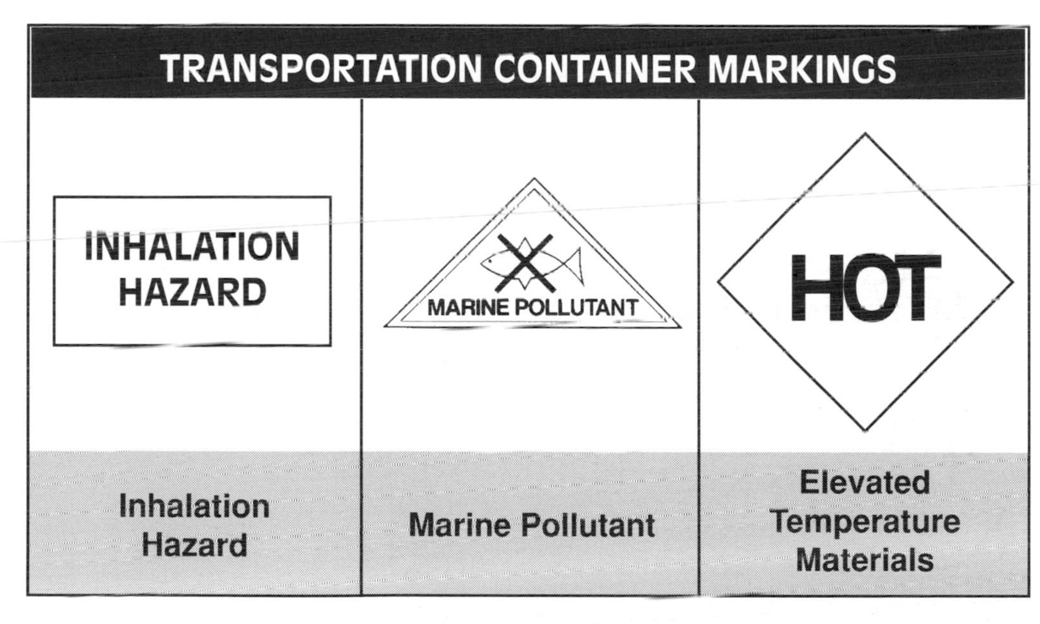

FIGURE 6-18: TRANSPORTATION CONTAINER MARKINGS FOR SPECIAL SITUATIONS

Railroad Tank Car Markings and Colors. There are a number of markings on railroad tank cars that can be used to gain knowledge about the tank itself and its contents. This information would be useful in evaluating the condition of the container. These markings include the following:

- *Commodity Stencil.* Tank cars transporting anhydrous ammonia, ammonia solutions with more than 50% ammonia, Division 2.1 material (flammable gas), or a Division 2.3 material (poison gas) must

NOTES

have the name of the commodity marked on both sides of the tank in 4-inch (102 mm) minimum letters.

- ❑ *Reporting Marks and Number.* Railroad cars are marked with a set of initials and a number (e.g., GATX 12345) stenciled on both sides (left end as one faces the tank) and both ends of the car. These markings can be used to obtain information about the contents of the car from the railroad, the shipper, or CHEMTREC. The last letter in the reporting marks has special meaning:
  - "X" indicates a rail car is not owned by a railroad (for a rail car, the lack of an "X" indicates railroad ownership).
  - "Z" indicates a trailer.
  - "U" indicates a container.

  Some shippers and car owners are also stenciling reporting marks and numbers on top of tank cars to assist in identification if the car is laying on its side as a result of an accident.
- ❑ *Capacity Stencil.* Shows the volume of a tank car in gallons (and sometimes liters), as well as in pounds (and sometimes kilograms). These markings are found on the sides of the car under the reporting marks. For certain tank cars (e.g., DOT 105, DOT 109, DOT 112, DOT 114, and DOT 111A100W4), the water capacity/water weight of the tank car is stenciled near the center of the car.
- ❑ *Specification Marking.* The specification marking indicates the standards to which a tank car was built. These markings will be on both sides of the tank car (right end as one faces the tank) and provide the following information:
  - Approving Authority
  - Class Number
  - Separator/Delimiter Character (significant in certain tank cars)
  - Tank test pressure
  - Type of material used in construction
  - Type of weld used
  - Fittings/material/lining

  See Figure 6-20 for additional information.
- ❑ *Hydrogen Cyanide Tank Cars.* Pay particular attention to railroad tank cars painted white with a red stripe running the length and around the ends of the car. This tank car transports hydrogen cyanide (HCN), a flammable and poisonous liquid. These color and design schemes are an industry standard and are not DOT requirements.

Cargo Tank Truck Markings and Colors. Like tank cars, cargo tank trucks also have various markings, including company names and logos, vehicle identification numbers, and the manufacturer's specification plate.

- ❑ *Manufacturer's Specification Plate.* Mounted at the left front of the tank since 1985, although units built prior to 1985 will have it on the right front of the container. The plate will provide the DOT container specification number (e.g., MC-306/DOT-406), date of manufacture

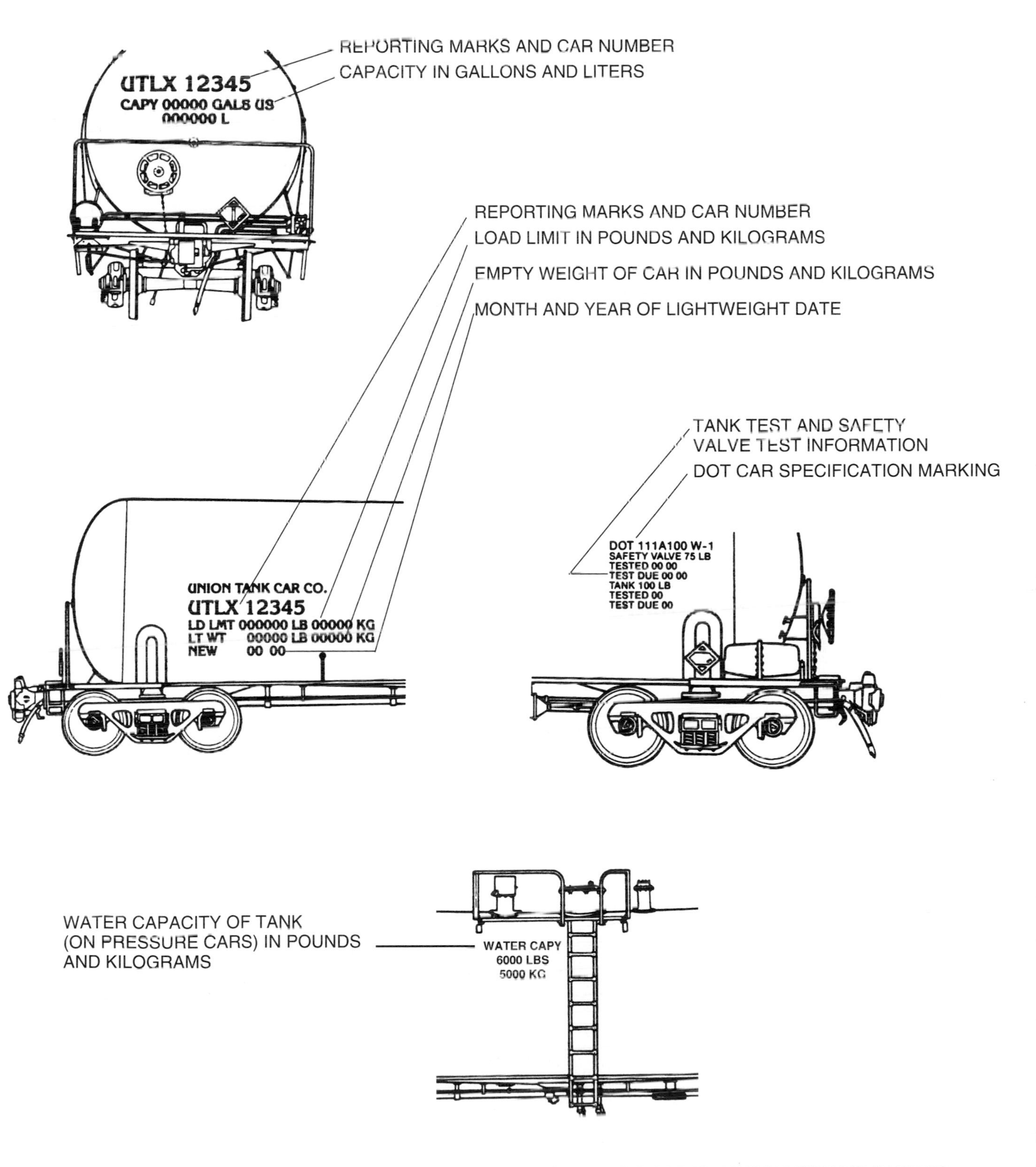

FIGURE 6-19: RAILROAD TANK CAR MARKINGS
UNION PACIFIC RAILROAD

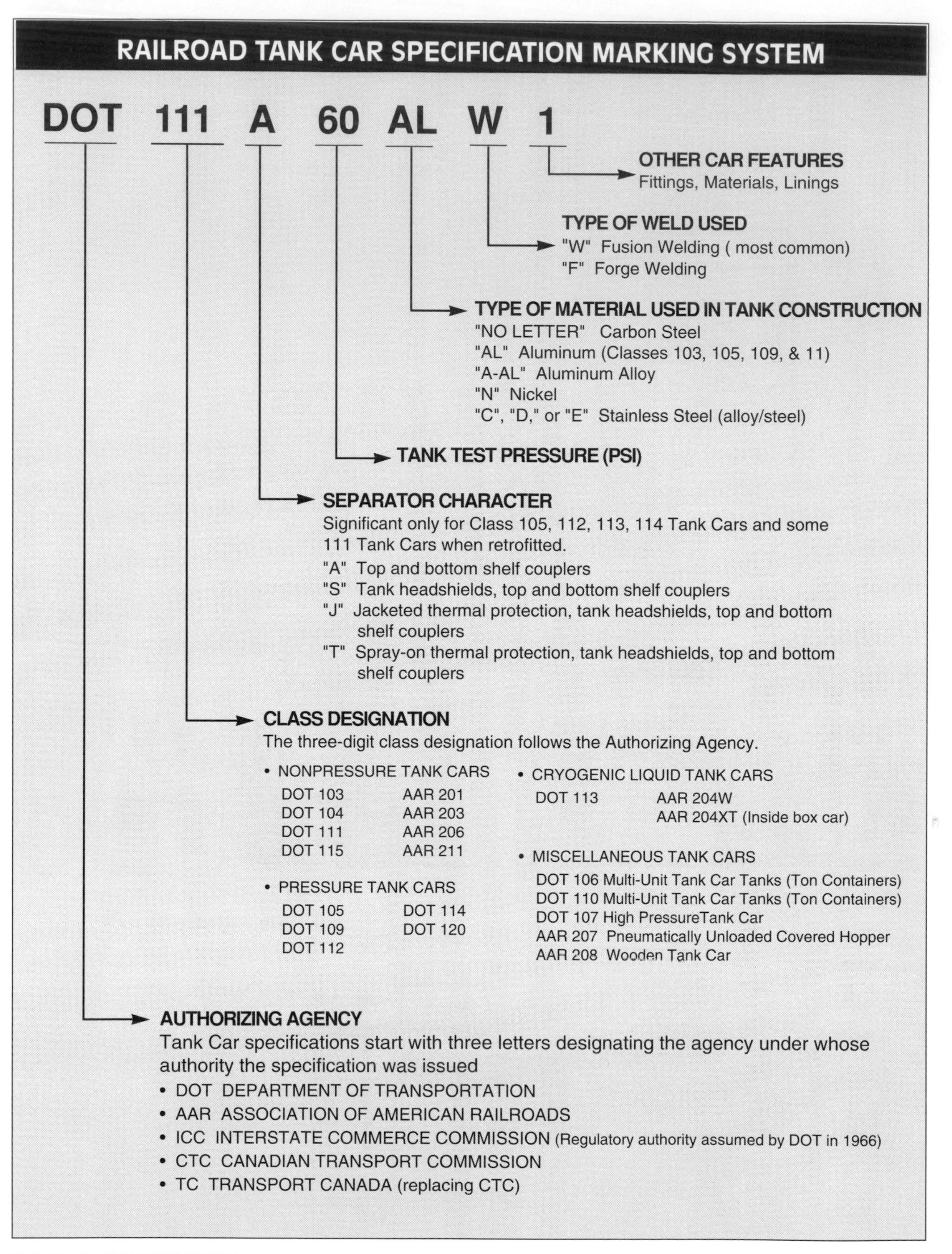

FIGURE 6-20: DOT SPECIFICATION MARKINGS FOR RAILROAD TANK CARS

NOTES

and test, shell material, number of compartments and their capacity, and the maximum product load (see Figure 6-21). Codes for shell materials include the following:

AL	Aluminum
CS	Carbon Steel
HSLA	High Strength Low Alloy Steel
HSLA-QT	High Strength Low Alloy, Quench Tempered Steel
MS	Mild Steel
SS	Stainless Steel

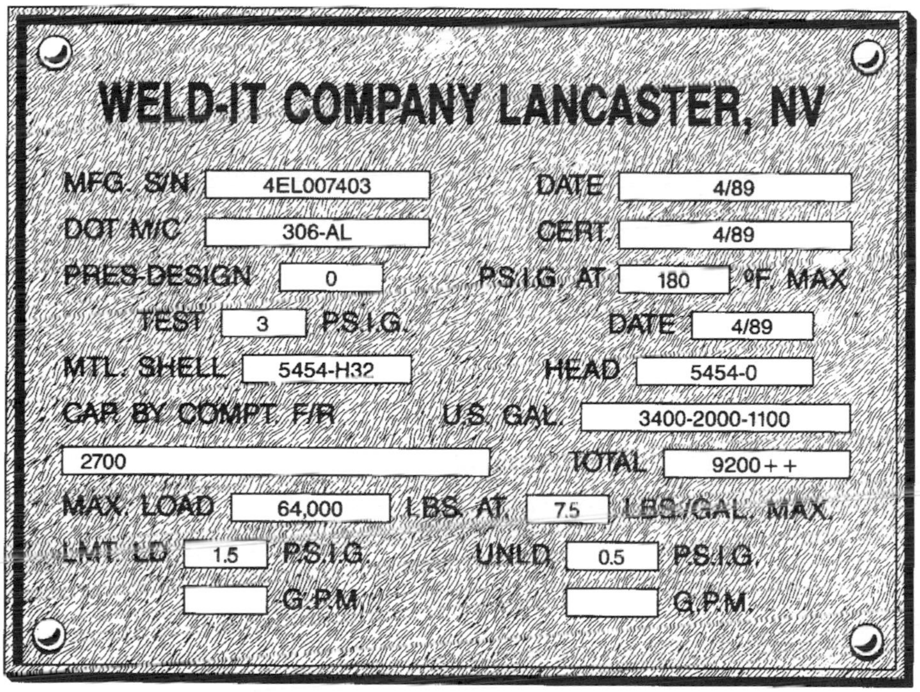

WELD-IT COMPANY LANCASTER, NV
MFG. S/N 4EL007403 DATE 4/89
DOT M/C 306-AL CERT. 4/89
PRES-DESIGN 0 P.S.I.G. AT 180 °F. MAX
TEST 3 P.S.I.G. DATE 4/89
MTL. SHELL 5454-H32 HEAD 5454-0
CAP BY COMPT. F/R U.S. GAL. 3400-2000-1100
2700 TOTAL 9200++
MAX. LOAD 64,000 LBS. AT 7.5 LBS./GAL. MAX.
LMT. LD 1.5 P.S.I.G. UNLD 0.5 P.S.I.G.
G.P.M. G.P.M.

DON SELLERS

FIGURE 6-21: THE MANUFACTURER'S SPECIFICATION PLATE WILL BE MOUNTED AT THE FRONT RIGHT OR LEFT SIDE OF A CARGO TANK TRUCK

Some cargo tanks are designed to multiple container specifications which allow them to transport more than one type of commodity. Common multi-purpose configurations include combination MC 306/DOT 406 and MC 307/DOT 407 units, or combination MC 307/DOT 407 and MC 312/DOT 412 units.

In addition to the manufacturer's specification plate, these tanks have a second "multi-purpose" plate that identifies the specification under which the cargo tank is being operated. These plates are color-coded, as are the fittings that are added to make the cargo tank meet the respective specifications. A sliding shield exposes the specification plate currently in use. However, these plates may move as a result of an accident or may not be properly positioned by the vehicle driver. The respective color code is:

NOTES

MC 306/DOT 406	Red Plate and Fittings
MC 307/DOT 407	Green Plate and Fittings
MC 312/DOT 412	Yellow Plate and Fittings
Nonspecification Tank	Blue Plate and Fittings

- *Tank Color.* MC 331 specification tanks are required to have the upper two-thirds of the tank painted white. Corrosive tank trucks (MC 312/DOT 412) will often have a contrasting color band on the tank in line with the dome covers and overturn protection. This band is usually a corrosive-resistant paint or rubber material.

Intermodal Portable Tank Markings. There are a number of markings on intermodal tank containers that can be used to gain knowledge about the tank design and construction features. These markings include the following:

- *Reporting Marks and Number.* Intermodal portable tank containers are registered with the International Container Bureau in France. They must be marked with reporting marks and a tank number. The initials indicate ownership of the tank, and the tank number identifies the specific tank. These markings are generally found on the right-hand side of the tank (as you face it from either side), and on both ends (see Figure 6-22).
- *Specification Marking.* The specification marking indicates the standards to which a portable tank was built. These markings will be on both sides of the tank, generally near the tank's reporting marks and number (see Figure 6-22). Examples of specification markings are:
    - IM 101
    - IM 102
    - Spec. 51
- *DOT Exemption Marking.* Exemptions are sometimes authorized from DOT regulations. In these cases, the outside of each package/container must be plainly and durably marked "DOT-E" followed by the exemption number assigned (e.g., DOT E8623). On intermodal tanks, these markings must be in 2-inch letters (see Figure 6-22).
- *AAR-600 Marking.* For interchange purposes in rail transportation, intermodal containers should conform to the requirements of Section 600—"Specification for Acceptability of Tank Containers" of the Association of American Railroads (AAR) Specifications for Tank Cars. Tanks meeting these requirements will display the "AAR 600" marking in 2-inch letters on both sides near the tank's reporting marks and number (see Figure 6-22). The "AAR 600" marking indicates tanks that can be used for regulated materials.
- *Country, Size, and Type Markings.* The tank will display a size/type code (see Figure 6-22—US-2275). The country code (two or three letters) indicates the tank's country of registry.

NOTES

The four-digit size/type code follows the country code. The first two numbers jointly indicate the container's length and height. The second pair of numbers is the type code which indicates the pressure range of the tank.

Common Country Codes

BM (BER)	Bermuda	NLX	Netherlands
CH (CHS)	Switzerland	NZX	New Zealand
DE	West Germany	PA (PNM)	Panama
DKX	Denmark	PIX	Philippines
FR (FXX)	France	PRC	People's Republic of China States
GB	Great Britain	RCX	People's Republic of China (Taiwan)
HKXX	Hong Kong	SGP	Singapore
ILX	Israel	SXX	Sweden
IXX	Italy	US (USA)	United States
JP (JXX)	Japan		
KR	Korea		
LIB	Liberia		

Common Size Codes

20 = 20 ft. (8 ft. high)
22 = 20 ft. (8 ft. 6 inches high)
24 = 20 ft. (> 8 ft. 6 inches high)

Common Type Codes—Maximum Allowable Working Pressure

Nonhazardous Commodities

70 = < 0.44 (6.4 psig) Bar test pressure
71 = 0.44 (6.4 psig) to 1.47 (21.3 psig) Bar test pressure
72 = 1.47 (21.3 psig) to 2.94 (42.6 psig) Bar test pressure
73 = spare

Hazardous Commodities

74 = < 1.47 (21.3 psig) Bar test pressure
75 = 1.47 (21.3 psig) to 2.58 (37.4 psig) Bar test pressure
76 = 2.58 (37.4 psig) to 2.94 (42.6 psig) Bar test pressure
77 = 2.94 (42.6 psig) to 3.93 (57.0 psig) Bar test pressure
78 = > 3.93 (57.0 psig) Bar test pressure
79 = spare

- *Dataplate.* Additional technical, approval, and operational data can be found on the dataplate which is permanently attached to the tank or frame.

FIGURE 6-22

Pipelines. Pipelines are the second largest hazmat transportation mode and often cross over or under roads, waterways, and railroads. At each of these crossover locations, a marker should identify the right-of-way. Although its format and design may be different, all markers are required to provide the pipeline contents (e.g., natural gas, propane, liquid petroleum products, etc.), the pipeline operator, and an emergency telephone number (see Figure 6-23).

The pipeline emergency telephone number goes to a control room where an operator monitors pipeline operations and can start emergency shutdown procedures. It should be stressed that even when a pipeline is immediately shut down, product backflow may continue for several hours until the product drains to the point of release.

Most gas pipelines are dedicated to one product (e.g., natural gas, butadiene, anhydrous ammonia, etc.). However, liquid petroleum transmission pipelines may carry several different petroleum products simultaneously. Figure 6-24 shows a typical liquid products shipping cycle. Liquid pipeline personnel normally refer to product flows in terms of "barrels" rather than gallons (NOTE: 1 barrel equals 42 gallons).

Product flows through many transmission pipeline systems are monitored through a computerized pipeline SCADA System (Supervisory Control and Data Acquisition System). The exact injection date and time of the particular product into the pipeline is noted and its delivery date/time is projected. As the product gets close to its destination, a sensor in the line signals the arrival of the shipment. The SCADA System provides pipeline personnel with the ability to monitor pipeline flows and pressures and initiate emergency shutdown procedures in the event of a release.

NOTES

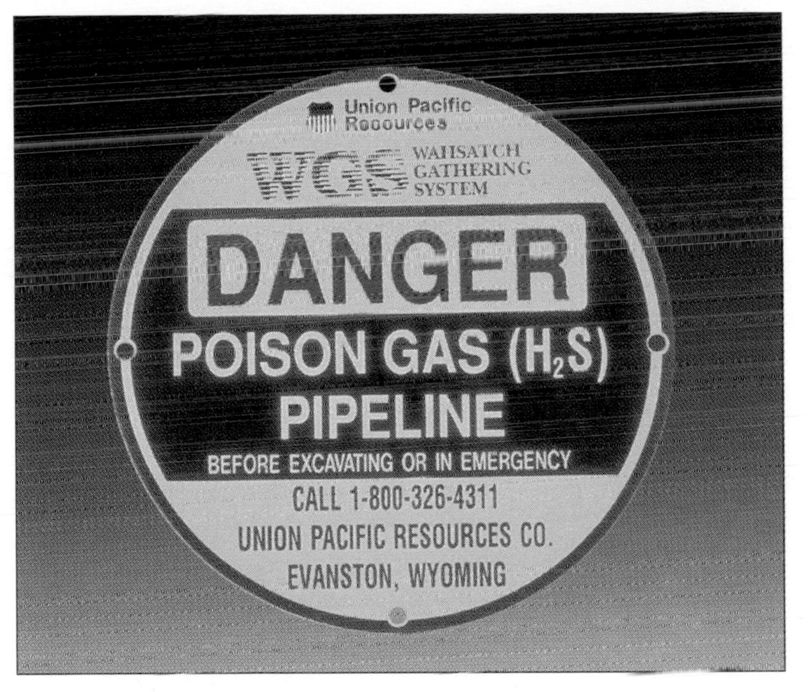

FIGURE 6-23: PIPELINE MARKERS MUST PROVIDE THE PIPELINE CONTENTS, THE PIPELINE OPERATOR, AND AN EMERGENCY TELEPHONE NUMBER

For liquid petroleum pipelines, there is usually no physical separator (e.g., sphere or pig) between different products. Rather, the products are allowed to "co-mingle." This interface can range from a few barrels to several hundred, depending upon the pipeline size and products involved. Verification of the arrival of the shipment is made by examining a sample of the incoming batch through color, appearance, and/or chemical characteristics.

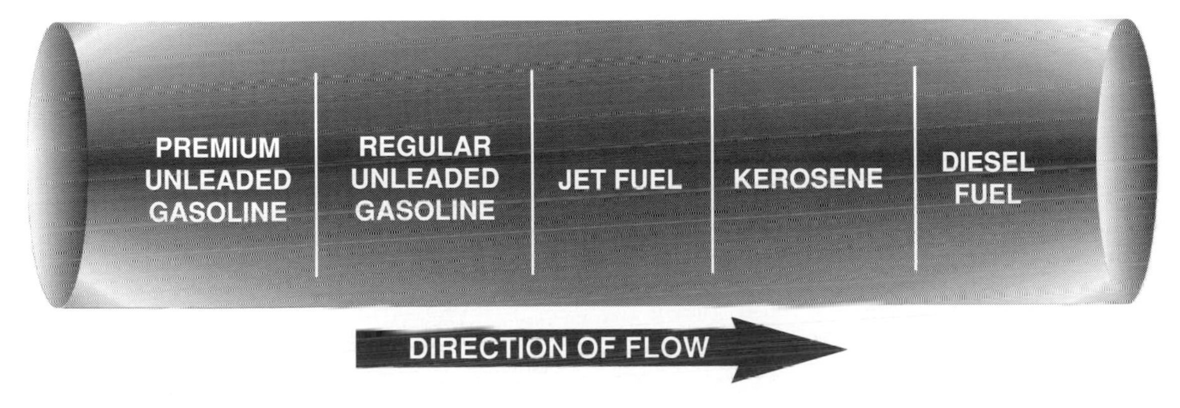

FIGURE 6-24: TYPICAL PRODUCT CYCLE FOR A LIQUID PETROLEUM PIPELINE

NOTES

NONBULK PACKAGE MARKINGS

Agricultural Chemicals and Pesticide Labels.

Individual nonbulk packages, particularly those storing pesticides and agricultural chemicals, will display useful information for identification and hazard assessment. These container markings include:

- **Toxicity Signal Words**—The signal word indicates the relative degree of acute toxicity. Located in the center of the front label panel, it is one of the most important label markings. The three toxic categories are high toxicity, moderate toxicity, and low toxicity (see Figure 6-25).
- **Statement of Practical Treatment**—Located near the signal word on the front panel, it is also referred to as the "first aid statement" or "note to physician." It may have precautionary information as well as emergency procedures for exposures. Antidote and treatment information may also be added.
- **Physical or Chemical Hazard Statement**—A statement displayed on a side panel, as necessary. It will list special flammability, explosion, or chemical hazards posed by the product.
- **Product Name**—The brand or trade name is printed on the front panel. If the product name includes the term "technical," as in Parathion-Technical, it generally indicates a highly concentrated pesticide with 70% to 99% active ingredients.
- **Ingredient Statement**—All pesticide labels must have statements which break down the chemical ingredients by their relative percentages or as pounds per gallon of concentrate. Active ingredients are the active chemicals within the mixture. They must be listed by chemical name, and their common name may also be shown. Inert ingredients have no pesticide activity and are usually not broken into specific components, only total percentage. A number of agricultural chem products, particularly those used in the home, use flammable products (e.g., propane) only as the inert ingredients and may not be easy to identify.
- **Environmental Information**—The label may provide information on both the storage and disposal of the product, as well as environmental or wildlife hazards that could occur. This information can be most useful when planning clean-up and disposal after a fire or spill.
- **EPA Registration Number**—This number is required for all ag chems and pesticide products marketed in the United States. It is one of three ways to positively identify a pesticide. The other ways are by the product name or by the chemical ingredient statement. The registration number will appear as a two- or three-section number and indicates the data described in Figure 6-26. When relaying this number, make sure you include each dash. A U.S. Department of Agriculture number may appear on products registered before 1970.
- **EPA Establishment Number**—The location where the product was manufactured. This number has little significance to ERP.

NOTES

TOXICITY SIGNAL WORDS	
LEVEL OF TOXICITY	SIGNAL WORD
HIGH  (skull and crossbones)	DANGER, POISON Skull and Crossbones Symbol
MODERATE	WARNING
LOW	CAUTION

FIGURE 6-25: TOXICITY SIGNAL WORDS FOUND ON AG CHEM AND PESTICIDE CONTAINERS

EPA REGISTRATION NUMBER

thoroughly before discarding in trash.
NOTICE: Buyer assumes all responsibil
accordance with directions.

Chevron Chemical
Company © 1989
Ortho Consumer
Products Division
P.O. Box 5047
San Ramon CA
94583-0947
Product 5467
Form C217-C
EPA Reg. No. 239-2491-AA
EPA Est. 239-MO-1
Made in USA

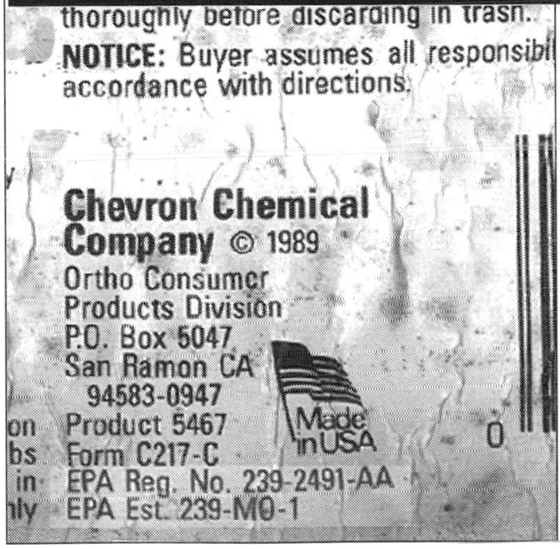

12345-6789-11

GROUP 1 GROUP 2 GROUP 3

GROUP 1 – Manufacturer

GROUP 2 – Specific Product

GROUP 3 – Locations Where Product May Be Used

FIGURE 6-26: EPA REGISTRATION NUMBER

When faced with an incident involving poisonous or toxic containers, obtain an uncontaminated container or label for reference whenever possible. When civilian or emergency services personnel are transported to a medical facility as a result of a suspected exposure, ensure proper identification and medical care by also forwarding an uncontaminated label or container.

Cylinder Color Codes. A number of gases are stored and transported in compressed gas cylinders. Although there are several voluntary color schemes, none are mandatory. The only reliable method to identify cylin-

NOTES

der contents is to check the attached label. Unfortunately, if cylinders are exposed to fire, the labels may burn off and further hamper identification.

Regulatory Markings. Under DOT regulations, the packaging manufacturer shall mark every container that is required to conform to UN standards in a durable and clearly visible manner (single line or multiple lines). Figure 6-27 outlines the sequence of markings, depending upon the type and use of the packaging.

FIGURE 6-27

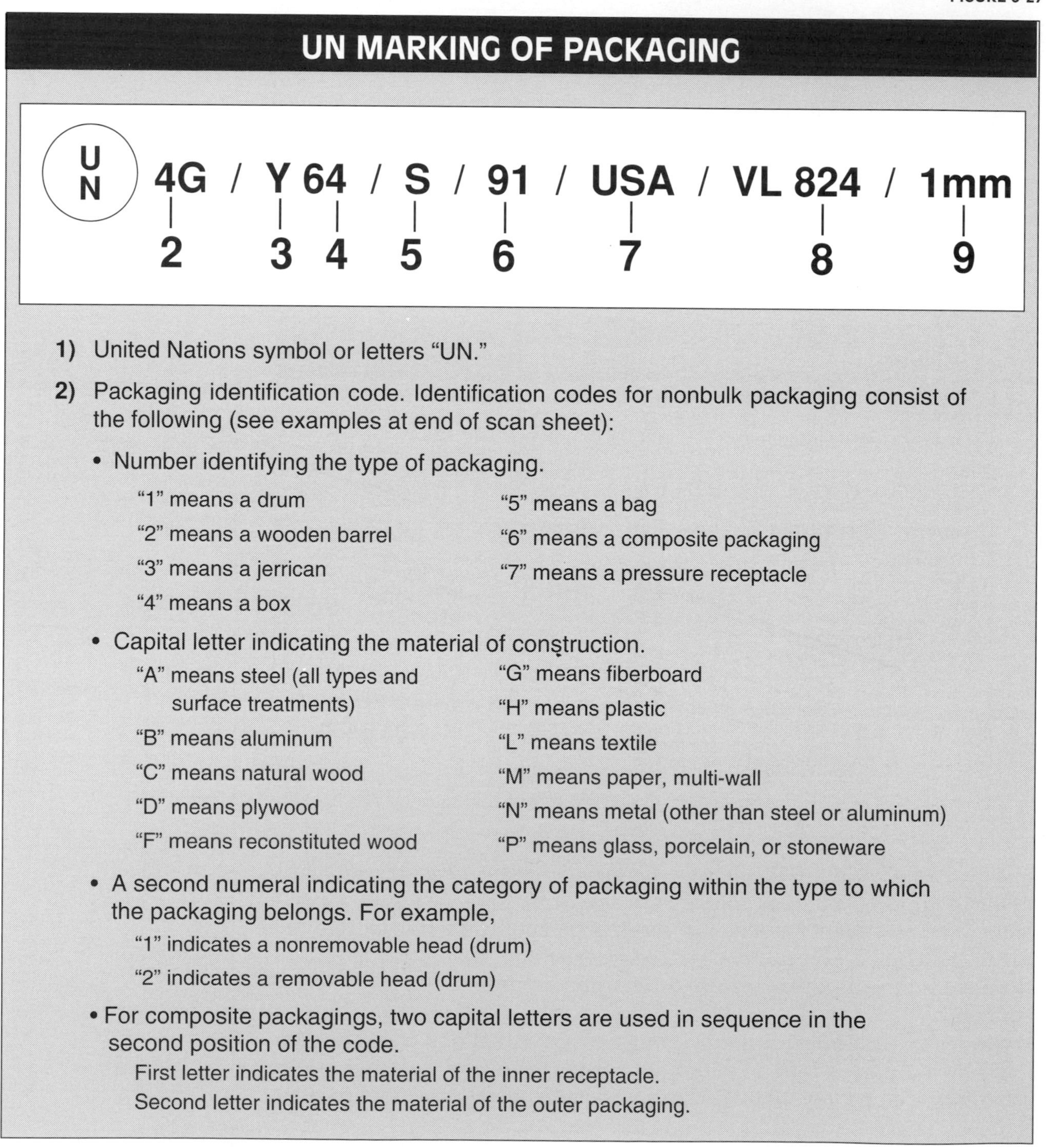

UN MARKING OF PACKAGING

UN 4G / Y 64 / S / 91 / USA / VL 824 / 1mm

2 3 4 5 6 7 8 9

1) United Nations symbol or letters "UN."

2) Packaging identification code. Identification codes for nonbulk packaging consist of the following (see examples at end of scan sheet):

- Number identifying the type of packaging.

  "1" means a drum
  "2" means a wooden barrel
  "3" means a jerrican
  "4" means a box
  "5" means a bag
  "6" means a composite packaging
  "7" means a pressure receptacle

- Capital letter indicating the material of construction.

  "A" means steel (all types and surface treatments)
  "B" means aluminum
  "C" means natural wood
  "D" means plywood
  "F" means reconstituted wood
  "G" means fiberboard
  "H" means plastic
  "L" means textile
  "M" means paper, multi-wall
  "N" means metal (other than steel or aluminum)
  "P" means glass, porcelain, or stoneware

- A second numeral indicating the category of packaging within the type to which the packaging belongs. For example,

  "1" indicates a nonremovable head (drum)
  "2" indicates a removable head (drum)

- For composite packagings, two capital letters are used in sequence in the second position of the code.

  First letter indicates the material of the inner receptacle.
  Second letter indicates the material of the outer packaging.

3) Performance Standard

"X" for packagings meeting Packing Group I, II, and III tests.
"Y" for packagings meeting Packing Group II and III tests.
"Z" for packagings meeting Packing Group III tests.

4) Specific gravity/mass designation (in kilograms).

5) Hydrostatic test pressure, in kilopascals (kPa) to the nearest 10 kPa. "S" indicates use for solids or inner packagings.

6) Year of manufacture (last two digits).

7) State authorizing the mark (e.g., USA).

8) Name and address or symbol of the manufacturing agency or approval agency.

9) Minimum thickness, in millimeters (mm), for metal or plastic drums or jerricans intended for re-use or re-conditioning.

The following are examples of Identification Codes (item #2) for nonbulk packages:

DRUMS

1A1 Nonremovable head steel drum
1A2 Removable head steel drum
1B1 Nonremovable head aluminum drum
1B2 Removable head aluminum drum
1N1 Nonremovable head metal drum (other than steel or aluminum)
1N2 Removable head metal drum (other than steel or aluminum)
1D Plywood drum
1G Fiber drum
1H1 Non-removable head plastic drum
1H2 Removable head plastic drum

BARRELS

2C1 Bung type wooden barrel
2C2 Slack type (removable head) wooden barrel

JERRICANS

3A1 Nonremovable head steel jerrican
3A2 Removable head steel jerrican
3H1 Nonremovable head plastic jerrican
3H2 Removable head plastic jerrican

BOXES

4A1 Unlined and uncoated steel box
4A2 Steel box with inner liner or coating
4B1 Unlined and uncoated aluminum box
4B2 Aluminum box with inner liner or coating
4C1 Ordinary natural wooden box
4C2 Natural wood box with sift-proof walls
4D Plywood box
4F Reconstituted wood box
4G Fiberboard box
4H1 Expanded plastic box
4H2 Solid plastic box

BAGS

5H1 Unlined and uncoated woven plastic bag
5H2 Sift-proof woven plastic bag
5H3 Water resistant woven plastic bag
5H4 Plastic film bag
5L1 Unlined or uncoated textile bag
5L2 Sift-proof textile bag
5L3 Water-resistant textile bag
5M1 Multi-wall paper bag
5M2 Multi-wall water-resistant paper bag

COMPOSITE PACKAGINGS	
6HA1	Plastic receptacle within a protective steel drum
6HA2	Plastic receptacle within a protective steel crate or box
6HB1	Plastic receptacle within a protective aluminum drum
6HB2	Plastic receptacle within a protective aluminum crate or box
6HC1	Plastic receptacle within a protective wooden box
6HD1	Plastic receptacle within a protective plywood drum
6HD2	Plastic receptacle within a protective plywood box
6HG1	Plastic receptacle within a protective fiber drum
6HG2	Plastic receptacle within a protective fiberboard box
6HH	Plastic receptacle within a protective plastic drum
6PA1	Glass, porcelain, or stoneware receptacle within a protective steel drum
6PA2	Glass, porcelain, or stoneware receptacle within a protective steel crate or box
6PB1	Glass, porcelain, or stoneware receptacle within a protective aluminum drum
6PB2	Glass, porcelain, or stoneware receptacle within a protective aluminum crate or box
6PC	Glass, porcelain, or stoneware receptacle within a protective wooden box

FIGURE 6-27: UN MARKING OF PACKAGING

PLACARDS AND LABELS

DOT Hazardous Materials Regulations outline the hazmat placarding and labeling requirements within the United States. Placarding and labeling can be used as a clue for hazmat recognition and classification. Although designed to be used for transportation purposes, they may also be found on both bulk and nonbulk packaging at fixed locations.

Placards and labels can provide recognition and general hazard classification by way of:

- Colored background.
- Respective hazard class symbol.
- Hazard class/division number (found at the bottom of the placard).
- Hazard class designation or the four-digit identification number (found in the center of the placard).

Labels are approximately 4-inch (100 mm) by 4-inch (100 mm) markings applied to individual hazardous materials packages. They are generally placed or printed near the contents names or are printed on the manufacturing label. When labels cannot be applied directly to the container because of its nonadhesive surface, they are placed on tags or cards attached to the package. The proper label(s) are determined by the product's hazard class. (See Figure 6-28).

Some chemicals have more than one hazard, so multiple labels may be required. For example, red fuming nitric acid nonbulk packages are labeled "CORROSIVE," "OXIDIZER," and "POISON" to indicate that it

NOTES

is a poison inhalation hazard (PIH) material. The subsidiary hazard label may not have a hazard class/division number at the bottom of the label.

Placards are approximately 10.75 inch (273 mm), diamond-shaped markings applied to both ends and each side of freight containers, cargo tanks, and portable tank containers. Factors such as the individual package labels, the size of individual packages, and the total quantity of the product will determine the correct placard to be used. Figure 6-29 shows the DOT placard requirements.

FIGURE 6-28: LABELS AND PLACARDS PROVIDE GENERAL HAZARD COMMUNICATION INFORMATION IN SEVERAL MANNERS

SPECIAL LABELS

In addition to those labels associated with a specific hazard class, there are others which may provide more specific hazmat information. They are shown in Figure 6-30.

SPECIAL PLACARDS

Special placards required on transport containers that may provide additional information during hazmat incidents include those shown in Figure 6-31.

HAZARDOUS MATERIALS PLACARDING GUIDELINES

Placard motor vehicles, freight containers, and rail cars containing any quantity of hazardous materials listed in Table 1.

Placard motor vehicles, freight containers, and rail cars containing 1,001 lbs. or more gross weight of hazardous materials listed in Table 2.

Placard freight containers 640 cubic feet or more containing any quantity of hazardous material classes listed in Tables 1 and/or 2 when offered for transportation by air or water.

TABLE 1

Placard	Hazard Class/ Division	Placard Name	Placard	Hazard Class/ Division	Placard Name
 EXPLOSIVES 1.1A 1	1.1	EXPLOSIVES 1.1	 POISON GAS 2	2.3	POISONOUS GAS
 EXPLOSIVES 1.2B 1	1.2	EXPLOSIVES 1.2	 DANGEROUS WHEN WET 4	4.3	DANGEROUS WHEN WET
 EXPLOSIVES 1.3C 1	1.3	EXPLOSIVES 1.3	 POISON 6	6.1 PG1 (Inhalation hazard only)	POISON
 RADIOACTIVE III 7	7 (Yellow III Label Only)	RADIOACTIVE			

FIGURE 6-29: HAZARDOUS MATERIALS PLACARDING REQUIREMENTS

TABLE 2

Hazard Class/ Division	Placard Name	Hazard Class/ Division	Placard Name
 1.4	EXPLOSIVES 1.4	 4.1	FLAMMABLE SOLID
 1.5	EXPLOSIVES 1.5 BLASTING AGENTS	 4.2	SPONTANEOUSLY COMBUSTIBLE
 1.6	EXPLOSIVES 1.6	 5.1	OXIDIZER
 2.1	FLAMMABLE GAS	 5.2	ORGANIC PEROXIDE
 2.2	NON-FLAMMABLE GAS	 6.1 PGI or PGII	POISON
 3	FLAMMABLE	 6.1	KEEP AWAY FROM FOOD
 COM-BUSTIBLE LIQUID	COMBUSTIBLE	 8	CORROSIVE
		9	CLASS 9 (optional) photo not shown

FIGURE 6-29: HAZARDOUS MATERIALS PLACARDING REQUIREMENTS

FIGURE 6-30

SPECIAL LABELS		
NAME	LABEL	APPLICATION
CARGO AIRCRAFT ONLY	DANGER	Materials not accepted for passenger-carrying aircraft.
MAGNETIZED MATERIALS	MAGNETIZED MATERIAL KEEP AWAY FROM AIRCRAFT COMPASS DETECTOR UNIT	Articles that contain magnets which may cause deviations in aircraft compasses.
POISONOUS SUBSTANCES	HARMFUL STOW AWAY FROM FOODSTUFFS 6	Shipments of poisonous substances. The text may be in the language of the country of origin.
INFECTIOUS SUBSTANCES	INFECTIOUS SUBSTANCE 6	Shipments of infectious substances. The text may be in the language of the country of origin.
POISON-INHALATION HAZARD	ORGANIC PEROXIDE 5.2 POISON	Chemicals which have a poison-inhalation hazard and are flammable, oxidizable, or an organic peroxide. When shipped in containers with a volume greater than 1 liter but less than 110 gallons, the label "poison," as well as the label for any other hazard, must also be on the container.

NAME	LABEL	APPLICATION
RADIOACTIVE MATERIALS		Labels will carry additional information (contents-name of radionuclide, radioactive activity, transport index). All labels will show vertical bars indicating radioactive levels (I, II, III) overprinted in red on lower half of each label.
MISCELLANEOUS DANGEROUS GOODS (CLASS 9)		Material that presents a hazard during transportation, but does not meet definition of any other hazard class (e.g., PCB's, adipic acid).
CANADIAN TRANSPORT LABEL CORROSIVE GAS		Canada classifies certain gases as "Corrosive Gases" (e.g., anhydrous ammonia).

FIGURE 6-30: SPECIAL LABELS AND THEIR APPLICATIONS

FIGURE 6-31

SPECIAL PLACARDS		
NAME	LABEL	APPLICATION
EXPLOSIVE PLACARD COMPATIBILITY GROUP LETTER		The designated alphabetical letter (e.g., K) is used to categorize different types of explosive substances and materials for the purpose of stowage and segregation.

SPECIAL RAIL PLACARDING		Rail shipments that require special handling. Each car placarded as Explosives 1.1 and 1.2, Poison Gas (Division 2.3), Zone A (loaded or residue), and Division 6.1, PG 1, Zone A (loaded or residue), must have a placard mounted on a 15-inch square white background.
RESIDUAL		Identifies packaging that contains residue of a hazmat and has not been cleaned, purged, or reloaded with a material not subject to hazmat regs. Not used for explosives, poison gas, and radioactive materials. Cargo Tank trucks display appropriate placard whether empty or full.
SUBSIDIARY HAZARD		Used for hazmats which meet definition for one hazard class, but have subsidiary/multiple hazards. Examples include POISON, CORROSIVE, and DANGEROUS WHEN WET placards. The second placard (subsidiary placard) may not display the identification number or the hazard class number.
MISCELLANEOUS DANGEROUS GOODS (CLASS 9)		Material that presents a hazard during transportation, but does not meet definition of any other hazard class. (e.g., PCB's, molten sulfur, adipic acid).
DANGEROUS		When total weight of 2 or more Table 2 materials is 1,000 lbs. (454 kg) or more, a Dangerous placard may be used. If 5,000 lbs. (2,268 kg) or more of any Table 2 materials are loaded at one location, the hazard class placard is used.
CANADIAN TRANSPORT PLACARD CORROSIVE GAS		Canada classifies certain gases as "Corrosive Gases" (e.g., anhydrous ammonia).

FIGURE 6-31: SPECIAL PLACARDS AND THEIR APPLICATIONS

NOTES

Placards and labels are simply another clue and should not be considered as a definitive source of hazmat information. Although enforcement measures are constantly improving, experience shows that a substantial number of vehicles are improperly placarded or not marked at all.

SHIPPING PAPERS AND FACILITY DOCUMENTS

SHIPPING PAPER REQUIREMENTS

Shipping papers are required to accompany each hazmat shipment. Response personnel must be familiar with the information noted on shipping papers, their location on each transport vehicle, as well as the individual responsible for them. Figure 6-32 summarizes this information for each transportation mode.

SHIPPING PAPERS INFORMATION

Mode of Transportation	Title of Shipping Papers	Location of Shipping Papers	Responsible Person(s)
Highway	Bill of Lading	Cab of Vehicle	Driver
Rail*	Waybill/Consist	With Crew (conductor)	Crew (conductor)
Water	Dangerous Cargo Manifest	Wheelhouse or Pipe-like Container on Barge	Captain or Master
Air	Air Bill with Shipper's Certification for Restricted Articles	Cockpit	Pilot

*STCC (Standard Transportation Commodity Code) number is used extensively on rail transportation shippings papers.

FIGURE 6-32: TYPES AND SOURCES OF SHIPPING PAPERS BY TRANSPORTATION MODE

Each transport mode has its own terms for shipping papers. Shipping papers are required to contain the following entries, with all information printed in the English language:

- ❑ **Proper Shipping Name.** Identifies the name of the hazmat as found in the DOT Hazardous Material Regulations. The word "WASTE" will precede the proper shipping name for those shipments that are classified as hazardous wastes.
- ❑ **DOT Hazard Class/Division Number.** Indicates the material's primary and secondary (as appropriate) hazard as listed in the DOT

NOTES

Hazardous Materials Regulations. A division is a subset of a hazard class. NOTE: A hazmat may meet the criteria for more than one hazard class, but is assigned to only one hazard class.

- ❑ **Identification Number(s).** The four-digit identification number assigned to each hazardous material. The identification number may be found with the prefix "UN"—United Nations or "NA"—North America.
- ❑ **Packing Group.** Further classifies hazardous materials based on the degree of danger represented by the material. There are three groups:
  - Packing Group I indicates great danger.
  - Packing Group II indicates medium danger.
  - Packing Group III indicates minor danger.

  Packing Groups may be shown as "PG I", etc. Packing Groups are not assigned to Class 2 materials (compressed gases), Class 7 materials (radioactives), some Division 6.2 materials (infectious substances), and ORM-D materials.
- ❑ **Total Quantity.** Indicates the quantity by net or gross mass, capacity, etc. May also indicate the type of packaging. The number and type of packaging (e.g., 1 TC, 7 DRM) may be entered on the beginning line of the shipping description. Carriers often use abbreviations to indicate the type of packaging. The following are examples of packaging abbreviations used by the Union Pacific Railroad:

BA = Bale	CH = Covered Hopper	KIT = Kit
BG = Bag	CL = Carload	KL = Container Load
BOX = Box	CY = Cubic Yard	PA = Pail
BC = Bucket	CYL = Cylinder	PKG = Package
CA = Case	DRM = Drum	SAK = Sack
CAN = Can	JAR = Jar	TB = Tube
CR = Crate	JUG = Jug	TC = Tank Car
CTN = Carton	KEG = Keg	TL = Trailer Load

- ❑ **Emergency Contact.** Indicates the telephone number for the shipper or shipper's representative that may be accessed 24 hours a day, 7 days a week in the event of an accident. If the shipper is registered with CHEMTREC, the phone number for CHEMTREC may be displayed as the emergency contact.

SHIPPING PAPERS—ADDITIONAL ENTRIES

Additional shipping paper entries may be required for some hazardous materials. They include the following:

- ❑ **Compartment Notation.** Identifies the specific compartment of a multi-compartmented rail car or cargo tank truck in which the material is located. On rail cars, compartments are numbered sequentially from the "B" end (the end where the hand brake wheel is located), while cargo tank trucks are numbered sequentially from the front.

NOTES

- **Residue (Empty Packaging).** Identifies packaging that contains a hazmat residue and has not been cleaned and purged or reloaded with a material that is not subject to the DOT Hazardous Materials Regulations. Residue is indicated by the words "Residue: Last Contained" before the proper shipping name. It is only used in rail transportation.
- **HOT.** Identifies elevated temperature materials other than molten sulfur and molten aluminum.
- **Technical Name.** Identifies the recognized chemical name currently used in scientific and technical handbooks, journals, and texts. Generic descriptions may be found provided that they identify the general chemical group. With some exceptions, trade names may not be used as technical names. Examples of acceptable generic descriptions are organic phosphate compounds, tertiary amines, and petroleum aliphatic compounds.
- **Not Otherwise Specified (N.O.S.) Notations.** If the proper shipping name of a material is an "N.O.S." notation, the technical name of the hazardous material must be entered in parentheses with the basic description. If the material is a mixture or solution of two or more hazardous materials, the technical names of at least two components that most predominantly contribute to the hazards of the mixture/solution must be entered on the shipping paper. For example, "Flammable Liquid, Corrosive Liquid, n.o.s., Corrosive n.o.s., 3, UN 2921, PGII (contains Methanol, Potassium Hydroxide)."
- **Subsidiary Hazard Class.** Indicates a hazard of a material other than the primary hazard assigned.
- **Reportable Quantity (RQ) Notation.** Indicates the material is a hazardous substance by the EPA. The letters "RQ" (reportable quantity) must be shown either before or after the basic shipping description entries. This designation indicates that any leakage of the substance above its RQ value must be reported to the proper agencies (e.g., the National Response Center). Regardless of which agencies are involved, the legal responsibility for notification still remains with the spiller.
- **Marine Pollutant.** Indicates that the material meets the definition of a marine pollutant.
- **EPA Waste Stream Number.** Indicates the number assigned to a hazardous waste stream by the U.S. EPA to identify that waste stream.
- **EPA Waste Characteristic Number.** Indicates the general hazard characteristics assigned to a hazardous waste by the U.S. EPA. Waste characteristics include EPA corrosivity, EPA toxicity, EPA ignitability, and EPA reactivity.
- **Radioactive Material Information.** Indicates the following:
    - "Radioactive Material"—if not part of the proper shipping name
    - Name of each radionuclicide
    - Physical/chemical form

NOTES

- Activity in curies
- Label applied
- Transport Index (if applicable)
- U.S. Department of Energy Approval Number (if applicable)
- Fissile Exempt (if applicable)
- Fissile Class (if applicable)

❑ **Poison Notation.** Indicates that a liquid or solid material is poisonous when the fact is not disclosed in the shipping name.

❑ **Poison-Inhalation Hazard (PIH) Notation.** Indicates gases and liquids that are poisonous by inhalation.

❑ **Hazard Zone.** Indicates relative degree of hazard in terms of toxicity (only appears for gases and liquids that are poisonous by inhalation):
- Zone A—LC50 less than or equal to 200 ppm (most toxic).
- Zone B—LC50 greater than 200 ppm and less than or equal to 1,000 ppm.
- Zone C—LC50 greater than 1,000 ppm and less than or equal to 3,000 ppm.
- Zone D—LC50 greater than 3,000 ppm and less than or equal to 5,000 ppm (least toxic).

❑ **Dangerous When Wet Notation.** Indicates a material that, by contact with water, is liable to become spontaneously flammable or give off flammable or toxic gas at a rate greater than 1 liter per kilogram of the material, per hour.

❑ **Limited Quantity (LTD QTY).** Indicates a material being transported in a quantity for which there is a specific labeling and packaging exception.

❑ **Canadian Information.** Indicates information required for hazardous materials entering or exiting Canada in addition to that required in the United States (e.g., ERG reference number, the 24-hour Canadian emergency telephone number, Canadian class, etc.).

❑ **Placard Notation.** Indicates the placard applied to the container. Where placards are not required (e.g., ORM commodities or non-flammable cryogenic gases), the notation "MARKED" is followed by the four-digit identification number.

❑ **Placard Endorsement.** Indicates the presence of a hazardous material requiring a placard. Found inside a rectangle made with any symbol (e.g., asterik (*), dollar sign ($), or the symbol for number (#)).

❑ **Trade Name.** A name that enables organizations, such as emergency responders and CHEMTREC®, to access the MSDS for additional information.

❑ **Hazardous Materials STCC Number.** A seven-digit Standard Transportation Commodity Code (STCC) number will be found on all shipping papers accompanying rail shipments of hazmats. It will also be found when intermodal containers are changed from rail to highway movement. Look for the first two digits—"49"— as the key iden-

NOTES

tifier for a hazmat. The STCC number will follow the notation "HAZ-MAT STCC -."

- **Shipper Contact.** Indicates the identity of the producer or consolidator of the materials described.

SHIPPING PAPERS—EMERGENCY RESPONSE INFORMATION

Emergency response information must also be included with shipping papers. An emergency telephone number for the shipper or shipper's representative is required on the shipping paper. Emergency response information must provide the following:

- Brief product description.
- Emergency actions involving fire.
- Emergency actions involving release only.
- Personnel protective measures.
- Environmental considerations, as appropriate.
- First aid measures.

Several common sources of emergency response information requirements are an MSDS, a copy of the DOT Emergency Response Guidebook, a copy of the specific page from the ERG for the hazmat being transported, or railroad emergency response information sheets which are cross-referenced with the train consist. With most major railroads, the railroad emergency response information sheets will be a part of the train consist.

FACILITY DOCUMENTS

Various types of facility documents are available to assist in the information process. They can be a source for hazmat recognition, identification, and classification at an emergency.

The specific type and nature of information provided will vary based upon pertinent federal, state, and/or local reporting requirements. Examples include hazmat inventory forms, shipping and receiving forms, Risk Management Plans and supporting documentation, MSDS's, and Tier II reporting forms required to be submitted to the LEPC and the fire department under SARA Title III.

Both the Risk Management Plans and the Tier II reporting forms can be used as part of the hazards analysis process. For example, the Tier II reporting forms provide information such as chemicals on-site which exceed the reporting thresholds, physical and chemical hazards, average and maximum amounts on-site, and types of storage containers and location.

MONITORING AND DETECTION EQUIPMENT

If you are unable to identify the hazardous material from the previously discussed methods, monitoring and detection equipment can often provide data and information concerning the overall nature of the problem you face as well as the specific materials involved. They are essential tools for identifying, verifying, or classifying the hazmat(s) involved.

NOTES

Although considered here as an identification tool, monitoring and detection equipment is also a critical tool for evaluating "real-time" data. It helps one to:

- Determine the appropriate levels of personal protective clothing and equipment.
- Determine the size and location of hazard control zones.
- Develop protective action recommendations and corridors.
- Assess the potential health effects of exposure.

The selection, application, and use of monitoring instruments is addressed in Chapter Seven—Hazard and Risk Evaluation.

SENSES

Senses are not a primary identification tool. In most cases, if you are close enough to smell, feel, or hear the problem, you are probably too close to operate safely.

Nonetheless, senses can be valuable assets and can offer immediate clues to the presence of hazardous materials. For our purposes, "senses" refers to any personal physiological reaction to or visual observation of the release of a hazmat. Smells, dizziness, unusual noises (i.e., relief valve actuations), and destroyed vegetation are some examples.

While sight and hearing can be extremely valuable, senses are among the most difficult clues to teach because of the many products, variations in amounts released, and differing individual physiological reactions. For example, a certain chemical might have characteristics which a second material in the same chemical family does not have, even though they are both equally hazardous. Hydrogen sulfide, for example, can deaden the sense of smell and lead ERP to believe they are operating in a safe haven when they are not.

SUMMARY

The evaluation of hazards and the assessment of the risks builds upon the timely identification and verification of the hazardous materials involved. A problem well-defined is half-solved. Identification and verification of the hazmats involved are critical to the safe and effective management of a hazmat incident.

The seven basic clues for recognition, identification, and classification are:

- Occupancy and location
- Container shapes
- Markings and colors
- Placards and labels
- Shipping papers and facility documents
- Monitoring and detection equipment
- Senses

REFERENCES AND SUGGESTED READINGS

Bowen, John, "PCB"—An Update." AMERICAN FIRE JOURNAL (June, 1987), pages 32–37.

Chemical Manufacturers Association and the Association of American Railroads, PACKAGING FOR TRANSPORTING HAZARDOUS AND NON-HAZARDOUS MATERIALS, Washington, DC: Chemical Manufacturers Association (June, 1989).

Code of Federal Regulations, TITLE 49 CFR PARTS 100–199 (TRANSPORTATION), Washington, DC: U.S. Government Printing Office.

Compressed Gas Association, HANDBOOK OF COMPRESSED GASES (3rd Edition). Arlington, VA: Compressed Gas Association (1990).

Federal Emergency Management Agency—National Fire Academy, Hazardous Materials Operating Site Practices Course—Student Manual, Emmitsburg, MD: FEMA (1993).

Federal Emergency Management Agency—National Fire Academy and the Union Pacific Railroad Company, Recognizing and Identifying Hazardous Materials Course, Emmitsburg, MD: FEMA (1993).

General American Transportation Corporation, GATX TANK CAR MANUAL (5th edition), Chicago, IL: General American Transportation Corporation (1985).

Hazardous Materials Advisory Council, AN OVERVIEW OF INTERMODAL PORTABLE TANKS, Washington, DC: Hazardous Materials Advisory Council (1986).

Hazmat World, "Safety Upgrades Plug Tank Car Leaks." HAZMAT WORLD (August, 1993), pages 42–43.

Isman, Warren E. and Gene P. Carlson, HAZARDOUS MATERIALS, Encino, CA: Glencoe Publishing Company (1980).

National Fire Protection Association, FIRE PROTECTION HANDBOOK (17th edition), Boston, MA: National Fire Protection Association (1991).

National Fire Protection Association, HAZARDOUS MATERIALS RESPONSE HANDBOOK (2nd edition), Boston, MA: National Fire Protection Association (1992).

National Fire Protection Association, LIQUEFIED PETROLEUM GASES HANDBOOK (3rd edition), Boston, MA: National Fire Protection Association (1992).

National Fire Protection Association, NATIONAL FLAMMABLE AND COMBUSTIBLE LIQUIDS CODE HANDBOOK (4th edition), Boston, MA: National Fire Protection Association (1990).

NIOSH/OSHA/USCG/EPA, OCCUPATIONAL SAFETY AND HEALTH GUIDANCE MANUAL FOR HAZARDOUS WASTE SITE ACTIVITIES, Washington, DC: U.S. Government Printing Office (1985).

Sea Containers Limited, GENERAL GUIDE TO TANK CONTAINER OPERATION, London, England: Sea Containers Group (1983).

Sea Containers Limited, INSPECTION, REPAIR AND TEST REQUIREMENTS FOR TANK CONTAINERS, London, England: Sea Containers Group (undated).

Union Pacific Railroad Company, A GENERAL GUIDE TO TANK CARS, Omaha, NE: Union Pacific Railroad Company, Technical Training (May, 1990).

Union Pacific Railroad Company, A GENERAL GUIDE TO TANK CONTAINERS, Omaha, NE: Union Pacific Railroad Company, Technical Training (December, 1988).

U.S. Department of Transportation—Research and Special Programs Administration, RESPONSE TO RADIOACTIVE MATERIAL TRANSPORT ACCIDENTS (DOT/RSPA/MTB-79/8), Washington, DC: DOT/RSPA (1980)

U.S. Environmental Protection Agency, "Review of the EPA Rule Regulating Polychlorinated Biphenyl (PCB) Transformer Fires" (Report E1E57-11-0024-80780), Washington, DC: EPA, Office of the Inspector General (March, 1988).

Wright, Charles J., "DOT's Hazardous Materials Regulations and Emergency Response." Student handout presented at the Kansas Hazardous Materials Symposium, Wichita, KS (November 7, 1993).

Wright, Charles J., "Handling Rail Emergencies Involving Hazardous Materials." Student handout presented at the Kansas Hazardous Materials Symposium, Wichita, KS (November 7, 1993).

Wright, Charles J. and William T. Hand, "Intermodal Tank Containers—Parts I and II," FIRE COMMAND, (June and July, 1988), pages 17, 36–37.

York, Kenneth J. and Gerald L. Grey, HAZARDOUS MATERIALS/WASTE HANDLING FOR THE EMERGENCY RESPONDER, Tulsa, OK: Fire Engineering Books and Videos (1989).

HAZARD AND RISK EVALUATION

OBJECTIVES

1) Describe the concept of hazard assessment and risk evaluation.
2) Describe the common hazard terms found in hazard information sources and their significance in a hazardous materials incident.
3) Identify the types of hazard and response information available from each of the following resources and explain the advantages and disadvantages of each resource:
   a) Reference manuals
   b) Technical information specialists
   c) Technical information centers
   d) Hazardous materials data bases
   e) Material data safety sheets (MSDS)
   f) Monitoring instruments
4) Identify the types of monitoring equipment used to determine the following hazards:
   a) Corrosivity
   b) Flammability
   c) Oxidizing potential
   d) Oxygen deficiency
   e) Radioactivity
   f) Toxicity

"A leader who doesn't hesitate before he sends his nation into battle is not fit to be a leader."

Golda Meir,
Israeli Prime Minister

NOTES

5) Identify the limiting factors associated with the selection and use of the following monitoring equipment:
   a) Carbon monoxide meter
   b) Colorimetric detector tubes
   c) Combustible gas indicator (CGI)
   d) Oxygen meter
   e) pH paper, meters, and strips
   f) Passive dosimeters
   g) Radiation detection instruments
6) Identify the steps in an analysis process for identifying unknown materials.
7) Identify and describe the components of the General Hazardous Materials Behavior Model (GEBMO).
8) Identify the guidelines for performing a damage assessment of a pressurized container.
9) Describe the factors which influence the underground movement of hazmats in soil and through groundwater.
10) Identify the hazards associated with the movement of hazmats in the following types of sewer collection systems:
    a) Storm sewers
    b) Sanitary sewers
    c) Combination sewers
11) List five site safety procedures for handling an emergency involving a hydrocarbon spill into a sewer collection system.
12) Describe the steps for determining strategic goals and tactical response objectives for a hazardous materials incident.

INTRODUCTION

The evaluation of hazard information and the assessment of the relative risks is one of the most critical decision-making points in the successful management of a hazardous materials incident. The decision to intervene, or more often to not intervene, is not easy. While most responders recognize the initial need for isolating the area, denying entry, and identifying the hazardous materials involved, many overlook the need for developing effective analytical and problem-solving skills.

This chapter will discuss the third step in the **Eight Step Process©—Hazard and Risk Evaluation**. The chapter is based upon the premise that ERP have (1) successfully implemented site management procedures, and (2) identified the nature of the problem and the hazmats potentially involved. Primary topics within this chapter will be describing common hazard terminology, outlining the common sources of hazard information, evaluating risks, and determining response objectives.

CASE STUDY

HAZARDOUS MATERIALS TRAIN DERAILMENT AKRON, OHIO FEBRUARY 26, 1989

On February 26, 1989, a CSX Transportation, Inc. freight train derailed in Akron, Ohio. Twenty-one freight cars derailed, including nine butane tank cars. These nine tank cars came to rest adjacent to the B. F. Goodrich Chemical Company plant, and butane released from two breached tank cars immediately caught fire. The fire required the evacuation of approximately 1,750 residents from a square mile area.

After the Akron Fire Department (AFD) received and verified the shipping information from the train crew, on-scene activities were accomplished in a timely and professional manner. These activities included controlling the butane tank car fires, controlling the fire at the B. F. Goodrich plant, and the timely evacuation of the area residents.

Both the Fire Department and the City of Akron depended upon the expertise of CSX for the removal of the wreckage from the derailment site. The AFD Operations Section Chief considered it unsafe to off-load the butane tank cars because of a continuing fire from one of the derailed tank cars. After discussions with CSX, he agreed with CSX's plan to rerail the tank cars, move them to the Akron Junction Yard where the tank cars would be more permanently secured, and finally transport them to Canton, Ohio where the product would be off-loaded. However, the railroad did not discuss alternatives with the Fire Department, nor did the railroad advise the Fire Department of the possible risks associated with rerailing the tank cars.

On February 28th, while the rerailed tank cars were being moved from the accident site, a butane tank car rolled off its tracks and forced the evacuation of approximately 25 families from the area. Only after this second derailment were alternative plans and the risks associated with each potential course of action thoroughly discussed with the Fire Department and City of Akron officials.

In their final report on the accident, the National Transportation Safety Board (NTSB) stated that it recognized the limited technical resources that may be available to local communities regarding train wreckage clearing operations and the communities' reliance on the railroad to take the appropriate course of action. Because of these reasons, NTSB believed that it is necessary for the railroad to discuss with local emergency response personnel (1) the severity of known damage to hazmat tank cars; (2) the relative dangers posed to public safety; (3) all possible courses of action; and (4) any associated risks.

NTSB also recognized the need for the Incident Commander to play an active role during the hazard and risk assessment process. Tasks should include searching out information on the severity of known tank car damage and the dangers posed, potential alternatives and solutions, and the risks involved.

Hazard and risk evaluation is the cornerstone of decision-making at a hazmat incident. The NTSB Accident Investigation Report is one of the first to address the issue of hazard and risk evaluation from the perspective of emergency responders.

NOTES

HAZARD AND RISK ASSESSMENT

The concept of hazard and risk assessment is routinely used by ERP in fire and rescue operations. Firefighters are generally able to size up and attack structural fires effectively because of their training and experience and the availability of information (e.g., water supply). In contrast, the lack of experience, information, and adequate training skills sometimes places response personnel in an unsafe position at a hazmat incident.

What are hazards and risks?

Hazards refer to a danger or peril. In hazardous materials operations, hazards generally refer to the physical and chemical properties of a material. You can obtain "hard data" on hazards through emergency response guidebooks, computer data bases, and the like. Examples include flash point, toxicity levels (LD-50, LC-50), exposure values (TLV's), protective clothing requirements, compatibility, etc.

Risk refers to the probability of suffering harm or loss. Risks can't be determined from books or pulled from computerized data bases—they are those intangibles that are different at every hazmat incident and must be evaluated by a knowledgeable Incident Commander. Although the risks associated with hazmat response will never be completely eliminated, they can be successfully managed. The objective of ERP operations is to minimize the level of risk to ERP, the community, and the environment. Hazmat responders must see their role as RISK EVALUATORS, not RISK TAKERS!

Risk levels are variable and change from incident to incident. Factors which influence the level of risk include:

- **Hazardous nature of the material(s) involved.** For example, toxicity, flammability, and reactivity.
- **Quantity of the material involved.** Risks will often be greater when dealing with bulk quantities of hazardous materials as compared to limited-quantity, individual containers. However, quantity must also be balanced against the hazardous nature of the material(s) involved—small quantities of highly toxic or reactive materials can create significant risks.
- **Containment system and type of stress applied to the container.** Containers may be either pressurized or nonpressurized. Risks are inherently higher for pressurized containers as compared to low pressure and atmospheric pressure containers. In addition, the type of stressors involved (thermal, chemical,mechanical, or combination) and the ability of the container to tolerate that stress will influence the level of risk.
- **Proximity of exposures.** This would include both distance and the rate of dispersion of any chemical release. Exposures include emergency response personnel, the community, property, and the environment.
- **Level of available resources.** The availability of resources and their response time will influence the level of risk. This includes both the training and knowledge level of responders.

NOTES

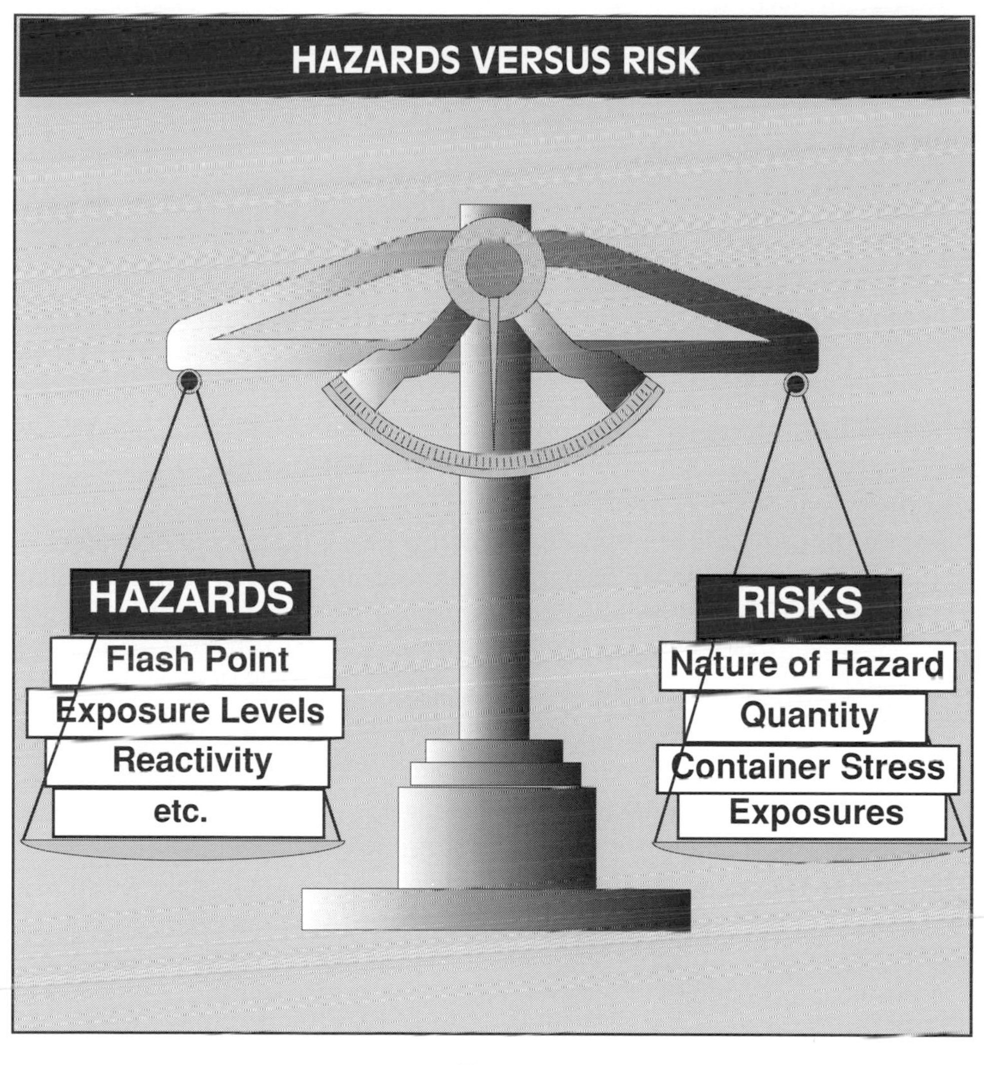

FIGURE 7-1: HAZARDS MUST BE WEIGHED AGAINST RISKS

EVALUATING HAZARDS: TERMS AND DEFINITIONS

In order to effectively evaluate risks, ERP must be able to identify and verify the materials involved and determine their hazards and behavior characteristics. This includes collecting and interpreting available hazard data.

Hazards are normally categorized based upon health, flammability, and reactivity. Considering SARA Title III and Right-to-Know regulations, responders should be able to read, understand, and interpret MSDSs. Common hazard terms and definitions are outlined in the following scan sheets.

NOTES

HAZARD TERMS AND DEFINITIONS

GENERAL TERMS AND DEFINITIONS

- **Element**—pure substance that cannot be broken down into simpler substances by chemical means.
- **Compound**—chemical combination of two or more elements, either the same elements or different ones, that is electrically neutral. Compounds have a tendency to break down into their component parts, sometimes explosively.
- **Mixture**—substance made up of two or more compounds, physically mixed together. A mixture may also contain elements and compounds mixed together.
- **Solution**—mixture in which all of the ingredients are completely dissolved. Solutions are composed of a solvent (water or another liquid) and a dissolved substance (known as the solute).
- **Slurry**—pourable mixture of a solid and a liquid.
- **Sludge**—solid, semi-solid, or liquid waste generated from a municipal, commercial, or industrial waste treatment plant or air pollution control facility, exclusive of treated effluent from a waste water treatment plant.
- **Salt**—chemical compound which results when metal elements chemically react with nonmetal elements. Occurs through ionic bonding.
- **Organic Materials**—materials which contain two or more carbon atoms. Organic materials are derived from materials that are living or were once living, such as plants or decayed products. Most organic materials are flammable. Examples include methane (CH_4) and propane (C_3H_8).
- **Inorganic Materials**—compounds derived from other than vegetable or animal sources which lack carbon chains but may contain a carbon atom (e.g., sulfur dioxide—SO_2).
- **Hydrocarbons**—compounds primarily made up of hydrogen and carbon. Examples include LPG, gasoline, and fuel oils.
- **Saturated Hydrocarbons**—a hydrocarbon possessing only single covalent bonds. All of the carbon atoms are saturated with hydrogen. Examples include methane (CH_4), propane (C_3H_8), and butane (C_4H_{10}).
- **Unsaturated Hydrocarbons**—a hydrocarbon with at least one multiple bond between two carbon atoms somewhere in the molecule. Generally, unsaturated hydrocarbons are more active chemically than saturated hydrocarbons, and are considered more hazardous. May also be referred to as the alkenes and alkynes. Examples include ethylene (C_2H_4), butadiene (C_4H_6), and acetylene (C_2H_2).
- **Aromatic Hydrocarbons**—a hydrocarbon which contains the benzene "ring" which is formed by six carbon atoms and contains resonant bonds. Examples include benzene (C_6H_6) and toluene (C_7H_8).
- **Halogenated Hydrocarbons**—a hydrocarbon with halogen atom (e.g., chlorine, fluorine, bromine, etc.) substituted for a hydrogen atom. They are often more toxic than naturally occurring organic chemicals, and

NOTES

they decompose into smaller, more harmful elements when exposed to high temperatures for a sustained period of time.

- **Cryogenic Liquids**—a gas with a boiling point of minus 150°F or lower. Cryogenic liquid spills will vaporize rapidly when exposed to the higher ambient temperatures outside of the container. Expansion ratios for common cryogenics range from 694 (nitrogen) to 1,445 (neon) to 1.

IDENTIFICATION TERMS

- **Shipping Name**—the proper shipping name or other common name for the material; also any synonyms for the material.
- **DOT Hazard Class**—the hazard class designation for the material as found in the Department of Transportation regulations, 49 CFR. There are currently 9 DOT hazard classes which are divided into 22 divisions.
- **ID Number**—the four-digit identification number assigned to a hazardous material by the Department of Transportation; on shipping documents may be found with the prefix "UN" (United Nations) or "NA" (North American). The ID numbers are not unique, and more than one material may have the same ID number.
- **STCC Number**—the Standard Transportation Commodity Code number used in the rail industry; a seven-digit number assigned to a specific material or group of materials and used in determination of rates. For a hazardous material, the STCC number will begin with the digits "49." Hazardous wastes may also be found with the first two digits being "48."
- **CAS Number**—the Chemical Abstract Service number. Often used by state and local Right-to-Know regulations for tracking chemicals in the workplace and the community. Sometimes referred to as a chemical's "social security number." Sequentially assigned CAS numbers identify specific chemicals and have no chemical significance.

PHYSICAL DESCRIPTION

- **Normal Physical State**—the physical state or form (solid, liquid, gas) of the material at normal ambient temperatures (68°F to 77°F).
- **Color**—the color of the material under normal conditions (i.e., room temperature and at atmospheric pressure).
- **Odor**—the odor of the material upon its release.
- **Temperature**—the temperature of the material within its container. The material's temperature will influence both the range of hazards and potential countermeasures. Temperatures are usually measured in Fahrenheit (F) or Centigrade (C).

PHYSICAL PROPERTIES

- **Specific Gravity**—the weight of the material as compared with the weight of an equal volume of water. If the specific gravity is less than one, the material is lighter than water and will float. If the specific gravity is greater than one, the material is heavier than water and will sink. Most insoluble hydrocarbons are lighter than water and will float on the surface. Significant property for determining spill control and clean-up procedures for waterborne releases.

NOTES

- **Vapor Density**—the weight of a pure vapor or gas compared with the weight of an equal volume of dry air at the same temperature and pressure. The molecular weight of air is 29. If the vapor density of a gas is less than one, the material is lighter than air and may rise. If the vapor density is greater than one, the material is heavier than air and will collect in low or enclosed areas.

  An easy way to remember those gases which are lighter than air is the acronym HA HA MICEN, where:

  H = Hydrogen

  A = Anhydrous Ammonia

  H = Helium

  A = Acetylene

  M = Methane

  I = Illuminating Gas

  C = Carbon Monoxide

  E = Ethlylene

  N = Nitrogen

  This is a significant property for evaluating exposures and where hazmat gas and vapor will travel.

- **Boiling Point**—the temperature at which a liquid changes its phase to a vapor or gas. The temperature at which the vapor pressure of the liquid equals atmospheric pressure. This is a significant property for evaluating the flammability of a liquid because flash point and boiling point are directly related. A liquid with a low flash point will also have a low boiling point, which translates into a large amount of vapors being given off.

- **Melting Point**—the temperature at which a solid changes its phase to a liquid. This temperature is also the freezing point depending on the direction of the change. For mixtures, a melting point range may be given. This is a significant property in evaluating the hazards of a material, as well as the integrity of a container (e.g., frozen material may cause its container to fail).

- **Sublimation**—the ability of a substance to change from the solid to the vapor phase without passing through the liquid phase. An increase in temperature can increase the rate of sublimation. Significant in evaluating the flammability or toxicity of any released materials which sublime. The opposite of sublimation is deposition (changes from vapor to solid).

- **Critical Temperature and Pressure**—critical temperature is the minimum temperature at which a gas can be liquefied no matter how much pressure is applied. Critical pressure is the pressure that must be applied to bring a gas to its liquid state. Both terms relate to the process of liquefying gases. A gas cannot be liquefied above its critical temperature. The lower the critical temperature, the less pressure required to bring a gas to its liquid state.

- **Volatility**—the ease with which a liquid or solid can pass into the vapor state. The higher a material's volatility, the greater its rate of evapora-

NOTES

tion. This is a significant property in that volatile materials will readily disperse and increase the hazard area.

- **Expansion Ratio**—the amount of gas produced by the evaporation of one volume of liquid at a given temperature. This is a significant property when evaluating liquid and vapor releases of liquefied gases and cryogenic materials. the greater the expansion ratio, the more gas that is produced and the larger the hazard area.
- **Vapor Pressure**—the pressure exerted by the vapor within the container against the sides of a container. This pressure is temperature dependent; as the temperature increases, so does the vapor pressure. Consider the following three points:
  1) The vapor pressure of a substance at 100°F is always higher than the vapor pressure at 68°F.
  2) Vapor pressures reported in millimeters of mercury (mm Hg) are usually very low pressures. 760 mm Hg is equivalent to 14.7 psi or 1 atmosphere. Materials with vapor pressures greater than 760 mm Hg are usually found as gases.
  3) The lower the boiling point of a liquid, the greater vapor pressure at a given temperature.
- **Solubility**—the ability of a solid, liquid, gas, or vapor to dissolve in water or other specified medium. The ability of one material to blend uniformly with another, as in a solid in liquid, liquid in liquid, gas in liquid, or gas in gas. This is a significant property in evaluating the selection of control and extinguishing agents, including the use of water and firefighting foams
- **Degree of Solubility**—an indication of the solubility and/or miscibility of the material.

  Negligible—less than 0.1%

  Slight—0.1 to 1.0%

  Moderate—1 to 10%

  Appreciable—greater than 10%

  Complete—soluble at all proportions
- **Viscosity**—measurement of the thickness of a liquid and its ability to flow. High viscosity liquids, such as heavy oils, must first be heated to increase their fluidity. Low viscosity liquids spread more easily and increase the size of the hazard area.
- **Other**—any additional or pertinent data found.

HEALTH HAZARDS

- **Acute**—marked by a single dose or exposure, generally having a sudden onset for a course of time (e.g., acute toxicity, acute exposure).
- **Chronic**—marked by a long or permanent duration, consistent or continuous (e.g., chronic toxicities are usually permanent or irreversible); often occurs from repeated exposures over a period of time.
- **Exposure Hazard**—hazards existing from the inhalation, ingestion, or absorption of the material involved.

NOTES

- **Odor Threshold (TLV_{Odor})**—the lowest concentration of a material's vapor in air that is detectable by odor. If the TLV Odor is below the TLV/TWA, odor may provide a warning as to the presence of a material.
- **Lethal Dose, 50% Kill (LD-50)**—the amount of a dose which, when administered to lab animals, kills 50% of them. Refers to an oral or dermal exposure and is expressed in terms of mg/kg. Significant in evaluating the toxicity of a material; the lower the value, the more toxic the substance.
- **Lethal Concentration, 50% Kill (LC-50)**—concentration of a material, expressed as parts per million (ppm) per volume, which kills half of the lab animals in a given length of time. Refers to an inhalation exposure; the LC-50 may also be expressed as mg/liter or mg/cubic meter. This is significant in evaluating the toxicity of a material; the lower the value, the more toxic the substance.
- **Lethal Dose Low (LD_{LOW})**—the lowest amount of a substance introduced by any route, other than inhalation, reported to have caused death to animals or humans.
- **Lethal Concentration Low (LC_{LOW})**—the lowest concentration of a substance in air reported to have caused death in humans or animals. The reported concentrations may be entered for periods of exposure that are less than 24 hours (acute) or greater than 24 hours (sub-acute and chronic).
- **Threshold Limit Value/Time Weighted Average (TLV/TWA)**—the airborne concentration of a material to which an average, healthy person may be exposed repeatedly for 8 hours each day, 40 hours per week, without suffering adverse effects. The young, old, ill, and naturally susceptible will have lower tolerances and will need to take additional precautions. TLV's are based upon current available information and are adjusted on an annual basis by organizations such as the American Conference of Governmental Industrial Hygienists (ACGIH). As TLV's are time weighted averages over an eight-hour exposure, they are difficult to correlate to emergency response operations. The lower the value, the more toxic the substance.
- **Threshold Limit Value/Short-Term Exposure Limit (TLV/STEL)**—the 15-minute, time weighted average exposure which should not be exceeded at any time or repeated more than four times daily with a 60-minute rest period required between each STEL exposure. The lower the value, the more toxic the substance.
- **Threshold Limit Value/Ceiling (TLV/C)**—the maximum concentration that should not be exceeded, even instantaneously. The lower the value, the more toxic the substance.
- **Threshold Limit Value/Skin (Skin)**—indicates a possible and significant contribution to overall exposure to a material by absorption through the skin, mucous membranes, and eyes by direct or airborne contact.
- **Permissible Exposure Limit (PEL) and Recommended Exposure Levels (REL)**—the maximum time weighted concentration at which 95% of exposed, healthy adults suffer no adverse effects over a 40-hour work

NOTES

week and are comparable to ACGIH's TLV/TWA. PEL's are used by OSHA and are based on an eight-hour, time weighted average concentration. REL's are used by NIOSH and are based upon a 10-hour, time weighted average concentration.

- **Carcinogen**—a material that can cause cancer in an organism. May also be referred to as "cancer suspect" or "known carcinogens."
- **Mutagen**—a material that creates a change in gene structure which is potentially capable of being transmitted to the offspring.
- **Teratogen**—a material that affects the offspring when the embryo or fetus is exposed to that material.
- **Sensitizer**—a chemical that causes a substantial proportion of exposed people or animals to develop an allergic reaction in normal tissue after repeated exposure to the chemicals. Skin sensitization is the most common form.
- **Other**—any additional or pertinent data found.
- **Decontamination Procedures**—the process of making personnel, equipment, and supplies safe by the elimination of harmful substances. Methods available for decontamination of the materials involved should be determined.
- **Emergency Medical Procedures**—the medical procedures for dealing with individuals contaminated with the hazardous materials should be outlined.

FIRE HAZARDS

- **Flash Point**—minimum temperature at which a liquid gives off enough vapors that will ignite and flash over but will not continue to burn without the addition of more heat. Significant in determining the temperature at which the vapors from a flammable liquid are readily available and may ignite.
- **Ignition (Autoignition) Temperature**—the minimum temperature required to ignite gas or vapor without a spark or flame being present. Significant in evaluating the ease at which a flammable material may ignite.
- **Flammable (Explosive) Range**—the range of gas or vapor concentration (percentage by volume in air) that will burn or explode if an ignition source is present. Limiting concentrations are commonly called the "lower flammable (explosive) limit" and the "upper flammable (explosive) limit." Below the lower flammable limit, the mixture is too lean to burn; above the upper flammable limit, the mixture is too rich to burn. If the gas or vapor is released into an oxygen-enriched atmosphere, the flammable range will expand. Likewise, if the gas or vapor is released into an oxygen-deficient atmosphere, the flammable range will contract.
- **Toxic Products of Combustion**—the toxic byproducts of the combustion process.
- **Possible Extinguishing Agents**—the extinguishing agents suitable for the control and extinguishment of a fire involving this material.

NOTES

REACTIVITY HAZARDS

- **Reactivity/Instability**—the ability of a material to undergo a chemical reaction with the release of energy. It could be initiated by mixing or reacting with other materials, application of heat, physical shock, etc.
- **Oxidation Ability**—the ability of a material to (1) either give up its oxygen molecule to stimulate the oxidation of organic materials (e.g., chlorate, permanganate, and nitrate compounds), or (2) receives electrons being transferred from the substance undergoing oxidation (e.g., chlorine and fluorine).
- **Water Reactive Materials**—materials which will react with water and release a flammable gas or present a health hazard.
- **Pyrophoric Materials**—materials that ignite spontaneously in air without an ignition source.
- **Chemical Interactions**—the interaction of materials in a container may result in a build-up of heat and pressure, which may cause a container to fail. Similarly, the combined materials may be more corrosive than the material the container was originally designed to withstand and lead to container failure.
- **Polymerization**—a reaction during which a monomer is induced to polymerize by the addition of a catalyst or other unintentional influences, such as excessive heat, friction, contamination, etc. If the reaction is not controlled, it is possible to have an excessive amount of energy released.
- **Catalyst**—used to control the rate of a chemical reaction by either speeding it up or slowing it down. If used improperly, catalysts can speed up a reaction and cause a container failure due to pressure or heat build-up.
- **Inhibitor**—added to products to control their chemical reaction with other products. If the inhibitor is not added or escapes during an incident, the material will begin to polymerize, possibly resulting in container failure.
- **Self-Accelerating Decomposition Temperature (SADT)**—property of organic peroxides. When this temperature is reached by some portion of the mass of an organic peroxide, irreversibe decomposition will begin.

CORROSIVITY HAZARDS

- **Corrosivity Hazards**—a material that causes visible destruction of, or irreversible alterations to, living tissue by chemical action at the point of contact.
- **pH**—acidic or basic corrosives are measured to one another by their ability to dissociate in solution. Those that form the greatest number of hydrogen ions are the strongest acids, while those that form the hydroxide ion are the strongest bases. The measurement of the hydrogen ion concentration in solution is called the pH (power of hydrogen) of the compound in solution. The pH scale ranges from 0 to 14, with strong acids having low pH values and strong bases or alkaline materials having high pH values. A neutral substance would have a value of 7.
- **Strength**—the degree to which a corrosive ionizes in water. Those that form the greatest number of hydrogen ions are the strongest acids (e.g.,

NOTES

pH < 2), while those that form the hydroxide ion are the strongest bases (pH > 12).

- **Concentration**—the percentage of an acid or base dissolved in water. Concentration is NOT the same as strength.
- **Neutralizing Agents**—those materials which can be used to neutralize the effects of a corrosive material.
- **Other**—any additional or pertinent data found.

RADIOACTIVE HAZARDS

- **Radioactivity Hazards**—the ability of a material to emit any form of radioactive energy.
- **Type of Radiation Emitted**—the type of radioactive energy emitted, either alpha particles, beta particles, or gamma radiation.
- **Activity**—the number of radioactive atoms that will decay and emit radiation in 1 second of time. Measured in curies (1 curie = 37 billion disintegrations per second), although it is usually expressed in either millicuries or microcuries. Activity indicates how much radioactivity is present and not how much material is present.
- **Transportation Index (TI)**—the number found on radioactive labels which indicates the maximum radiation level (measured in milli-roentgens/hour—mR/hr) at 1 meter from an undamaged package. For example, a TI of 3 would indicate that the radiation intensity that can be measured is no more than 3 mR/h at 1 meter from the labeled package.
- **Other**—any additional or pertinent data found.

SOURCES OF HAZARD DATA AND INFORMATION

Two primary considerations within the hazard and risk assessment process are to (1) gather hazard data and information on the materials involved and to (2) compile that data in a useful manner so that it can be evaluated by either the Hazmat Branch Officer or the Incident Commander in assessing the level of risk.

In this section, we will discuss the various sources of hazard data and information and some methods for compiling that data into a logical and useful format.

For ease of discussion, sources of hazard data and information can be broken into the following categories:

- Reference Manuals
- Technical Information Centers
- Hazardous Materials Data Bases
- Technical Information Specialists
- Hazard Communication and Right-to-Know regulations
- Monitoring Instruments

NOTES

REFERENCE MANUALS AND GUIDEBOOKS

A variety of emergency response guidebooks and reference manuals currently exist for hazmat responders. While response personnel have often been hindered by the lack of adequate information during an emergency, there have also been situations where they have been overwhelmed by the amount of data and information available. In addition, there have also been situations where conflicting information has been obtained from different sources.

DON SELLERS

FIGURE 7-2: ERP BEING OVERWHELMED WITH INFORMATION

Despite the large number of written resources available, most ERP rely upon three to five basic response guidebooks for their primary data and information. These sources are generally selected based upon personal preferences and experience, ease of use, a large chemical listing, and a large, effective data bank. Figure 7-3 lists some of the most common reference guidebooks.

COMMON EMERGENCY RESPONSE GUIDEBOOKS AND MANUALS

FIRST RESPONDER GUIDEBOOKS

EMERGENCY RESPONSE GUIDEBOOK FOR HAZARDOUS MATERIALS
Office of Hazardous Materials Research and Special Programs Administration, U.S. Department of Transportation
Washington, DC 20036
(202) 366-4488

DANGEROUS GOODS INITIAL EMERGENCY RESPONSE GUIDE
Transport Canada
Regulatory Affairs—Transport Dangerous Goods
Ottawa, Ontario, Canada
K1A ON5
(613) 992-4624

THE FIREFIGHTERS HANDBOOK OF FIRE PROTECTION GUIDE ON HAZARDOUS MATERIALS
Maltese Enterprises, Inc.
P. O. Box 31009
Indianapolis, IN 46231
(317) 243-2211

GENERAL INFORMATION

EMERGENCY HANDLING OF HAZARDOUS MATERIALS IN SURFACE TRANSPORTATION
Association of American Railroads
Hazardous Materials Systems
50 F Street, NW
Washington, DC 20001
(202) 639-2100

HAZARDOUS MATERIALS EMERGENCY ACTION GUIDESHEETS
Association of American Railroads
Hazardous Materials Systems
50 F Street, NW
Washington, DC 20001
(202) 639-2100

FIRE PROTECTION GUIDE ON HAZARDOUS MATERIALS
National Fire Protection Association
1 Batterymarch Park
Quincy, MA 02269
(617) 770-3000

CHRIS—CHEMICAL HAZARDS RESPONSE INFORMATION SYSTEM
Superintendent of Documents
U.S. Government Printing Office
Washington, DC 20402
(202) 783-3238

NIOSH POCKET GUIDE TO CHEMICAL HAZARDS
Superintendent of Documents
U.S. Government Printing Office
Washington, DC 20402
(202) 783-3238

DANGEROUS PROPERTIES OF INDUSTRIAL MATERIALS (SAX)
Lab Safety Supply Company
P.O. Box 1368
Janesville, WI 53547-1368
(800) 356-0783

HAWLEY'S CONDENSED CHEMCAL DICTIONARY
Lab Safety Supply Company
P.O. Box 1368
Janesville, WI 53547-1368
(800) 356-0783

THE MERCK INDEX
Lab Safety Supply Company
P.O. Box 1368
Janesville, WI 53547-1368
(800) 356-0783

SPECIALIZED INFORMATION

FARM CHEMICALS HANDBOOK
Meister Publishing Company
37733 Euclid Avenue
Willoughby, OH 44094
(216) 942-2000
Subject Area: Agricultural Chemicals

RECOGNITION AND MANAGEMENT OF PESTICIDE POISONINGS
Superintendent of Documents
U.S. Government Printing Office
Washington, DC 20402
(202) 783-3238
Subject Area: Agricultural Chemicals

GATX TANK CAR MANUAL
General American Transportation Corporation
120 South Riverside Plaza
Chicago, IL 60606
(312) 621-6200
Subject Area: Railroad Tank Cars

GUIDELINES FOR THE SELECTION OF CHEMICAL PROTECTIVE CLOTHING, VOLUMES I AND II
National Technical Information Service
Springfield, VA 22161
(703) 487-4600
Subject Area: Chemical Protective Clothing

CHEMICAL PROTECTIVE CLOTHING PERMEATION AND DEGRADATION COMPENDIUM
Lewis Publishers
2000 Corporate Blvd., NW
Boca Raton, FL 33431
(407) 994-0555
Subject Area: Chemical Protective Clothing

HAZARDOUS MATERIALS INJURIES: A HANDBOOK FOR PRE-HOSPITAL CARE
Bradford Communications Corporation
10742 Tucker Street
Beltsville, MD 20705
(301) 345-0100
Subject Area: Emergency Medical Care

EMERGENCY CARE FOR HAZARDOUS MATERIALS EXPOSURE
The C. V. Mosby Company
11830 Westline Industrial Drive
St. Louis, MO 63146
(314) 872-8370
Subject Area: Emergency Medical Care

HAZARDOUS MATERIALS EXPOSURE: EMERGENCY RESPONSE AND PATIENT CARE
Lab Safety Supply Company
P.O. Box 1368
Janesville, WI 53547-1368
(800) 356-0783
Subject Area: Emergency Medical Care

BROTHERICIKS HANDBOOK OF REACTIVE CHEMICAL HAZARDS
Lab Safety Supply Company
P.O. Box 1368
Janesville, WI 53547-1368
(800) 356-0783
Subject Area: Reactive Chemicals

HANDBOOK OF COMPRESSED GASES
Lab Safety Supply Company
P. O. Box 1368
Janesville, WI 53547-1368
(800) 356-0783
Subject Area: Compressed Gases

FIGURE 7-3

NOTES

As with all resources, guidebooks are an information tool which have both advantages and limitations. Several operational considerations should be kept in mind when using them:

1. You must know HOW to use emergency response guidebooks before the incident in order to use them effectively. Evaluate reference materials before use and make sure that all references are using the same definitions for hazard terms. A good guidebook should have a well-written "How to Use" section.
2. Many responders will evaluate a minimum of two or three independent information sources and reference guidebooks before permitting personnel to operate within the Hot Zone. With experience, ERP develop a preference for those manuals and information sources they feel most comfortable with.
3. In some instances, there may be conflicting information between guidebooks. This is often due to the testing methods used (e.g., closed cup vs. open cup test for flash point), or differences in terminology. Always select the most conservative values or recommendations to ensure the greatest margin of safety.
4. Be realistic in your evaluation of the data contained in the guidebooks. For example, the ambient temperature of a liquid, and whether it is flammable or combustible, is much more critical than a slight discrepancy in flash points between references.
5. Conflicts may exist in protective clothing compatibility charts. Responders should initially rely upon the protective clothing manufacturers' chemical resistance tables. If conflicts are present, ALWAYS choose the most conservative recommendation.
6. During extended operations, it may be necessary to obtain extra copies of the pertinent reference guide information for distribution to the operating companies. Some response teams carry portable copying machines for this purpose.
7. Although reference guidebooks contain data on those chemicals most commonly encountered during hazmat incidents, they are not a complete listing of the chemicals found in your community. There is no replacement for hazard analysis and contingency planning at both the plant and community levels.

TECHNICAL INFORMATION CENTERS

A number of private and public sector hazardous materials emergency "hotlines" exist. Their functions include providing immediate chemical hazard information, accessing secondary forms of expertise for additional action and information, and acting as a clearinghouse for spill notifications.

In the United States, the most recognized emergency telecommunications center is CHEMTREC™ (Chemical Transportation Emergency Center). CHEMTREC™ is a public service operated by the Chemical Manufacturers Association (CMA) in Washington, DC. CHEMTREC™ can be contacted 24 hours daily at (800)424-9300 from anywhere with-in the United States, as well as Puerto Rico, the Virgin Islands, and Canada.

Callers outside of the United States and ships at sea can contact the Center using CHEMTREC's international and maritime number at (202) 483-7616 (NOTE: Collect calls are accepted). The U.S. Department of Transportation has recognized CHEMTREC™ as a central emergency response information center for hazardous materials transportation incidents.

The CHEMTREC™ Center provides a number of emergency and non-emergency resources, including the following:

1) **Emergency Response Information.** First, CHEMTREC™ provides immediate advice to callers anywhere on how to cope with chemicals involved in a transportation or fixed facility emergency. Secondly, it can access shippers, manufacturers, or other forms of expertise for additional and appropriate follow-up action and information. For example, CHEMTREC™ emergency specialists can initiate a conference phone call between on-scene responders and company representatives, as well as fax MSDS's from their data base directly to on-scene personnel.

2) **Chemical Industry Mutual Aid Network.** CHEMTREC™ is the coordination point for a chemical industry mutual aid network. The mutual aid network consists of chemical company ERT's as well as response teams from commercial contractors under contract to CMA. The primary objective of the network is to provide a chemical industry presence on-scene as soon as possible. **Chemical mutual aid teams cannot be activated by emergency responders; they must be requested by chemical companies that are members of the network.**

   The network works like this—if a mutual aid member is unable to respond to an incident involving one of their products in a timely manner, CHEMTREC™ links the company with the member response team closest to the scene and able to provide assistance. The member team would then respond to the incident scene and render assistance until the shipper's personnel arrive on-scene.

   CHEMTREC™ also serves as the point of contact for several product-specific industry mutual aid programs. These include products such as chlorine, sulfur dioxide, hydrogen fluoride, hydrogen cyanide, phosphorus, vinyl chloride, compressed gases, and swimming pool chemicals.

3) **Nonemergency Information.** The CHEMTREC™ Center also operates a toll-free nonemergency telephone number to provide ERP, chemical users, transporters, and the general public with access to health, safety, and environmental information on chemicals and chemical products. This nonemergency hotline is staffed weekdays from 9:00 am to 6:00 pm (Eastern Standard Time) and can be contacted at (800) 262-8200.

   The center will either answer the caller's question or will reference a data base of more than 2,500 manufacturers who have provided the CHEMTREC™ Center with pre-designated health and safety contacts who can be reached during normal business hours to answer questions regarding their products.

NOTES

Other emergency response and general information numbers that may be useful for both planning and response purposes include:

- **CANUTEC (Canadian Transport Emergency Center) at (613) 996-6666. The general information number is (613) 992-4624.** The Canadian counterpart of CHEMTREC™, it provides assistance in identification and establishing contact with shippers and manufacturers of hazardous materials that originate in Canada.
- **U.S. Coast Guard and the Department of Transportation National Response Center (NRC) at (800) 424-8802 or 267-2675 in the Washington, DC area.** The NRC is the federal government's central reporting point for emergencies which involve hazardous materials. The NRC must be notified by the responsible party (i.e., the spiller) if hazardous materials releases exceed the reportable quantity provisions of CERCLA. In addition, the NRC is used for the required reporting of hazardous materials transportation incidents that cause death, serious injury, property damage in excess of $50,000, or a continuing threat to life and property.
- **Agency for Toxic Substances and Disease Registry (ATSDR) at (404) 633-5313.** ATSDR is the leading federal public health agency for hazmat incidents and operates a 24-hour emergency number for providing advice on health issues involving hazmat releases. The following health professionals are available for consultation and advice:
  - Within 10 minutes, an Emergency Response Coordinator.
  - Within 20 minutes, a Preliminary Assessment Team consisting of a toxicologist, a chemist, an environmental health specialist, a physician, and other health personnel, as required.
  - Within 8 hours, an On-Site Response Team (if necessitated by the incident).
- **National Animal Poison Control Center at (800) 548-2423.** Operated by the University of Illinois, this number provides 24-hour consultation in the diagnosis and treatment of suspected or actual animal poisonings or chemical contamination. In addition, it staffs an emergency response team to investigate such incidents in North America and performs laboratory analysis of feeds/animal specimens/environmental materials for toxicants and chemical contaminants.
- **National Pesticide Telecommunications Network at (800) 858-7378.** Operated by Texas Tech University, the Network provides information on pesticide-related health/toxicity/minor clean-up to physicians, veterinarians, response personnel, and the general public. Hours of operation are 8:00 am to 6:00 pm (Central Standard Time), Monday through Friday.
- **Association of American Railroads, Hazardous Materials Systems (Bureau of Explosives) at (202) 639-2222.** A nonemergency number for assistance in handling railroad-related hazardous materials incidents.

NOTES

- **Environmental Protection Agency (EPA) RCRA Hotline at (800) 424-9346.** A nonemergency number for assistance regarding RCRA and CERCLA regulations. It serves as the central contact point and attempts to route inquiries to the proper, responsible parties for guidance and action.
- **TSCA, Public Information Desk at (202) 554-1404.** The Toxic Substance Control Administration handles questions regarding the Toxic Substance Control Act.
- **Department of Transportation (DOT) at (202) 366-4488.** Provides information and assistance concerning the hazardous materials transportation regulations found in the Code of Federal Regulations, Title 49.

In addition to these numbers, response personnel should develop a telephone roster of those individuals and agencies at the state and local level who can offer technical assistance. Examples include environmental and health departments, local chemical industry personnel, local hazmat spill cooperatives and clean-up contractors, and regional poison control centers.

HAZARDOUS MATERIALS DATA BASES

Portable computers and modem hook-ups provide easy access to both public- and private-sector data bases during an incident. In addition, a number of responders have acquired software packages with chemical emergency planning and response capabilities (e.g., CAMEO, Emergency Information Systems—Chemical version [EIS-C], the GlovES+™ chemical compatibility program, PlantSafe™) and CD-ROM disks (e.g., TOMES) containing various data bases, reference materials, resource listings, and specialized emergency response recommendations.

When evaluating computer software packages and electronic data bases, consider:

- The number of chemicals, including the number of fields of information available
- The frequency at which the data base is updated
- The subscription and user access fees
- The hardware requirements for both fixed and portable applications
- User friendly applications

A good source of information and networking is the Hazardous Materials Information Exchange (HMIX), which is jointly sponsored by the U.S. Department of Transportation (DOT) and the Federal Emergency Management Agency (FEMA). HMIX is a computerized bulletin board designed especially for the distribution and exchange of hazmat information. HMIX is available 24 hours daily, 7 days a week, and can be accessed through (708) 972-3275.

NOTES

TECHNICAL INFORMATION SPECIALISTS

One of the most likely sources of hazard information will be personnel who either work with the chemical(s) or their processing or who have some specialized knowledge, such as container design, toxicology, or chemistry. When evaluating these "specialist employees" and the information they provide, consider these observations:

- Many individuals who are specialists in a narrow, specific technical area may not have an understanding of the broad, multi-disciplined nature of hazmat response. Some information sources will provide extensive data on, for example, container design, yet may be unfamiliar with personal protective equipment used by ERP.
- Each individual source will have his or her own advantages and limitations. It is a good idea to simply remove the term "expert" from your vocabulary during a hazardous materials response.
- You will often interact with individuals with whom you have had no previous contact. Before relying on their recommendations, ascertain their level of expertise and job classification by asking specific questions. More than one responder has been disappointed or embarrassed to find that the "expert" they have been waiting for is actually a truck driver or part of a maintenance crew.
- When questioning outside information sources, consider yourself as playing the role of a detective. REMEMBER, FINAL ACCOUNTABILITY RESTS WITH THE INCIDENT COMMANDER. While this is certainly not an interrogation process, you must be confident of the individual's expertise and authority. In some cases, responders will ask questions for which they already know the answer in order to evaluate that person's competency and knowledge level.
- Local ERP and facility personnel must get out into their communities and make personnel contacts with the individuals with whom they will regularly communicate. These include state, regional, and federal environmental response personnel, law enforcement, clean-up contractors, industry representatives, wrecking and rigging companies, etc.
- Investigate the existence of local and state "Good Samaritan" legislation which may cover outside representatives as they assist you on the scene.

HAZARD COMMUNICATION AND RIGHT-TO-KNOW REGULATIONS

Today there are many state and local worker and community Right-to-Know laws across the country. In addition, OSHA has issued its Hazard Communication Standard (OSHA 1910.1200) for chemical markings and worker exposures to chemicals in the workplace.

While the scope of these regulations varies, local emergency response agencies have both access to MSDS's as well as regulatory requirements mandating the development of facility pre-incident plans and community hazardous materials plans.

Most concerns with Right-to-Know regulations center upon managing the MSDS information data base. From the ERP's perspective, the most effective Right-to-Know regulations are generally those which are maintained and updated by the manufacturer or user yet directly accessible by emergency services.

Responders must be able to both read and interpret material safety data sheets. OSHA requires that certain data and information be provided on each MSDS. Every MSDS must list the following basic information:

- **General Information**—includes manufacturer's name, address and emergency phone number, chemical name and family, and all synonyms.
- **Hazardous Ingredient Statement**—breaks out the active ingredients by percentage. Trade secret restrictions may sometimes minimize the amount of information available on an MSDS, although responders should have access to this data during an emergency.
- **Physical Data**—includes physical properties.
- **Fire and Explosion Data**—includes control and extinguishment measures, proper extinguishing agents, etc.
- **Health and Reactivity Hazard Data (as necessary)**—includes toxicology information, signs and symptoms of exposure, emergency care, chemical incompatibilities, decomposition products, etc.
- **Spill and Leak Control Procedures**—include procedures and precautions for handling chemical releases, as well as waste disposal methods.
- **Special Protection Information**—includes protective clothing and respiratory protection requirements.
- **Other Special Precautions (as necessary).**

Although MSDS's provide response personnel with a significant amount of information, you should realize that:

- There is currently no uniform format or layout. The only regulatory requirements are that the specified data be provided. However, the Chemical Manufacturers Association (CMA) has recently developed a voluntary consensus standard entitled *Guidelines for the Preparation of Material Safety Data Sheets (ANSI Z400.1)*. The standard outlines requirements for a uniform MSDS format and terminology.
- Computer-generated MSDS's may be difficult to initially interpret because of their layout.
- There are no regulatory requirements concerning the language used. MSDSs for the same chemical that are produced by different manufacturers may appear differently and, in some instances, may use different terminology.

NOTES

NOTES

MONITORING INSTRUMENTS

Monitoring and detection equipment are critical tools for evaluating "real-time" data to:

- Determine the appropriate levels of personal protective clothing and equipment.
- Determine the size and location of hazard control zones.
- Develop protective action recommendations and corridors.
- Assess the potential health effects of exposure.
- Determine when the incident scene is safe so that the public and/or facility personnel may be allowed to return.

Monitoring is an integral part of site safety operations. Numerous emergency response organizations have been issued regulatory citations for their failure to identify hazardous and IDLH conditions, to constantly evaluate the incident site for changes, and to verify the accuracy of hazard control zone locations.

Hazardous materials concentrations can be identified, quantified, and/or verified in two ways: on-site use of direct-reading instruments, and laboratory analysis of samples obtained through several collection methods. Both tools will be discussed in this section.

FIGURE 7-4

SELECTING DIRECT-READING INSTRUMENTS

Direct-reading instruments provide information at the time of sampling. This enables rapid, on-scene decision-making. They are used to detect and monitor flammable or explosive atmospheres, oxygen deficiencies, certain toxic and hazardous gases and vapors, and ionizing radiation. The selection of types of monitoring instruments should be based upon local/facility hazards and anticipated emergency responses.

NOTES

Many direct-reading devices were designed for specific needs within industry. When evaluating survey instruments for emergency response use in the field, consider the following criteria:

- ❑ **Portability and "User Friendliness"**—Ease to carry, weight, etc. Consider who will be using the instrument and their ability to consistently use it safely and correctly.
- ❑ **Instrument Response Time**—Period of time between when the instrument senses a product and when a monitor reading is produced. Depending upon the monitoring instrument, can range from several seconds to minutes.
- ❑ **Sensitivity and Selectivity**—The ability of the instrument to select slight changes in product concentrations and select a specific chemical or group of chemicals that react similarly. Monitoring instruments are calibrated on specific materials (e.g., methane, hexane). Increased selectivity widens the relative response of an instrument and can increase its accuracy; however, it may not be possible to determine the exact contaminant present.

  NOTE: Amplifiers are used in some monitoring instruments to widen their response to more hazmats and increase accuracy. However, other electrical equipment in the area, including radios, other types of monitoring instruments, power lines, transformers, etc., may interfere with the amplifier in the instrument being used.
- ❑ **Lower Detection Limit (LDL)**—The lowest concentration to which a monitoring instrument will respond. The lower the LDL, the quicker contaminant concentrations can be evaluated. Many instruments have several scales of operation for monitoring both very low and very high concentrations.
- ❑ **Calibration**—The process of adjusting a monitoring instrument so that its readings correspond to actual, known concentrations of a given material. Instruments are initially calibrated by the manufacturer; consideration should be given to the ability to calibrate instruments in the field without having to return the instrument to the manufacturer. Conditions for calibration include atmosphere, humidity, temperature, atmospheric pressure, etc.
- ❑ **Relative Response Curves**—Relative response is the difference between a calibrated response and a response from a product for which the meter is not calibrated. This is a common issue with combustible gas indicators (CGI's). The manufacturer should provide a comparison or response curve table to adjust the readings for the product being evaluated.
- ❑ **Inherent Safety**—The inherent safety and the ability of the device to operate in hazardous atmospheres must be evaluated (see Figure 7-5). In addition, it should be confirmed that the instrument has been certified by an approved testing laboratory for expected operations.

NEC HAZARDOUS CLASSIFICATIONS

When evaluating direct-reading monitoring instruments, responders must have a basic understanding of certain terms and classifications which relate to the ignition potential of electrical instruments in hazardous environments. Hazardous locations are defined in Article 500 of NFPA 70—The National Electrical Code.

Three simultaneous conditions create a hazardous location:

1) Vapors, dusts, or fibers present in sufficient quantity to ignite (i.e., within the flammable range).
2) Source of ignition may be present (e.g., the instrument).
3) The resulting exothermic reaction could propagate beyond where it started.

There are three classes, two divisions, and various sub-groups within the NEC classification structure.

Classes—Describe the type of flammable materials that produce the hazardous atmosphere.

Class I Locations—Flammable gases or vapors may be present in quantities sufficient to produce explosive or ignitible mixtures.

Class II Locations—Concentrations of combustible dusts may be present (e.g., coal or grain dust).

Class III Locations—Areas concerned with the presence of easily ignitible fibers or flyings (e.g., cotton milling).

Groups. Groups are products within a Class. Class I is divided into four groups (Groups A–D) on the basis of similar flammability characteristics. Class II is divided into three groups (Groups E–G). There are no groups for Class III materials.

Group A Atmospheres

Acetylene

Group B Atmospheres (not sealed in conduit 1/2 inch or larger)

1,3-butadiene
Ethylene Oxide
Formaldehyde (gas)
Hydrogen
Manufactured Gas (containing > 30% hydrogen by volume)
Propylene Oxide
Propyl Nitrate
Allyl Glycidyl Ether
n-Butyl Glycidyl Ether

Group C Atmospheres (selected chemicals)

Acetaldehyde	Carbon Monoxide	Dicyclopentadiene
Diethyl Ether	Epichlorohydrin	Ether Acetate
Ethylene	Ethylene Glycol	Ethyl Mercaptan
Hydrazine	Hydrogen Cyanide	Hydrogen Selenide
Hydrogen Sulfide	Methylacetylene	Monoethyl Ether
Nitropropane	Tetraethyl Lead	Tetrahydrofuran

Group D Atmospheres (selected chemicals)

Acetone	Acetonitrile	Acrylonitrile
Ammonia	Aniline	Benzene
Butane	Chlorobenzene	Cyclohexane
Dichloroethane	Ethane	Ethyl Alcohol
Ethylene Glycol	Fuel Oils	Gasoline
Hexane	LPG	Methane
Methyl Alcohol	Methyl Ethyl Ketone	Monomethyl Ether
Naptha	Propane	Styrene
Vinyl Chloride	Xylene	

Group E Conductive Dusts

Atmospheres containing metal dusts, including aluminum, magnesium, and their commercial alloys, and other metals of similarly hazardous characteristics.

Group F Semi-Volatile Dusts

Atmospheres containing carbon black, coal, or coke dust with more than 8% volatile material.

Group G Nonconductive Dusts

Atmospheres containing flour, starch, grain, carbonaceous, chemical thermoplastic, thermosetting, and molding compounds.

Divisions—Describe the type of location that may generate or release a flammable material.

Division 1—Location where the vapors, dust, or fibers are continuously generated and released. The only element necessary for a hazardous situation is a source of ignition.

Division 2—Location where the vapors, dusts, or fibers are generated and released as a result of an emergency or failure in the containment system.

Division 1 areas have a greater probability of generating a hazardous atmosphere than in Division 2. Although Division 1 devices are permitted for use in Division 2 areas, instruments approved for Division 2 areas are not usable in Division 1 areas. At a minimum, it is recommended that direct-reading instruments be rated for operations in Class I, Division 2 hazardous classifications.

To reduce the ignition potential of both fixed and portable monitoring instruments, manufacturers use several different methods. They include:

Explosion-Proof Construction. Encases the electrical equipment (i.e., potential ignition source) in a rigidly built container so that (1) it withstands the internal explosion of a flammable mixture, and (2) it prevents propagation to the surrounding flammable atmosphere. Used in Class I, Division 1 atmospheres at fixed installations.

Intrinsically Safe Construction. Equipment or wiring is incapable of releasing sufficient electrical energy under both normal and abnormal conditions to cause the ignition of a flammable mixture. Commonly used in portable direct-reading instruments for operations in Class I, Division 2 hazardous classifications.

Purging. Totally enclosed electrical equipment is protected with an inert gas under a slight positive pressure from a reliable source. The inert gas provides positive pressure within the enclosure and minimizes the development of a flammable atmosphere. Used in Class I, Division 1 atmospheres at fixed installations.

Explosion-proof construction and purging are primarily used at fixed facilities and for protecting stationary instrumentation. In contrast, direct-reading instruments used for field applications will rely upon intrinsically safe construction. Users should ensure that the instrument has been certified by an approved testing laboratory (e.g., Underwriters' Laboratories or Factory Mutual) for operations within your required class, group, and division.

NATIONAL FIRE PROTECTION ASSOCIATION

FIGURE 7-5: NEC HAZARDOUS CLASSIFICATIONS

In addition to the above criteria, the following operations, storage, and use considerations should be evaluated:

- Where and in what type of storage container will the instruments be stored? This is especially critical when placing monitoring instruments inside or outside of vehicles.
- Can field maintenance be easily performed? For example, are calibration kits available and can sensors be easily changed?

NOTES

- Can buttons, switches, etc. be easily manipulated while wearing chemical gloves and double gloves?
- How long does it take for the monitoring instruments to "warm up" before they can be used in the field?
- What types of alarms does the instrument have? Can meters be easily read while wearing respiratory protection? Is there a glare problem during daytime operations and a lighting problem for operations at night?
- What types of batteries are required for the instrument—off-the-shelf batteries or rechargeable batteries? While rechargeable batteries work very well, when the batteries are dead they usually cannot be replaced with off-the-shelf batteries.

TYPES OF DIRECT-READING INSTRUMENTS

All direct-reading instruments have inherent limitations. Many detect and/or measure only specific classes of chemicals. As a general rule, they are not designed to measure and/or detect airborne concentrations below 1 ppm. Also, many direct-reading instruments designed to detect one particular substance may detect other substances (interference) and give false readings.

Figure 7-6 outlines the common types of direct-reading instruments used in the emergency response field.

When using direct-reading instruments, interpret instrument readings conservatively and consider the following guidelines:

- Calibrate instruments according to the manufacturer's instructions before and after every use.
- Develop chemical response curves if they are not provided by the manufacturer.
- Remember that instrument readings have limited value when dealing with unknown substances. Report readings of unknown contaminants as "needle deflection" or "positive instrument response" rather than specific concentrations (i.e., ppm). Conduct additional monitoring at any location where a positive response occurs.
- A reading of zero should be reported as "no instrument response" rather than "clean," since quantities of chemicals may be present that are not detectable by that particular instrument.
- Repeat the initial survey using several types of detection devices to maximize the number of chemicals detected.
- After the initial survey, continue frequent monitoring throughout the incident.

Monitoring instruments help responders to (1) verify information regarding the type(s) of materials involved in an emergency and their associated hazards; (2) determine the appropriate levels of personal protective clothing and equipment necessary for entry operations; (3) determine the size and location of control zones; and (4) develop protective action recommendations and corridors (e.g., evacuation vs. sheltering-in-place).

DIRECT-READING INSTRUMENTS AND RELATED EQUIPMENT

COMBUSTIBLE GAS INDICATOR CGI (Also referred to as Explosivemeter)

HAZARD MONITORED
Combustible Gases and Vapors

APPLICATION
Measures the concentration of a combustible gas or vapor in air.

METHOD OF OPERATION
Operate by catalytic combustion, where a sample is drawn across a surface of heated platinum. If the sample is flammable, it will burn on the platinum catalyst.

Catalytic combustion will produce heat in direct proportion to the concentration of combustible gas present.

Heated surface is connected to an electrical circuit (known as Wheatstone Bridge) so that increases in temperature give corresponding indication on the meter.

Meter may read in either percent of the LEL, ppm, or percent of gas by volume.

GENERAL COMMENTS
Intended for use in normal atmospheres (i.e., not oxygen deficient or enriched). Will affect readings if oxygen levels are deficient (lower readings).

Often require an initial warm-up period to heat up the platinum catalyst.

Does not respond to all combustible gases/vapors in the same manner as the reference gas (e.g., methane, hexane). Response curves will be required.

Operating filament is damaged by certain materials, including silicone, tetraethyl lead, and acid gases. Some problems can be reduced through use of filters and water traps between instrument and sampling tube.

Often found as a combination CGI/oxygen meter or as a triple combination meter with a third gas (e.g., carbon monoxide, hydrogen sulfide).

OXYGEN MONITORS

HAZARD MONITORED
Oxygen Deficient and Enriched Atmospheres

APPLICATION
Measures the percentage of oxygen in air.

Should measure both oxygen-deficient and oxygen-enriched atmospheres.

METHOD OF OPERATION
Operate by diffusion process, in which air diffuses into the sensor.

May have either a passive sensor or a battery or bulb operated pump which draws in the air sample.

Oxygen reacts with electrolytes in a cell, which generates a current flow in the meter. Meter will normally read percent of oxygen in the sample (e.g., 21% etc.).

GENERAL COMMENTS
Some materials (e.g., chlorine, fluorine) will indicate a high or normal level of oxygen, when actual atmosphere may be oxygen deficient.

Extreme cold temperatures can often result in sluggish, delayed movement of the meter.

Must be calibrated prior to use to compensate for altitude and barometric pressure.

High CO_2 concentrations can shorten the life of the oxygen sensor.

Can be individual unit or combined with a CGI.

COLORIMETRIC INDICATOR TUBES (Also referred to as Detector Tubes)

HAZARD MONITORED

Specific Gases and Vapors

APPLICATION

Measures the concentration of specific gases and vapors in air.

Used in hazard categorization systems for testing for unknowns; enable the user to classify the hazard class or chemical family of the unknown contaminant.

METHOD OF OPERATION

Use tubes which are filled with different reagents which react with the chemical being tested. When that chemical is present, the reagent changes color or produces a stain.

Sample is drawn through the tube by either a bellows or piston pump.

Readings can be direct or interpolated based upon the number of strokes or pumps. Wrong number of strokes/pumps can lead to incorrect readings.

Read the instructions for each tube prior to use.

GENERAL COMMENTS

Primarily used to determine if a specific chemical is present or not, as compared to specific quantitative results.

May be found in pre-packaged Hazmat Kits with a sampling matrix for the identification of unknowns.

The measured concentration of the same compound may vary between different manufacturers' tubes. In addition, cannot use different manufacturers' tubes interchangeably.

Many similar chemicals can interfere with the sampling and give "false positive" readings.

The greatest sources of error are (1) how the user judges the color change or stain's endpoint, and (2) the tube's limited accuracy. Tubes have a margin of error up to 35%.

Tubes have a limited shelf life. Can also be affected by high humidity and temperature extremes.

Response times may vary greatly from chemical to chemical.

SPECIFIC CHEMICAL MONITORS

HAZARD MONITORED

Specific Gases and Vapors

APPLICATION

Designed to detect either a large group of chemicals or a specific chemical.

Most common examples are carbon monoxide and hydrogen sulfide.

Monitors may also be found for chlorine, ammonia, sulfur dioxide, and hydrogen cyanide.

METHOD OF OPERATION

Utilize electrochemical cells or metal oxide semi conductors (MOS) for detecting specific chemicals.

MOS detectors change electrical conductivity when exposed to certain gases or vapors.

GENERAL COMMENTS

More accurate than detector tubes, but they are limited to approximately a dozen different chemicals.

Often found as a combination meter with a CGI and/or oxygen monitor.

FLAME IONIZATION DETECTOR (FID) (Also referred to as an OVA, Organic Vapor Analyzer)

HAZARD MONITORED

Organic Vapors and Gases

APPLICATION

Operates in two modes—survey mode and gas chromatograph (GC).

Survey Mode—Detects total concentration of many organic vapors and gases. All organic compounds are ionized and detected at the same time.

GC Mode—Identifies and measures specific compounds. Volatile compounds are separated.

METHOD OF OPERATION

Survey Mode—Ionizes any chemical with an Ionization Potential (IP) of less than 15.4 eV. Sample is exposed to a hydrogen flame which ionizes the sample. Ions are collected, amplified, and produce a current which is read on a display as total organic vapors present (in ppm).

GC Mode—Sample is drawn into a column with an inert material. Hydrogen gas is passed through the tube and picks up various components of the sample. Each chemical takes a period of time to exit the tube. As the sample passes into the detector the energy is measured on a strip chart recorder as a peak. Identification can be determined by comparing the strip chart results to known standards.

GENERAL COMMENTS

Not all FID's are intrinsically safe.

Used in survey mode for determining toxic levels of contaminants. Does not detect inorganic vapors and gases.

In survey mode, readings are reported relative to calibration standard (e.g., methane).

High concentration of contaminants or oxygen-deficient atmospheres may require system modification.

Response affected by temperatures below 40°F, as gases condense in the pump and column.

Not commonly used by public safety responders. Requires experience to both operate and interpret data correctly. Interpretation of the data is difficult, and each chemical requires specific calibration.

PHOTOIONIZATION DETECTOR (PID)

HAZARD MONITORED

Organic and Some Inorganic Gases and Vapors

APPLICATION

Detects total concentration of many organic and some inorganic gases and vapors.

Is normally a general measuring/survey device, but if the sample is known, the unit can be calibrated for specific chemicals.

Being increasingly used for both emergency response and remedial operations as a general survey instrument.

METHOD OF OPERATION

Sample is exposed to ultraviolet light which ionizes the sample. Ions are collected, amplified, and produce a current which is read on a display as total organic vapors present (in ppm).

The Ionization Potential (IP) of the ultraviolet lamp is critical. The 10.2 or 10.6 electron volts (eV) lamps are often used for emergency response purposes because of their durability.

If the IP of the sample is above the eV rating of the lamp, no reading will be detected.

GENERAL COMMENTS

IP's can be found in the NIOSH *Pocket Guide to Chemical Hazards.*

Reading is dependent upon calibration. Most commonly calibrated on isobutylene.

Does not detect methane.

Readings may change when gases are mixed.

Response may be affected by radio frequencies, power lines, and transformers. Dust and high humidity can block the transmission of UV light, causing a reduction in instrument reading.

Good for surveys of organics. However, when surveying for unknowns, don't only use a PID—use it in conjunction with other meters.

RADIATION MONITORS

HAZARD MONITORED

Alpha, Beta, or Gamma Radiation Sources

APPLICATION

Measure accumulated radiation exposure.

METHOD OF OPERATION

Ionization detectors which collect and count ions electronically.

Readings can be provided in a number of formats, including

cpm = counts per minute

rem = roentgen equivalent man

mR/hr = milliroentgens per hour

Lowest level of detection is 0.01 mR/h.

Typical background radiation is 0.01–0.05 mR/h.

GENERAL COMMENTS

Number of different probes available, with the Geiger–Mueller being most common. Designed to differentiate between gamma and beta—will not detect alpha sources.

No response does not equal clean!

GENERAL COMMENTS

Electromagnetic fields can give "false positive" readings.

Measurements can be affected by wind, shielding, etc.

Many detectors are battery operated. Unless units are frequently used, batteries should not be stored in unit. Weak batteries will produce inaccurate or erratic readings.

Factory calibration should take place at least annually. Consider need for radiation check sources (e.g., Cesium 137) to ensure unit is working.

pH METERS, PAPERS, AND STRIPS

HAZARD MONITORED

Corrosivity—Acidity or Alkalinty

APPLICATION

Measure the acidity or alkalinity of a corrosive material.

METHOD OF OPERATION

pH Meters

pH Paper/Strips—chemical reaction changes the color of the detection paper. Acids are normally shades of red and purple, while bases/caustics are shades of blue.

GENERAL COMMENTS

Readings less than 2 or more than 12.5 are cause for extreme caution.

Do not provide the specific concentration of the corrosive material.

pH meters must be calibrated before each use. Probes must be thoroughly rinsed with distilled water before and after each calibration and use.

pH papers should be pre-moistened with water prior to use.

INDICATOR PAPERS

HAZARD MONITORED

Test for Specific Hazards, such as Oxidizers, Peroxide, and Hydrogen Sulfide in Air.

Indicating papers are usually part of a hazmat identification system, such as the HazCat® Chemical Identification System.

APPLICATION

Oxidizer indicating paper tests for the presence of oxidizing materials in excess of normal oxygen in air.

Peroxide indicating paper tests for presence of peroxides.

Sulfide indicating paper tests for the presence of hydrogen sulfide and related sulfide compounds.

METHOD OF OPERATION

Oxidizer Test—uses potassium iodide starch paper (KI paper) wetted with 2–3 drops of 3N hydrochloric acid and touched to the unknown. Color change to blue/black, purple, or black indicates oxidizers. The faster the color change, the greater the oxidizing ability.

Peroxide Test—uses peroxide test paper touched to the unknown. Use and sequence of wetting the paper depends upon whether the unknown is a solid, aqueous solution, or organic solvent. Color change to blue after 15 seconds indicates a peroxide.

Sulfide Test—uses lead acetate paper wetted with 2–3 drops of 3N hydrochloric acid, and touched to the unknown. Color change to brown/black indicates presence of sulfides. May also get a sulfur odor.

GENERAL COMMENTS

Oxidizer Test—positive reading indicates chlorine, nitrogen dioxide, oxygen, ozone, etc.

Peroxide Test—positive reading indicates presence of organic peroxides (e.g., hydrogen peroxide). NOTE: peroxide paper both expensive and sometimes difficult to obtain.

Sulfide Test—positive reading indicates hydrogen sulfide.

Indicator paper readings should be confirmed by other instruments (e.g., colorimetric tubes).

HALOGEN LEAK DETECTOR

HAZARD MONITORED

Halogenated Hydrocarbons at Low Levels

APPLICATION

Refrigeration gun, which selectively identifies halogenated hydrocarbons at low levels (ppm).

Detect materials such as freons, methyl bromide, perchloroethylene, trichloroethylene, etc.

METHOD OF OPERATION

Commercial equipment which is commonly used in the refrigeration industry for locating refrigerant leaks.

GENERAL COMMENTS

Not intrinsically safe and should not be used in flammable atmospheres.

Only capable of providing qualitative reading (yes/no).

Acetone will produce a false positive reading.

PASSIVE DOSIMETERS

HAZARD MONITORED

Specific Chemical and Radiation Hazards

APPLICATION

Used to determine the chemical or radiological exposure that an individual receives while working around a specific material or area.

METHOD OF OPERATION

Passive monitor which measures an individual's exposure.

May provide either immediate results or required to be sent to a laboratory for analysis.

GENERAL COMMENTS

Must acquire the specific dosimeter for the materials in question (e.g., organic vapors, ethylene oxide, mercury).

Commonly used to monitor for the TLV/TWA and the TLV/STEL.

FIGURE 7-6: COMMON MONITORING INSTRUMENTS

NOTES

Initial air monitoring and reconnaissance operations pose the greatest threat to emergency responders. Remember the following safety considerations:

- Air monitoring personnel have the greatest risk of exposure, so protective clothing must be sufficient for the expected hazards. The situation should determine the level of protective clothing used. If there is no potential for skin exposure, chemical splash protective clothing (EPA Level B) and positive-pressure self-contained breathing apparatus (SCBA) may be used. However, if there is a chance for skin exposure or there is a high potential for an unknown vapor exposure, then chemical vapor protective clothing (EPA Level A) should be used.
- The air monitoring team should consist of at least two personnel, with a back-up team wearing an equal level of protection.
- Protect the instruments as appropriate. Instruments may be wrapped in clear plastic to protect against liquid, vapor, and dust contamination; however, make sure that the intake and exhaust ports are open.
- Approach the hazard area from upwind whenever possible. The initial site survey should begin upwind and then move to the flanks of the release. As soon as any positive indication is received, proceed with caution.
- Priority areas should include confined spaces, low-lying areas, and behind natural or artificial barriers (e.g., hills, structures, etc.) where heavier than air vapors can accumulate.

MONITORING STRATEGY

The Incident Commander and/or the Hazmat Branch Officer must establish a monitoring strategy. In developing this strategy, the following operational issues must be considered:

- Always use the appropriate monitoring instrument(s) based upon dealing with known or unknown materials. The instrument(s) should be able to detect the anticipated hazard(s), measure appropriate concentrations, and operate under the given field conditions.
- Monitoring personnel should have a good idea of what readings to expect. In the event that abnormal or unusual readings are encountered, the possibility of instrument failure should be considered. Try to confirm the initial reading with another instrument.
- The absence of a positive response or reading does not necessarily mean that contaminants are not present. A number of factors can affect contaminant concentrations, including wind, temperature, and moisture.
- Never assume that only one hazard is present.
- Interpret the instrument readings in more than one manner (i.e., always play "devil's advocate").
- Establish action levels based upon instrument readings.

NOTES

Monitoring operations at long-term incidents may be performed by different responders, using different monitoring instruments at different locations. Reliable monitoring operations can only occur by using consistent monitoring procedures. Various systems are available to indicate the locations where monitoring is conducted. For example, in Figure 7-7 a clock system is used to refer to locations in and around the incident site. Small traffic cones may also be numbered and used to indicate the relative positions. The entry point is at the 6 o'clock position; points away from the source of the spill would be lettered at approximate distances (e.g., every 10 ft, 20 ft, etc.), based upon the incident. For example, initial monitoring was conducted at B-6.

Monitoring results should be documented as follows:

- Instrument—the type of monitoring instrument being used.
- Location—location where the monitoring is conducted (e.g, traffic cone #1, flag, etc.).
- Time—time at which the monitoring is conducted.
- Level—level where the monitoring reading is taken (e.g., foot, waist, head).
- Reading—the actual reading given by the monitoring instrument.

Monitoring priorities will be dependent upon whether ERP have identified the hazmat(s) involved. Priorities should be systematic and continuously evaluated throughout the course of the emergency.

Unknowns will create the greatest challenge for emergency responders. The suspected physical state of the unknown (i.e., solid, liquid, or gas) and the location of the emergency (e.g., outdoors, indoors, confined space) will influence the monitoring strategy. It should be emphasized that the role of hazmat personnel when dealing with unknown hazardous materials is much like that of a detective. At the conclusion of the testing process, responders may still be unable to specifically identify the material(s) involved; however, they should be able to rule out a number of hazard classes and shorten the list of possibilities.

The following monitoring priority is used by many HMRT's when dealing with unknowns. **Although radiation is listed as the initial priority, most responders choose to initially monitor for flammability. Reasons include the high probability of encountering a flammable atmosphere (over 50% of hazmat incidents involve flammable liquids and gases), the fact that flammable atmospheres pose an immediate threat to the safety of responders, and the need to assess the safety of the "work area" as soon as possible.** Initial efforts should be toward determining if IDLH concentrations are present and providing an initial base to confirm or refute the existence of specific hazards.

1) **Radiation**—if there is any doubt, radiation detection should be the first priority. Remember—gamma rays travel the greatest distance and are the primary hazard for external exposure. A positive reading above background level would confirm the existence of a radiation hazard. If initial gamma radiation readings are negative but clues are present indicating the possible presence of radioactive materials, additional testing for beta and alpha sources should be conducted.

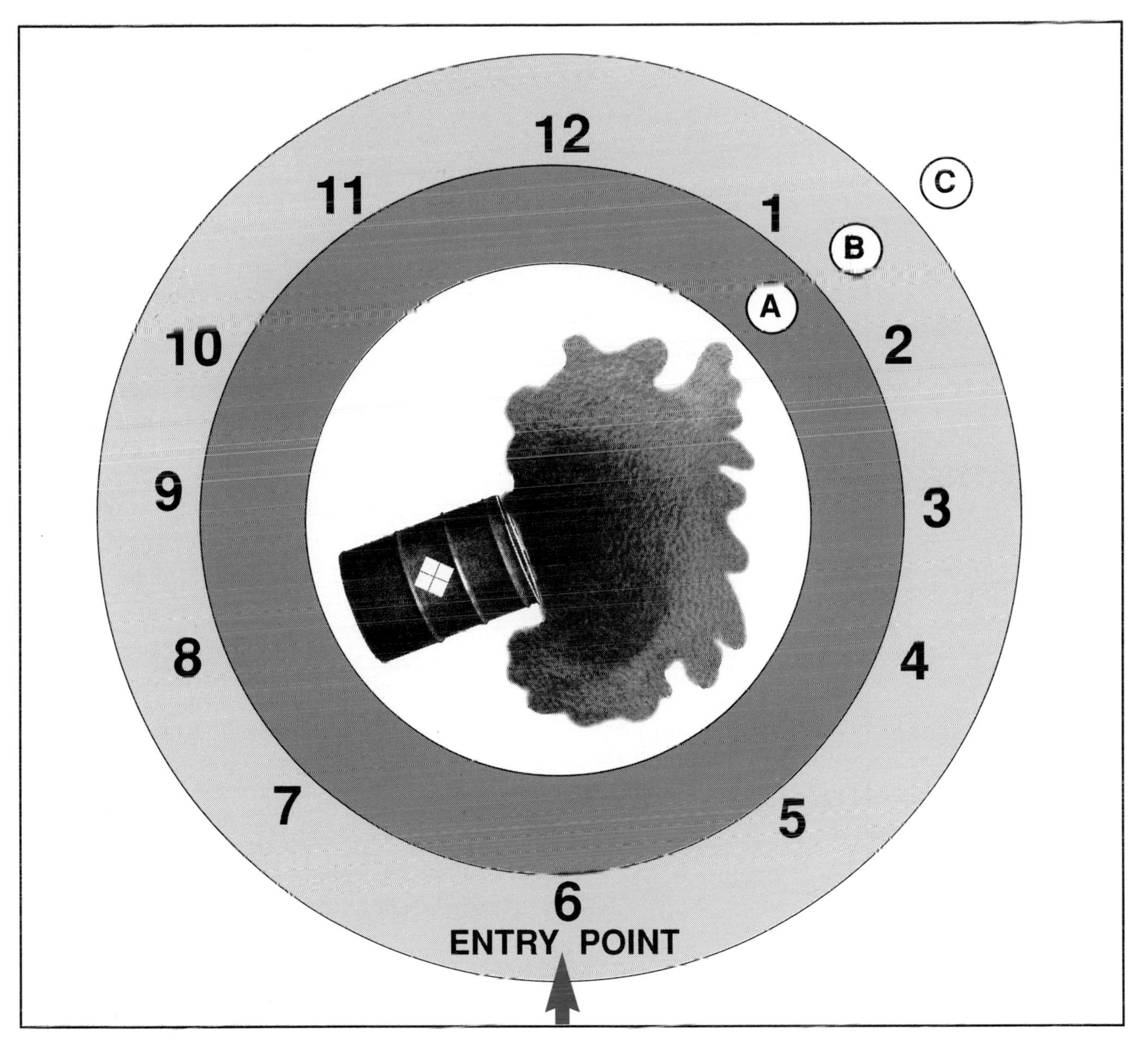

FIGURE 7-7: MONITORING RESULTS SHOULD BE DOCUMENTED USING A CONSISTENT PROCESS

2) **Flammability**—since flammability and oxygen levels are directly related, monitoring for flammability and oxygen are usually implemented simultaneously through combination CGI/O_2 meters. Remember that an oxygen-deficient atmosphere will shorten the flammable range, while an oxygen-enriched atmosphere will expand the flammable range.

3) **Oxygen**—monitoring should evaluate the presence of both oxygen deficient and enriched atmospheres, particularly when dealing with confined spaces. ERP must consider that the level of oxygen may also be influenced by the level of contaminants (i.e., increasing level of contaminants displaces available oxygen).

4) **Toxicity**—the level and sophistication of toxicity monitoring will depend upon available instrumentation. Resources range from the simple to the sophisticated and may include the following:

NOTES

- *Chemical Classifier™ Kits,* which can test liquid spills for corrosivity, oxidizers, fluoride, organic solvents/petroleum distillates, and iodine/bromine/chlorine.
- *Specific or combination air monitors,* which detect toxic gases such as hydrogen sulfide or carbon monoxide. Carbon monoxide monitors are useful in fire and post-fire situations. Hydrogen sulfide monitors are useful when dealing with confined spaces and when working around petroleum facilities where "sour" gas is handled.
- *Colormetric detector tubes* can be used for both known and unknown substances. Commercial detector tube kits are available for identifying unknown airborne hazards. By conducting a series of measurements with various pre-established detector tubes, responders can often determine the chemical class (e.g., organic gases, alcohols, acidic gases, etc.) which is present. These kits only indicate the presence of several classes of gases and vapors; they provide a gross estimate of concentration and will not differentiate between specific gases within a certain class.
- *Survey instruments,* such as flame ionization detectors (FID's) and photoionization detectors (PID's).

5) **Samples**—if air monitoring instruments provide no information on the identity of the unknown, ERP may have to collect a sample and conduct several field chemical tests to determine the hazard(s) present. Both locally developed and commercial chemical identification kits (e.g., HazCat™ Chemical Identification System) can be found in the field. These systems rely upon a series of chemical tests which follow a logic sequence to identify unknown materials.

EVALUATING MONITORING RESULTS—ACTIONS LEVELS AND GUIDELINES

Initial monitoring efforts should be directed toward determining if IDLH concentrations are present. Decisions regarding protective clothing recommendations, establishing control zones, and evaluating any related public protective actions should be based upon the following parameters:

1) Radioactivity—any positive reading above background level would confirm the existence of a radiation hazard and should be used as the basis for initial actions.
2) Flammability—if dealing with a confined space or indoor release, the IDLH/action level is 10% of the lower explosive limit (LEL). If dealing with an open-air release, the initial action level is 20% of the LEL.
3) Oxygen—an IDLH oxygen deficient atmosphere is 19.5% oxygen or lower, while an oxygen-enriched atmosphere contains 23.5% oxygen or higher. In evaluating an oxygen deficient atmosphere, consider that the level of available oxygen may be influenced by contaminants which are present.
4) Toxicity—unless a published action level or similar guideline (e.g., ERPG-2) is available, the STEL or IDLH values should initially be used. If there is no published IDLH value, ERP may consider using

an estimated IDLH of ten times the TLV/TWA. Control zones can be established for toxic materials using the following guideline:

- Hot Zone—monitoring readings above STEL or IDLH exposure values.
- Warm Zone—monitoring readings equal to or greater than TLV/TWA or PEL exposure values.
- Cold Zone—monitoring readings less than TLV/TWA or PEL exposure values.

COLLECTING SAMPLES

If air monitoring instruments provide no information on the identity of an unknown, responders may have to collect a sample and conduct several field chemical tests to categorize the hazards of the material. These tests can place unknown materials in a specific hazard class, such as flammables, corrosives, oxidizers, etc. Both locally and commercial identification kits (e.g., HazCat™ Chemical Identification System) are used in the field. These systems rely upon a series of chemical tests which follow a logic sequence to identify unknown solid and liquid materials.

The following supplies and equipment will be required for sample collecting. These materials can be obtained through both laboratory and safety supply stores, and hardware stores. The equipment includes:

- Nonsparking bung wrench.
- Glass tube or disposable polypropylene/PVC bailer.
- Coliwasa waste samplers (Composite Liquid Waste Sampler).
- Loosely woven fiberglass cloth.
- Turkey baster (resembles a huge eyedropper).
- Nonsparking sample pole, extendible to 10 feet.
- Plastic or glass cup.
- Glass and plastic bottles.
- Bomb sampler or weighted bottle sampler.

When collecting samples, the following procedures should be followed:

- Collect the samples from an upwind position. In most emergency response situations, a minimum of chemical liquid splash protective clothing (EPA Level B) should be used.
- When collecting the sample, record the type of container, any markings on the container, and any reactions or other relevant information concerning the site.
- Once the sample is properly collected, take it to a safe testing location in the warm zone.
- If the sample may become part of a criminal or regulatory investigation, ensure that chain of custody procedures are completely followed and documented.

Sampling Methods and Procedures. The following are accepted methods used for collecting samples:

NOTES

- **Drums**—When necessary to open a drum to collect a sample, use a non-sparking bung wrench. Manual drum opening should be performed only with structurally sound drums having contents that are known to be not shock sensitive, nonreactive, nonexplosive, and nonflammable. Monitor the bung hole with a CGI prior to and during opening. There are destructive nonsparking drum openers that can be operated remotely.

  Use a glass tube (stinger), a coliwasa, or a polyethylene bailer to collect a sample from the drum. The glass tube is simple, cost-effective, and quick, and is the most widely used implement for sampling. By inserting the glass tube into the drum and then removing it, a cross section of the drum's contents can be collected. This is similar to taking milk out of a glass by holding a finger over the end of a straw.

  The coliwasa is designed to collect a sample from drums and other containerized wastes. Some coliwasa kits have coupler sets which allow different coliwasa sets to be combined to increase their length. Drums may "layer" or stratify after sitting for extended periods. The bottom, middle, and top samples within the sampling tube should be placed into separate containers for analysis.
- **Sumps and Wells**—Disposable polypropylene bailers were made for groundwater monitoring wells but also work well for retrieving samples from other static water sources. The open top bailer has a ball at the bottom that allows liquids to enter but prevents the liquids from escaping. Bailers can be lowered into the water by a suspension cord.
- **Puddles**—Use a turkey baster or a small plastic/glass cup attached to the end of a pole to collect a sample from the puddle.
- **Slick on Top of Water**—Use a piece of loosely woven fiberglass cloth attached to a string. Place the fiberglass on top of the water; the floating material will collect on the cloth. Two or three samples may be necessary to obtain a sufficient amount for testing.
- **Heavier than Water Unknowns (from underwater)**—Lower a bomb sampler or a weighted bottle sampler into the water and open it when the bomb enters the layer to be sampled.
- **Deep Holes**—Lower a glass or plastic bottle on a string or fishing line into the hole to collect the sample.
- **Dry Piles of Solids**—Surface samples from piles can be collected with a plastic or stainless steel scoop or cup attached to the end of a sample pole. This allows for sample collection from a safe distance. A lab spoon, plastic spoon, or shovel will also work, but avoid the use of devices plated with chrome or other materials.

COMPILING HAZARD INFORMATION

In the process of evaluating risks, response personnel will be gathering and updating all kinds of information from all types of sources. Not only must responders question the accuracy and validity of the information

NOTES

being collected, but equally important is the management of that information. Responders have sometimes been overwhelmed by the volume of hazard information collected during the course of an incident and have been unable to effectively or efficiently manage the data and information base.

To minimize these problems and concerns, responders should prioritize their information requirements—what do I need to know right now, in 1 hour, and in 8 hours? For example, accurate hazard information, chemical reactivities, protective clothing recommendations, and initial control measures need to be determined by the IC as soon as possible. However, equipment disposal and clean-up information are not needed in the initial stages of the incident and can be delayed until later.

Many responders rely upon printed data forms and checklists to ensure that all information requirements have been prioritized and addressed. Regardless of their exact design and format, some logical, functional, and "user friendly" system must be used.

EVALUATING RISKS

Previous sections have laid the foundation for gathering and compiling hazard information. We will now discuss evaluating the relative risks at a hazardous materials incident. The basis of this evaluation is an analysis of the "inputs," specifically the hazmat involved, its container, the environment in which the emergency occurs, and; finally, the actions and resources of emergency response personnel. The output of the risk evaluation process will then be the establishment of strategical response objectives and tactical options.

All emergencies consist of a series of events that occur in a logical sequence. For example, a fire in a confined room or building will move from the incipient to the free-burning phase and eventually to the smoldering phase, if there is no fire department intervention. Similarly, a hazmat emergency will also follow a logical sequence of events arriving at some stabilized outcome if there is no intervention by ERP. Figure 7-8 describes the Hazardous Materials Emergency Model and its components.

The overall objective of emergency response personnel at a hazmat emergency is to favorably change or influence the outcome. These outcomes are defined in terms of fatalities, injuries, property and environmental damage, and systems disruption.

To determine whether or not to intervene, you must first estimate the likely harm that will occur without intervention. Simply, what will happen if you do nothing? This requires you to: (1) visualize the likely behavior of the hazardous material and/or its container, along with the likely harm associated with that behavior, and (2) describe the outcome of that behavior.

HAZARDOUS MATERIAL INFORMATION CHECKLIST

HAZARDOUS MATERIAL# NAME ____

HEALTH:

SOURCE:

FLAMMABILITY:

SOURCE:

REACTIVITY:

SOURCE:

PHYSICAL PROPERTIES

SOURCE:

ISOLATION/EVACUATION, DISTANCES:	PROTECTIVE CLOTHING REQ'D:
SOURCE:	SOURCE:

HOT ZONE: DIMENSIONS	WARM ZONE: DIMENSIONS
PROTECTIVE CLOTHING REQ'D	PROTECTIVE CLOTHING REQ'D

☐ HOT AND WARM ZONES ESTABLISHED TIME: ____

WEATHER INFORMATION CURRENT CONDITIONS:	FORECAST CONDITIONS:
SOURCE:	

DECONTAMINATION INFORMATION

LOCATION OF DECON AREA ____

☐ HAZMAT ID CARDS POSTED

☐ ON-SCENE REPRESENTATIVES CARDS POSTED

☐ SHIPPING DOCUMENTS AND/OR MSDS OBTAINED AND KEPT

☐ INFORMATION SOURCES REFERENCED (MINIMUM OF 3)
- ☐ DOT GUIDEBOOK
- ☐ NFPA MANUAL
- ☐ AAR MANUAL
- ☐ AAR ACTION GUIDES
- ☐ CHRIS MANUAL
- ☐ FARM CHEM HANDBOOK
- ☐ OTHER ____
- ☐ COMPUTER DATABASES: ____
- ☐ MANUFACTURER
- ☐ CHEMTREC
- ☐ ON-SCENE REPS.

☐ ISOLATION/ EVACUATION DISTANCES RECOMMENDED TO HAZMAT OFFICER AND SAFETY TIME ____

☐ EXPOSURES IDENTIFIED
LIFE ____
PROPERTY ____
ENVIRONMENTAL ____
(STORM DRAINS, WATER) ____

☐ CONTENTS OF ALL EXPOSED CONTAINERS/VEHICLES/ STRUCTURES IDENTIFIED

☐ MANUFACTURER CONTACTED FOR INFORMATION ON:
- ☐ PROPERTIES
- ☐ HAZARDS
- ☐ DECONTAMINATION
- ☐ PROTECTIVE CLOTHING
- ☐ HANDLING

☐ MANUFACTURER RESPONSE TEAM/REPRESENTATIVE ENROUTE

YES ____ NO ____ ETA ____ TIME ____

☐ PROTECTIVE CLOTHING RECOMMENDED TO HAZMAT OFFICER, TEAM LEADER AND SAFETY

☐ DECONTAMINATION METHOD RECOMMENDED TO TEAM LEADER, DECON AND SAFETY TIME ____

☐ EMERGENCY CARE FOR EXPOSURE RESEARCHED
- ☐ REFERENCE LIBRARY
- ☐ MANUFACTURER
- ☐ COMPUTER DATABASE
- ☐ POISON CONTROL CENTER
- ☐ CENTER FOR DISEASE CONTROL

TIME ____

☐ EMERGENCY CARE INFORMATION GIVEN TO EMS AND DECON

☐ SHIPPER CONTACTED FOR INFO ON CONTAINER(S)/VEHICLE
SHIPPER REPRESENTATIVE ENROUTE

YES ____ NO ____ ETA ____ TIME ____

☐ HAZARD AND RISK ASSESSMENT COMPLETED/REEVALUATED

☐ OTHER INFORMATION SOURCES REFERENCED

☐ RECON SECTOR DATA COLLECTED
- ☐ CHECKLIST
- ☐ SITE DRAWINGS
- ☐ PHOTOGRAPHS
- ☐ PERSONNEL DEBRIEFED

☐ PRODUCT CONTROL/CONFINEMENT OPTIONS RECOMMENDED TO HAZMAT OFFICER, TEAM LEADER, AND RESOURCES

☐ AGENCIES CONTACTED
- ☐ STATE HEALTH — CONTACT PERSON ____ ETA ____ TIME ____
- ☐ WATER RESOURCES ADMINISTRATION — CONTACT PERSON ____ ETA ____ TIME ____
- ☐ COUNTY HEALTH — CONTACT PERSON ____ ETA ____ TIME ____
- ☐ OEP — CONTACT PERSON ____ ETA ____ TIME ____
- ☐ OTHER (e.g., PUBLIC WORKS) — AGENCY/DEPT ____ CONTACT PERSON ____ ETA ____ TIME ____

☐ CONTRACTOR COMPANIES CONTACTED (e.g., CRANES)

COMPANY ____
CONTACT PERSON ____
ETA ____ TIME ____
COMPANY ____
CONTACT PERSON ____
ETA ____ TIME ____

☐ NATIONAL RESPONSE CENTER (NRC) NOTIFIED TIME ____

☐ TERMINATION ACTIVITIES COMPLETED TIME ____

☐ MEDICAL EVALUATIONS SIGNS/SYMPTOMS OF
- ☐ EXPOSURE
- ☐ CHECKLISTS/NOTES COLLECTED FROM ALL SECTORS

____ SAFETY ____ RECON ____ DECON
____ RESOURCE ____ OTHER
- ☐ DEBRIEFING

FIGURE 7-8

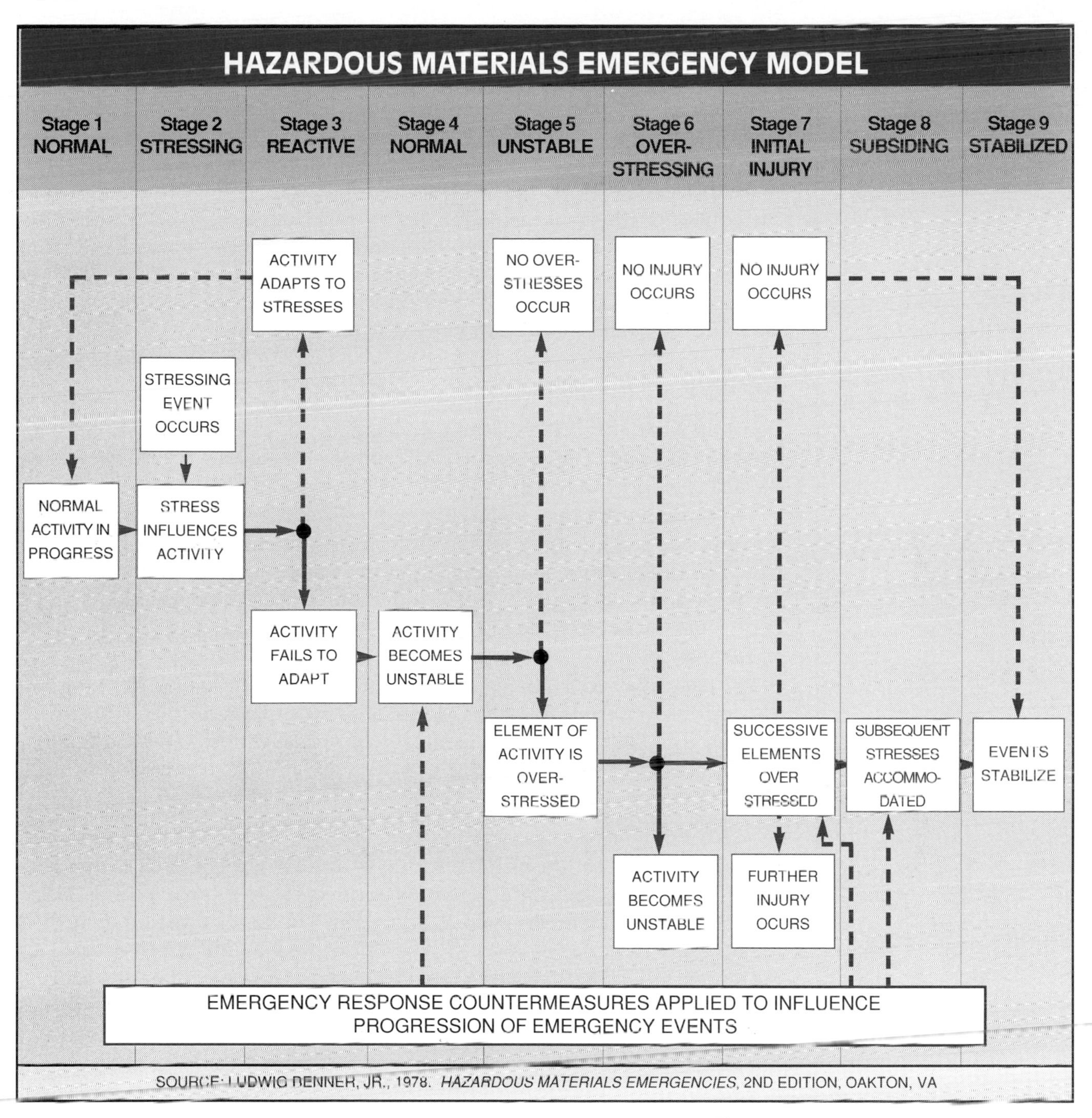

FIGURE 7-9: ALL EMERGENCIES CONSIST OF A SERIES OF EVENTS THAT OCCUR IN A LOGICAL SEQUENCE

To properly visualize likely behavior, five basic questions must be addressed:

1. Where will the hazardous material and/or its container go when released?
2. Why is the hazardous material and/or its container likely to go there?
3. How will the hazardous material and/or its container get there?
4. When will the hazardous material and/or its container get there?
5. What harm will the hazardous material and/or its container do when it gets there?

NOTES

When answering these questions, you must recognize that there are several factors which will affect hazmat behavior. They include the inherent properties and quantities of the materials involved (i.e., toxicity, flammability, corrosivity, etc.), the built-in design and construction features of the container (thermal insulation, safety relief devices, fixed water spray systems), the natural laws of physics and chemistry (influence dispersion patterns), and the pertinent environmental factors (i.e., terrain, weather and atmospheric conditions, wind direction and speed, and the physical surroundings [i.e., rural location vs. downtown]).

BEHAVIOR OF HAZMATS AND CONTAINERS

Since all incidents follow a logical sequence of events, hazmat releases will do likewise. This sequence will remain consistent regardless of the hazard class encountered.

To visualize hazmat behavior and what is likely to occur, use the concept of "events analysis," which is simply the process of breaking down complex actions into smaller, more easily understood parts. Events analysis helps responders to (1) understand, track, and predict a given sequence of events and (2) decide when and how to change that sequence.

An easy way to visualize hazmat behavior is by using the General Hazardous Materials Behavior Model or GHBMO, pronounced "gebmo." As you look at these events, consider how they interrelate at a hazmat emergency.

STRESS EVENT

Under normal conditions, hazardous materials are controlled by some type of container or containment system, such as a tank, piping, bags, and bottles. For an emergency to occur, this container must be disturbed or stressed in some fashion. Stress is defined as an applied force or system of forces that tend to either strain or deform a container (external action) or trigger a change in the condition of the contents (internal action). It is important to recognize that stress can affect the container and/or its contents.

Three types of stress—thermal, mechanical, and chemical—are quite common. Less likely, though still possible, are etiological and radiation stresses. These stressors may be present alone or in combination with each other.

- **Thermal Stress**—generally indicates temperature extremes, both hot or cold. Examples include fire, sparks, friction or electricity, and ambient temperature changes. Extreme or intense cold, such as found with liquefied gases and cryogenic materials, may also act as a stressor. Clues of thermal stress include the operation of safety relief devices or the bulging of a container.
- **Mechanical Stress**—the result of a transfer of energy when one object physically contacts or collides with another. Indications would be punctures, gouges, scores, breaks, or tears in the container. A container may also be weakened but have no visible

GENERAL HAZARDOUS MATERIALS BEHAVIOR MODEL					
EVENT					
STRESS	BREACH	RELEASE	ENGULF	IMPINGE	HARM
EVENT CATEGORIES					
THERMAL MECHANICAL CHEMICAL IRRADIATION ETIOLOGICAL	DISINTEGRATION RUNAWAY CRACKING ATTACHMENTS OPEN UP PUNCTURES SPLITS OR TEARS	DETONATION VIOLENT RUPTURE RAPID RELIEF SPILL OR LEAK	CLOUD PLUME CONE STREAM IRREGULAR DEPOSITS	SHORT TERM MEDIUM TERM LONG TERM	THERMAL RADIATION ASPHYXIATION TOXIC CORROSIVE ETIOLOGIC MECHANICAL
EVENT INTERRUPTION PRINCIPLES					
INFLUENCE APPLIED STRESSES	INFLUENCE BREACH SIZE	INFLUENCE QUANTITY RELEASED	INFLUENCE SIZE OF DANGER ZONE	INFLUENCE EXPOSURE IMPINGED	INFLUENCE SEVERITY OF INJURY
REDIRECT IMPINGEMENT SHIELD STRESSED SYSTEM MOVE STRESSED SYSTEM	CHILL CONTENTS LIMIT STRESS LEVEL ACTIVATE VENTING DEVICES	CHANGE CONTAINER POSITION MINIMIZE PRESSURE DIFFERRENTIAL CAP OFF BREACH	INITIATE CONTROLLED IGNITION ERECT DIKES OR DAMS DILUTE	PROVIDE SHIELDING BEGIN EVACUATION	RINSE OFF CONTAMINATION INCREASE DISTANCE FROM SOURCE PROVIDE SHIELDING
SOURCE: LUDWIG BENNER, JR., 1978. *HAZARDOUS MATERIALS EMERGENCIES*, 2ND EDITION, OAKTON, VA					

FIGURE 7-10: THE GHMBO PROVIDES A METHOD FOR VISUALIZING AND PREDICTING HAZARDOUS MATERIALS BEHAVIOR

signs of potential release. Assessment of damaged containers will be discussed later in this section.

- **Chemical Stress**—the result of a chemical reaction of two or more materials. Examples include corrosive materials attacking a metal, the pressure or heat generated by the decomposition or polymerization of a substance, or any variety of corrosive actions.

All three types can occur in combination with each other. The combination of thermal and mechanical stresses at the Waverly, Tennessee railroad incident in 1978 resulted in the deaths of 15 individuals, including several firefighters, the fire chief, and the police chief. Always look beyond the obvious for damaged or stressed containers.

If the potential types of stressors can be identified, preventive measures can be installed or implemented to minimize their impact in the event of an emergency. For example, headshields and shelf couplers have been installed on railroad tank cars to minimize the potential for a mechanical stressor in the event of a

NOTES

train derailment, while sprayed-on thermal insulation or jacketing are installed to minimize the impact of a thermal stressor. Likewise, installed water spray systems are used within the petrochemical industry to minimize thermal stress upon exposed bulk pressure vessels and containers.

BASIC PRINCIPLE: ALWAYS LOOK FOR STRESSED CONTAINERS. IF THE HAZMAT HAS ESCAPED BEFORE YOUR ARRIVAL, LOOK FOR MORE THAN ONE STRESSED CONTAINER.

BREACH EVENT

If the container is able to adapt to the stress, the incident will be stabilized at that point. However, when the container is stressed beyond its limits of recovery (its design strength or ability to hold contents), it will open up or "breach." Different containers breach in different ways; glass bottles shatter, bags tear, pressure cylinders split, and drums tear.

There are five basic types of breach behaviors:

- **Disintegration**—the total loss of container integrity. Although usually associated with explosives, it can be easily visualized as the shattering of a glass bottle or carboy.
- **Runaway Cracking**—occurs in closed containers such as liquid drums or pressure vessels. A small crack in a closed container suddenly develops into a rapidly growing crack which encircles the container. As a result, the breach will generally break apart into two or more pieces in a violent manner. Linear cracking is commonly associated with catastrophic BLEVE scenarios, such as Kingman, Arizona and Livingston, Louisiana.
- **Failure of Container Attachments**—attachments open up or break off the container, such as safety relief valves, frangible disks, fusible plugs, discharge valves, or other related appliances.
- **Container Punctures**—usually related to mechanical stressors and result in a breach of the container. Examples include 55-gallon drums being punctured by fork lift trucks and coupler punctures of railroad tank cars.
- **Container Splits or Tears**—examples include torn fiber or plastic bags such as those used for certain oxidizers and agricultural chemicals, split 55-gallon drums, and seam or weld failures.

Containers breach differently. The breach will depend on (1) the type of container and (2) the type of stress applied to it. When you are unsure how a container is likely to breach, get technical assistance.

BASIC PRINCIPLE: BECOME FAMILIAR WITH HOW CONTAINERS BREACH IN AN EMERGENCY. IF YOU DON'T KNOW, GET HELP FROM SOMEONE WHO DOES.

RELEASE EVENT

Once a container is breached, the hazardous material is free to escape in the form of energy, matter, or a combination of both. The rate of this release is critical since it will directly determine your ability to control it.

NOTES

Generally, the faster the release, the greater likelihood of harm.

There are four types of release:

- **Detonation**—an explosive chemical reaction with a release rate less than 1/100th of a second. This gives responders NO time to react. Examples include military munitions, mining explosives such as PETN and Tovex, and organic peroxides.
- **Violent Rupture**—associated with chemical reactions having a release rate of less than one second (e.g., deflagration). Again, there is no time to react. This behavior is commonly associated with runaway cracking and overpressures of closed containers.
- **Rapid Relief**—ranges from several seconds to several minutes, depending on the size of the opening, type of container, and the nature of its contents. Responders may often have time to reach a safe location or develop tactical countermeasures. This behavior is associated with releases from containers under pressure, through relief valve actuations, broken or damaged valves, punctures, splits, tears, or broken piping.
- **Spills or Leaks**—release rates vary from minutes to days. They are generally a low-pressure, nonviolent flow through broken or damaged valves and fittings, splits, tears, or punctures. Because of the slower release rate, responders will often have adequate time to develop prolonged countermeasures.

Hazardous materials and/or energy WILL be released when a container is breached. Responders must visualize not only how a release will occur but also how quickly. The speed of the release will depend on the nature of the hazmat as well as the stored energy—such as pressure. The greater the stored energy, the faster the release.

BASIC PRINCIPLE: THERE ARE ONLY TWO THINGS JUMPING OUT AT YOU—ENERGY AND MATTER. LOOK FOR ENERGY AND MATTER TO BE RELEASED WHEN A CONTAINER IS BREACHED.

ENGULFING EVENT

Once the hazardous material and/or energy is released, it is free to travel or disperse, engulfing an area. The farther the contents move outward from their source, the greater the level of problems. How quickly they move and how large of an area they engulf will depend upon the type of release, the nature of the hazmat, the chemical and physical laws of science, and the environment.

To visualize the area the hazmat and/or energy is likely to engulf, consider the following questions:

1. **What is jumping out at you?** Energy or matter?
2. **What form is it in?** Pressure waves, liquids, organisms, etc.?
3. **What is making it move?** Wind, pressure, contaminated individuals?
4. **What path will it follow?** Linear, random, ground contour, etc.?
5. **What dispersion pattern will it create?** Cloud, plume, etc.?

NOTES

These answers will help ERP to predict and define (visualize) where the hazardous material and/or its container will go when released. In turn, responders can determine the primary danger zone and their exposures.

BASIC PRINCIPLE: PREDICT YOUR DISPERSION PATTERNS.

HAZARDOUS MATERIALS BEHAVIOR—DISPERSION CONSIDERATIONS				
What Is Jumping Out At You?	**What Is Its Form?**	**What Is Making It Move?**	**What Path Will It Follow?**	**What Dispersion Pattern Could It Form?**
ENERGY	INFRA- RED RAYS	THERMAL DIFFERENTIAL	LINEAR OR RADIAL	CLOUD
	GAMMA RAYS (NUCLEAR)	SELF-PROPELLED	LINEAR OR RADIAL	CONE OR CLOUD
	PRESSURE WAVES	SELF-PROPELLED	LINEAR	CLOUD
MATTER				
SOLIDS	DUSTS OR POWDERS	AIR ENTRAINMENT PERSONAL TRANSPORT	WITH WIND (LINEAR) RANDOM	PLUME, IF UNCONFINED IRREGULAR DEPOSITS
	FRAGMENTS, SHRAPNEL, OR CHUNKS	SELF-PROPELLED	LINEAR	CLOUD
	ORGANISMS	AIR ENTRAINMENT PERSONAL TRANSPORT	WITH WIND (LINEAR) RANDOM	PLUME, IF UNCONFINED IRREGULAR DEPOSITS
	ALPHA AND BETA	SELF-PROPELLED	LINEAR	CONE OR CLOUD
LIQUIDS	POURABLE LIQUIDS	GRAVITY	FOLLOW CONTOUR	STREAM
GASES	VAPORS	GRAVITY AIR ENTRAINMENT VAPOR DIFFUSION	FOLLOW CONTOUR WITH WIND (LINEAR) UPWARD FROM SURFACE	STREAM PLUME, IF UNCONFINED CLOUD ABOVE LIQUID
	GASEOUS	GASEOUS DIFFUSION	OUTWARD FROM SURFACE	PLUME, IF UNCONFINED OR SHAPE OF CONFINED AREA
LIQUEFIED GASES	VAPORIZING	SELF-PROPELLED & BOILING	LIQUID FOLLOWS CONTOUR, GAS MOVES WITH WIND (LINEAR)	LIQUID FORMS STREAM, GAS FORMS PLUME

SOURCE: LUDWIG BENNER, JR., 1978. *HAZARDOUS MATERIALS EMERGENCIES*, 2ND EDITION, OAKTON, VA

FIGURE 7-11: HAZARDOUS MATERIALS FORM A DISPERSION PATTERN ONCE RELEASED FROM THEIR CONTAINER

IMPINGEMENT (CONTACT) EVENT

As the hazardous material and/or its container engulf an area, they will impinge, or come in contact with, exposures. Exposures include people (civilian and emergency response personnel), property (physical and environmental), and systems (transportation, community water supply, production shutdown, etc.). They may also impinge upon other hazardous materials containers, producing additional problems.

Impinged exposures may or may not suffer any harm. The level of harm is directly dependent on the dose received (remember—dose makes

NOTES

the poison!); therefore, you must consider factors such as the type of exposure, the duration of the exposure, and the ability of the impinged exposure to minimize the effects of the exposure.

Impingements are categorized based upon their duration. Short-term impingements (i.e., a transient vapor cloud) have durations of minutes to hours. Medium-term impingements may extend over a period of days, weeks, and even months. Examples include lingering pesticide residues resulting from fires or spills and asbestos remediation following a process unit fire or explosion.

Long-term impingements extend over years and perhaps even generations. The clean-up operations associated with the radioactive accidents at Three Mile Island and Chernobyl are good examples.

Therefore, estimating likely harm within an engulfed area includes consideration of all of the following factors:

- Harmful characteristics of the material released.
- Concentration of the hazardous material.
- Duration of the impingement.
- Characteristics of the exposure (i.e., vulnerability).

For example, a toxic vapor cloud is released from a fixed facility. It will impinge upon all people and objects which fall within its path. However, if the concentration of the toxic chemical is low enough or if the length of impingement is short enough, little harm will be done.

BASIC PRINCIPLE: IDENTIFY THE EXPOSURES LIKELY TO BE IMPINGED UPON.

HARM EVENT

Before responders can favorably influence the outcome of a hazmat incident, they must first understand what harm is likely to occur within the engulfed area if they do NOT intervene. Harm simply refers to the effects of exposure to the hazardous material and/or the container.

Harm can be categorized into the following types:

- **Thermal**—harm resulting from exposure to hot and cold temperature extremes.
- **Radiation**—harm resulting from exposure to radioactive materials.
- **Asphyxiation**— harm resulting from exposure to simple asphyxiants (e.g., nitrogen, natural gas) and chemical asphyxiants (e.g., carbon monoxide, methylene chloride).
- **Toxicity**—harm resulting from exposure to poisons and toxic materials.
- **Corrosivity**—harm resulting from exposure to corrosive materials (acids and bases).
- **Etiologic**—harm resulting from exposure to etiologic agents and infectious substances.
- **Mechanical**—harm resulting from contact with scattering fragments from an explosion, BLEVE, etc.

NOTES

There are three factors which affect or determine the level of harm: the timing of the release (speed of escape and travel, length of exposure), the size of the dispersion pattern and the area covered, and the lethality of the chemicals involved (concentration of the chemical or dosage received).

In general, the faster the rate of release, the larger the area covered, and the more lethal the materials involved, the less ERP can do to minimize the level of harm once the hazmat escapes from its container. Finally, by understanding the type and nature of the potential harm, ERP can start to evaluate the appropriate level of protective clothing required.

BASIC PRINCIPLE: EMERGENCY RESPONSE PERSONNEL MUST IDENTIFY THE HARM THAT CAN OCCUR.

PRACTICAL APPLICATION OF THE GHMBO

Now that all six events of the General Hazardous Materials Behavior Model have been discussed, how do they relate to "what happens on the street"? As part of the incident size-up process, ERP should initially determine exactly where, in the sequence of events, this particular incident is. Knowing where you are in this sequence will assist you in formulating a practical set of incident priorities and strategical goals.

In a complex incident, such as a major train derailment or a major fire in a petrochemical process area, containers may be at different stages of the hazmat behavior sequence simultaneously. One or more of the containers has already released its contents and poses an immediate threat. However, there are other containers that may be stressed, waiting only for the proper conditions which will cause them to breach. Visualize and predict the behavior of all of the containers.

EVALUATING RISKS—SPECIAL PROBLEMS

Three situations which responders commonly deal with are the potential behavior of stressed pressurized containers, the behavior of hazmats underground, and the behavior of hazmats in sewer collection systems. Responders often have difficulty in both acquiring and interpreting information in these areas. In this section, we will build upon the basic concepts of the GHMBO and examine these three areas more closely.

DAMAGE ASSESSMENT OF PRESSURIZED CONTAINERS

Pressurized containers have sustained extensive damage in accidents and have not released their contents. However, mechanical container damage without a release of the contents has also led to delayed, catastrophic releases (e.g., Waverly, Tennessee). During this delay, ERP are likely to become involved in emergency response and transfer operations and could sustain death or serious injury should the container fail.

Responders can be confronted with a variety of pressurized containers, including cylinders, cargo tank trucks (MC-331), and railroad tank cars (e.g., DOT-105, 112, and 114 tank cars). Much of the available literature on damage assessment is based upon testing and experience with railroad

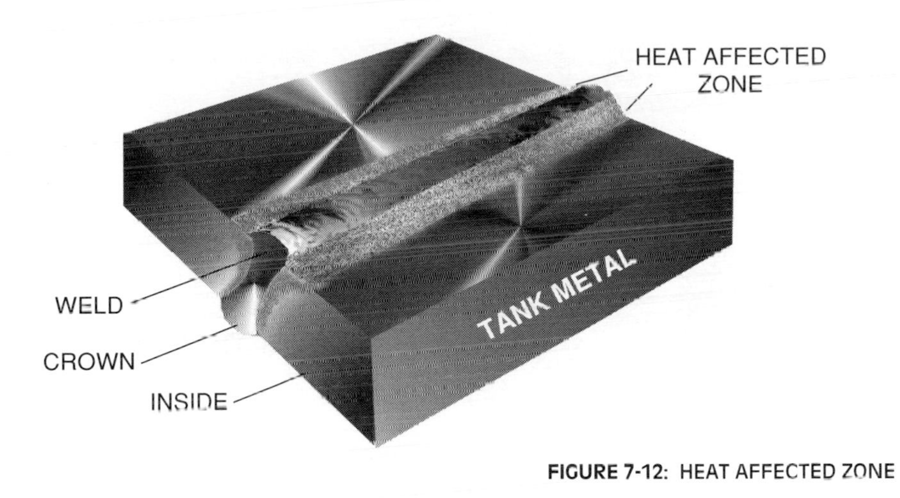

FIGURE 7-12: HEAT AFFECTED ZONE

NOTES

tank cars. It should be recognized that design and construction standards for tank cars are substantially greater than those for cylinders and tank trucks, and provide a greater margin of safety. Much of the information contained in this section was referenced from the Union Pacific Railroad Company booklet written by Charles Wright entitled *Determining the Extent of Tank Damage to Pressure Tank Cars.*

The violent rupture of pressurized containers can be triggered by two related conditions: (1) the presence of a crack in the container shell associated with dents, and (2) the thinning of the tank shell as a result of scores, gouges, and thermal stress. Factors which can affect the severity of the container damage include the ductility of the container metal (i.e., ability of the metal to bend or stretch without cracking) and the internal pressure causing stress on the container metal.

Emergency response personnel should have a basic understanding of ductility. Ductility is the relative ability of a metal to bend or stretch without cracking. Ductile materials have a fine grain structure and tend to bend but not crack. Brittle materials have a coarse-grain structure and tend to crack rather than bend. When a ductile steel container cracks, the crack tends to be small; in contrast, a crack in a brittle steel container tends to run linearly and cause the container to fail.

The ductility of a container metal is affected by the following:

1) **Specification of the steel**—particularly important in evaluating the integrity of damaged railroad tank cars, as different types of steel with different characteristics have been used over time.

2) **Temperature of the steel**—the higher the temperature of the steel at the time of damage, the more ductile (less brittle) the steel will be and the less risk there is for failure. For many liquefied gases which are loaded cold, it will take time for the shell temperature to rise to ambient temperature levels.

3) **Cold work**—deformation of steel when it is bent at ambient temperatures or results from an impact or static load. Cold work reduces the ductility of the metal.

4) **Heat affected zone**—area in the undisturbed tank metal next to the actual weld material. This area is less ductile than either the weld or the steel plate due to the effect of the heat of the welding process. This zone is most vulnerable to damage as cracks are likely to start there (see Figure 7-12).

MECHANICAL STRESS AND DAMAGE

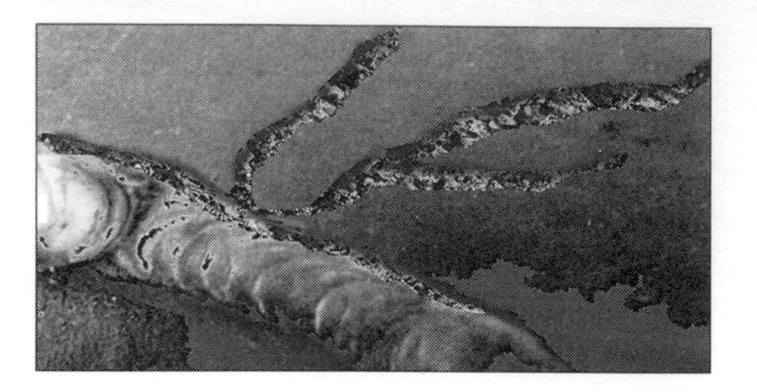

CRACK—narrow split or break in the container metal which may penetrate through the container metal (may also be caused by fatigue). It is a major mechanism which could cause catastrophic failure.

SCORE—reduction in the thickness of the container shell. It is an indentation in the shell made by a relatively blunt object. A score is characterized by the reduction of the container or weld material so that the metal is pushed aside along the track of contact with the blunt object.

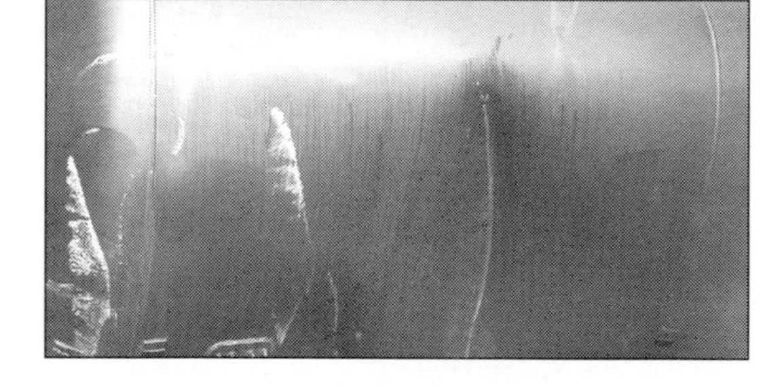

GOUGE—reduction in the thickness of the tank shell. It is an indentation in the shell made by a sharp, chisel-like object. A gouge is characterized by the cutting and complete removal of the container or weld material along the track of contact.

WHEEL BURN—reduction in the thickness of a railroad tank shell. It is similar to a score but is caused by prolonged contact with a turning railcar wheel.

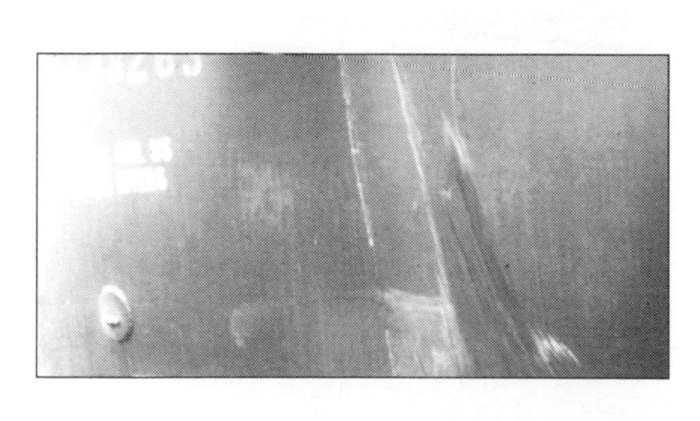

DENT—a deformation of the tank head or shell. It is caused from impact with a relatively blunt object (e.g., railroad coupler, vehicle). If the dent has a sharp radius, there is the possibility of cracking.

RAIL BURN—a deformation in the shell of a railroad tank car. It is actually a long dent with a gouge at the bottom of the inward dent. A rail burn can be oriented circumferentially or longitudinally in relation to the tank shell. The longitudinal rail burns are the more serious because they have a tendency to cross a weld. A rail burn is generally caused by the tank car passing over a stationary object, such as a wheel flange or rail.

STREET BURN—a deformation in the shell of a highway cargo tank. It is actually a long dent that is inherently flat. A street burn is generally caused by a container overturning and sliding some distance along a cement or asphalt road.

FIGURE 7-13: EXAMPLES OF PRESSURIZED CONTAINER DAMAGE

Guidelines for damage assessment of pressurized containers include:

1) Gather information concerning the type of container (e.g., DOT specification number), material of construction (e.g., aluminum, steel), and internal pressure. Methods for determining the internal pressure include pressure gauges and temperature gauges with pressure conversion charts.
2) Determine the type of stress applied to the container (e.g., thermal or mechanical).
3) Evaluate the stability of the container. Do not inspect an unstable container as it may move or shift during the inspection process. It may be necessary to stabilize the container with blocks, cribbing, or other means.
4) Examine the surface of the container, paying attention to the types of damage and the radius (i.e., sharpness) of all dents. Railroad personnel will often use a Tank Car Dent Gauge as a "go/no go" device for comparing the radius of curvature of a tank car dent to accepted standards to determine the severity of damage.

Experience shows that the most dangerous situations will include the following:

- Cracks in the base metal of a tank or cracks in conjunction with a dent, score, or gouge. Both of these situations justify off-loading the container as soon as safely possible.

NOTES

- Sharply curved dents or abrupt dents in the cylindrical shell section which are parallel to the long axis of the container. The container should be off-loaded without moving it when a dent is considered critical.
- Dents accompanied with scores and gouges.
- Scores and gouges across a container's seam weld. If the score crosses a welded seam and removes no more than the weld reinforcement (i.e., that part of the heal which sticks above the base metal), the stress is considered noncritical. However, if the score removes enough of the base metal at the welded seam, the stress is considered critical.

5) If you are unsure of the container damage or how the container is likely to breach, get technical assistance. This may include railroad personnel, gas industry representatives, and cargo tank truck specialists.
6) In summary, damage assessment is, at best, a qualitative assessment process (e.g., stable vs. nonstable, go/no go, etc.).

MOVEMENT AND BEHAVIOR OF HAZMATS UNDERGROUND

When hazmats are released into the ground, their behavior will depend upon their physical and chemical properties (e.g., liquid vs. gas, hydrocarbon vs. polar solvent), the type of soil (e.g., clay vs. gravel vs. sand), and the underground water conditions (e.g., location and movement of the water table). While such incidents may involve any hazmat class, flammable liquids, gases, and vapors are the most common.

As with hazmat containers and their behavior, responders need to have a basic understanding of geology, groundwater and groundwater movement to evaluate the underground dispersion of hazmat releases and potential exposures and to determine potential outcomes. The ultimate clean-up and remediation of underground spills and releases is not the responsibility of ERP.

GEOLOGY AND GROUNDWATER

Soil consists of loose, unconsolidated surface materials, such as sand, gravel, silt, and clay. Bedrock is the hard, consolidated material that lies under the soil, such as sandstone, limestone, or shale. Most areas have a soil cover ranging from a few feet to several hundred feet.

Generally, rocks and soils consist of small fragments or sand grains. When compressed together, they may form small voids or pores. Measurement of the total volume of these voids is called the "porosity" of the rock or soil. If these pores are interconnected, the rock or soil is "permeable" (i.e., fluid can pass through it).

The size of these voids will vary from large (e.g., gravel) to small (e.g., sand and topsoil) to essentially zero (e.g., dense clay). Rock almost never

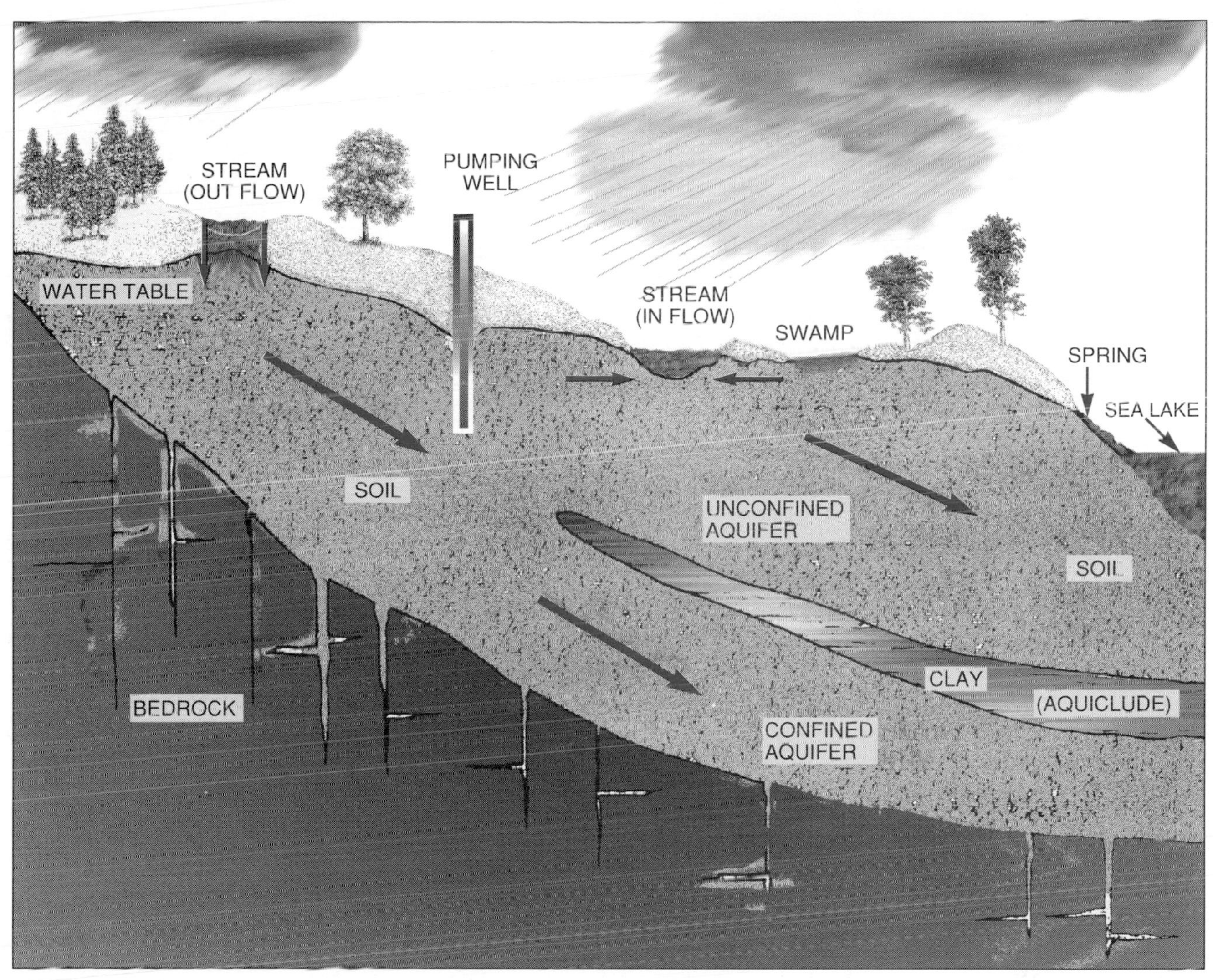

AMERICAN PETROLEUM INSTITUTE

FIGURE 7-14: HYPOTHETICAL GROUNDWATER SYSTEM

has large voids, but sandstone and limestone have voids which are similar to a fine sand. Aquifers are permeable sections of soil or rock capable of transmitting water. In contrast, silt and shale have many, but extremely small, pores that are poorly interconnected. Since fluids cannot readily pass through such materials, they are known as "impermeable" materials.

In most areas, water exists at some depth in the ground. The source of most groundwater is precipitation over land, which percolates into porous soils and rocks at the surface. Rivers and streams that seep water into the sub-surface are a second source of groundwater. Groundwater accounts for the majority of the drinking water supply for the United States.

In most areas, groundwater moves extremely slowly. The rate of flow depends on the permeability of the underground aquifer and the slope or "hydraulic gradient" of the water table. Flows can range from 6 feet per year in fine clays to 6 feet per day in gravels. In addition, the location and relative production rates of groundwater wells can significantly disrupt normal flow patterns. Figure 7-14 shows a hypothetical ground water system.

NOTES

BEHAVIOR OF HAZMATS IN SOIL AND GROUNDWATER

Hazmats may be absorbed into the soil through either surface spills or leaks from underground pipelines or storage tanks. Flammable liquids and toxic solvents can create significant problems if they migrate into confined areas or are allowed to flow into waterways.

Flammable and toxic gases, such as natural gas or hydrogen sulfide, can also accumulate in underground pockets or confined areas. These can occur naturally or may be released from a pipeline failure. When dealing with natural gas, recognize that some soils (e.g., clay) can "scrub" or remove the methyl mercaptan odorant commonly used on natural gas distribution lines, thereby removing any physical clues of smell or odor.

The underground movement of hazmats follows the most permeable, least resistant path. For example, the backfill in trenches carrying utility conduits, sewers, or other piping is often much more permeable than the undisturbed native soil. In urban areas, this can facilitate the rapid and easy movement of liquids and gases to nearby basements, sewers, or other below-grade structures. Identifying these conduits is critical in identifying potential exposures. Utilities and public works agencies, including regional "Miss Utility" and "One Call Systems," can often provide assistance in locating these underground conduits. ERP should also seek technical information specialists, such as geologists, who have either maps or knowledge of the local water tables, soil types and densities, underground rock formations, and so forth.

Liquid hazmats which are spilled into soil will tend to flow downward with some lateral spreading (see Figure 7-15). The rate of hazmat movement in soil will depend upon the viscosity of the liquid, soil properties, and the rate of release. For example, light hydrocarbon products, such as gasoline, will penetrate rapidly while heavier oils, such as #4 fuel oil, will move more slowly. If the soil near the surface has a high clay content and very low permeability, the hazmat may penetrate very little or not at all. However, a porous, sandy soil may quickly absorb the product. Eventually, the downward movement of the hazmat through soil will be interrupted by one of three events: (1) it will be absorbed by the soil; (2) it will encounter an impermeable bed; or (3) it will reach the water table.

Hazmats that are absorbed by the soil may move again at some later time as the water table is elevated. For example, ERP will often receive a report of hydrocarbon or gasoline vapors in an area, with the source of the odor being unknown. In other situations, recent rains may cause the water table to rise and bring hydrocarbon liquid and vapors to the surface. With these scenarios, monitoring readings should be "mapped," as they can assist emergency responders in identifying the location of the problem and its direction of movement.

Although combustible gas indicators (CGI) are excellent tools for evaluating flammable atmospheres, they may not be very effective for assessing low level flammable concentrations such as found with subsurface and sewer spills. To register a positive reading, many CGI's require a concentration of up to 10,000 ppm (1% in air). While this may not be a flammable concentration, it often represents a significant environmental problem.

FIGURE 7-15: MOVEMENT OF HAZMATS IN SOIL

Hazmats which encounter an impermeable layer will spread laterally until becoming immobile or until the hazmat comes to the surface where the impermeable layer outcrops. If the hazmat comes in contact with the water table, there is a high potential for the water supply to become polluted and for the hazmat to move and accumulate in an underground structure (e.g., basement, sewer system). Remember that groundwater supplies can become contaminated by concentrations as small as 200 parts per billion (ppb). Preventing spills and releases from entering the soil is a critical element in many areas. While ERP can do little to influence the underground movement of the material once it has entered the soil, exposures must still be identified and protected.

Hydrocarbon liquids will not mix with water and will simply float on the surface of the water table. Many oils and refined petroleum products, however, contain certain components which are slightly soluble in water. Gasoline is high in water-soluble components which, when dissolved in water, produce odors and taste that can be detected at levels of only a few parts per million (e.g., methanol, toluene, methyl tertiary butyl ether [MTBE]).

MOVEMENT AND BEHAVIOR OF SPILLS INTO SEWER COLLECTION SYSTEMS

Sewers, manholes, electrical vaults, french drains, and other similar underground structures and conduits can be critical exposures in the event of a hazmat spill. If the hazmat release penetrates the sub-structure walls or enters through surface sewers and manholes, significant quantities of liquid can be expected to flow into the sewer collection system. Depending upon the hazmat involved, this situation can pose an

NOTES

immediate fire problem, as well as significant environmental concerns regardless of whether ignition occurs or not.

Most sewer emergencies involve flammable and combustible liquids. The probability of an explosion within an underground space will depend upon two factors: (1) that a flammable atmosphere exists, and (2) that an ignition source is present. The severity of an explosion and its consequences will depend upon the type of sewer collection system, the process and speed at which the hazmat moves through the system, and the ability of emergency responders to confine the release and implement fire and spill control procedures.

TYPES OF SEWER SYSTEMS

Sewer systems can be categorized based upon their application:

- **Sanitary Sewers.** This is a "closed" system which carries wastewater from individual homes, together with minor quantities of storm water, surface water, and groundwater that are not admitted intentionally. Sanitary sewers also collect wastewater from industrial and commercial businesses in the service area. The collection and pumping system will transport the wastewater to a wastewater treatment plant, where various liquid and solid treatment systems are employed to process the wastewater. Sewer diameters from 8 inches to 60 inches are common.
- **Storm Sewers.** This is an "open" system which collects storm water, surface water, and groundwater from throughout a community but excludes domestic wastewater and industrial wastes. A storm sewer system may dump runoff directly into a retention area which is normally dry or into a stream, river, or waterway without treatment. Storm sewers are generally much larger than sanitary sewers, with diameters ranging from 2-foot pipes to greater than 20-foot tunnels. They can be used by "illegal dumpers" as a means of hazardous waste disposal.
- **Combined Sewers.** Carries domestic wastewater as well as storm water and industrial wastewater. Although separate sanitary sewers are being constructed today, it is quite common for older cities to have an extensive amount of combined sewers. Combined sewers are often very large and can be as much as 20 feet in diameter. Combined sewers may also have regulators or diversion structures that allow overflow directly to rivers or streams during major storm events.

WASTEWATER SYSTEM OPERATIONS

There are three primary elements of a wastewater system: (1) collection and pumping, (2) liquid treatment systems, and (3) solid treatment systems. The highest risks of a fire or explosion are associated with collection and pumping operations, and with the early stages of liquids and solid processing. Similarly, the greatest potential for either environmental damage or shutdown of a wastewater treatment plant operation will take place at the liquid and solid stream treatment processes.

NOTES

Wastewater, storm water, and surface water initially enter the collection and pumping system through a series of collectors and branch lines which tie together small geographic areas. These collectors and branch lines are eventually tied into a trunk sewer (also known as main sewer), which then carries the wastewater to its final destination for either treatment or disposal.

Where the terrain is flat, the collection system may consist solely of gravity piping. However, in most areas the collection system will require pumping or lift stations. Most pumping stations will have two parts—a dry well and a wet well. The wet well receives and temporarily stores the wastewater. Wet wells often contain electrical equipment such as fans, pumps, motors, and other accessories. In some instances, proper management of the wet well may provide an opportunity for the collection and removal of a flammable liquid. The dry well provides isolation and shelter for the controls and equipment associated with pumping the wastewater. They are designed to completely exclude wastewater and wastewater-derived atmospheres, although there may be accidental leakage from pumpshafts or occasional spills.

Depending upon the type of sewer system and the specific location, most areas are classified by the *National Electrical Code* as Class I, Division 2 areas. However, pumping stations should be viewed as potential ignition sources in case of hydrocarbon spills into the sewer collection system. Pumping stations are sometimes equipped with hydrogen sulfide or fixed combustible gas detectors to detect the presence of flammable vapors and gases.

PRIMARY HAZARDS AND CONCERNS

There are two basic scenarios involving releases into a sewer collection system. The first is an aboveground release where a spill flows into the sewer collection system through catch basins, manholes, etc. Clearly, a problem exists. The second scenario involves underground tank and pipeline leaks where the product migrates through the sub-surface structure into the sewer collection system. This type of emergency is usually not obvious from the surface and presents a greater challenge in identifying the source of the problem and controlling the release within the sewer system.

With the sub-surface scenario, emergency responders will often receive a report of hydrocarbon or gasoline vapors in an area, with the source of the odor being unknown. In other situations, recent rains may cause the water table to rise and bring hydrocarbon liquid and vapors to the surface. Monitoring readings should be "mapped," as they can assist emergency responders in identifying the location of the problem and its direction of movement.

Rules of thumb for evaluating monitoring readings are: (1) if readings are high and then drop off or dissipate in a relatively short period of time, the source of the problem is often a spill or dumping directly into the sewer collection system; and (2) if readings are consistent over a period of time, the source of the problem is often a sub-surface release, such as a underground storage tank or pipeline.

NOTES

Spills and releases into the sewer collection system will create both fire and environmental concerns.

- **Fire Concerns.** Flammable liquids, such as gasoline and JP-4 jet fuel, will create the greatest risk of a fire or explosion. The potential for ignition within a sewer collection system will be greatest at points where liquids may enter or where entry is possible. Manholes, storm sewers, catch basins, and pumping station wet wells are likely to present the greatest areas of concern.

  If a flammable liquid enters a *sanitary or combined sewer*, the probability of ignition will be high. If floor traps are not filled with water, flammable liquid in either a sanitary or combined sewer collection system will back up vapors into building basements and other low-lying areas where there are multiple ignition sources, such as pilot lights, hot water heaters, electrical equipment, sparking and arcing, etc.

  When a flammable liquid enters a *storm sewer* collection system, manholes, catch basins, and pumping station wet wells are likely points of ignition. In some instances, vapors have flowed out of the sewer collection system and accumulated in low lying areas, only to be ignited by an ignition source completely outside of the sewer system (e.g., passing vehicle).

  In the event of a fire or explosion, secondary and tertiary problems will likely be created by the emergency as the explosion will affect all utilities that occupy the same or nearby utility corridors. Natural gas leaks, electric and telephone utility outages, and a loss of both water and water pressure should be anticipated in areas which suffer a major fire or explosion.

- **Environmental Concerns.** Environmental concerns will be greatest when dealing with poisons, environmentally sensitive materials, or when there is no ignition of flammable liquids following their release into the sewer system. Depending upon the type of sewer collection and treatment system, environmental impacts may range from a shutdown of the wastewater treatment facility and/or destruction of microorganisms necessary for the treatment process, to a spill impacting environmentally sensitive areas (e.g., wetlands, wildlife refuge, etc.) or threatening both potable and aquifer drinking water supplies.

  The selection of control agents, such as dispersants and firefighting foams, to control a fire or spill may also have secondary environmental impacts. Both sewer department and environmental personnel from the various on-scene governmental agencies should be consulted on any decision to apply firefighting foams or dispersants either into a sewer collection system or onto a waterway.

COORDINATION WITH SEWER DEPARTMENT

Pre-planning with the sewer department is critical. ERP should have a basic knowledge and understanding of the sewer system and its operations and a good working relationship with sewer department personnel.

NOTES

Responders should identify areas where there is a probability of hazmats entering the sewer collection system and discuss procedures and tactical options for handling such an emergency with the sewer department as part of its pre-incident planning and training activities.

When an emergency occurs, a sewer department representative must be immediately available on the scene. The evaluation and selection of tactical control options should be based upon input from the sewer department and the respective governmental environmental agency. Maps of the local sewer system will be a key element in identifying the direction in which a spill may potentially head and in identifying likely exposures. Effective use of sewer maps will require a sewer department representative who is familiar with the unique aspects of the local system, local construction techniques, etc. However, sewer maps may not always be accurate and up-to-date, particularly in identifying all lateral/domestic connections and branch connections on older combined sewer systems.

FIGURE 7-16

SITE SAFETY PROCEDURES FOR HYDROCARBON SPILLS INTO SEWER COLLECTION SYSTEMS

- Verify that the sewer department has been notified and is enroute. Identifying the direction of flow is critical in identifying exposures and establishing evacuation zones.
- Continuous air monitoring must be provided, particularly in low lying areas. ERP may have difficulty obtaining a sufficient number of monitoring instruments and trained personnel to perform monitoring over a relatively large geographic area.
- Monitoring readings should be "mapped" as they are taken as this will assist responders in identifying the location of the spill within the sewer system and its general speed and direction of movement. This information, in turn, will assist in establishing response priorities.
- Control all ignition sources in the area, including vehicles, traffic flares and smoking materials. Depending upon the nature of the emergency, large spills into a sanitary or combined sewer collection system may require the shutdown of gas and electric utilities until the situation is under control.
- If the spill is in a sanitary or combined sewer collection system and its speed and direction of movement is known, ERP may be able to notify homeowners and facilities ahead of the spill to pour water into their basement floor traps as a quick preventive measure to minimize hydrocarbon vapor migration and build-up.
- Do not allow any personnel to stand on or near manholes and catch basins. In the event of a fire or explosion, manhole lids can be blown into the air and fire can quickly emanate from catch basins and other sewer openings. Manhole lids can be blown into adjoining buildings and vehicles and represent a significant life safety hazard.

NOTES

- ERP should not enter a sewer collection system unless advised by representatives of the sewer system. In addition to the obvious fire and health hazards, sewer collection systems consist of piping and collection areas of various diameters and depths and pose significant physical hazards.
- There are many confined spaces within a wastewater collection and pumping system. ERP should also consider the presence of oxygen deficient atmospheres and other toxic and flammable gases, including hydrogen sulfide, methane, and sewer and sludge gases.
- When either flushing or applying control agents into a sewer collection system, the agent must be applied at multiple points along the projected flow path. If control agents are only applied at the source of the release, the agent will never "catch up" with the head of the flow, and there will be a continuous flammable atmosphere within the sewer system.
- The injection of some control agents into a sewer collection system, such as firefighting foams, dispersants, and water, may also introduce air and possibly move the environment into the flammable range.

FIGURE 7-16

SUMMARY

Hazard and risk assessment is one of the most critical functions in the successful management of a hazardous materials incident. The key elements of this analytical process are (1) identifying the materials involved, (2) gathering hazard information, (3) visualizing hazmat behavior and predicting outcomes, and (4) based upon the evaluation process, establishing response objectives. The system which is the "horizontal bridge" tying these elements together is the General Hazardous Materials Behavior Model.

An accurate evaluation of the real and potential problems will enable response personnel to develop accurate and informed strategical response objectives and tactical decisions. Remember—your job is to be a risk evaluator, not a risk taker. Bad risk takers get buried; effective risk evaluators come home.

REFERENCES AND SUGGESTED READINGS

American Petroleum Institute. API 1628—GUIDE TO THE ASSESSMENT AND REMEDIATION OF UNDERGROUND PETROLEUM RELEASES (Second Edition), Washington, DC: American Petroleum Institute (1989).

Andrews, Robert C., Jr., "The Environmental Impact of Firefighting Foam." INDUSTRIAL FIRE SAFETY (November/December, 1992), pages 26–31.

Berger, M., W. Byrd, C. M. West, and R. C. Ricks, TRANSPORT OF RADIOACTIVE MATERIALS: Q & A ABOUT INCIDENT RESPONSE, Oak Ridge, TN: Oak Ridge Associated Universities (1992).

Emergency Film Group, Air Monitoring (two videotape series), Plymouth, MA: Emergency Film Group (1991).

Fingas, Merv F. et. al. "The Behavior of Dispersed and Nondispersed Fuels in a Sewer System." AMERICAN SOCIETY OF TESTING AND MATERIALS—SPECIAL TECHNICAL PUBLICATION 1018 (1989).

Fingas, Merv F. et. al., "Fuels in Sewers: Behavior and Countermeasures." JOURNAL OF HAZARDOUS MATERIALS, 19 (1988), pages 289–302.

Maslansky, Carol J. and Steven P. Maslansky, AIR MONITORING INSTRUMENTATION, New York, NY: Van Nostrand Reinhold (1993).

National Fire Protection Association, HAZARDOUS MATERIALS RESPONSE HANDBOOK (2nd edition), Boston, MA: National Fire Protection Association (1992).

National Fire Protection Association, NATIONAL ELECTRICAL CODE HANDBOOK, Boston, MA: National Fire Protection Association (1993).

National Fire Protection Association, RECOMMENDED PRACTICE FOR THE CONTROL OF FLAMMABLE AND COMBUSTIBLE LIQUIDS IN MANHOLES, SEWERS AND SIMILAR UNDERGROUND STRUCTURES—NFPA 328, Boston, MA: National Fire Protection Association (1992).

National Fire Protection Association, RECOMMENDED PRACTICE FOR HANDLING UNDERGROUND RELEASES OF FLAMMABLE AND COMBUSTIBLE LIQUIDS—NFPA 329, Boston, MA: National Fire Protection Association (1992).

National Fire Protection Association, RECOMMENDED PRACTICE FOR FIRE PROTECTION IN WASTEWATER TREATMENT AND COLLECTION FACILITIES—NFPA 820, Boston, MA: National Fire Protection Association (1992).

National Transportation Safety Board, "Derailment of a CSX Transportation Freight Train and Fire Involving Butane in Akron, OH." (Report NTSB/HZM-90/2). Washington, DC: National Transportation Safety Board (February 26, 1989).

Plog, B. A. FUNDAMENTALS OF INDUSTRIAL HYGIENE (3rd Edition), Chicago, IL: National Safety Council (1988).

RIGHT-TO-KNOW POCKET GUIDE FOR LABORATORY EMPLOYEES, Schenectady, NY: Genium Publishing Corporation (1990).

Union Pacific Railroad Company, DETERMINING THE EXTENT OF TANK DAMAGE TO PRESSURE TANK CARS, Omaha, NB: Union Pacific Railroad Company—Hazardous Material Training (1990).

U.S. Environmental Protection Agency. DETECTING LEAKS: SUCCESSFUL METHODS STEP-BY-STEP (EPA 530/UST-89 1012), Washington, DC: EPA (1989).

Water Pollution Control Federation, EMERGENCY PLANNING FOR MUNICIPAL WASTEWATER FACILITIES (MOP SM-8), Arlington, VA: Water Pollution Control Federation (1989).

York, Kenneth J. and Gerald L. Grey., HAZARDOUS MATERIALS/WASTE HANDLING FOR THE EMERGENCY RESPONDER, Tulsa, OK: Fire Engineering Books and Videos (1989).

THE FAR SIDE By GARY LARSON

"Well, it's a delicate situation, sir. ... Sophisticated firing system, hair-trigger mechanisms, and Bob's wife just left him last night, so you *know* his mind's not into this."

PERSONAL PROTECTIVE CLOTHING AND EQUIPMENT

OBJECTIVES

1. Define the following terms and their impact and significance on the selection of chemical protective clothing:
   a) Degradation
   b) Penetration
   c) Permeation
   d) Breakthrough Time
   e) Permeation Rate
2. Identify three indications of material degradation of chemical protective clothing.
3. Describe the difference between limited-use and multi-use chemical protective clothing materials.
4. Identify the process and factors to be considered in selecting the proper level of protective clothing at a hazmat incident.
5. Identify the process and factors to be considered in selecting the proper level of respiratory protection at a hazmat incident.
6. Describe the advantages, limitations, and proper use of the following types of respiratory protection at hazmat incidents:
   a) Air purifying respirators
   b) Atmosphere supplying respirators (i.e., SCBA, airline respirator)

Fire is an inspirational thing...especially when your'e the one on fire!

Richard Pryor,
Comedian, Actor

NOTES

7. Identify the operational components of the air purifying respirators and supplied air respirators by name, and match the function to the component.
8. Describe the advantages, limitations, and proper use of structural firefighting clothing at a hazmat incident.
9. Identify the four levels of chemical protection (EPA/NIOSH), the equipment required for each level, and the conditions under which each level is used.
10. Identify three types of liquid chemical splash protective clothing and describe the advantages and disadvantages of each type.
11. Identify the three types of chemical vapor protective clothing and describe the advantages and disadvantages of each type.
12. Identify the two types of high temperature protective clothing and describe the advantages and disadvantages of each type.
13. Identify the safety and emergency procedures for personnel wearing liquid chemical splash and vapor protective clothing.
14. Identify the elements of a personal protective clothing and equipment (PPE) program.

INTRODUCTION

Protection against chemical hazards can be provided through engineering controls, by the design of safe work practices, and/or through the use of personal protective clothing and equipment (PPE). While engineering controls and safe work practices are preferred for personal protection in an industrial environment, emergency responders must typically rely upon the safe and effective use of personal protective clothing and equipment.

This chapter covers the fourth step in the **Eight Step Process©**. In emergency response applications, the selection of PPE cannot be safely and adequately addressed until tactical response objectives are determined as part of the Hazard and Risk Evaluation process. Likewise, emergency response operations cannot be safely performed if ERP are not provided with the proper level of personal protective clothing, and are trained in its proper application, limitations, and use.

This chapter will review the application and use of personal protective clothing and equipment (PPE) at hazmat emergencies. Topics will include the basic principles of chemical protective clothing, the types and levels of protective clothing available, criteria for PPE selection and use, operational issues and concerns, and preventive maintenance considerations.

Case Study

On August 12, 1983, a railroad tank car loaded with dimethylamine located on an industrial railroad track at A. J. Chemicals in Benecia, California began leaking product from its sample line. Dimethylamine is a corrosive and flammable gas shipped in pressurized containers. The owner of the company was immediately notified, and he called the shipper of the product for instructions on how to handle the leak. Because he lacked the necessary equipment, the owner then notified the Benecia Fire Department (BFD). Upon their arrival, the BFD Incident Commander requested assistance from the Hazmat Supervisor of the railroad which had transported the tank car and from the San Francisco Fire Department Hazardous Materials Response Team (HMRT).

Soon after his arrival, the railroad Hazmat Supervisor and three firefighters, all wearing chemical vapor protective clothing, climbed upon the tank car and installed a repair clamp over the leak in the sample line. Approximately 30 minutes later, the clamp began to leak and a second entry operation was undertaken. During this second entry, both the railroad supervisor and the firefighters began to experience visibility problems with the CPC facepieces. Problems included melting, clouding, reduced visibility, and lens crazing which further reduced visibility. Leaks also developed in the seams of several of the suits. The facepiece in one the chemical vapor suits shattered when the supervisor dropped about 3 ft. while getting off the tank car, subsequently exposing him to the hazardous environment.

Because of their CPC suit failures, the BFD Incident Commander prohibited further attempts to stop the leak. The sample line leak was finally stopped by a contract hazardous materials clean-up company. No difficulties were experienced with the chemical vapor suits used by the contractor; however, the exposure period was very short. All of the CPC used in this emergency were made by the same manufacturer but were not of the same design.

Chemical compatibility information initially available to emergency responders led them to believe that the CPC suits were adequate for protecting their personnel against the dimethylamine environment.

This incident vividly illustrates some of the personal protective clothing issues and concerns regularly expressed by hazmat emergency responders. The willingness of the individuals involved in this emergency to network and "share the lessons learned" eventually led to changes and improvements in both the hazmat emergency response and chemical protective clothing professions. Unlike 1983, today there are NFPA consensus standards which specify minimum documentation, design and performance criteria, and test methods for chemical protective clothing. In addition, the quality of CPC and chemical compatibility information has significantly improved.

NOTES

BASIC PRINCIPLES

When evaluating protective clothing for use at a hazmat incident, primary concerns should focus upon chemical resistance, the integrity of the entire protective clothing ensemble (including the garment, visor, zippers, gloves, boots. etc.), and the tasks to be performed. It is important that one be familiar with the following terminology and its significance in evaluating protective clothing chemical compatibility:

Degradation—the physical destruction or decomposition of a clothing material due to exposure to chemicals, use, or ambient conditions (e.g., storage in sunlight). Degradation is noted by visible signs such as charring, shrinking, swelling, color change or dissolving, or by testing the clothing material for weight changes, loss of fabric tensile strength, etc. Degradation can occur when chemical compatibility data are not properly interpreted or understood by responders, the wrong protective clothing material is used, or exposure recommendations are exceeded.

Although permeation testing is most common, chemical compatibility charts may be found which are based upon degradation. These data are normally based upon laboratory tests conducted with pure, undiluted test chemicals on clean, uncontaminated swatches of material over a pre-established time period (often two to six hours). Virtually all testing is done at room temperature (70° to 75°F). Degradation tests do not account for simultaneous exposures to two or more chemicals, exposures at elevated temperatures, or previous exposure to chemicals. Furthermore, degradation testing alone is insufficient to demonstrate the barrier performance of protective clothing materials.

Penetration—the flow or movement of a hazardous chemical through closures, seams, porous materials, and pinholes or other imperfections in the material. While liquids are most common, solid materials (e.g., asbestos) can also penetrate through protective clothing materials.

Penetration testing may be used to demonstrate the liquid barrier properties of protective clothing materials and is important for evaluating the performance of liquid splash chemical protective clothing. Penetration resistance data are provided as "pass" or "fail" relative to the specific chemical or mixture tested.

Causes of penetration include clothing material degradation, physical damage to the suit (e.g., punctures, abrasions, etc.), normal wear and tear, and PPE defects. Chemical protective clothing (CPC) can be penetrated at several locations, including the facepiece and exhalation valve, suit exhaust valves, and suit fasteners and closures. The potential for penetration generally increases at excessively hot or cold temperatures.

Permeation—the process by which a hazardous chemical moves through a given material on the molecular level. Permeation differs from penetration in that permeation occurs through the clothing material itself rather than through the openings in the clothing material. Permeation can lead to protective clothing failures when chemical compatibility data are not properly interpreted or understood by responders, or if breakthrough times (see below) are exceeded.

Because of its significance in evaluating the integrity of PPE, chemical contamination, and decontamination, permeation will be discussed more completely.

NOTES

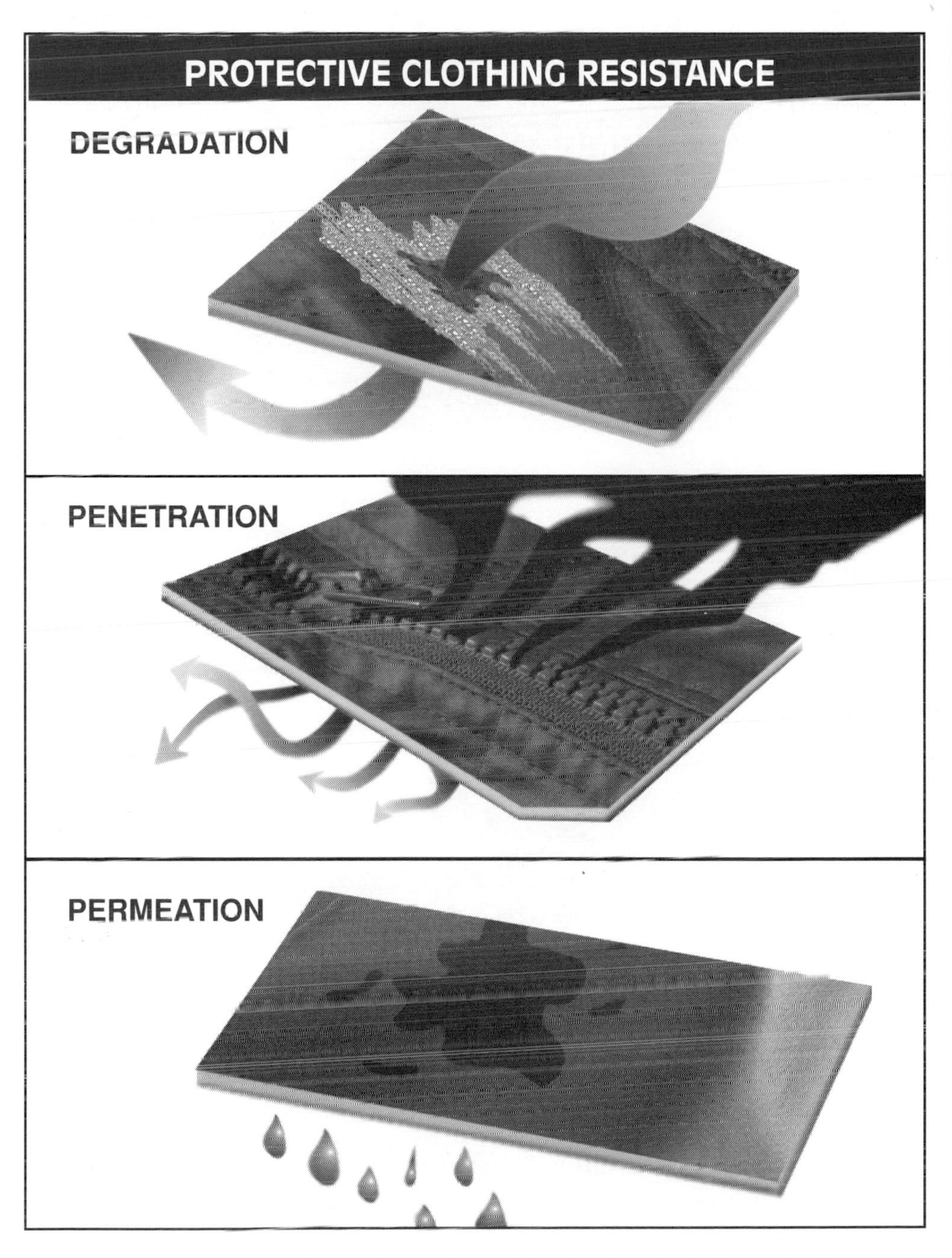

FIGURE 8-1: PROTECTIVE CLOTHING RESISTANCE TO CHEMICAL ATTACK IS DESCRIBED IN TERMS OF CHEMICAL DEGRADATION, PENETRATION, AND PERMEATION

NOTES

PERMEATION THEORY

The process of chemical permeation through an impervious barrier is a three-step process consisting of:

1. Adsorption of the chemical into the outer surfaces of the material; generally not detectable by the wearer.
2. Diffusion of the chemical through the material.
3. Desorption of the chemical from the inner surface of the material (toward the wearer).

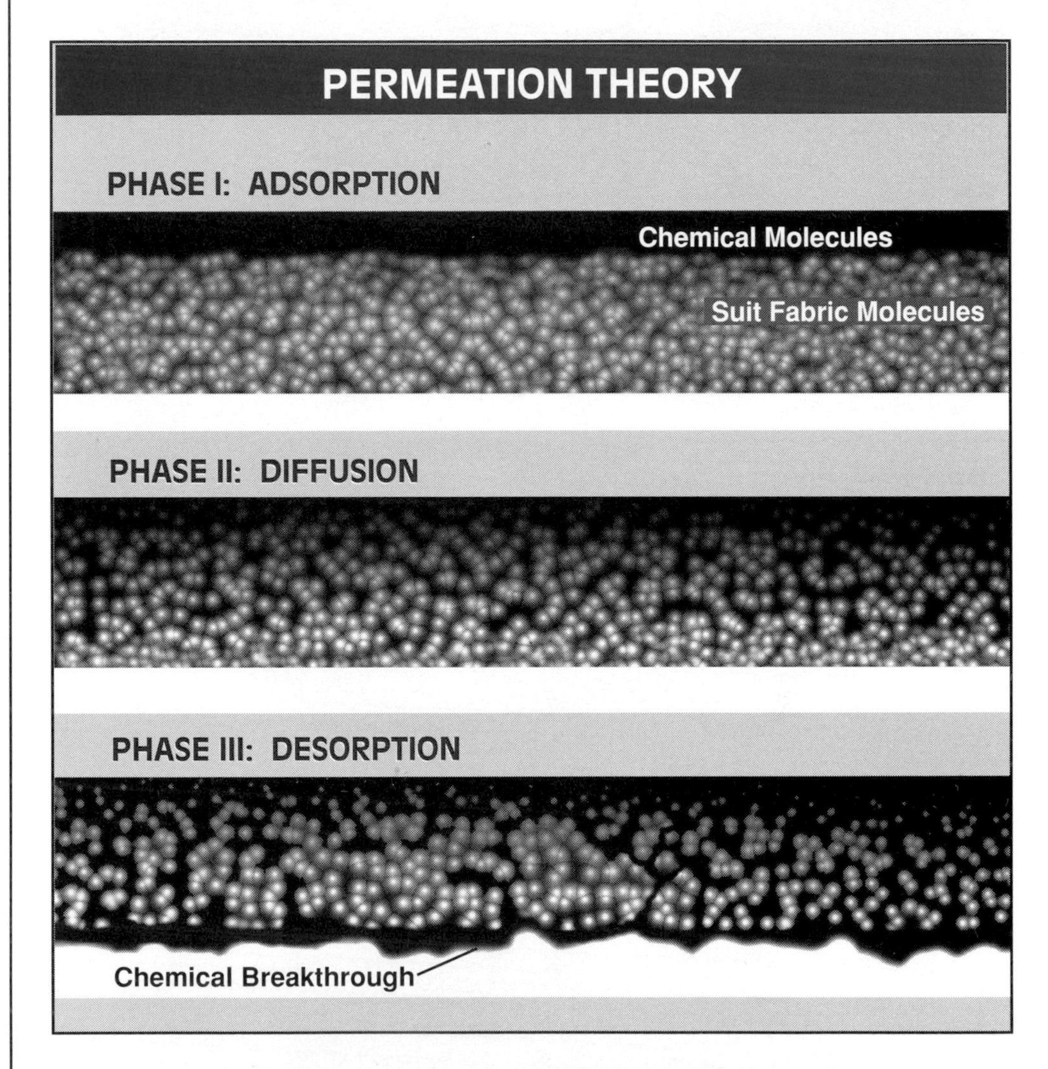

FIGURE 8-2: CHEMICAL PERMEATION THROUGH PROTECTIVE CLOTHING IS A THREE-STEP PROCESS: ADSORPTION, DIFFUSION, AND DESORPTION

Breakthrough time is defined as the time from the initial chemical attack on the outside of the material until its desorption and detection inside. The units of time are usually expressed in minutes or hours, and a typical test runs up to a maximum of 8 hours. If no measurable breakthrough is detected after 8 hours, the result is often reported as a breakthrough time of >480 minutes or >8 hours.

NOTES

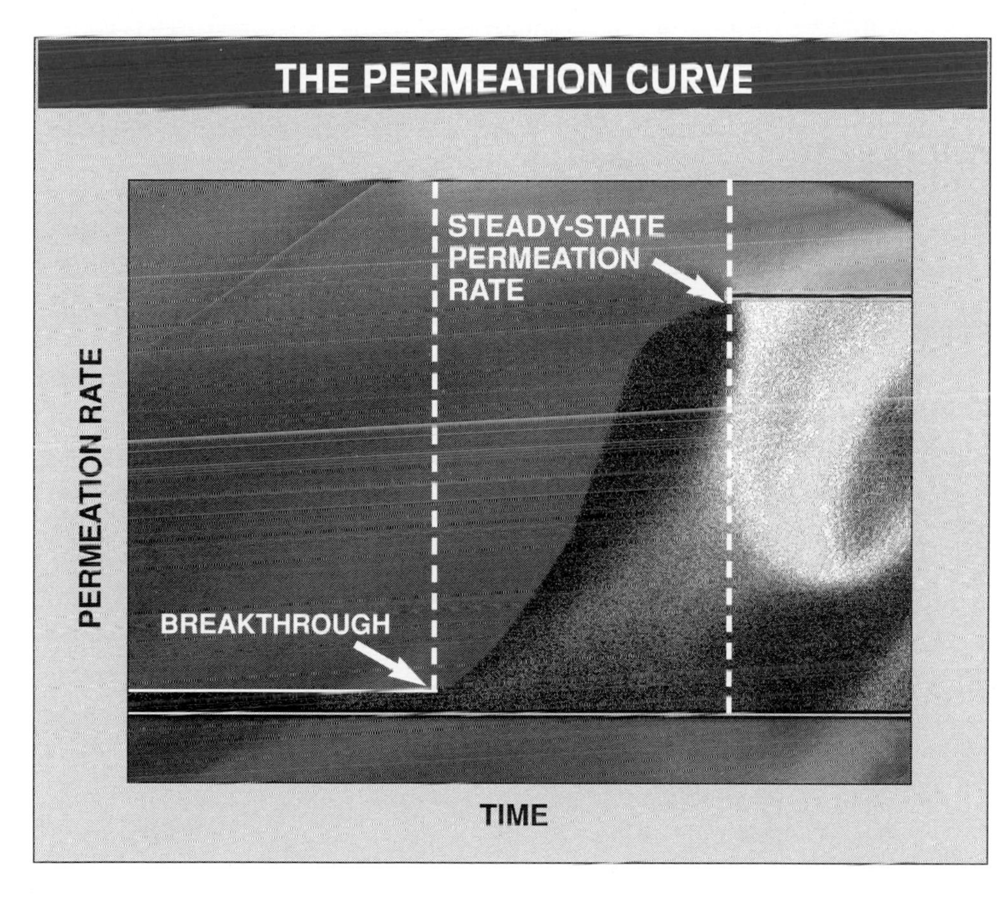

FIGURE 8-3: THE PERMEATION CURVE ILLUSTRATES THE RELATIONSHIP BETWEEN PERMEATION RATE AND TIME. BREAKTHROUGH IS THE INITIAL POINT AT WHICH A CHEMICAL IS DETECTED ON THE INSIDE OF A PROTECTIVE CLOTHING MATERIAL

Permeation rate is the rate at which the chemical passes through the CPC material and is generally expressed as micrograms per square centimeter per minute ($\mu g/cm^2/min$). For reference purposes, .9 $\mu g/cm^2/min$ is equal to approximately 1 drop/hour. The higher the rate, the faster the chemical passes through the suit material. Most chemical compatibility charts will contain both the breakthrough time and the permeation rate.

Measured breakthrough times and permeation rates are determined by laboratory permeation testing procedures against a list of chemicals outlined in ASTM F 1001, Standard Guide for Chemicals to Evaluate Protective Clothing Materials. NFPA 1991, Standard of Vapor Protective Suits for Emergency Response, uses the ASTM F 1001 list as part of a battery of chemicals for determining CPC material permeation resistance. Use of the breakthrough times from this testing allows responders to estimate the duration of maximum protection under a worst-case scenario of continuous chemical contact.

In evaluating permeation resistance and breakthrough times, several other terms may be found in the CPC manufacturer's compatibility information:

NOTES

- **Actual Breakthrough Time**—Breakthrough time as previously defined.
- **Normalized Breakthrough Time**—A calculation, using actual permeation results, to determine the time at which the permeation rate reaches 0.1 $\mu g/cm^2/min$. Normalized breakthrough times are useful for comparing the performance of several different protective clothing materials. Note that in Europe, breakthrough times are normalized at 1.0 $\mu g/cm^2/min.$, a full order of magnitude less sensitive.
- **Minimum Detectable Permeation Rate (MDPR)**—The minimum permeation rate that can be detected by the laboratory analytical system being used for the permeation test.
- **System Detection Limit (SDL)**—The minimum amount of chemical breakthrough that can be detected by the laboratory analytical system being used for the permeation test. Lower SDL's result in lower (or earlier) breakthrough times.

Chemical permeation rates are a function of many factors, including:

- **Temperature**—Most chemical compatibility tests are conducted at ambient temperatures (68° to 72°F). However, as temperature increases, permeation rates increase and breakthrough times shorten.
- **Thickness**—Permeation is inversely proportional to the thickness of the clothing material. In other words, doubling its thickness will theoretically cut the permeation rate in half. The breakthrough time will, therefore, become longer.
- **Chemical Mixtures and Their Effects Upon Chemical Resistance Are Relatively Unknown**—It is impossible to test for every possible chemical combination. Tests have also shown that combining data concerning exposures to individual chemicals has limited value. For example, Viton™/chlorobutyl laminate resists hexane for over three hours and acetone for one hour. However, any mixture of hexane and acetone will permeate Viton™/chlorobutyl laminate in under 10 minutes. Chemical mixtures may result in a stronger attack on protective clothing than with individual exposures.
- **Previous Exposures**—Once a chemical has begun the diffusion process, it may continue to diffuse even after the chemical itself has been removed from the outside surface of the material. This is significant when considering the re-use of any protective clothing. DECONTAMINATION IS NO ASSURANCE THAT PERMEATION HAS STOPPED. Although it is possible to test for permeation, the testing process also destroys the clothing material.

PROTECTIVE CLOTHING MATERIALS

NOTES

Several parameters must be evaluated when choosing protective clothing materials:

1. **Chemical Resistance**—This is most critical as the clothing material must maintain its integrity and protection qualities when it comes in contact with a hazardous material. Chemical resistance (either permeation or penetration) must be evaluated. Permeation resistance test data should be used for vapor chemical protective clothing, particularly for chemicals which are toxic by skin absorption or present other hazards in vapor form.

   Chemical penetration data should be used for liquid splash chemical protective clothing. However, penetration data should not be used when chemicals are highly toxic or give off hazardous vapors.

2. **Flammability**—The CPC garment should not contribute to the fire hazard and should maintain its protective capabilities when exposed to elevated temperatures. When burning, the clothing material should not melt or drip.

   Be aware that chemical protective clothing is not appropriate for firefighting operations, or for protection in flammable or explosive environments. Although some clothing manufacturers offer "flash" protective garments for use with CPC, they offer only limited protection for rapid, short-duration fires and should not be considered for firefighting applications.

3. **Strength and Durability**—These characteristics reflect a material's ability to resist cuts, tears, punctures, abrasions, and other physical hazards found at the incident. The breaking strength, seam strength, and closure strength also relate to how well clothing materials withstand repeated use or the stresses of use.

4. **Overall Integrity**—Chemical protective clothing should provide complete protection to the wearer. Vapor chemical protective clothing should provide gas tight integrity as demonstrated by passing a "pressure" or inflation test. Similarly, liquid splash chemical protective clothing should be tested for liquid-tight integrity to verify that the suit does not allow liquids onto the wearer's skin.

5. **Flexibility**—This is the ability of the user to move and work in protective clothing and, generally, a factor of how the ensemble is fabricated. Emphasis is on the garment weight, fabrication of seams (i.e., bonded, sewn, glued, sealed), and their resistance to chemical and wear exposures. Flexibility is particularly important with gloves and chemical protective ensembles.

   When dealing with laminated fabrics, "microcracking" may become a concern. Continuous flexing of a laminate material may cause microcracks to develop on the inner laminate layers and lead to a CPC failure with little indication of potential failure.

6. **Temperature Characteristics**—Temperature characteristics affect the ability of protective clothing to maintain its protective capacity in temperature extremes. Higher temperatures increase the effects

NOTES

of all chemicals upon polymers. A material suitable for chemical exposures at ambient temperatures may fail at elevated temperatures. Exposure to low temperatures may cause materials to stiffen, crack, flake, or separate.

7. **Shelf Life**—Exposure to ultraviolet light causes many CPC materials to age and deteriorate. Some materials, particularly rubber-like materials, deteriorate over time, much like automobile tires. These changes may also occur without use. Shelf life information should be obtained from the manufacturer for the specific CPC products and materials being used.
8. **Decontamination and Disposal**—The ability of the protective clothing material to be cleaned and decontaminated must be evaluated against potential chemical exposures. Limited-use garments often represent cost-effective options. Although decon may reduce the level of contamination, in many instances it will not be completely eliminated.

Protective clothing materials can be classified by use into two categories:

- **Limited-Use**—Limited-use materials are protective clothing materials which are used and then discarded. They eliminate many health and safety concerns regarding CPC decontamination and their return to service. If contaminated, disposal must be in accordance with local, state, and federal environmental regulations.

  Most limited-use materials consist of laminates of plastics with nonwoven fabrics. In general, these materials are less durable and not as strong as re-usable rubber or thermoplastic-based materials. The chemical resistance of these garments may be compromised by the physical breakdown of the material which occurs with re-use.

  Limited-use garments are generally suitable for a single use and should be disposed of in accordance with local, state, and federal environmental regulations. Examples of limited-use garments include Tyvek™ QC, Tyvek™/Saranex™ 23-P, Barricade™, Kappler CPF™ III and CPF™ IV, Chemrel Max™, Lakeland Interceptor™, and the Lifeguard Responder™.

  Advantages of limited-use materials include lower costs, ability to stock a larger and more varied protective clothing inventory, and reduced inspection and maintenance requirements. They are often used for support functions, including decontamination, remedial cleanup of identified chemicals, and training.
- **Multi-Use**—Based upon the chemical exposure, multi-use materials are designed and fabricated to allow for decontamination

and re-use. Generally thicker and more durable than limited-use garments, they are used for liquid chemical splash and vapor protective suits, gloves, aprons, boots, and thermal protective clothing. The most common materials include butyl rubber, Viton, polyvinyl chloride (PVC), neoprene rubber, and Teflon™.

Although these garments are considered re-usable, certain chemical exposures may require the disposal of this clothing as well. Disposal must be in accordance with local, state, and federal environmental regulations.

NOTES

FIGURE 8-4

NFPA CHEMICAL PROTECTIVE CLOTHING STANDARDS

The NFPA Technical Committee on Fire Service Protective Clothing and Equipment has developed three consensus standards which specify minimum documentation, design and performance criteria, and test methods for chemical protective clothing. These standards are often referenced as minimum requirements in purchase specifications and cover the following:

- NFPA 1991—Vapor Protective Suits for Hazardous Chemicals Emergencies.
- NFPA 1992—Liquid Splash Protective Suits for Hazardous Chemical Emergencies.
- NFPA 1993—Support Function Protective Clothing for Emergency Response.

Each standard requires independent, third party certification to ensure that the protective clothing meets the respective design, performance, and documentation requirements. Compliant products carry a special label showing that the protective clothing meets the respective NFPA standard and is the best evidence of a certified suit/product.

NFPA 1991 requires that the protective clothing suit meet the following criteria:

- The primary suit materials shall not exhibit chemical detection times of one hour or less when tested against the battery of 21 chemicals as outlined in ASTM F 1001, Guide for Chemicals to Evaluate Protective Clothing Materials. The test battery includes 15 liquids and 6 gaseous chemicals which are representative of the range of chemicals encountered in both industry and transportation.
- The NFPA 1991 chemical permeation test battery for chemical vapor protective clothing is listed on following page.

NOTES

Chemical	Chemical Family	Chemical	Chemical Family
Acetone (liquid)	Ketone	Acetonitrile (liquid)	Nitrile
Anhydrous Ammonia (gas)	Basic Gas	1,3-Butadiene (gas)	Unsaturated Hydrocarbon Gas
Carbon Disulfide (liquid)	Organic Sulfur	Chlorine (gas)	Acid Gas
Dichloromethane (liquid)	Chlorinated Paraffin	Diethylamine (liquid)	Amine
Dimethyl Formamide (liquid)	Amide	Ethyl Acetate (liquid)	Ester
Ethylene Oxide (gas)	Heterocyclic Ether Gas	Hexane (liquid)	Saturated Hydrocarbon
Hydrogen Chloride (gas)	Inorganic Acid Gas	Methanol (liquid)	Primary Alcohol
Methyl Chloride (gas)	Chlorinated Hydrocarbon Gas	Nitrobenzene (liquid)	Nitro Compound
Sodium Hydroxide (liquid)	Inorganic Base	Sulfuric Acid (liquid)	Inorganic Mineral Acid
Tetrachloroethylene (liquid)	Chlorinated Olefin	Tetrahydrofuran (liquid)	Heterocyclic Ether
Toluene (liquid)	Aromatic Hydrocarbon		

- There are additional requirements that address overall suit or garment integrity, material flame resistance, material durability and strength, and the proper function of suit components (e.g., exhaust valves). Chemical resistance requirements are applied to all parts of the ensemble including the garment, gloves, boots, visor, seams, and closures. Each part of the clothing is tested for other properties as described above.
- Chemical vapor CPC certified to NFPA 1991 are recommended for clothing use with the Hot Zone.
- Vapor protective suits may also meet additional, but optional, requirements for chemical flash fire escape protection (flash cover) and liquefied gas protection (cryogenic cover).
- The manufacturer must provide a logbook for each chemical vapor protective suit. The logbook should provide for the following data:
  - Dates of each inspection, inspection findings, and name of inspector.

NOTES

- Dates of each use, duration of use, and user's name.
- Chemicals to which the suit is exposed, including the length and duration of exposure.
- Dates of all repairs, including description and name of person making repairs.
- Dates and types of decontamination the suit is subjected to, and the name of the person or facility responsible for the decon.
- Date(s) the suit is taken out of service and the reason for such action.
- Date(s) the suit is returned to the manufacturer and the reason for the return.

NFPA 1992 and 1993 require that the protective clothing suit meet the following criteria:

- The primary suit materials shall not exhibit chemical detection times of one hour or less when tested against the battery of 8 liquid chemicals as outlined below. The NFPA 1992 and 1993 penetration test battery for liquid splash and support function protective clothing is listed below.

Chemical	Chemical Family	Chemical	Chemical Family
Acetone	Ketone	Sodium Hydroxide	Inorganic Base
Diethylamine	Amine	Sulfuric Acid	Inorganic Mineral Acid
Ethyl Acetate	Ester	Tetrahydrofuran	Heterocyclic Ether
Hexane	Saturated Hydrocarbon	Toluene	Aromatic Hydrocarbon

- A smaller battery of chemicals is used for NFPA 1992 and 1993. Many of the chemicals required to be tested under NFPA 1991 are either suspected/known carcinogens or pose known skin absorption hazards. Liquid splash and support function protective clothing should not be worn against chemicals of this type.
- Liquid splash CPC certified to NFPA 1992 are recommended for clothing use within the Hot Zone, while liquid splash CPC certified to NFPA 1993 are recommended for Warm Zone operations, such as decontamination.

FIGURE 8-4: NFPA CHEMICAL PROTECTIVE CLOTHING STANDARDS

NOTES

CHEMICAL COMPATIBILITY AND SELECTION CONSIDERATIONS

No single protective clothing material offers total chemical protection. The selection of protective clothing should be based upon a hazard assessment of those chemicals found in the community or the facility. Unfortunately, there may be some chemicals for which there is NO adequate protection.

Chemical protective clothing will often be constructed of a combination of several materials. Be aware of which of these materials form the basis for chemical compatibility recommendations. Recognize that it may not be possible to determine the specific material(s) used in some limited-use laminate garments due to proprietary reasons.

The manufacturer should provide technical test data outlining the chemical compatibility of both the primary suit material and all secondary components (e.g., gloves, boots, closure assemblies, visors, and exhaust valves). When evaluating chemical vapor protective suits, acquire a complete inventory of all suit components and their construction materials. THE PERFORMANCE OF CHEMICAL PROTECTIVE CLOTHING IS ONLY AS STRONG AS ITS WEAKEST MATERIAL.

Manufacturers will publish quantitative chemical resistance data for particular chemicals. This data will normally be based upon standardized laboratory tests such as those established by the ASTM F 1001 Committee. Standard permeation and penetration tests often incorporate a very large safety factor in predicting failures. The size of this safety factor depends on the established testing criteria.

Quantitative ratings will be described in terms of either chemical permeation/breakthrough times and rates, or as "pass/fail" chemical penetration testing results. Remember that the longer the breakthrough time, the better the level of protection. Breakthrough rates are an indication of how quickly a chemical will permeate through a CPC material. If two CPC materials have comparable breakthrough times, the CPC with the lowest reported permeation rate should normally represent the safest option.

Chemical compatibility recommendations may also be provided in the form of qualitative chemical resistance ratings or use recommendations for a specific protective clothing material and particular chemicals. These ratings are often in the form of a four- to six-grade scale (e.g., excellent, good, fair, and poor/not recommended), or a color code (e.g., green, yellow, red). They often will not include performance specifications or quantitative data such as breakthrough times. Degradation resistance data may be misleading and should be avoided when selecting CPC materials.

When evaluating chemical compatibility recommendations, consider the following guidelines:

- The primary reference source for chemical compatibility recommendations should be the CPC manufacturer's technical documentation. Other credible sources will include CPC

NOTES

reference manuals and computer databases, such as Forsberg and Keith's *Chemical Protective Clothing Permeation and Degradation Compendium* and the National Toxicology Program's *GlovES+*™ computer expert system.

- Determine the basis of chemical compatibility recommendations. Degradation or immersion testing is not sufficient for compatibility assessment. Permeation test data should be sought since permeation of rubber or plastic fabrics can occur with little or no physical effect on the clothing material. PERMEATION IS AN INSIDIOUS PROCESS WHICH CAN OCCUR WITH NO SIGN OF DEGRADATION.
- Compatibility recommendations based upon immersion testing data may be quite old, and they may also be based on the subjective evaluations rather than quantitative measurements for swelling, weight, or strength changes. In some cases, the testing criteria and qualitative descriptions for defining "good," "excellent," and other key words may not be documented.
- Materials constructed of the same primary fabric or material (e.g., butyl rubber, PVC) are not necessarily equal in performance ("A rose by any other name..."). Variations in formulations, thickness, and coating and backing materials influence chemical exposure times.
- There may be a conflict in compatibility recommendations between sources. Responders should initially rely upon the protective clothing manufacturer's chemical resistance recommendations. ALWAYS SELECT THE MOST CONSERVATIVE DATA.

LEVELS OF PROTECTIVE CLOTHING

The need for proper protective clothing and equipment in a hostile environment is obvious. Unfortunately, there is no one type of PPE that satisfies our protection needs under all conditions. For example, chemical protection and thermal protection are very difficult to combine into one protective clothing material. The IC and the Hazmat Branch Director must be familiar with the various types and levels of protective clothing available.

There are three basic types of protective clothing:

1. **Structural Firefighting Clothing**—designed to protect against extremes of temperature, steam, hot water, hot particles, and ordinary hazards of firefighting.
2. **Chemical Protective Clothing**—designed to protect skin and eyes from direct chemical contact. Divided into two groups:
   - Liquid chemical splash protective clothing
   - Chemical vapor protective clothing

NOTES

3. **High Temperature Protective Clothing**—designed to protect against short-term exposures to high temperatures. Divided into two groups:
   - Proximity suits
   - Fire entry suits

The EPA has developed a classification scheme for the various levels of chemical protective clothing. Figure 8-5 outlines the equipment and its associated protection level. For our purposes, protective clothing and equipment will be discussed in terms of its use (respiratory protection, structural firefighting clothing, chemical protective clothing, high temperature protective clothing).

RESPIRATORY PROTECTION

Respiratory protection is the primary concern when wearing PPE. The respiratory system is the most exposed, direct, and, generally, the most critical exposure route. Protection can be provided by either air purification devices (APR's) or by atmosphere supplying respiratory equipment.

AIR PURIFICATION DEVICES

Air purification relies upon respirators or filtration devices which remove particulate matter, gases, or vapors from the atmosphere. These devices range from full facepiece, dual cartridge masks with eye protection to half-mask, facepiece-mounted cartridges with no eye protection.

Air purification devices do not have a separate source of air. They use ambient air, which is then filtered or purified before inhalation. When used for gases or vapors, they are commonly equipped with a sorbent material which absorbs or reacts with the hazardous gas. The proper cartridge must be used for the expected contaminants (e.g., acid gas, organic vapor, etc.). Particle-removing respirators use a mechanical filter to separate the contaminants from the air. Both devices require a minimum of 19.5% oxygen in the atmosphere.

Air purification devices present many problems when used in emergency response applications. They:

- Cannot be used in IDLH or oxygen-deficient atmospheres containing less than 19.5% oxygen. When used, they require constant monitoring for contaminant and oxygen levels.
- Should not be used in the presence or potential presence of unidentified contaminants. Not recommended for areas where contaminant concentrations are unknown or exceed the designated use concentrations. "Designated use concentrations" are based upon testing at a given temperature (usually room temperature) over a narrow range of flow rates and relative humidities. Therefore, the level of protection may be compromised in nonstandard conditions.
- Present logistical problems for storage and maintenance because of the variety of canisters and facepiece assemblies that are required. The shelf life of filtration and purification devices will

EPA U.S. ENVIRONMENTAL PROTECTION AGENCY
LEVELS OF CHEMICAL PROTECTION

LEVEL A PROTECTION SHOULD BE WORN WHEN THE HIGHEST LEVEL OF RESPIRATORY, SKIN, EYE, AND MUCOUS MEMBRANE PROTECTION IS NEEDED.

- PERSONAL PROTECTIVE EQUIPMENT
  - ☐ POSITIVE-PRESSURE (PRESSURE-DEMAND), SELF-CONTAINED BREATHING APPARATUS (MSHA/NIOSH APPROVED).
  - ☐ FULLY-ENCAPSULATING CHEMICAL RESISTANT SUIT.
  - ☐ GLOVES, INNER, CHEMICAL RESISTANT.
  - ☐ GLOVES, OUTER, CHEMICAL RESISTANT.
  - ☐ BOOTS, CHEMICAL RESISTANT, STEEL TOE AND SHANK (DEPENDING ON SUIT BOOT CONSTRUCTION, WORN OVER OR UNDER SUIT BOOT). UNDERWEAR, COTTON, LONG-JOHN TYPE.*
  - ☐ HARD HAT (UNDER SUIT).*
  - ☐ COVERALLS (UNDER SUIT).*
  - ☐ TWO-WAY RADIO COMMUNICATIONS (INTRINSICALLY SAFE).

LEVEL B PROTECTION SHOULD BE SELECTED WHEN THE HIGHEST LEVEL OF RESPIRATORY PROTECTION IS NEEDED, BUT A LESSER LEVEL OF SKIN AND EYE PROTECTION. LEVEL B PROTECTION IS THE MINIMUM LEVEL RECOMMENDED ON INITIAL SITE ENTRIES UNTIL THE HAZARDS HAVE BEEN FURTHER IDENTIFIED AND DEFINED BY MONITORING, SAMPLING, AND OTHER RELIABLE METHODS OF ANALYSIS, AND PERSONNEL EQUIPMENT CORRESPONDING WITH THOSE FINDINGS UTILIZED.

- PERSONAL PROTECTIVE EQUIPMENT
  - ☐ POSITIVE-PRESSURE (PRESSURE-DEMAND), SELF-CONTAINED BREATHING APPARATUS (MSHA/NIOSH APPROVED).
  - ☐ CHEMICAL RESISTANT CLOTHING (OVERALLS AND LONG SLEEVED JACKET, COVERALLS, HOODED TWO PIECE CHEMICAL SPLASH SUIT, DISPOSABLE. CHEMICAL RESISTANT COVERALLS).*
  - ☐ COVERALLS (UNDER SPLASH SUIT).*
  - ☐ GLOVES, OUTER, CHEMICAL RESISTANT.
  - ☐ GLOVES, INNER, CHEMICAL RESISTANT.
  - ☐ BOOTS, OUTER, CHEMICAL RESISTANT, STEEL TOE AND SHANK.
  - ☐ BOOTS, OUTER, CHEMICAL RESISTANT.*
  - ☐ TWO-WAY RADIO COMMUNICATIONS (INTRINSICALLY SAFE).
  - ☐ HARD HAT.*

LEVEL C PROTECTION SHOULD BE SELECTED WHEN THE TYPE OF AIRBORNE SUBSTANCE IS KNOWN, CONCENTRATION MEASURED, CRITERIA FOR USING AIR-PURIFYING RESPIRATORS MET, AND SKIN AND EYE EXPOSURE IS UNLIKELY. PERIODIC MONITORING OF THE AIR MUST BE PERFORMED.

- PERSONAL PROTECTIVE EQUIPMENT
  - ☐ FULL-FACE, AIR-PURIFYING RESPIRATOR (MSHA/NIOSH APPROVED). CHEMICAL RESISTANT CLOTHING (ONE PIECE COVERALL, HOODED TWO PIECE CHEMICAL SPLASH SUIT,
  - ☐ CHEMICAL RESISTANT HOOD AND APRON, DISPOSABLE CHEMICAL RESISTANT COVERALLS).
  - ☐ GLOVES, OUTER, CHEMICAL RESISTANT.
  - ☐ GLOVES, INNER, CHEMICAL RESISTANT.*
  - ☐ BOOTS, STEEL TOE AND SHANK, CHEMICAL RESISTANT.
  - ☐ CLOTH COVERALLS (INSIDE CHEMICAL PROTECTIVE CLOTHING).*
  - ☐ TWO-WAY RADIO COMMUNICATIONS (INTRINSICALLY SAFE).*
  - ☐ HARD HAT.*
  - ☐ ESCAPE MASK.*

LEVEL D IS PRIMARILY A WORK UNIFORM. IT SHOULD NOT BE WORN ON ANY SITE WHERE RESPIRATORY OR SKIN HAZARDS EXIST.

* Optional

FIGURE 8-5

NOTES

also vary depending upon the type and concentration of materials used.

- Have a limited protection duration. Once opened, sorbent canisters begin to absorb humidity and air contaminants whether in use or not. Their efficiency and service life will decrease dramatically.
- Protect against only specific chemicals and only to specific concentrations. Their effectiveness against two or more chemicals simultaneously is highly questionable.

For these reasons, air purification devices should not be used at hazmat releases unless qualified personnel (e.g., industrial hygienists) have monitored the atmosphere surrounding the environment and determined that such devices can be safely used. As a general rule, they should not be used by initial responders at hazmat incidents.

ATMOSPHERE SUPPLYING DEVICES

Respiratory protection devices with an air source are referred to as atmospheric supplying devices. There are two basic types:

1. **Self-Contained Breathing Apparatus (SCBA)**
2. **Supplied Air Respirators (SAR)**—supply air from a source away from the scene; connected to the user by an airline hose.

These devices provide the highest available level of protection against airborne contaminants and in oxygen-deficient atmospheres. Only positive-pressure devices which maintain positive pressure in the facepiece during both inhalation and exhalation should be used during hazmat response operations. Most active hazmat teams will use 45- to 60-minute air bottles to provide a sufficient back-up air supply for entry, exit, and decon operations.

Although SCBA is the most common, SAR's may be used when extended working times are required. Decontamination, clean-up, and remedial operations are good examples. Dual flow SCBA's which have the capability of being supplied by either an SCBA or an airline may provide additional flexibility for both entry and decon operations.

Operational considerations when using supplied air respirators (SAR's) include:

- NIOSH certification limits the maximum hose length from the source to 300 feet.
- Use of airlines in IDLH or oxygen-deficient atmosphere requires a secondary emergency air supply, such as an escape pack for immediate back-up protection in case of airline failure.
- Using airline hose will probably impair user mobility and slow the operation. The user will also have to retrace his or her entry path when leaving the work area.

- The airline hose is vulnerable to physical damage, chemical contamination, and degradation. Airline sleeves constructed of disposable materials can provide additional protection. Decontamination may be difficult.
- As the hose length is increased, the minimum approved airflow required at the facepiece may not be delivered.
- Respiratory protection devices which combine SCBA and airline hose units are now available. The user can operate in either the SCBA or airline hose modes by operating a manual or automatic switch.

NOTES

Air purification devices may be appropriate for operations involving volatile solids and for remedial clean-up and recovery operations where the type and concentration of contaminants is verifiable. However, air-supplied devices such as airline hose units will offer the greatest protection for exposures to gases and vapors.

STRUCTURAL FIREFIGHTING CLOTHING

ERP may respond to a structure fire or vehicle accident only to find themselves suddenly involved in a hazardous materials incident. Likewise, petrochemical facility emergency response teams (ERT's) often deal with hazmat emergencies involving flammable gases and liquids. Structural firefighting gear can offer sufficient protection to the wearer who is fully aware of the hazards being encountered and the limitations of the protective clothing.

Many reference books refer to the term "full protective clothing." This term does not have a uniform definition for ERP, EMS, law enforcement, and chemical industry personnel. The definition of full protective clothing for firefighting is not the same as full protective clothing for hazmat response.

For our purposes, structural firefighting clothing includes a helmet, positive-pressure self-contained breathing apparatus (SCBA), PASS device, turnout coat and pants, gloves and boots, and a hood made of a fire-resistant material. This ensemble should meet NFPA 1500—Fire Service Occupational Safety and Health requirements. This ensemble is shown in Figure 8-6.

FIGURE 8-6: STRUCTURAL FIREFIGHTING CLOTHING

NOTES

Structural firefighting gear may be used when the following conditions are met:

- Unlikely contact with splashes of extremely hazardous materials.
- Total atmospheric concentrations do not contain high levels of chemicals toxic to the skin. There are no adverse effects from chemical exposure to small areas of unprotected skin.

In some situations, duct tape or bands may be applied around the neck, waist, forearms, and ankles for additional vapor protection. These materials provide a false sense of security and do not increase the ability of structural firefighting clothing to provide either chemical splash or vapor protection. **REMEMBER—STRUCTURAL FIREFIGHTING CLOTHING IS NOT DESIGNED TO OFFER CHEMICAL PROTECTION.**

The increased use of plastics and other synthetic materials releases a variety of products of combustion in a structural fire, many of which are highly toxic or carcinogenic. This has led to increased concerns with the contamination and decontamination of structural firefighting clothing. Examples of products of combustion include inorganic gases (e.g., hydrogen sulfide, nitrogen oxides), acid gases (e.g., hydrochloric acid, sulfuric acid), hydrocarbons (e.g., benzene), metals, and polynuclear aromatic compounds (PNA's).

Hazardous chemicals can both penetrate and permeate firefighting protective fabrics. Certain areas are more likely to absorb hazmats than others. Consider the following points:

- Clothing and equipment materials are porous and are easily contaminated by chemical penetration. These include:
  - Turnout clothing outer shells, thermal liners, collars and wristlets.
  - Station/work uniforms.
  - Glove shells and liners.
  - Fire retardant hoods.
  - Boot linings.
  - Helmet straps and linings.
  - SCBA straps.
- Coated or rubber-like materials are more likely to be affected by chemical permeation. These include:
  - Moisture barriers.
  - Reflective trim.
  - Boot outer layers.
  - SCBA masks.
  - Hard plastics used in the helmet and SCBA components.
- Ash, resins, and other smoke particles can easily become trapped within the protective clothing fibers.

- Infectious bloodborne diseases, including the HIV, Hepatitis B, and Hepatitis C viruses, can be readily absorbed into the protective clothing fibers.

BODY PROTECTION

The most serious problem faced when using structural gear for hazmat operations is ensuring that all exposed skin surfaces are covered and protected. Utilizing ear flaps, neck flaps, gloves, and 3/4 boots alone is simply not acceptable.

A hood made of fire-resistant materials such as Nomex™ or Kevlar™ will provide some protection for the head, ears, neck, and throat. When worn properly, they do not interfere with the SCBA face seal. A disadvantage is that any chemical splashed onto the hood may be absorbed and remain in direct contact with the skin.

Turnout coats and pants should be constructed with a vapor barrier, usually neoprene, Goretex™, or similar materials. Note that the manufacturers of Goretex™ recommend that their fabric not be worn in any type of chemical atmosphere since it will not stop the passage of chemical vapors.

GLOVES

Gloves must be selected in reference to the tasks to be performed and the specific chemicals they will be exposed to. Because of the likelihood of physical contact, protective gloves should be considered as a critical element in the protective clothing ensemble. Factors to evaluate include chemical resistance, physical hazard resistance resistance, and temperature resistance.

Cotton, synthetic fiber, leather, and firefighting gloves will absorb liquids, keeping these chemicals in contact with the skin when working around poisons and agricultural chemicals. They will also deteriorate when exposed to corrosive liquids. In comparison, synthetic rubber and plastic gloves may melt when exposed to high temperatures associated with firefighting, or may deteriorate upon contact with certain petrochemical products. They also may not provide adequate protection against many corrosives and agricultural chemicals. Polyvinyl alcohol (PVA) gloves have excellent compatibility against certain petroleum solvents, but break down upon exposure to water

Products which penetrate natural rubber may create serious exposure problems for gloves, boots, and SCBA facepieces. Examples include methyl bromide and dichloropropene.

RESPIRATORY PROTECTION

Since toxic, corrosive, and flammable vapors along with the products of combustion are present, air supplying respiratory protection devices are required. Positive-pressure SCBA must be the minimum level of respiratory protection. Because of problems with decontamination before refilling, additional SCBA and air supply units are almost always required at hazmat incidents.

NOTES

GENERAL

It is not uncommon for exposure to a specific chemical or hazmat environment to require the complete discarding of all structural firefighting clothing. Leather and fibrous materials are easily permeated by many chemicals and make decontamination difficult at best. Polycarbonate helmets may be affected by solvents. Pesticides, PCB-related fires, and radioactive materials incidents may make any decon impossible. Disposal should be done in accordance with local, state, and federal environmental regulations.

LIQUID CHEMICAL SPLASH PROTECTIVE CLOTHING

There are many hazmat incidents where structural firefighting clothing will not offer an adequate level of protection. The hazards and potential harm of the released material may require that specialized clothing be worn. Liquid chemical splash protective clothing consists of several pieces of clothing and equipment designed to provide skin and eye protection from chemical splashes. **It does not provide total body protection from gases or vapors, and should not be used for protection against liquids which give off vapors known to affect or be absorbed through the skin.** For hazmat operations, SCBA or airline hose units must also be used for respiratory protection.

Liquid chemical splash protective clothing may be used under the following conditions:

- The vapors or gases present are not suspected of containing high concentrations of chemicals which are harmful to, or can be absorbed by, the skin.
- It is highly unlikely that the user will be exposed to high concentrations of vapors, gases, or liquid chemicals which will affect any exposed skin areas.
- Operations will not be conducted in a flammable atmosphere. However, in some situations, it is possible to wear liquid chemical splash protective clothing either over or under structural firefighting clothing to offer limited chemical splash and thermal protection.

Liquid chemical splash protective clothing can be best discussed in terms of skin, body, and respiratory/inhalation protection.

SKIN AND BODY PROTECTION

Liquid chemical splash protective clothing is routinely used in hospitals and laboratories where various chemicals, etiological agents, and infectious diseases are handled. They are also found in nuclear facilities and installations which handle or process radioactive materials, and are used

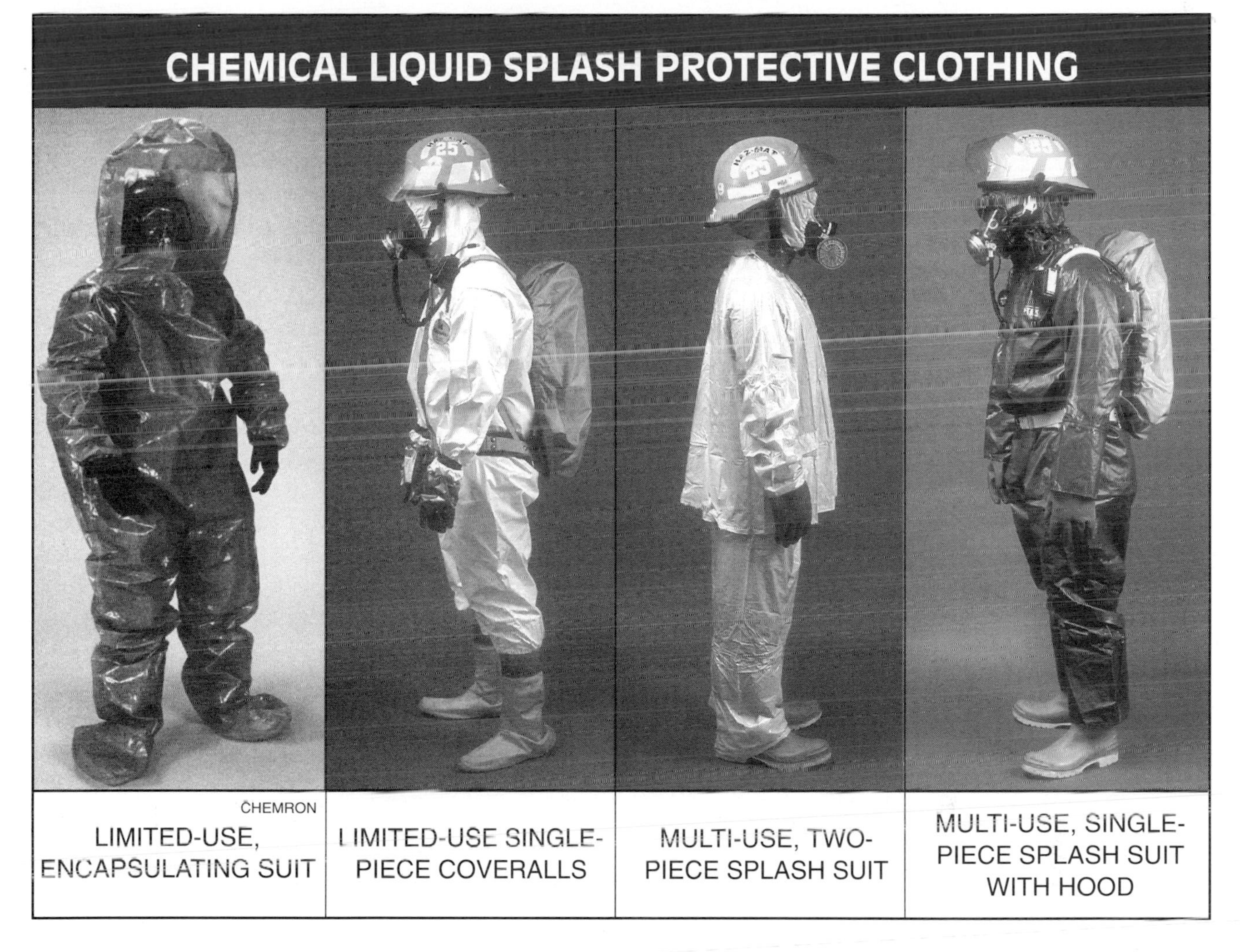

FIGURE 8-7: EXAMPLES OF LIQUID CHEMICAL SPLASH PROTECTIVE CLOTHING

in handling mild corrosives and PCBs, in protecting against asbestos fibers and lead dust, and in formulating and applying agricultural chemicals.

In emergency response, liquid chemical splash protective clothing is often used to enter the initial site, protect decon personnel, and provide limited chemical protection in combination with structural firefighting clothing.

Several common types include:

- **Single-piece Suits**—usually coveralls, a splash suit, or an encapsulating suit which is not vapor tight. Hoods and booties may be attached to coveralls and splash suits. Both limited-use and multi-use garments are available. Considering their low cost, absence of decon problems, and the varied needs of ERP, single-piece limited-use suits are an excellent alternative for the first responder.
- **Two-piece Suits**—usually consist of bib overalls or pants worn with a jacket. Some ensembles include an expanded back or

NOTES

"humpback" design which covers SCBA. They encapsulate the user but do not provide total vapor protection. Accessories such as gloves, boots, and hoods are available.

- **Head Protection**—hard hat, helmet, or hood. Some form of hard hat protection is recommended when using a hood. This gear is designed in various configurations and materials.
- **Gloves**—some coveralls and jackets have a sleeve mounted splash guard which prevents wrist exposure. Some manufacturers have also developed an O-ring glove and cuff assembly which ensures a leak-proof glove/cuff assembly. These designs allow gloves to be easily interchanged according to the hazard.

  For maximum hand protection, overgloving and doublegloving should always be used. Doublegloving involves the use of latex surgical gloves under a work glove. It permits doffing of the work glove without compromising exposure protection and also provides an additional barrier for hand protection. Doublegloving also reduces the potential for hand contamination when removing protective clothing during decon procedures. This process is discussed in Chapter Eleven, Decontamination.

  Overgloving is the wearing of a second glove over the work glove for additional chemical and abrasion protection during lifting and moving operations.

  Gloves should be acquired in at least two sizes—size 9 gloves are commonly used for single- or doublegloves, while sizes 11 and 12 are appropriate for overgloves for additional strength and chemical protection. Common types include Viton™, neoprene, butyl rubber, Chloropel™, Silver Shield™, and polyvinyl chloride (PVC).
- **Footwear and Shoe Covers**—foot protection may be boots, separate shoe covers or booties which are part of the ensemble. Boots should provide both chemical and mechanical protection (e.g., cuts, punctures, etc.). Shoe boots (i.e., firefighting boots) are usually used by ERP, but work boots over footwear are also found. Common boot materials include PVC, neoprene, PVC/nitrile, and nitrile.

  Many liquid chemical splash suits constructed from limited-use materials have integral or connected booties. Although booties and shoe covers may be worn over shoes, they offer more protection when worn inside chemical resistant boots. If worn over shoes, they should have abrasion ridges on the soles to minimize the slip and fall hazard. Shoe covers and booties are most often found in those areas where radioactive materials, etiological agents, and agricultural chemicals are handled, or where cleanliness is a concern.
- **Aprons and Body Coverings**—aprons, lab coats, sleeve guards, and other body coverings are designed for protection against spills and splashes that occur when physically handling chemicals and other hazardous materials. They are primarily used for

routine chemical handling operations rather than emergency response.

Duct tape is sometimes used to ensure that zippers remain closed, to accommodate size differences, secure gloves and boots, and secure the hood to the SCBA mask on some liquid chemical splash suits. **Duct tape provides a false sense of security and does not increase the ability of the garment to provide liquid chemical splash protection. In some instances, duct tape may actually damage the garment material.** It should be noted that the 3M Company has recently developed a tape specifically for protective clothing applications.

NOTES

CHEMICAL VAPOR PROTECTIVE CLOTHING

Many hazardous materials exposures will require the use of specialized protective clothing which provides full-body protection with gas-tight integrity. When used with air-supplied respiratory devices, chemical vapor protective clothing provides a gas-tight envelope around the wearer.

Chemical vapor suits should be used when the following conditions exist:

- Extremely hazardous substances are known or suspected to be present, and skin contact is possible (e.g., cyanide compounds, toxic and infectious substances).
- There is potential contact with substances that harm or destroy skin (e.g., corrosives).
- Anticipated operations involve a potential for splash or exposure to vapors, gases, or particulates capable of being absorbed through the skin.
- Anticipated operations involve unknown or unidentified substances and require intervention by ERP.

TYPES OF CHEMICAL VAPOR PROTECTIVE SUITS

Chemical vapor protective clothing is manufactured in several configurations. It can be categorized by the manner in which respiratory protection is incorporated with the suit. Figure 8-8 outlines the relative advantages and disadvantages of each type.

- **Type 1 Suit**—allows for SCBA to be worn underneath the suit. It is the most common design used in North America and provides total vapor protection by encapsulating the wearer. Easily identified by its "humpback" design.
- **Type 2 Suit**—SCBA is worn over or outside of the suit, exposing it to the environment. The SCBA facepiece is either incorporated into the suit hood or worn over the hood. The gas-tight integrity of the suit is dependent on the quality of the seal between the suit hood and the SCBA facepiece. In addition, the facepiece

NOTES

serves as the primary barrier for respiratory protection against chemical permeation. Very common design in the European countries.

- **Type 3 Suit**—completely encapsulates the user, but has the capability to use a supplied air respirator for respiratory protection. This suit is identical to the Type 1 design with the addition of an airline hose bulkhead connection. Requires a secondary emergency air source (e.g., SCBA or escape pak) for hazmat operations.

VAPOR SUIT ATTACHMENTS AND ACCESSORIES

- **Gloves**—can be permanently attached or detachable. Permanently attached gloves offer integral, vapor tight wrist protection. However, when the gloves must be replaced, the entire suit must be taken out of service and possibly returned to the manufacturer.

  Most manufacturers allow the interchange of gloves. This permits the user to select the glove material which offers the highest level of chemical and/or physical protection, yet facilitates maintenance and repair. Methods of attaching the glove to the suit include concentric rings, a ring/clamp system, an interlocking ring/pin system, and connecting rings. This ensures that the glove/suit seal is vapor tight and stops the penetration of vapors and/or liquids.

  Gloves must match the chemical resistance of the primary suit material. Overgloving and doublegloving should always be used.

- **Boots**—As with gloves, boots can be either an integral part of the suit or detachable. Their type and construction will vary and may be either standard sized shoe boots or overboots. With overboots, some responders use cheap sneakers in lieu of work shoes under the overboots for easy disposal/decontamination. Steel toe and shank protection should also be provided. Common boot materials include PVC, neoprene, PVC/nitrile, and nitrile. Detachable boots can be easily removed and replaced by assemblies similar to those found on detachable gloves.

  Many chemical vapor suits incorporate an integral bootie design which is constructed of the same material as the suit. Chemical boots are worn over the booties and a splash guard is then pulled down to prevent liquid product from entering the boot. This feature allows the user to wear footwear that fits, a task more difficult with other boot designs.

- **Suit Fit and Closure Assemblies**—Mobility is sacrificed whenever a chemical vapor suit is worn. The degree of restriction depends on the suit type, the primary suit material, and type of respiratory protection used. Most suit manufacturers offer several sizes. Unless suits can be provided for individual users, it is a good idea to order large and extra-large sizes. While a smaller person can always modify a larger suit, someone 6′2″ will rarely squeeze into a "medium."

  All chemical vapor suits are sealed by a closure assembly. The pressure sealing zipper is the most common. When closed, the zipper forms a gas-tight seal. A second type consists of outer extruded

CHEMICAL VAPOR PROTECTIVE CLOTHING

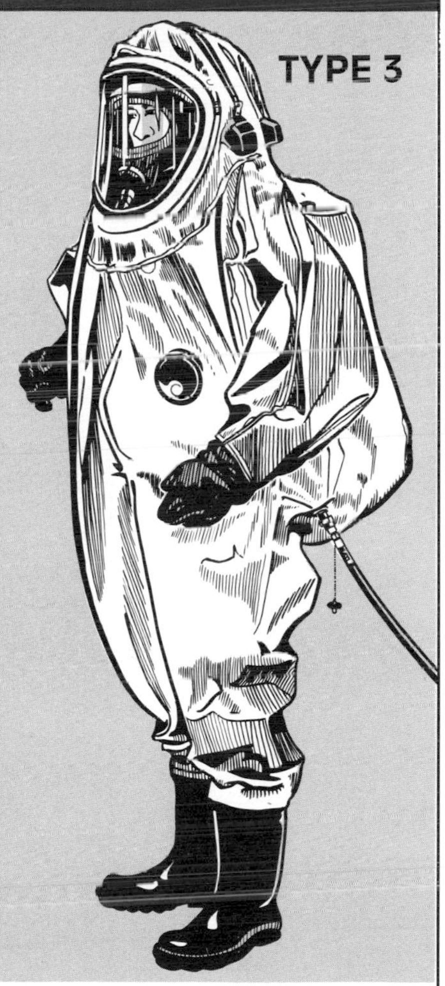

CHEMICAL VAPOR SUIT (SCBA INSIDE SUIT)

ADVANTAGES

1. Offers maximum level of protection to the user.
2. Positive internal pressure may help to minimize minor leaks.
3. If SCBA malfunctions, the user may have some time to reach a nonhostile environment.

LIMITATIONS

1. Reduced mobility and visibility, with some problems in confined space operations.
2. No easy method of re-supplying air cylinders—must decontaminate before opening the suit.
3. Problems in implementing SCBA emergency procedures if suit doesn't have "Batwing Sleeves."
4. Higher weight than similar Type 2 suits.

CHEMICAL VAPOR SUIT (SCBA OUTSIDE SUIT)

ADVANTAGES

1. Greater comfort and mobility than Type 1 suits because of close-fitting cut.
2. Air cylinders can be removed or changed without opening the suit.
3. Turnout coat or limited-use garment can be worn over the suit for additional protection.

LIMITATIONS

1. SCBA exposed to the atmosphere without protection.
2. Facepiece compatibility (lens and mask) may not be equivalent to suit compatibility.
3. Many facepiece and other SCBA materials are not tested for chemical permeation resistance.

CHEMICAL VAPOR SUIT (SUPPLIED AIR RESPIRATOR)

ADVANTAGES

1. Permits extended operations.
2. Positive pressure always maintained in the suit.
3. Airline hose may provide mechanism for minor body cooling.
4. May rely upon SCBA as primary air supply and airline hose or a second SCBA bottle as secondary air supply.

LIMITATIONS

1. Limited maneuverability due to hoseline. Creates a tripping hazard and may become tangled.
2. If using airline hose, distance is limited to the length of the airline hose (generally not greater than 300 ft.).
3. Construction of the airline hose in regards to chemical compatibility.

DON SELLERS

FIGURE 8-8: ADVANTAGES AND LIMITATIONS OF CHEMICAL VAPOR PROTECTIVE CLOTHING

NOTES

sealing lips with an inner restraint zipper. The inner zipper provides closure strength while the sealing lips provide the gas-tight seal. The extruded sealing lip assembly is similar in principle to the Ziploc™ closure of plastic bags.

Emergencies may arise when the suit integrity is compromised or when the SCBA malfunctions. Beware of the initial impulse to immediately get out of the suit, as this may endanger the user. Many suits have an expanded sleeve design (i.e., "Batwing sleeve") which allows the user to easily remove one or both arms from the suit to manipulate the SCBA valves. Should the suit integrity remain intact, several minutes of air should remain within the suit, allowing the user to reach a safe haven.

- **Overgarments**—Many chemical vapor protective clothing manufacturers offer flash protection and/or low temperature protective overgarments. Flash overgarments are not entry or proximity clothing; they lack any insulation and will provide only limited protection in the event of a flash fire. Likewise, low temperature overgarments offer only limited protection against splashes of liquefied gases, cryogenic liquids, and their associated vapors.
- **Head Protection**—May be provided through either a built-in hard hat, a separate hard hat, or a bicycle cap.
- **Cooling and Ventilation**—Both liquid chemical splash and vapor protective clothing seal the body in a manner which retains body heat and moisture. Heat stress becomes a concern even in moderate temperatures. The heat stress factor, in conjunction with added weight and restrictions in movement, results in a high level of physiological and psychological stress. Chemical protective clothing should only be worn by individuals in good health and physical condition, as profiled by a comprehensive medical surveillance program.

  Cooling methods are discussed in Chapter Two, Health and Safety.
- **Communication**—Verbal, person-to-person communications while wearing chemical vapor clothing is virtually impossible. Radio communications is a necessity for entry operations. In addition, other alternatives, such as hand signals and large flash cards, must also be evaluated.

  Communications systems for use within protective clothing ensembles include radio headsets and ear, mouth, and bone microphones. Any device should be intrinsically safe and approved for use in explosive atmospheres (see discussion on terms in Chapter Six, Identification). Three modes of operation exist: continuous transmit, push to transmit, and voice-activated.

  When radio systems fail or high noise levels impair verbal communications, it will be necessary to use hand signals.

  Visual identification of personnel in the Hot Zone is particularly important. It enables the Safety Officer and Hazmat Branch Director to differentiate between personnel operating in the hot and warm zones. Identification methods can be large numbers attached to the suit or color-coded (or numbered) traffic vests. Reflective tape is helpful when operating at night or in low light environments. Cyalume® lightsticks suspended from the suit can also be used to identify personnel during night operations. Check the manufacturer's recommendations before applying tape, as it may damage the suit material.

NOTES

HIGH TEMPERATURE PROTECTIVE CLOTHING

ERP may be required to operate in high temperature environments which exceed the protective capabilities of structural firefighting clothing. These types of hazards are often found during aircraft rescue firefighting (ARFF) operations and in petrochemical facilities. In these circumstances, special aluminized fabric protective clothing may be necessary.

Thermal energy can be encountered in the three forms listed below. You must recognize and understand their differences before selecting appropriate thermal protection.

- Ambient heat—the temperature of the surrounding atmosphere in a given scenario. For example, the ambient temperature in a public building is generally 65° to 70°F. During a structure fire, it may range from 120°F to 200°F at the floor level.
- Conductive heat—the heat generated by direct physical contact with a hot surface. For example, touching a metal valve which has been heated as a result of a flammable liquid spill can expose ERP to temperatures over 1,000°F.
- Radiant heat—the heat generated by a heat source such as a flammable gas or liquid fire. The heat is absorbed by materials which are struck by the radiant heat emitted by the heat source.

High temperature protective clothing is designed primarily for radiant heat exposures. Its principal application will be in those situations where there is a minimal likelihood of direct contact with chemical vapors or splashes. SCBA should always be considered part of the ensemble.

TYPES OF HIGH TEMPERATURE CLOTHING

The two types of high temperature protective clothing commonly used by HMRT's are designed for specific thermal environments.

- **Proximity Suits**—designed for exposures of short duration and close proximity to flame and radiant heat, such as in aircraft rescue firefighting (ARFF) operations. The outer shell is a highly reflective, aluminized fabric over an inner shell of a flame-retardant fabric such as Kevlar™ or Kevlar™/PBI™ blends. **These ensembles are not designed to offer any substantial chemical protection**. Design and performance criteria for proximity suits is outlined in NFPA 1976—Standard on Protective Clothing for Proximity Fire Fighting.

  Proximity suits are available as a separate coat and pants ensemble or as coveralls. In many instances, a hood is also used. The outer shell must be kept clean to ensure maximum reflection of radiant heat. The outer shell of aged suits will commonly begin to crack or flake off after several years of regular use. At this

NOTES

point, the protection factor drops significantly, and the suit should be replaced.

Older proximity suits may incorporate loose asbestos fibers in their design. Because of the asbestos hazard, they should be bagged and properly disposed of.

- **Fire Entry Suits**—offer complete, effective protection for short-duration entry into a total flame environment. They are designed to withstand exposures to radiant heat levels up to 2,000°F. Entry suits consist of a coat, pants, and separate hood assembly. They are constructed of several layers of flame-retardant materials, with the outer layer often aluminized.

  Because of the rapid speed with which burn injuries affect the body, entry suits are not effective for rescue operations in total flame environments. Unprotected individuals cannot be safely rescued from a fully involved area. Possible rescues are further impeded by the time necessary to don gear and enter the fire area.

  Entry suits may be useful for offensive operations such as valve shutdowns in a flammable gas or liquid facility. However, there is a lack of mobility and flexibility when attempting these manipulations. Fire entry suits are usually a low priority budget item for most fire departments and HMRT's. There are currently no standards specifying the performance of fire entry suits.

TYING THE SYSTEM TOGETHER

OPERATIONAL CONSIDERATIONS

The selection and use of specialized protective clothing at a hazmat emergency should be approached from a "systems" perspective. This system begins with the evaluation of four variables: (1) the hostile environment, (2) the tasks to be performed, (3) the type of protective clothing required, and (4) the capabilities of the user/wearer.

The most significant hazard in the system is often the user who has unrealistic expectations of the protective clothing. Consider the following guidelines when using protective clothing:

- The selection, maintenance, and use of protective clothing must be integral part of an overall safety program. It is the *last* line of defense which is activated only after site safety procedures have been completed.
- Chemical protection and thermal protection are different qualities that require different materials. While some ensembles combine flash protection garments with chemical protective clothing, they offer limited thermal protection beyond short-duration exposures.
- In some situations, structural firefighting clothing may be worn in combination with liquid chemical splash protective clothing, such as coveralls or a non-vapor-tight encapsulating suit. There are no hard and fast rules in this area—the response situation

DON SELLERS

FIGURE 8-9: TWO COMMONLY USED TYPES OF HIGH TEMPERATURE PROTECTIVE CLOTHING: PROXIMITY SUITS AND FIRE ENTRY SUITS

NOTES

will dictate what level of protection will be worn on the inside and outside. However, the combination of the two levels will increase the potential for heat stress upon responders.

- Always minimize direct contact with any chemicals, regardless of your level of protection. Use common sense—fingers, tongues, and toes are not effective monitoring instruments! Avoid walking into or touching chemicals whenever possible.
- Ensure that entry and back-up crews have equivalent levels of protection. When chemical vapor protection is required, at least four suits will be needed.
- Communications is a critical element in entry operations, especially when using chemical vapor suits. A radio system backed by hand signals is a minimum requirement. Where possible,

NOTES

entry, back-up, and safety personnel should operate on a radio channel separate from other on-scene units.

- Air supply will often be the most critical element of entry operations. Plan for the time needed for entry, exit, and decontamination procedures. Most response organizations use an entry limit of 20 to 30 minutes and primarily rely upon 60 minute SCBA's. It is good practice for the Hazmat Safety Officer to notify the entry crew of their time limits at five-minute intervals.
- Any plan to allow the same entry personnel to re-enter the Hot Zone should be approved by the Hazmat Branch Director, Safety Officer, Medical Officer, and, most importantly, the individual involved. Do not underestimate the physiological and psychological effects of wearing specialized protective clothing.
- Support staff are always needed to assist entry and back-up crews in donning and doffing protective clothing. As a rule, one support person is necessary for each entry person. Donning times in excess of 10 minutes should be considered as excessive. A pre-entry checklist for support personnel is shown in Figure 8-10.

SUPPORT CONSIDERATIONS

The use of protective clothing at a hazmat incident should be the final step of a comprehensive system which begins with an analysis of the plant or community's hazards. This system must include organized and documented training with all types of protective clothing ensembles used within the organization, as well as a regular, effective preventive maintenance and testing program.

Both classroom and hands-on training is essential. Aggressive, experienced firefighters and ERT personnel do not always make effective hazmat response personnel. As a result, many organizations have developed a protective clothing qualification system which requires initial qualification in each type of suit followed by regular retesting.

Protective clothing training evolutions are most effective when conducted on a "building block," or modular basis and combined with other manipulative skills.

Training with protective clothing is essential for the development of effective skills. To minimize damage to liquid chemical splash and vapor suits, many organizations purchase training suits, limited-use suits, or use older models which have been removed from service. When using older suits, ensure that they have been completely decontaminated and do not have a significant chemical exposure history. "Frontline" chemical vapor suits should only be used during actual emergencies.

MAINTENANCE AND INSPECTION PROCEDURES

Preventive maintenance is an integral element of a comprehensive personal protective equipment program. Unfortunately, it is also one of the most neglected.

ENTRY SUPPORT CHECKLIST

INCIDENT ______________________________

ENTRY TEAM MEMBER ______________________________

SUPPORT TEAM MEMBER ______________________________

LEVEL OF PROTECTION ______________________________

SUIT MATERIAL ____________________ SUIT NUMBER __________

DUTIES

- ❐ Coordinate with Medical to begin monitoring of entry and back-up personnel. Ensure that medical evaluation information is maintained and given to the HAZMAT SAFETY OFFICER.
- ❐ Obtain protective clothing and equipment information from INFORMATION.
  - Protective Clothing ______________________________
  - Gloves ______________________________
  - Boots ______________________________
  - Respiratory Protection ______________________________
- ❐ All personal items removed, tagged, and secured.
- ❐ Suit compatibility information double-checked with INFORMATION.
- ❐ Medical evaluation completed and entry/back-up personnel approved by HAZMAT SAFETY OFFICER.
- ❐ Final entry hazards and procedures briefing completed.
- ❐ Verify status of air supply.
- ❐ Communications checked (Channel______)
- ❐ Proper facepiece seal
- ❐ Visual check of Entry Suit:

  ____ All zippers and closures properly secured
  ____ No obvious suit damage
  ____ Overgloving, doublegloving, and footwear verified as appropriate
- ❐ Final communications check completed
- ❐ ENTRY OFFICER notified that Entry Team is ready.

FIGURE 8-10

NOTES

The manufacturer's maintenance and testing recommendations should be consulted for maintenance intervals and procedures. At a minimum, protective clothing should be inspected at the following benchmarks:

- Upon receipt from the manufacturer or distributor.
- After each use.
- Periodic inspections (i.e., monthly or quarterly).
- Whenever questions arise regarding selected protective equipment or when problems with similar equipment arise.

Each inspection will cover different areas in varying levels of thoroughness depending on the type of protective clothing. Detailed inspection procedures are usually available from the manufacturer.

Documentation and maintenance of all appropriate records is a top priority. Individual inventory or record numbers should be assigned to all re-usable pieces of protective clothing, including gloves. This will simplify the process of tracking gloves, liquid chemical splash suits, and chemical vapor suits over a period of time and facilitate monitoring for potential problems.

A logbook should be maintained for each type of protective clothing and equipment which documents its overall history. This documentation notes each time the clothing is worn, inspection and maintenance data, unusual conditions or observations, decontamination solutions and procedures, and dates with appropriate signatures. Periodic records review may pinpoint an item with excessive maintenance costs or out-of-service times.

All protective clothing must be stored properly to prevent damage caused by dust, moisture, sunlight, chemical exposures, temperature extremes, and impact. The manufacturer's storage guidelines should always be followed. Numerous equipment failures during actual use are attributed to improper storage procedures.

SUMMARY

Personal protective clothing and equipment are critical to the success of an organization's hazardous materials response program. A good personal protective clothing program should address six fundamental elements, including hazard identification, PPE selection and use, medical monitoring, training, inspection, and maintenance.

COMMON PROTECTIVE MATERIALS AND TERMINOLOGY

ACETATE—polymer of cellulose acetate; a clear, relatively inexpensive material used for face and eye protection.

ACRYLIC—polymer of methyl methacrylate; clear plastic used for face and eye protection.

NOTES

BARRICADE™—a multi-layered composite film laminated to a strong polypropylene substrate. Developed by DuPont, the limited-use fabric is commonly used for liquid chemical splash protection.

BUTYL RUBBER—material with good resistance to weathering and a wide variety of chemicals. Resists degradation to many contaminants except for halogenated hydrocarbons and petroleum products. Both supported and unsupported forms are used in protective clothing.

CHEMICAL PROTECTIVE FABRICS (CPF™ I, II, III, IV)—family of limited-use garment materials developed by Kappler USA for use with liquid chemical splash protective clothing. CPF™ I provides the lowest level of physical strength and chemical resistance, while CPF™ IV provides the highest level.

CHEMREL™—multi-layered composite material laminated over a polymer substrate. Developed by the Chemron Corporation, the limited-use clothing material is commonly used for liquid chemical splash protection. Provides a higher level of chemical resistance than the Chemtuff™ protective clothing material.

CHEMREL MAX™—multi-layered composite material laminated over a polymer substrate. Developed by the Chemron Corporation, the limited-use clothing material is used for both liquid chemical splash and chemical vapor protection. Provides a higher level of chemical resistance than both the Chemtuff™ and Chemrel™ protective clothing materials.

CHEMTUFF™—proprietary fabric material developed by the Chemron Corporation. The limited-use clothing material is primarily used for routine chemical exposures in controlled environments.

CHLORINATED POLYETHYLENE (CPE, Chloropel)—polyethylene elastomer with a chlorine content of 36% to 45%. Generally has better chemical resistance and physical properties than polyethylene. Considered a good all-around protective material.

FLOURINATED ETHYLENE PROPYLENE (FEP)—polymer with exceptional chemical resistance. Used for protective clothing in film and coating forms.

GORE-TEX™—a proprietary fabric in which microporous polytetraflururoethylene (PTFE), such as Teflon™, is laminated on one or both sides of a fabric. The fabric allows moisture vapor to pass through, reducing heat stress. It prevents penetration by many liquids and solids but does not provide vapor protection.

KEVLAR™—noncombustible, flame-resistant fabric manufactured by DuPont. Used primarily in firefighting protective clothing and coveralls.

LAMINATED—joining two or more sheets or fabrics together by means of heat or adhesive.

LATEX—stable dispersion of polymer or rubber particles in water. Latex gloves and coated fabrics are prepared by coagulating and cross-linking the particles on a form or cloth substrate. Most natural rubber, neoprene, and nitrile gloves are prepared from latices.

LATEX DIPPED—glove prepared by dipping a glove form or fabric glove into a rubber latex bath. The entire glove is formed in one dip.

NOTES

NATURAL RUBBER—highly flexible and conforming material used primarily for gloves. Has high elasticity.

NEOPRENE—synthetic rubber having chemical and wear-resistant properties superior to natural rubber. Easily available and inexpensive. Resists degradation by caustics, acids, and alcohols.

NITRILE RUBBER (Hycar™, Kyrnac™, Paracril™, NBR)—used for supported and unsupported gloves and coated fabric. Resists degradation by petroleum products, alcohols, acids, and caustics.

NOMEX™—noncombustible, flame-resistant fabric manufactured by DuPont. Used primarily in firefighting protective clothing and coveralls.

PACESETTER™—limited-use fabric material developed by Lion Apparel for use as liquid chemical splash protective clothing. The trilaminate material consists of 7.5 oz. 100% Nomex™ poplin, jersey, and Gore-Tex™.

PBI™—noncombustible, flame-resistant fabric manufactured by the Hoechst Celanese Company. Used primarily in firefighting protective clothing and coveralls.

POLYCARBONATE—hard, transparent plastic used for face and eye protection. Exceptionally impact resistant. Often used for safety glass lenses.

POLYETHYLENE—chemically resistant material used as a freestanding film or fabric coating (i.e., polyethylene coated Tyvek™). Provides increased resistance to acids, bases, and salts.

POLYVINYL ALCOHOL (PVA)—water-soluble polymer with exceptional resistance to many organic solvents that permeate most rubbers. Material is stiff, which limits manual dexterity. Will break down upon exposure to water.

POLYVINYL CHLORIDE (PVC)—material used for gloves, splash suits, aprons, and other body coverings, as well as fabric coatings. Clear forms are used for flexible face shields. Resists degradation by acids and caustics.

RESPONDER™—a limited-use garment material developed by Kappler USA for use with liquid chemical splash and chemical vapor protective clothing.

SARANEX™—multi-layered laminate of polyethylene and Saran™ which is a widely used limited-use garment material. Generally applied as a laminate onto Tyvek™.

SOLVENT DIPPED—glove preparation prepared by repeatedly dipping a glove form or substrate into a solution of rubber in a solvent. The rubber is then cured.

SUPPORTED—materials containing a substrate such as cotton, polyester, or nylon fabric which is coated, laminated, or impregnated with a rubber or polymer.

TFE (PTFE)—polytetrafluoroethylene. For example, Teflon™.

TRELLCHEM™—family of multi-use clothing materials developed by Trelleborg Viking Inc. for liquid chemical splash and vapor protection.

NOTES

Includes Trellchem™ VPS, a blend of rubber and polymer barrier materials, and Trellchem™ HPS, a blend of rubber and polymer barrier materials coated on the outer surface with Viton™.

TYCHEM™—family of limited-use clothing materials developed by DuPont for liquid chemical splash protection. Includes Tychem™ 7500 and Tychem™ 9400.

TYVEK™— proprietary, nonwoven fabric used for limited-use (disposable) clothing developed by DuPont. Excellent protection against particulate contaminants.

TYVEK™ QC—Tyvek™ which is coated with 1.25 mils polyethylene. Used for liquid chemical splash protection against many inorganic acids, bases, and liquid agricultural chemicals.

VITON™—proprietary fluoroelastomer. Excellent resistance to degradation and permeation by aromatic and chlorinated hydrocarbons and petroleum compounds. Also very resistant to oxidizers. Very expensive synthetic elastomer used primarily in gloves and chemical vapor suits.

REFERENCES AND SUGGESTED READINGS

Carroll, Todd R. "Contamination and Decontamination of Turnout Clothing." Emmitsburg, MD: Federal Emergency Management Agency—U.S. Fire Administration (April 29, 1993).

Forsberg, Krister and Lawrence H. Keith, CHEMICAL PROTECTIVE CLOTHING PERMEATION AND DEGRADATION COMPENDIUM, Boca Raton, FL: Lewis Publishers (1993).

Glenn, Stephan, W. David Eley, and Paul A. Jensen, "Evaluation of Three Cooling Systems Used in Conjunction with the USCG Chemical Response Suit." HAZARDOUS MATERIALS CONTROL (September–October, 1990), pages 54–59.

Keith, L. H., "Is Your Method of Selecting CPC Based on Reliable or Liable Data?" OCCUPATIONAL HEALTH & SAFETY (August, 1992), pages 35–48.

LeMaster, Frank, "Why Protective Clothing Must Be Cleaned." THE VOICE (August/September, 1993), pages 19–21.

National Fire Protection Association, HAZARDOUS MATERIALS RESPONSE HANDBOOK (2nd edition), Boston, MA: National Fire Protection Association (1992).

National Fire Protection Association, LIQUID SPLASH PROTECTIVE SUITS FOR HAZARDOUS CHEMICAL EMERGENCIES—NFPA 1992, Boston, MA: National Fire Protection Association (1992).

National Fire Protection Association, SUPPORT FUNCTION PROTECTIVE GARMENTS FOR HAZARDOUS CHEMICAL OPERATIONS—NFPA 1993, Boston, MA: National Fire Protection Association (1992).

National Fire Protection Association, VAPOR-PROTECTIVE SUITS FOR HAZARDOUS CHEMICAL EMERGENCIES—NFPA 1991, Boston, MA: National Fire Protection Association (1992).

National Transportation Safety Board. NTSB Safety Recommendations and Report on the Benecia, California Dimethylamine Railroad Tank Car Incident. Washington, DC: National Transportation Safety Board (April 23, 1984).

Schwope, A. D. et. al., GUIDELINE FOR SELECTION OF CHEMICAL PROTECTIVE CLOTHING (3rd edition), Cambridge, MA: Arthur D. Little, Inc. (1992).

Stull, Jeffrey O., "Chemical Protective Clothing." OCCUPATIONAL HEALTH & SAFETY (November, 1992), pages 49–52.

Stull, Jeffrey O., "Using, Applying Chemical Permeation Data." SAFETY & PROTECTIVE FABRICS (November, 1993), pages 10–11.

U.S. Environmental Protection Agency, STANDARD OPERATING SAFETY GUIDELINES, Washington, DC: EPA (1988).

York, Kenneth J. and Gerald L. Grey, HAZARDOUS MATERIALS/WASTE HANDLING FOR THE EMERGENCY RESPONDER, Tulsa, OK: Fire Engineering Books and Videos (1989).

CHAPTER 9

INFORMATION MANAGEMENT AND RESOURCE COORDINATION

OBJECTIVES

1. Define the terms "information," "data," and "facts" as related to the function of information management.
2. Describe the types of information required to safely and effectively manage a hazmat incident.
3. Describe the criteria for evaluating hazardous materials information management systems for field applications.
4. Identify and describe the Hazmat Branch functions required to manage information at a hazmat incident.
5. Define the term "resources" as related to the function of resource coordination.
6. Describe the process and procedures for coordinating internal and external resource groups at a hazmat incident.
7. List and describe three techniques for improving coordination and communications with internal or external resource groups.

"You don't manage people; you manage things. You lead people."

Admiral Grace Hooper (USN)

INTRODUCTION

This chapter will describe the fifth step in the **Eight Step Process©—Information Management and Resource Coordination**. We've previously mentioned that managing a hazmat emergency is much like managing

NOTES

a war. In both instances, the timely access to and effective distribution of information and resources are critical to the overall success of the mission.

The function of Information Management and Resource Coordination can be viewed from two perspectives. From a strategical perspective, it is a "constant" that starts with the notification and dispatch of emergency responders and ends with the termination of the incident. From a tactical perspective, it is the transition point from the "size-up" phase of the hazmat incident to the implementation of the incident action plan. Both perspectives share a common "bottom line"—failure to get the right information to the right people at the right time can jeopardize both the safety of responders and the overall success of the emergency response effort.

If you have been reading this text in sequence from Chapter One, it should be obvious that it is absolutely critical that the "right hand knows what the left hand is doing." Among the basic principles which we have emphasized in previous chapters include:

- Information and resources cannot be effectively coordinated if an incident management organization is not in place.
- Information which is poorly coordinated among the players at the emergency scene can politically damage the Incident Commander's credibility and ultimately undermine the response operation.
- Safe operations cannot be carried out unless information and resources are coordinated in a timely and effective manner.
- Decisions must be made from a solid technical basis.

Pulling all the pieces together to develop and implement a coordinated incident action plan may seem like an overwhelming problem, and it often is! In this chapter, we will review several structured and systematic ways that responders can use to organize information and resources to their fullest advantage.

INFORMATION MANAGEMENT

First we need to understand some basic terms. They are:

- **Data**—Individual elements which are gathered and organized for analysis. In emergency response, the most common data elements are the physical and chemical properties of a material. For example, specific gravity, flash point, exposure values, and vapor density.
- **Facts**—Things which are known to be true, or statements about something which has occurred. In emergency response, facts are typically based upon objective observations. For example, observations reported by a RECON team concerning mechanical damage to a container, or the concentration of airborne contaminants as indicated by monitoring instruments.
- **Information**—The communication of knowledge. From an emergency response perspective, information is knowledge that is

based upon data or facts which can be used to support decision-making.

In summary, the available data and facts collected at the scene of an emergency make up the information that the Incident Commander will use to make decisions.

In the 1970's and the early 1980's, there was often a lack of timely and accurate technical information to support the Incident Commander's decision-making process. However, the growth and development of the portable computer field dramatically increased our field information capabilities. Today, portable computers are common electronic items for many households. And tomorrow? There will be electronic notebooks with integrated cellular telephone and fax capabilities linked to central databases by satellite.

We can summarize the issue of information access and management with this simple statement—"We have some good news and we have some bad news." The good news is that today virtually everyone can access reliable and quality technical information. The bad news is that the there is now *so much* information available that we sometimes have a difficult time interpreting, managing, and prioritizing it at the emergency scene. See Figure 9-1.

For example, a freight train derailment involving hazardous materials can generate a computer consist profiling how the train is made up, emergency response information sheets on each hazmat car, and individual waybills for each car in the train. If the train consists of 80 cars and 20 of those are carrying hazardous materials, the Incident Commander could potentially have access to over 100 different documents. This doesn't include other sources of information which emergency response personnel may generate on their own, such as through CHEMTREC®, reference manuals, computer data bases, or pre-plans. Sorting, evaluating, interpreting, and communicating this information to people who need it can be a perplexing problem.

Information management must begin well before the incident. Some important decisions which must be made include:

1. What type of information will be needed at the emergency scene? How should the information be compiled?
2. What is the priority of information? What do I need immediately versus an hour into the incident?
3. How will the information be stored for quick recovery at the incident scene—manually or electronically?
4. Are the information and retrieval systems suitable for field applications?
5. Who will be responsible for managing and coordinating information at the incident scene? Are they properly trained and equipped for the job?

In the subsequent sections we will address each of these issues individually.

NOTES

TONY EXLINE & JEFF CALLAWAY/GROUND ZERO GRAFIX

FIGURE 9-1: THE INCIDENT COMMANDER CAN BE OVERWHELMED WITH INFORMATION AND RESOURCES AT THE COMMAND POST

INFORMATION REQUIREMENTS

Thinking is the basis of problem solving and good decision-making. The dilemma is that both thinking and decisions cannot be made without reliable information. Likewise, responders never have enough time to sift through and understand all of the data and information that modern technology has made available.

While it is difficult to operate safely when little or no information is available, it can be even more dangerous to initiate offensive hazmat operations when the IC is overwhelmed with data, facts, and information.

The information overload problem can be minimized by determining what type of information is absolutely essential for emergency decision-making. Essentially, this process involves identifying what you need to know versus what is nice to know. For example, what are the first things you want to know about your "enemy"? Your list should read something like this:

1. What are its hazards?
2. What are the PPE requirements (or what must I do to protect myself)?
3. What are the health concerns (e.g., exposure values, signs and symptoms of exposure, etc.)?

4. What are the initial tactical recommendations (e.g., spill control, leak control, fire control, public protective actions)?
5. What type of decontamination procedures and methods will be required?

Trying to collect and carry every source of hazard and response information in an emergency response vehicle just isn't practical. (Imagine rolling the county library's bookmobile to all hazmat incidents... "Command to communications, give me a second alarm and add the bookmobile to the assignment!").

Responder information needs will be different based upon local or facility conditions. Nonetheless, there are some basic sources of reliable information which should be available either on the scene of an emergency or accessed from on-site communications.

NOTES

Emergency Response Plans

Emergency Response Plans typically include a Hazard Analysis section which identifies special problems that may exist within the plant or community, the potential risk and consequences of a hazmat incident, and the level of available emergency response resources. Process safety and risk management documents can also be good sources of hazard analysis information.

While an Emergency Response Plan does not typically have much use in the field ("If we need the Plan to figure out what to do at 3 o'clock in the morning, we've got bigger problems than the Plan can fix!"), it is a good beginning point to identify target hazards for which a more detailed Pre-Incident Plan should be developed. This process should ultimately allow one to determine the types of information to be stored in the information management system.

Plans and planning documents may also help to determine the types of information needed at the scene of an emergency. For example, if the Hazard Analysis section of a plan identifies that poison gas is used in a particular process at a chemical plant, specific toxicity/exposure data and evacuation information would be very useful for immediate retrieval at an actual incident.

Chapter One describes several types of hazard analysis tools and sources of planning information that are available. Familiarize yourself with these different techniques so that you know what to ask for when you write or visit a facility for information.

Pre-Incident Plans

Pre-Incident Plans (or pre-plans) are individually prepared documents which focus on a specific problem or location, such as a railyard or a bulk storage facility. In fixed facilities, such as a refinery or chemical plant, pre-plans may concentrate on a particular process or tank with special hazards.

The larger the facility or jurisdiction, the more essential it becomes that the pre-planning process be practical and prioritize problem areas based upon the hazard analysis process. Criteria for developing special pre-plans in these situations may include the following.

NOTES

- ❑ Type of hazards and risks present. High consequence scenarios are good candidates for further analysis.
- ❑ Environmentally sensitive exposures. Scenarios and locations in close proximity to waterways or sole source aquifers.
- ❑ Unusual or poor water supply requirements. Examples include facilities located in an area with a reputation for water supply problems or installations that may require unusually high fire flows (e.g., large diameter flammable liquid storage tank).
- ❑ Locations which will require large quantities of foam concentrate such as bulk petroleum storage facilities, pipelines, etc. The quantity requirements need to be known as well as application rates, availability of foam supplies, etc.
- ❑ Restricted or delayed response routes. For example, single approach and access corridors, railroad tracks or draw bridges which are frequently blocked, etc.
- ❑ Poor accessibility. Examples include restricted entrance corridors, secure installations, obstacles, unusual ground slope, etc.

Pre-plans can provide valuable information if a field survey form is used to record key information. Although informal site visits may be instructive for the personnel participating in them, they often provide little long-term improvement if there is no mechanism for compiling and maintaining key information.

To be useful, pre-planning documents must be completed using a standardized format. A well-designed survey form assures that essential information is gathered and consistently recorded. Field testing a survey form before adopting a standard format will pay big dividends for your planning, training, and response programs. Bigger is not necessarily better; a simple form can reduce the maintenance time required to keep pre-plans current.

A well prepared pre-plan should include a simple plot plan which shows the basic details of the facility but is not cluttered with extraneous information. The pre-plan should also indicate the availability of any special plans prepared by the facility which may be referenced during an emergency, (for example, foam calculations, tactical checklist, etc.).

Published References

There are a wide variety of published emergency response references available to the emergency responder. They are generally divided into the following categories:

- ❑ Reference Manuals and Guidebooks (e.g., DOT ERG, AAR Emergency Action Guides).
- ❑ Technical Information Centers (e.g., CHEMTREC®, CANUTEC).
- ❑ Hazardous Materials Data Bases (e.g., TOXNET, TOMES).

Chapter Seven provides a good overview of the different types of widely recognized published references. Many of these are in their third generation (electronic) or third printing (written). Familiarize yourself with these sources of information.

NOTES

Learn to be selective in what you decide to purchase and carry to the scene of the emergency. It is much better to have a few published references that you have trained on and are comfortable with than to have a large and unfamiliar reference library.

Information is no substitute for thinking! People have been trained to believe things in writing without questioning their source or authenticity. Just because it is written down on paper or shows up on a computer screen does not mean that you have to buy it, subscribe to it, or believe what it says. Be inquisitive and challenge information you may have to rely on in the field.

INFORMATION STORAGE AND RECOVERY

Many HMRT's have access to portable computer systems. Their use is growing as more and better types of software applications become available. However, don't overlook the simplicity of a three-ring binder or a small on-board filing cabinet. Pre-incident plans, site drawings, and process flow diagrams are especially well suited for "manual" retrieval.

A computer system, no matter how user friendly it is, does not guarantee an improvement to your information management system. Don't expect much from your portable computer system if you haven't read the manual and had some realistic training that simulates field conditions.

The key to successfully managing and retrieving hazmat information under emergency conditions is good organization and simplicity. The "acid test" for deciding whether one type of information management system is better than another should be: "Will it work in the field in a consistent and reliable manner?" As previously noted, there will always be some situations where manual information management systems may be better suited than computerized systems.

When evaluating systems, consider the following:

User Friendly: Stressful situations call for simple solutions. Beware of hardware and software that requires a lot of brains and experience to operate. Almost every HMRT has an in-house computer wizard (You know the type...60% of the time they're worth their weight in gold. Unfortunately, they also operate on a different orbit and frequency the remaining 40% of the time!). **Don't build your system around one individual!** The same case can be made for every piece of equipment in your inventory.

Durability. Experience shows that equipment which works well in an office environment often won't hold up in the emergency response field. Junk yards are full of equipment that was rattled to pieces by the *smooth ride* of a 10-ton hazmat squad. Dust, cold weather, Mongo the Hazmat Tech, and a little spilled coffee on the keyboard often equals a dead computer.

Remember—a computer's outstanding attributes of speed, storage capacity, consistency, and the ability to process complex logical instructions are of no value unless they are applied within a good management process.

NOTES

MANAGING INFORMATION IN THE FIELD

Managing information in the field becomes particularly important as the Incident Commander, Hazmat Branch Officer, and others evaluate options concerning protective clothing, decontamination requirements, and public protective actions. Bad news also doesn't get better with time; information must be moved throughout the incident management organization in a timely manner.

Information management is a dynamic process which must adjust its scale over time to provide the right information to the right people at the right time. The larger and more complex the incident, the greater the need for a formal structure to manage the data and facts that will flow between Command and the various individuals and organizations at the emergency scene. Information must also flow freely to and from the incident scene to off-site support facilities, such as the Emergency Operations Center, CHEMTREC®, etc. See Figure 9-2.

Hazmat Branch Functions

An effective way to organize and manage information within the Hazmat Branch is to sub-divide Branch functions into specific Groups or Units, as described in the Hazmat Branch Operations section of Chapter Three. Following this general approach, the Hazmat Branch Officer assigns different functions to the first available and qualified personnel as the situation requires. Basic functions include:

- ❑ **Safety:** Ensuring safe and accepted practices and procedures are followed throughout the incident.
- ❑ **Site Control:** Establishing hazard control zones, establishing and monitoring access routes, etc.
- ❑ **Information:** Gathering, compiling, coordinating, and disseminating data and information relative to the incident.
- ❑ **Entry:** Entry and back-up operations in the hot zone including reconnaissance, monitoring, sampling, and mitigation.
- ❑ **Decontamination:** Development and implementation of the decon plans and procedures.
- ❑ **Medical:** Pre- and post-entry medical monitoring for entry teams, as well as providing technical medical guidance to the Hazmat Branch.
- ❑ **Resources:** Controlling and tracking supplies and equipment used by the Hazmat Branch.

See Figure 3-4 for an example of how each of these functions is organized within the Hazmat Branch.

CREW RESOURCE MANAGEMENT (CRM) PROGRAMS

C. R. THOMAS, UNITED AIRLINES

There are many parallels between aircraft cockpit operations during an in-flight emergency and command post operations during a hazmat emergency. The cockpit functions as the "command post," the pilot is the Incident Commander, and the crew are the members of the Incident Command staff. Information coordination and resource management are directly influenced by the crew's ability to handle conflict, communications, and interpersonal dynamics.

To improve cockpit management and operations, the commercial airline industry has developed crew training programs which focus upon topics such as command, leadership, and resource management. Commonly referred to as Crew Resource Management (CRM), these programs can serve as a model in helping emergency responders improve their incident management capabilities. Several examples are listed below.

LEADERSHIP

Leadership can directly influence the quality of the IMS organization. While the exercise of leadership is typically associated with the Incident Commander, every officer has a certain amount of authority, responsibility, and opportunity to exercise leadership. Leadership styles will vary depending upon the situation. Skills which can be used to improve leadership effectiveness include problem definition, inquiry, and advocacy.

- **Problem Definition.** The first and most critical step in problem solving is defining the problem ("A problem well-defined is half-solved."). Once the problem is understood, alternative solutions can be evaluated. Involving the IMS organization in the "diagnosis" increases both decision-making effectiveness and organizational "buy in."
- **Inquiry.** Human error is inevitable. Unfortunately, the cost of human error at a hazmat emergency is often measured in lives. Inquiry is a mental process that involves constantly evaluating and re-evaluating everything that can be anticipated during an incident. Commonly known as "playing the devil's advocate," this process can help the IC sense a difference between what actually is happening or about to happen, and what should be occurring. Inquiry is the responsibility of every individual within the IMS organization.
- **Advocacy.** Advocacy is the responsibility of personnel to speak out in support of a course of action different than that currently being planned or implemented (e.g., " I have a problem with that..."). Reasons may include safety issues, technical concerns, etc.

Advocacy is constructive questioning and is not a resentment of authority. Inquiry and advocacy are essential to each other. Inquiry which results in the detection of potential safety problems is of little use unless the information is advocated so others can react.

RESOURCE MANAGEMENT

Resource management includes the effective use of emergency responder skills, knowledge, and expertise. Communications, coordination, conflict resolution, and critique are critical elements which can enhance the effectiveness of both individuals and the IMS organization.

- *Communication* involves both talking and listening. To talk when no one is listening is worthless. To have deaf ears to what is being said is equally ineffective. One must always be sensitive to the "atmosphere" of the discussion. When the atmosphere is open, it allows for a free flow of communication and encourages input.
- *Coordination* is the process by which information, plans, and operational activities are considered and shared throughout the IMS organization. Coordination minimizes the likelihood of confusion because the players understand the tactical action plan. It also reduces the potential for error because of overlooked or disregarded information.
- *Conflict Resolution.* Conflict is inevitable. Differences in thoughts, opinions, values, or actions—whether actual or perceived—can lead to disagreements and disputes. Differences in personality alone may create conflict.

  Conflict is not necessarily negative or destructive. What makes conflict "bad" is the inability to constructively cope with it, such as when the conflicting positions are passively given up or aggressively suppressed. When effectively channeled by the IC, conflict can be transformed into an effective comparison of viewpoints, problem definition, tactical options, and sound solutions.
- *Critique.* Many have said that, "experience is the best teacher." This may be true, but only if one takes full advantage of the lessons learned. Critique involves studying a plan of action both before, during, and after its implementation. Feedback is at the heart of a critique, but don't confuse critique with criticism.

FIGURE 9-2

Checklist System

The most simple and reliable method of coordinating information between the various Hazmat Branch functions is to use the checklist system. Formal checklists have several distinct advantages as they relate to information management in the field. These include:

- Identifying the tasks assigned to each Hazmat Branch function. It assigns duties and responsibilities to each member of the team so that work is not duplicated and important information gathering activities are not overlooked.
- Listing critical activities and action items required for each function. As elements are addressed, they are formally "checked off" as a method of verifying that the activity has been completed.
- Prioritizing actions so that important activities are completed early in the incident.

NOTES

- Providing a framework for development of the Incident Action Plan.
- Identifying which Hazmat Branch functions or individuals need to be formally contacted to coordinate information.
- Providing the required documentation of the incident.

See Figure 9-3 for an example of an INFORMATION checklist.

In order for the Checklist System to be effective, checklists must be updated on a regular basis using the critique methods described in Chapter Twelve–Termination. It is important that responders take ownership of the checklists and make gradual improvements to the system over time.

While there is nothing wrong with "borrowing" an established checklist system from another organization as a beginning point, checklists should be adapted and customized to suit local needs.

Information Officer

The person designated as the Information Officer (INFO) will play a key role in the successful mitigation of any hazmat incident. This individual should be chosen because of his or her ability to communicate, comprehend and manage information, work effectively under stress, and coordinate activities with individuals from different backgrounds.

Extended incidents may require that an Information Group be formed. This allows the workload to be split up into areas such as on-scene library research, contacting CHEMTREC®, the manufacturer, computer operations, and on-scene data collection.

In situations where there is not sufficient room for everyone to comfortably function, one or more coordinators can move to nearby offices or houses in the area to complete their assignments. Access to comfortable surroundings with telephones, office supplies, bathrooms, and other amenities of life make the job much more endurable and the work product more reliable. Good lighting and an uninterrupted power supply are also essential.

RESOURCE COORDINATION

RESOURCES DEFINED

What are RESOURCES and how do they differ from INFORMATION? As previously described, information is knowledge that is based upon data or facts which can be used to support decision-making. In contrast, resources are made up of the people, equipment, and supplies required to manage a hazardous materials emergency. As with military operations, resources are the basis of strategy and tactics.

Human Resources include Emergency Response Personnel, Support Personnel, Technical Information Specialists, and Specialist Employees. Coordinating human resources is important because people provide the thinking power and manual labor required to bring the situation under control. People also represent the greatest financial, legal, political, and

FIGURE 9-3

REFINERY TERMINAL FIRE COMPANY
HAZARDOUS MATERIALS RESPONSE PROCEDURES

INFORMATION OFFICER WORKSHEET "INFORMATION"

I. GENERAL INFORMATION

The Information Officer Worksheet is designed to be used by those designated individuals who will function as the Haz Mat Information Officer. The objective of the worksheet is to assist the Haz Mat Safety Officer in assembling and analyzing technical reference materials, incident data, and other resources.

The Information Officer is responsible for developing, documenting, and coordinating all data and information relevant to the incident. The data gained will be used in hazard and risk assessment, protective action recommendations, selection of protective clothing and equipment, and development of incident management considerations.

The radio designation for the Information Officer will be "INFORMATION."

II. INCIDENT INFORMATION

DATE ______________________ TIME ______________________

LOCATION ____

____

____

NATURE OF THE INCIDENT ____

____

____

____

HAZ MATS INVOLVED____

____

____

____

____

❑ HAZ MAT BRANCH OFFICERS (identified by Command Vest)

- Haz Mat Branch Officer (HAZ MAT BRANCH) ______________________
- Haz Mat Safety Officer (HAZ MAT SAFETY) ______________________
- Information Officer (INFORMATION) ______________________
- Entry Officer (ENTRY) ______________________
- Resources Officer (RESOURCES) ______________________
- Decontamination Officer (DECON) ______________________

- [ ] HAZ MAT BRANCH PERSONNEL

____

____

____

____

____

____

III. INFORMATION FUNCTIONS

INFORMATION sector activities shall be conducted in accordance with the following:

- [ ] A minimum of three (3) information sources shall be utilized while developing hazard and risk recommendations.
- [ ] All data and information gathered shall be coordinated with the HAZ MAT SECTOR OFFICER, the SAFETY OFFICER, DECONTAMINATION, RESOURCES, and the MEDICAL SECTOR, as appropriate.
- [ ] INFORMATION personnel shall complete the Hazardous Materials Data Sheet for each chemical involved, or shall utilize similar data sheets or information sources.
- [ ] Coordinate the development of recommendations for the use of personal protective clothing within the hot, warm, and cold hazard control zones, including chemical compatibility concerns.
- [ ] Contact technical information sources, as necessary. These shall include CHEMTREC, facility chemists, shippers and manufacturers, Poison Control Centers, etc.

NOTES:

____

____

____

- [ ] HAZARD CONTROL ZONES ESTABLISHED

- Hot and Warm Zones Established Time ______________

- Hot Zone
  - Dimensions
  - Protective Clothing Required
    1. Suit
    2. Gloves
    3. Boots
    4. Respiratory Protection

- Warm Zone
  - Dimensions
  - Protective Clothing Required
    1. Suit
    2. Gloves
    3. Boots
    4. Respiratory Protection

❑ WEATHER INFORMATION

CURRENT CONDITIONS	FORECAST CONDITIONS

❑ DECONTAMINATION INFORMATION (see Haz Mat Data Sheet)

❑ EMERGENCY MEDICAL INFORMATION REFERENCED

❑ TECHNICAL INFORMATION SOURCES REFERENCED (MINIMUM OF 3)

❑ DOT Guidebook

❑ NFPA Haz Mat Manual

❑ AAR Manual

❑ Computer databases

❑ ____________________

❑ ____________________

- ❑ AAR Emergency Action Guides
- ❑ CHRIS Manual
- ❑ Sax's Handbook to Industrial Chemicals
- ❑ ______________________
- ❑ ______________________
- ❑ ______________________
- ❑ ______________________
- ❑ Facility Information
- ❑ CHEMTREC® (800) 424-9300
- ❑ Manufacturer
- ❑ Poison Control Center
- ❑ On-Scene Representatives
- ❑ ______________________
- ❑ ______________________

❑ AGENCY CONTACTS AND NOTIFICATIONS

Agency	Contact
Texas Water Commission 851-8484	Contact Person ____________ ETA ________ Time ________
Texas Air Control Board 882-5828	Contact Person ____________ ETA ________ Time ________
Nueces County health Department 851-7200	Contact Person ____________ ETA ________ Time ________
Nueces County LEPC 888-0444	Contact Person ____________ ETA ________ Time ________
U.S. Coast Guard Corpus Christi MSO 888-3163	Contact Person ____________ ETA ________ Time ________
U.S. Environmental Protection Agency (214) 655-2277	Contact Person ____________ ETA ________ Time ________
National Response Center (NRC) (800) 424-8802	Contact Person ____________ ETA ________ Time ________
Other ____________ ____________	Contact Person ____________ ETA ________ Time ________
Other ____________ ____________	Contact Person ____________ ETA ________ Time ________

❑ CONTRACTOR COMPANY CONTACTS

Company	Contact
Company ____________ ____________	Contact Person ____________ ETA ________ Time ________
Company ____________ ____________	Contact Person ____________ ETA ________ Time ________

FIGURE 9-3: INFORMATION CHECKLIST

NOTES

technical exposure for the Incident Commander. Chapter Three lists some of the different types of human resources ("players") which may be involved in a hazmat incident.

Equipment Resources include items which are re-usable, such as hand tools, generators, pumps, monitoring instruments, etc. Hazmat equipment can represent a substantial cost outlay whether it is rented, leased, or owned. Very few organizations are self-sufficient when it comes to the equipment required to resolve a large-scale hazmat incident. Consequently, good coordination is required to assure that the right equipment resources are available when they are needed, that they are tracked throughout the operation, their use and costs are monitored, and they are returned to their owner in a timely manner.

Supply Resources differ from equipment resources in that they are usually considered expendable. In other words, you use them once or twice, then dispose of them. Examples include foam concentrate, decontamination solutions, limited-use protective clothing, product control materials, and most medical supplies. Hazmat supplies require special coordination because they are usually consumed in bulk quantities, may require special decontamination procedures, may take a long time until inventories are replenished, and can be expensive.

COORDINATING RESOURCES

Hazardous materials incidents are unique to the emergency response business because of the resources required to mitigate the problem. There are few technical emergencies which involve such a broad spectrum of private and public services. For example, the same train derailment previously described may require extensive resources to bring the situation under control. These could include:

- Railroad operations and hazmat specialists.
- Product specialists representing a variety of different companies.
- Wreck clearing contractors.
- Environmental specialists and contractors.
- Firefighters, police officers, EMS and rescue personnel from multiple jurisdictions.
- State and County Emergency Management officials.
- Local, state, and federal environmental officials.
- National Transportation Safety Board (NTSB) investigators.
- Federal Railroad Administration (FRA) inspectors.

This is just the short list! If special hazardous materials are involved like military munitions or radioactive materials, or if there are multiple fatalities, the list could grow to 20 or 30 different agencies and organizations. Naturally, they all bring their own equipment and supplies (toys) to the incident, and thus there is a greater need for the Incident Commander to be well organized and coordinated.

NOTES

Internal Resource Coordination

Coordinating resource requirements within the "internal" structure of the emergency response organization can be an easy process if your organization understands and regularly operates within an Incident Management System framework. As information flows between the Incident Commander and Hazmat Branch, resource needs and requirements will begin to evolve. It is important that the Incident Commander implement a system of identifying which resources (people, equipment, and supplies) are needed early in the incident to reduce response time.

Special resource requirements are funnelled through the chain of command to the Logistics Section, which coordinates the requirements through the overall command structure and gets the needed resources to the incident scene. Once these resources arrive, they are assigned to a Staging Area and held in reserve until they are required or assigned to the Branch or organization needing them.

Within the Hazmat Branch, resources are coordinated by the Resource Group. The Resource Group is responsible for controlling and tracking all equipment and supplies used by the Hazmat Branch during the course of the emergency. This includes all expendable supplies and materials. Obviously, the Resource Group must coordinate closely with the Logistics Section Chief.

What are some of the attributes of an effective Logistics Section Chief and/or Hazmat Resources Officer? Key traits and characteristics include:

- Self-starter—looks at the scope and nature of the incident and starts to determine the necessary type and level of resources. Needs little direction.
- "Scrounger"—knows everyone and has the ability to get anyone and everything!
- Good listener and prioritizer—can determine what we need immediately versus 3 hours from now.

As is the case with coordinating information, a formal checklist can provide some structure to coordinating resources. Figure 9-4 provides an example.

External Resource Coordination

Coordinating "external" resource needs and requirements can present a real challenge if the participating agencies and players are not familiar with the Incident Management System and cannot play the game by using Unified Command Structure rules. (See Chapter Three for a more detailed discussion on how Unified Command works.)

Outside agencies, whether public or private sector, will almost always be involved in providing some type of resources at a major incident. Each and every agency with a legitimate "piece of the action" must be coordinated through a single control point (Unified Command) on the scene.

As resources from external agencies arrive at the scene, a leader or representative from the organization should be directed to the Command Post. Some important items that should be addressed by the IC include:

FIGURE 9-4

REFINERY TERMINAL FIRE COMPANY
HAZARDOUS MATERIALS RESPONSE PROCEDURES

RESOURCES OFFICER WORKSHEET "RESOURCES"

I. GENERAL INFORMATION

The Resources Worksheet is designed to be used by those designated individuals who will function as the Resources Officer. The Resources Sector should be established by the Haz Mat Branch Officer as needed.

The Resources Officer shall assume responsibility for the control and documentation of tools and equipment utilized during hazardous materials operations. The Resources Officer shall report to the Haz Mat Branch Officer.

The radio designation for the Resources Officer will be "RESOURCES."

II. INCIDENT INFORMATION

DATE ______________________________ TIME ____________________________

LOCATION ____

____

____

NATURE OF THE INCIDENT ____

____

____

____

HAZ MATS INVOLVED ____

____

____

____

____

❑ HAZ MAT BRANCH OFFICERS (identified by Command Vest)

- Haz Mat Branch Officer (HAZ MAT BRANCH) ____________________
- Haz Mat Safety Officer (HAZ MAT SAFETY) ____________________
- Information Officer (INFORMATION) ____________________
- Entry Officer (ENTRY) ____________________
- Resources Officer (RESOURCES) ____________________
- Decontamination Officer (DECON)

❑ RESOURCES SECTOR PERSONNEL

____

____

III. RESOURCES FUNCTIONS

INFORMATION sector activities shall be conducted in accordance with the following:

❑ Coordinate with the HAZ MAT BRANCH OFFICER to determine the type of response operation to be conducted and possible equipment and resources required.

❑ Supplies, tools, and equipment required for entry operations shall be located in the warm zone, at or near the entry point into the hot zone. An equipment staging area utilizing a salvage cover shall be set up at this location.

❑ Expendable items must be monitored and replaced or re-supplied.

❑ Contaminated items remaining within the hot zone must be identified.

❑ Ensure that any equipment utilized throughout the incident is appropriately handled (i.e., isolated, decontaminated, etc.).

IV. RESOURCES TRACKING

NOTE: RTFC supplies and equipment can be referenced on attached equipment inventory.

ITEM	AVAILABLE	USED
PROTECTIVE CLOTHING		

ITEM	AVAILABLE	USED
SPILL AND LEAK CONTROL EQUIPMENT		
DECONTAMINATION MATERIALS		
HAND TOOLS		
OTHER		

V. POST-INCIDENT RESOURCE CONCERNS

- ❑ All tools and equipment accounted for.
- ❑ All tools and equipment decontaminated.
  Decontamination Method(s) ____

  ____
- ❑ Any protective clothing or equipment required to be isolated for further analysis or disposal?
  YES _____NO _____

  Items:____

  ____

  ____
- ❑ Any contractor equipment requiring decontamination? YES _____NO _____

  Items:____

  ____

  ____

NOTES:

FIGURE 9-4· RESOURCE CHECKLIST

NOTES

- ❑ Make sure that the external players understand that you are running the operation using a Unified Command structure. If you get a blank stare, a little ICS-101 indoctrination may be necessary.
- ❑ Determine at what capacity the agency or organization is going to participate. If you requested their resources, this may seem obvious. However, resources sometimes just show up. Also, some agencies will simply turn out for "fact finding missions" or as observers.
- ❑ Physically identify each agency or organization representative as they arrive. Request a business card and post it in the command post so that you are sure of the individual's name, title, rank, etc. Extended incidents may require a flow chart with these cards attached in their appropriate boxes to establish "who's on first." If the organization has its own communications system, list cellular telephone numbers and radio frequencies on the business card so that you can contact representatives directly.
- ❑ Assign all external resources to a Section, Branch, or Sector within the command structure. Don't allow external resources to "free-lance" or do the "end-run." Remember that final responsibility for site safety of external resources rests with the Incident Commander.
- ❑ If external agencies have a problem, make sure it is brought to your immediate attention. Remember—bad news doesn't get better with time. If there's a problem, the earlier you know about it, the sooner you can start to fix it!

The best way to guarantee that external resources are properly coordinated and integrated into your command structure is to develop written guidelines or Memorandums of Understanding (MOU) between your respective organizations. Certain types of incidents (e.g., railroad accidents) have unique problems which may bring new players to the emergency. However, most emergency response organizations call on the same cast of external players on a regular basis. Get to know them in advance.

RESOURCE COORDINATION PROBLEMS

Fortunately, the majority of hazmat incidents you will encounter won't require major resource commitments from outside agencies. Nevertheless, you should be prepared to deal with the influx of resources large incidents may generate.

Full-scale emergencies, like train derailments, plane crashes, and petrochemical plant fires, can shift the Incident Commander's role from one of a strategist and tactician to a politician and diplomat. Chapter Four, The Politics of Incident Management, provides an in-depth look at the political aspects of command. However, it is worth reviewing some basic problem-solving techniques as they apply to resource coordination.

Most resource coordination problems fall into three categories:

1. Failure to understand or work within the structure of the Incident Management System.
2. Given the type and nature of the incident, failure to anticipate potential problems and "gaps" in information or resources.

NOTES

3) Communications and personality problems between the players.

Emergencies don't permit much time for resolving long-standing personality conflicts; they only intensify under stress. If you must bring uncooperative people into the command structure, assign a qualified person as a "chaperone" to keep them out of trouble. One way to keep uncooperative and unwanted external resources out of your hair is to team them up with someone in your command structure who knows what they are doing.

The Incident Commander can resolve many communications and personality problems by using a little psychology and some group leadership techniques at the command post. Some useful techniques that can be effectively applied in stressful situations include:

Listening: Pay attention to others as they communicate. Don't just listen to what people are saying, pay attention to their body posture, gestures, mannerisms, and voice inflections. Identify angry players and resolve the problem before the situation gets out of hand and eats up your valuable time.

Clarifying: Clarify what the person's gripe is. Identify and sort out individual problems. Big problems are more manageable when they are broken into smaller individual issues. If you can identify what the person's problem is, you can address one issue at a time. Issues can also be handed off to subordinates for resolution.

Summarizing: If large groups of people are involved in a Unified Command setting, the decision-making process can get bogged down or fragmented. If several individuals have a problem, summarizing where you are can be helpful. For example, if the discussion is turning into a debate, the Incident Commander can interrupt and ask each agency representative to briefly state how he or she feels about the issue. The IC can then summarize by identifying some common ground and turn the discussion toward acceptable alternatives. Look for issues to agree on, then build on those as a platform to work through the unresolved issues.

Empathizing: If a special interest emerges and becomes a problem, try empathizing with the individual to reassure him or her that the concern is valid. For example, protecting sea birds from a massive oil spill created by a burning barge at the dock is a high priority to fish and wildlife officials and environmental activists; however, under the circumstances, life safety is a bigger and more immediate concern. Empathize and show respect for the individual's concern, then get a "buy in" to the fact that your alternatives are limited. A little well-placed empathy goes a long way to building support for your plan for managing the resources available.

Develop a good working relationship with supporting agencies, suppliers, contractors, and consultants before the incident and most of the personality issues will dissolve, or at a minimum they won't get in the way of on-scene operations. The Local Emergency Planning Committee is a good place to start laying the foundation.

NOTES

SUMMARY

The function of Information Management and Resource Coordination can be viewed from two perspectives. From a strategical perspective, it is a "constant" that starts with the notification and dispatch of emergency responders and ends with the termination of the incident. From a tactical perspective, it is the transition point from the "size-up" phase of the hazmat incident to the mitigation and termination phases. Both perspectives share a common "bottom line"—failure to get the right information to the right people at the right time can jeopardize both the safety of responders and the overall success of the emergency response effort.

As much time, thought, and effort must be put into information management and resource coordination as into any other hazmat function. The way other agencies and the public perceive how the incident was handled generally depends on the way information was managed and people were handled on the scene.

Information and resources must be managed within the framework of the Incident Management System. The checklist system is one of the most effective tools for assuring that information and resources are effectively coordinated both internally and externally.

REFERENCES AND SUGGESTED READINGS

Buskirk, Richard, HANDBOOK OF MANAGEMENT TACTICS, Boston, MA: Cahners Books (1976).

Chemical Manufacturers Association, EVALUATING PROCESS SAFETY IN THE CHEMICAL INDUSTRY: A MANAGER'S GUIDE TO QUANTITATIVE RISK ASSESSMENT, Washington, DC: Chemical Manufacturers Association (1989).

Chemical Manufacturers Association, SITE EMERGENCY RESPONSE PLANNING GUIDEBOOK, Washington, DC: Chemical Manufacturers Association (1992).

Corey, Gerald, and Marianne Schneider, GROUPS: PROCESS AND PRACTICE (2nd edition), Monterey, CA: Brooks/Cole Publishing Company (1982).

Drucker, Peter F., THE EFFECTIVE EXECUTIVE, New York, NY, Harper and Row (1985).

Engels, Donald W., ALEXANDER THE GREAT AND THE LOGISTICS OF THE MACEDONIAN ARMY, Berkeley and Los Angeles, CA: University Press (1978).

Hersey, Paul and Ken Blanchard, MANAGEMENT OF ORGANIZATIONAL BEHAVIOR: UTILIZING HUMAN RESOURCES (4th Edition), Englewood Cliffs, NJ, Prentice-Hall (1982).

Hildebrand, JoAnne Fish, "Stress Research: Solutions to the Problem, Part-3." FIRE COMMAND (July 1984). Page 23-25.

Isman, Warren E. and Carlson, Gene P., HAZARDOUS MATERIALS. Encino, CA: Glencoe Publishing Company, Inc. (1980).

Nasbitt, John, MEGATRENDS, New York: Warner Books (1982).

Nasbitt, John, and Patricia Aburdene, RE-INVENTING THE CORPORATION, New York: Warner Books (1985).

Pagonis, William, P., Lt. General, and Cruikshank, Jeffrey L., MOVING MOUNTAINS: LESSONS LEARNED IN LEADERSHIP AND LOGISTICS IN THE GULF WAR, Boston, MA: Harvard Business School Press (1992).

Siu, R. G. H., THE CRAFT OF POWER, New York: John Wiley and Sons, Inc. (1979).

Van Horn, Richard L., "Don't Expect Much from Your Computer System." THE WALL STREET JOURNAL ON MANAGEMENT, New York, NY: Dow Jones and Company (1985).

IMPLEMENTING RESPONSE OBJECTIVES

OBJECTIVES

1) Describe the process of size-up as a method of determining the strategic and tactical options available to produce a favorable outcome at a hazardous materials incident.
2) Define the terms:
   - Strategic Goals
   - Tactical Objectives
3) Describe the objectives and dangers of search and rescue operations at a hazmat incident.
4) Identify and describe the factors for selecting and evaluating response objectives to achieve the following strategic goals:
   - Rescue
   - Public Protective Actions
   - Spill Control
   - Leak Control
   - Fire Control
   - Recovery
5) Identify and describe the application, advantages, and limitations of the following methods of spill control:
   - Absorption
   - Adsorption
   - Covering

"It's not that I'm afraid to die. I just don't want to be there when it happens."

Woody Allen,
Director and Actor

NOTES

- Damming
- Diking
- Dilution
- Diversion
- Dispersion
- Retention
- Vapor Dispersion
- Vapor Suppression

6) Identify and describe the application, advantages, and limitations of the following methods of leak control:
   - Neutralization
   - Overpacking
   - Plugging and Patching
   - Solidification
   - Vacuuming
7) Identify and describe the application, advantages, and limitations of fire control operations for the following emergencies:
   - Flammable liquids
   - Flammable gases
   - Reactive chemical fires and reactions
8) Identify the factors to be considered in evaluating a confined space rescue.
9) Identify the safety considerations for product removal and transfer operations, including:
   - Site safety guidelines
   - Bonding
   - Grounding
   - Elimination of ignition hazards
   - Chemical compatibility issues
10) Identify and describe the application, advantages, and limitations of the following product transfer methods:
    - Portable pumps (gasoline, diesel, power-take-off, electric, air, water)
    - Pressure differential
    - Vacuum trucks

INTRODUCTION

This chapter discusses the sixth step in the **Eight Step Process©–Implementing Response Objectives**. It represents the phase in a hazmat emergency where the Incident Commander implements the best available strategic goals and tactical objectives which will produce the most favorable outcome. Remember that "outcomes" are measured in terms of fatalities, injuries, property and environmental damage, and systems disruption.

NOTES

In *Chapter Seven, Hazard and Risk Evaluation,* we explained that all hazmat incidents follow a logical sequence of events along a predictable timeline. In other words, every hazmat incident has a beginning and an end. A good Incident Commander must have the ability to quickly size-up what events have already occurred, determine what events are happening now, and predict which events will occur in the future. If an event can be predicted, it may be able to be prevented. This is the essence of the risk evaluation process. In summary, the operational strategy for the incident is developed based upon the IC's evaluation of the current conditions and forecast of future conditions.

This chapter will focus on the various strategic goals and tactical objectives available to the IC to influence outcomes. Topics will include basic principles of decision-making, guidelines for determining and implementing strategic goals and tactical objectives, and special tactical problems. We will not use a "how to do" approach which focuses upon the specific tasks involved in implementing respective tactical objectives; rather, we will focus upon the management criteria and guidelines associated with selecting and implementing the appropriate response action.

The authors would like to thank Ludwig Benner, Jr. for permission to reproduce materials from his copyrighted works in the Basic Principles discussion of this chapter.

BASIC PRINCIPLES

UNDERSTANDING EVENTS

In order to determine which strategy and tactics are best suited to change the outcome for hazmat emergencies, the IC must be able to understand what has already occurred, what is occurring now, and what will occur in the future. In other words, if the IC can visualize what has already happened and understand what is happening now, it may be possible to interrupt the chain of events and favorably change what will be happening in the future.

If you think about every emergency you ever responded to in your career, it could be plotted along a timeline. At some point on that time-line you arrived on the scene, sized up the situation, took some action, and began influencing the outcome in some positive way (e.g., you rescued a victim, extinguished a fire, etc.). See Figure 10-1.

The options available to the IC has a lot to do with where you are on the timeline when you arrive at the scene. If containers have already breached and produced fatalities and injuries, the options available to influence the outcome in a favorable way are fairly limited. On the other hand, if you arrive on the scene while containers are still being stressed, it may be possible to prevent containers from breaching.

Experience over the last thirty years has been marked by many significant hazmat incidents which produced bad outcomes. Unfortunately many of these incidents occurred after emergency responders intervened in an incident and tried to change the outcome using offensive tactics. In other words, by intervening they made the outcome worse.

NOTES

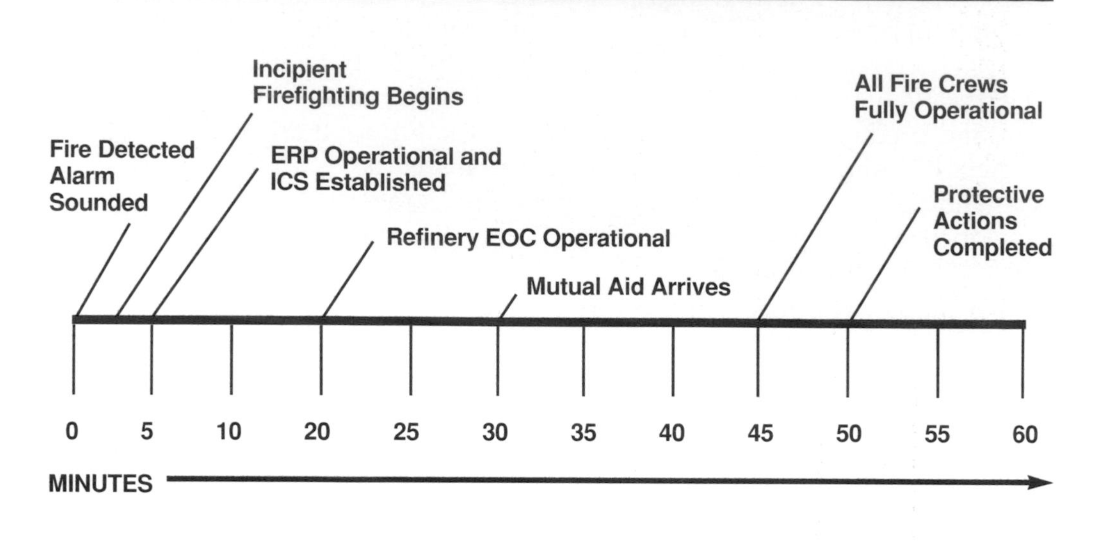

FIGURE 10-1: EVERY EMERGENCY CAN BE PLOTTED ALONG A TIMELINE

One study conducted by the National Transportation Safety Board in 1979 noted that between 1968 and 1978 there were 55 fatalities from hazardous materials accidents involving firefighters and police officers. Many of these incidents involved multiple fatalities.

Fortunately, recent history shows that there are fewer incidents which produce the multiple fatalities and mass casualties experienced in the 1970's and 1960's. One of the primary reasons for this is that emergency responders are better trained to understand how hazardous materials behave and what responders can realistically achieve with the resources available.

PRODUCING GOOD OUTCOMES

Producing good outcomes at hazmat emergencies depends a great deal on the IC's ability to visualize the emergency in a chronological sequence of events. The IC should analyze and visualize events at the emergency scene in three phases in time. These include what has already taken place, what is taking place now, and what will be taking place in the immediate future. Stated another way, the IC must evaluate the incident in terms of past, present, and future events.

PAST EVENTS

During the initial size-up, the IC should develop a mental picture of what has already occurred before the arrival of ERP. Examples include whether containers have already breached, which containers have already been stressed, the dispersion patterns created, and the exposures which have already been contacted and harmed.

PRESENT EVENTS

NOTES

Understanding what has already happened can help the IC focus on what is occurring now and what may be happening in the next 10 to 15 minutes. Getting the "Big Picture" as quickly as possible is an important step in (1) predicting what will be happening in the future, and (2) determining which options are available to influence those events. How quickly the IC develops this "Mental Movie" of what is happening has a lot to do with how quickly command is established and how information is managed to support decision making.

FUTURE EVENTS

If the IC has a good mental picture of what is occurring along a timeline, it may be possible to influence the outcome of the emergency in a variety of different ways. These can be divided into five basic options and can be remembered by the acronym MOTEL:

Magnitude—Trying to keep the incident as little as possible. For example, rotating a one-ton chlorine container 180° can change a liquid leak to a vapor leak and reduce the size of the vapor cloud.

Occurrence—Preventing a future event from occurring by influencing the current event. For example, activating a water spray system or extinguishing a ground fire to minimize thermal stress upon surrounding containers.

Timing—Trying to change when an event happens and/or how long it lasts. For example, product transfer operations can be initiated to reduce the quantity of product remaining in a fixed facility storage tank while resources are being assembled to implement leak control tactics.

Effects—Trying to reduce the size and/or effects of an event. For example, using vapor dispersion tactics (e.g., water monitors and water spray systems) to reduce the size or magnitude of an anhydrous ammonia release.

Location—Trying to change where the next event occurs. For example, a tank truck or railroad tank car with a minor leak at the dome cover could be moved to a safer, less congested area and handled.

STRATEGIC GOALS

STRATEGY DEFINED

Uncertainty is a reality of emergency response. Yet one of the IC's most critical tasks is to minimize uncertainty by using a structured decision-making process to size up the problem and select the safest strategy to make the problem go away. Adopting a "Strategic Goal" to manage the incident is one of the IC's first priorities before intervening in an incident. Hazmats are the true equalizer—acting on instinct without facts and without predicting the hazmats' behavior can get people killed.

NOTES

The terms "strategy" and "tactics" are sometimes used interchangeably, but they actually have very different meanings. A strategy is a plan for managing resources. It becomes the IC's overall goal or game plan to control the incident. Like most goals, a strategy is usually very broad in nature and is selected at the Command Level. Several strategic goals may be pursued simultaneously during an incident. Examples of common strategic goals include the following:

- Rescue
- Public Protective Actions
- Spill Control (Confinement)
- Leak Control (Containment)
- Fire Control
- Recovery

In contrast, tactics are the specific objectives the IC uses to achieve strategic goals. Tactics are normally decided at the Section and Branch Levels in the command structure (see Figure 10-2). For example, tactical objectives to achieve the strategic goal of spill control (confinement) would include absorption, diking, damming, and diversion.

If the IC hopes to have strategic goals understood and implemented, they must be packaged and communicated in simple terms. If strategic goals are unclear, tactical objectives will become equally muddied. Hazmat strategic goals and tactical objectives can be implemented from three distinct operational modes:

STRATEGIC GOALS VS. TACTICAL OBJECTIVES		
STRATEGIC GOALS	COMMAND LEVEL	• Rescue • Protective Actions • Confinement (Spill Control) • Containment (Leak Control) • Fire Control • Recovery Offensive, defensive, or nonintervention mode
TACTICAL OBJECTIVES	BRANCH LEVEL	• Containment • Neutralization • Patching/Plugging • Solidification • Vacuuming • Vapor Suppression
TASK	COMPANY OR INDIVIDUAL LEVEL	Example: • Specific procedures for applying Chlorine B Kit on 1-ton cylinder

FIGURE 10-2

NOTES

Offensive Mode—Offensive-mode operations commits the IC's resources to aggressive leak, spill, and fire control objectives. An offensive strategic goal/tactical objective is achieved by implementing specific types of offensive operations which are designed to quickly control or mitigate the problem. Although offensive operations can increase the risk to emergency responders, the risk may be justified if rescue operations can be quickly achieved, if the spill can be rapidly confined or contained, or if the fire can be quickly extinguished.

Defensive Mode—Defensive-mode operations commits the IC's resources (people, equipment, and supplies) to less aggressive objectives. A defensive strategic goal/tactical objective is achieved by using specific types of defensive tactics, such as diverting or diking the hazmat. The IC's defensive plan may require "conceding" certain areas to the emergency while directing response efforts toward limiting the overall size or spread of the problem (e.g., concentrating all efforts on building dikes in advance of a spill to prevent contamination of a fresh water supply). As a general rule, defensive operations expose ERP to less risk than offensive operations.

Nonintervention Mode—Nonintervention means taking no action other than isolating the area. The basic plan calls for ERP to wait out the sequence of events underway until the incident has run its course and the risk of intervening has been reduced to an acceptable level (e.g., waiting for an LPG container to burn off). This strategy usually produces the best outcome when the IC determines that implementing offensive or defensive strategic goals/tactical objectives will place ERP at an unacceptable risk. In other words, the potential costs of action far exceed any benefits (e.g., BLEVE scenario).

SELECTING STRATEGIC GOALS

Once the IC has run a "mental movie" for each of the options being considered, Command must make a decision to do something or to do nothing—both constitute a decision. Selecting the best strategic goal involves weighing what will be gained against the "costs" of what will be lost—a process often easier said than done.

Determining what will be "gained" involves weighing many different variables, including:

- **Potential Casualties and Fatalities**—Will lives be saved by pursuing aggressive rescue operations? For example, if a civilian is trapped in a confined area contaminated with poison gas, can his or her life be saved by committing personnel to a rescue? Likewise, does the rescue environment present an unreasonable risk to responders?
- **Potential Property Damage or Financial Loss**—What will be the financial cost of implementing one option over another? For example, a small chemical spill occurs in an auto assembly facility. Should the entire building be evacuated and production operations be halted ($$$$$)? Of course, if you don't evacuate and the problem quickly worsens, the potential risk to both employees and emergency responders may be unacceptable.

NOTES

- **Potential Environmental Damage**—What will be the impact on the environment as a result of your actions? For example, if offensive fire control tactics are implemented at an agricultural chemical warehouse fire, water runoff and pollution will likely result. However, allowing the fire to burn out with no application of water may result in widespread air pollution over a much larger area. This may require additional public protective actions downwind. Which option is the most acceptable?
- **Potential Disruption of the Community**—Will the community be disrupted to an unacceptable level? For example, a gasoline tank truck has overturned and is burning on a major freeway just before rush hour. Is it an acceptable risk to the community to let the tanker burn for three hours and consume the product and destroy bridges and roadway, or will extinguishment be required sooner? Which option will actually open the freeway sooner?

DECISION-MAKING TRADE-OFFS

Every decision the IC makes involves making trade-offs between two or more factors that are usually in conflict with one another (e.g., life safety vs. property or environmental damage). Decision-making at a hazmat incident is not a "black and white" process; rather, it is a mine field full of varying shades of grey.

Conflicting information and values can also add to the uncertainty of the decision-making process. Consider the following:

- **Conflicting or uncertain information.** The more critical the life safety situation, the less likely accurate and reliable information will be available to make an informed decision. Experience shows that the IC is sometimes pressured into making split-second decisions based on incomplete or inadequate information. For example, firefighters have been killed and seriously injured pursuing aggressive rescue operations when told that there is "someone" trapped inside a building, only to discover later that the information was incorrect.

  The most difficult decisions the IC must make are those related to life safety. Experience clearly shows that the greater the risks to people, the greater the risks ERP are willing to take to save them.

  A corollary to this fact is that the worse the potential outcome, the greater the level of uncertainty the IC should be willing to accept in implementing response goals and objectives. Understanding this point and incorporating it in your decision-making process will save lives!
- **Conflicting or competing values.** Every individual has a set of values that he or she brings to the incident. Regardless of one's background, these values influence the decision-making process. Even among similar groups with the same types of values, people have different opinions about the perceived risk involved in carrying out an option as well as the perceived value of what will be gained.

NOTES

> When lives are at stake, each person can be expected to select a different option as the "best" one in his or her judgment. If you doubt this, conduct an exercise in which you give responders a specific situation and ask them to rate the risks involved in protecting life, property, environment and in disrupting the community. You may be surprised what you learn about how people think and how much risk they are willing to take!

There is no single best way to evaluate life safety decisions. Pure risk taking (the black area) can lead to self-destruction, while pure safety (the white area) can lead to taking no action at all. Somewhere between these two extremes is the right decision (the grey area).

Getting buy-in from the risk takers (i.e., the people in the suits) is an important step in getting the IC's strategic game plan implemented. The success of obtaining agreement depends on:

1. The IC's ability to understand the differences in how ERP perceive risks.
2. The IC's ability to explain the options available to the risk takers.
3. Getting the other agencies and organizations involved in the incident to understand the big picture concerning what will be lost vs. what will be gained.

The more you understand how different organizations and individuals perceive risk and the value of implementing different options before an incident, the quicker you can implement these decisions at the incident.

TACTICAL OBJECTIVES

TACTICS DEFINED

Once the Command Level has committed to a strategy, subordinate Branches and Sectors will implement the IC's general game plan by establishing specific tactical objectives. There is no way to make a hazmat emergency go away without tactics!

A well-defined tactic has a stated objective which can be implemented using specific procedures and tasks within a reasonable period of time. For example, the IC selects an offensive leak control strategy for a leaking 2,000-gallon storage tank of nitric acid. In turn, the Hazmat Branch will determine the specific tactical objectives in order to achieve the strategic goal. A variety of tactical procedures can be used to achieve this goal, including product transfer, patching and plugging, neutralization, etc. The specific tactic or technique used is ultimately determined by the people closest to the problem.

The tactical action plan must be easy to understand and laid out with straightforward objectives. A good standard for evaluating a tactical objective is that it should be able to be communicated by radio without providing too much detail. Simply stating that you want the leak stopped and the problem to go away doesn't provide much direction.

NOTES

A IC who gets involved in making detailed tactical decisions like which type of plug to use loses the broad perspective of the incident and develops tunnel vision. This can be a serious problem with new IC's who have worked their way up through the system and just can't let go of the "hands on" details (like the fire chief on the nozzle going down the hallway). Our military experience has taught us that Generals are good at strategy, Captains are good at tactics, and Sergeants are good at making problems go away. Things get pretty screwed up when the process works in reverse. Just as a military commander cannot effectively implement tactics from Washington, a sergeant deep in enemy territory cannot get the "Big Picture" and develop sound strategy.

TACTICAL DECISION-MAKING

Tactical decision-making is about whether to use one tactic over another. Most hazmat incidents cannot be resolved using just one tactic. Effective tactical decision-making requires thinking ahead and planning various tactical options so that the right people, with the right training and the right equipment, are available at the appropriate time.

When deciding which tactic to use, consideration should be given to how long it will take to accomplish the specific objective. Conditions can change rapidly as the incident progresses along a natural timeline. What seemed like a good tactical objective early in the incident may no longer be an option as conditions change. Spills can grow larger and leaks can get worse as ERP gather their equipment and prepare for entry. Valuable time can be lost if the entry team has to shift from Plan A to Plan B. The trick is to keep ahead of what the hazmat will be doing when the entry team is ready to go to work.

The length of time required to implement tactical objectives at the task level must be compared to how long the "window of opportunity" will be open. As the clock ticks, tactical options to deal with the problem will often become more limited. Remember, all hazmat incidents follow a natural timeline. Leaks and spills usually get worse all by themselves before the situation gets better.

Some tactics can be employed to delay events or slow down the clock until entry teams are ready to implement the "final solution." In other words, responders may buy time with less effective, but easy to implement, tactics until the most effective tactic can be implemented.

Examples of tactical options which can be used to buy time include:

- **Barriers**—Put a physical barrier between the hazmat, its container, and surrounding exposures. For example, building dikes, retention ponds, or diversion dams well in advance of an oncoming spill can confine the hazmat to a limited area or slow it down until entry teams are ready to contain the leak.
- **Distance**—Separate people from the hazmat. The further away you are from the problem, the lower the risk. Increasing the size of control zones or moving the problem to an isolated area can further reduce the life safety risk until the entry team is ready to enter the hot zone to resolve the problem.

NOTES

- **Time**—Reduce the duration of the release, or trade off or rotate persons exposed to the hazmat. For example, if you can reduce the pressure on a leaking pipe or vessel, the magnitude of the problem can be significantly reduced. This may allow responders to buy some time to work out a more complete solution to the problem.
- **Techniques**—These are specific tasks and procedures performed by responders to stop the leak and control the problem (e.g., uprighting a leaking liquefied gas cylinder so that vapors rather than liquid product will be released).

Tactical decision-making sometimes involves trial and error. What sounded like a great idea after recon operations may not be very practical after the entry team begins on-site operations. Not everything works in the field the way you planned it. No amount of training and simulation can prepare you for actual field conditions—always have options available. Remember—surprises are nice on your birthday, but not on the emergency scene.

In the next part of this chapter, we will discuss specific strategic goals and the tactics associated with each option.

RESCUE AND PROTECTIVE ACTIONS

RESCUE

Life safety is always the IC's highest priority. One of the first concerns after sizing up the incident is search and rescue of victims. Hazmat rescue problems fall into three general categories. These include:

1) **Searching for and relocating people who will be immediately exposed and harmed by the hazmat as the situation gets worse.**

   This group consists of everyone inside of the Hot Zone who is not wearing protective clothing and equipment rated for the hazards. This can consist of civilians and employees who have left the immediate hazard area on their own and believe that they are now in a safe location. It may also include the curious, the "authorized," or the just plain stupid. This includes people who need to be rescued but just don't know it yet! Normally, all that is needed is a little organization and direction to move these people back and away from the hazard area. This group may also include people who were initially exposed but have not yet shown the signs or symptoms of exposure.

2) **Rescuing victims who have been disoriented or disabled by the hazmat.**

   This group includes individuals or groups of people who have been exposed to the hazmat and are suffering from its harmful effects. Examples include victims who have been burned, poisoned, blinded, etc. Normally, rescue involves packaging and removing the victim following standard operating procedures. Chapter Eleven provides some specific guidelines on how to handle and treat these contaminated patients.

NOTES

3) **Planning and executing technical rescue.**

This category includes rescues of one or more victims who have been exposed to the hazmat and require physical extrication. Examples of technical hazmat rescue situations include:

- High angle situations, including injured or disabled victims found in high areas (e.g., cooling towers and elevated structures in refineries or chemical plants, on top of a large diameter storage tank, scaffolding, etc.).
- Victims pinned and trapped inside wreckage or debris (e.g., train crew pinned in a train locomotive cab or a tank truck driver inside of an overturned vehicle).
- Confined space rescue situations (e.g., inside storage tanks, on lowered floating roof tanks, underground vaults, sewers, etc.).

TECHNICAL RESCUES

Technical rescues are extremely dangerous operations which must be adequately planned using highly trained individuals. Consequently, many fire departments and industrial organizations have formed specialty teams to provide the expertise required to handle these unique situations.

Technical rescue problems involving hazardous materials have several common elements which make them difficult to plan for and execute. These include:

- ❑ **Hazardous atmospheres,** which typically involve multiple combinations of flammability, toxicity, and oxygen deficiency or enrichment that change over time. The longer the victim is exposed to these atmospheres, the less likely the chance for survival.
- ❑ **Hazardous work areas**, including slippery or uneven walking surfaces, missing or damaged catwalks, ladders, and handrails, as well as sharp and jagged metal edges. Under normal working conditions, these types of physical hazards require special precautions. Add the restricted vision and motion problems created by wearing SCBA and specialized protective clothing, and these damaged work areas become ultra-hazardous.
- ❑ **Limited access to the rescue site**, including areas with a single access point and confined/narrow walkways or ladders. If the route to or from the rescue site is blocked or cut off, the rescue team will be trapped.

Although it is improving, the track record of making effective technical rescues has not been very good. Many rescuers have become victims because they (1) underestimated the hazards and risks, (2) took action without the proper tools or equipment, or (3) were not properly trained for the tasks at hand.

The most significant problem involved in performing hazmat-related technical rescues is time. Simply, there isn't enough of it. Remember—without an oxygen supply to the brain, clinical death occurs in three minutes, and biological death occurs in five minutes. Exposure to hazardous atmospheres accelerates the timeline.

NOTES

RISK TAKING

The IC must weigh the risk of conducting a technical rescue against the probability of success. Stated bluntly, are we rescuing a victim or recovering a body? Making the decision to attempt or abandon a rescue under extremely hazardous conditions will be one of the most stressful and difficult decisions an IC has to make. Before implementing rescue operations, the IC should be satisfied that some basic questions have been adequately addressed. These include:

1) **Do we actually have a real victim?** In other words, has anyone actually seen the victim(s) to confirm that they are really trapped? If so, was there any movement, etc.? This sounds like a stupid question, but more than one rescue team has searched an area to rescue someone sitting in the coffee shop.
2) **What type of hazardous material has the victim been exposed to and what are its hazardous characteristics?** Specifically, what are its health hazards? The more toxic the material and the longer the exposure time, the lower the likelihood of a successful rescue.
3) **How long has the victim been exposed or trapped?** Try to determine when the incident initially occurred. For example, check with your emergency communications center and determine what time the initial 911 call was placed. Was there a delay in placing the initial call? If you are 20 minutes into a confined space incident and the victim has been exposed to 5,000 ppm of hydrogen sulfide, the chances for survival would be remote at best.
4) **Do you have the appropriate number of personnel, with the right training and the proper equipment, to conduct a technical rescue?** Anyone who has ever participated in a high angle rescue or confined spaces rescue course can tell you that talking about it and being able to do it are two different things. Don't let unprepared rescue teams lay their life on the line without the right equipment.
5) **Do we have control of the situation and do we have a coordinated action plan?** Once the frantic pace of a rescue begins, it is *very difficult* to stop the operation. ERP's eagerly do what they do best, save lives. Be sure that you are handling the rescue following Standard Operating Procedures and are not "winging it."

PUBLIC PROTECTIVE ACTIONS

While we have already discussed public protective actions in detail in Chapter Five, Site Management and Control, some of the basics are worth reviewing here as they relate to implementing response objectives.

Public protective actions (i.e., evacuation or protection-in-place) must be continuously monitored during the course of an incident. Incidents are dynamic events—weather conditions may change, the problem may grow, resources may be used up, etc. Initial protective actions will often be insufficient as we gather more information, get smarter about the incident, and fully assess the level of hazards and risks. Likewise, the Incident Commander will often be under tremendous political pressure

NOTES

to allow civilians and employees who have been evacuated to return to their homes and work stations.

The IC must maintain strict control of the scene throughout the entire incident. The following activities can help maintain site discipline until the problem has been eliminated and the incident safely terminated:

- ❑ Maintain a site Safety Officer throughout the incident. On campaign operations, rotate this position at regular intervals to maintain alertness. A Safety Officer is especially important during hazardous operations such as leak control and technical rescue.
- ❑ Use formal site safety checklists. If you have read this text from the beginning, this statement is probably getting a little old. However, checklists don't make mistakes, tired ERP's do. The longer you work at the hazmat scene, the more likely you will overlook something critical, like making sure your suit is zipped up.
- ❑ Enforce perimeter security and the use of Hazard Control Zones. A weak perimeter often leads to loss of site control, which increases the potential for an accident.
- ❑ Establish a crew rotation schedule. This is especially important during extreme weather conditions. ERP remove protective clothing because they are uncomfortable. Don't contribute to the problem by holding personnel in forward positions for an unreasonable time period. Rested people are more alert.

SPILL CONTROL /CONFINEMENT OPERATIONS

Spill control strategies and confinement tactics are the actions taken to confine a product release to a limited area. These actions usually occur remote from the spill or leak site and are, therefore, defensive in nature. As a general rule, confinement tactics expose ERP to less risk than containment tactics. If the IC can accomplish the same objective using defensive confinement tactics such as diking or remotely closing a valve, then they should be implemented before attempting higher risk, offensive-oriented options.

Confinement operations present several advantages over containment options, including the following:

- Avoid direct personnel exposure.
- Can often be performed without special equipment other than some shovels and dirt. Even the most animalistic truck company can build a dike!
- Can usually be performed by first responders with minimal supervision.

The decision to use confinement tactics is based on the availability of time, personnel, equipment, and supplies. It also should be made with a review of the potential harmful effects the leaking material will have on ERP downwind of the spill, where most of the spill control operations normally take place.

NOTES

For example, a decision to divert a flowing diesel fuel spill from a storm drain to a roadside ditch may be based on the observation that the fuel is flowing too fast and sufficient personnel and equipment are not available to construct a dike. Finally, the fuel will cause substantially less potential damage in the ditch than in the storm system.

Confinement tactics such as diversion can usually begin immediately upon the arrival of ERP. Diking can be started with basic first-responder equipment as more ERP arrive. Retention techniques will then follow as specialty teams and equipment become available.

Don't make the mistake of concentrating all resources on one tactic. It is easy to assign all ERP to the construction of a dike, for example, which may fail and force everyone to move to a safer location to begin again. Recognize that virtually all confinement tactics are "first aid" measures and will eventually fail over time.

FIGURE 10-3: CONFINEMENT TACTICS ARE PRIMARILY DEFENSIVE OPERATIONS

CONFINEMENT TACTICS

There are a number of tactical options available to achieve the spill control strategic goal. These include both physical and chemical methods. A summary of the various tactical options is given below.

Absorption—This is the physical process of absorbing or "picking up" a liquid hazardous material (sorbate) to prevent enlargement of the contaminated area. As the material is picked up, the sorbent will often swell and expand in size. Depending upon the absorbent, it can be used for liquid spills on both land and water.

Operationally, absorbents are effective when dealing with liquids of less than 55 gallons. Larger spills are more difficult to absorb, and often

NOTES

the cost and time exceed the benefits. Materials used as absorbents include clay, sawdust, charcoal, absorbent particulate, socks, pans, pads, and pillows. Absorbent socks and tubes can also be deployed as a circular dike around small spills. When using absorbents, compatibility must be considered (e.g., sawdust used on an oxidizer could start a fire).

Adsorption—This is the chemical process in which a sorbate (liquid hazardous material) interacts with a solid sorbent surface. Since the sorbent surface is solid, the sorbate adheres to the surface and is not absorbed, as with absorbents. An example is activated carbon.

Characteristics of this chemical interaction include:

- Sorbent surface is rigid and there is no increase/swelling in the size of the adsorbent.
- The adsorption process is accompanied by the heat of adsorption, whereas absorption is not. As a result, spontaneous ignition may be a possibility with some liquid chemicals.
- Adsorption can only occur when the sorbent has an activated surface, such as activated carbon.

Adsorbents are primarily used for liquid spills on land and should be nonreactive to the spilled material.

Covering—This is a physical method of confinement. It is typically a temporary measure until more effective control tactics can be implemented. Depending upon the product involved, it may be necessary to first consult with a product specialist.

Examples of covering include:

- Placing a plastic cover or tarp over a spill of dust or powder.
- Placing a cover or barrier over a radioactive source, normally (alpha or beta) to reduce the amount of radiation being emitted.
- Covering a flammable metal or pyrophoric material with the appropriate dry powder agent.

Damming—This is a physical method of confinement by which barriers are constructed to prevent or reduce the quantity of liquid flowing into the environment. Damming consists of constructing a barrier across a waterway to stop/control the product flow and pick-up the liquid or solid contaminants.

There are two types of dams—overflow and underflow.

- **Overflow Dam.** Used to trap sinking heavier-than-water materials behind the dam (specific gravity >1). With the product trapped, uncontaminated water is allowed to flow unobstructed over the top of the dam. Operationally, this is most effective on slow moving and relatively narrow waterways.
- **Underflow Dam.** Used to trap floating lighter-than-water materials behind the dam (specific gravity <1). Using PVC piping or hard sleeves, the dam is constructed in a manner that allows uncontaminated water to flow unobstructed under the dam while keeping the contaminant behind the dam. Operationally, this is most effective on slow moving and relatively narrow waterways.

NOTES

If the pipes are not deep enough on the upstream side of the dam, a whirlpool may be created and pull the hazardous substance through the pipes. This problem can be overcome through the use of a t-siphon on the upstream side. To be effective, several overflow or underflow dams should be placed downstream to catch what was missed by the first dam.

Diking—This is a physical method of confinement by which barriers are constructed on ground used to control the movement of liquids, sludges, solids, or other materials. Dikes prevent the passage of the hazmat to an area where it will produce more harm.

Dikes are most effective when they can be built quickly. Although any available material will do the job, the best quickly acquired supplies are dirt, tree limbs, boards, roof ladders, pike poles, and salvage covers. Bagged materials such as tree bark, sand, and kitty litter can be found at hardware and garden supply stores when more substantial control is required. However, when really large spills occur, dump-truck-sized deliveries will be required.

Dikes will usually be constructed by first responders using whatever equipment is available on the scene. When considering building a dike, quickly compare your resources to the quantity of the spilled material. Most ERP overestimate the amount of spill and underestimate the personnel required to complete a dike.

Slow-moving or heavy materials should be confined by use of a circle dike. Faster moving products will require a "V"-shaped dike located in the best available low-lying area. Always use the land to your advantage.

Dike construction should begin by choosing large, heavier materials for reinforcement, followed by an outer layer of lighter material such as dirt. Operationally, dikes are normally a temporary measure and can begin to leak after a while. Seepage can be minimized by using plastic sheets or tarps at the dike base and within the dike by placing a final layer of dirt along the leading edge between the plastic and the ground. Be aware that plastic sheets may be degraded by certain chemicals

Factors which can limit dike construction include situations in which:

- The surrounding area is concrete or asphalt with no available soil. Either sacrifice the area for better turf or truck in necessary materials.
- The ground is frozen. Snow may be used in conjunction with materials such as plastic and ladders. Otherwise, truck in necessary materials.
- Essential equipment is unavailable. At least three pointed, long-handled shovels are necessary. When possible, construct dikes upwind in safe areas. Be sure to consider the need for SCBA.

Dilution—This is a chemical method by which a water-soluble solution, usually a corrosive, is diluted by adding large volumes of water to the spill. There are four important criteria that must be met before dilution is attempted. These include determining in advance that (1) the substance is *not water reactive*, (2) will not generate a toxic gas upon contact with water, (3) will not form any kind of solid or precipitate, and (4) is totally water soluble.

NOTES

As a general rule, dilution should only be attempted on liquid and solid substances that are corrosives, and only when all other reasonable methods of mitigation and removal have proven unacceptable. In other words, dilution tactics are a last resort.

Dilution can be effective for small corrosive spills of one quart or less, especially for concentrated corrosives with a pH of 0–2 (acidic) or 12–14 (alkaline). In outdoor situations, local water department or fish and wildlife representatives should be consulted for their approval to use dilution tactics. Federal and most state regulations limit corrosive entries into storm drains and drainage canals to a pH of 6 to 8 as long as no other pollutants are involved which may be harmful to the environment or wildlife.

The major disadvantage to dilution is that it is not well understood by emergency response personnel. It is not a straight linear one-to-one process. It is important to recognize that dilution is actually a logarithmic process (i.e., on a one to ten scale). For example, a 1-gallon spill of an acid with a pH of zero will require one million gallons of water just to bring its pH up to six! That is a lot of water just to dilute 1 gallon of acid. This same rule applies to the full range of corrosives, from a pH of 0 to 14. The following chart provides some guidelines which will help the IC determine if dilution is the best tactical option.

Acid Spill of 1 Gallon (pH of 0)

Water to add	Brings pH to
10 gallons	1
100 gallons	2
1,000 gallons	3
10,000 gallons	4
100,000 gallons	5
1,000,000 gallons	6

Alkaline Spill of 1 Gallon (pH of 14)

Water to add	Brings pH to
10 gallons	13
100 gallons	12
1,000 gallons	11
10,000 gallons	10
100,000 gallons	9
1,000,000 gallons	8

NOTES

A Rule of Thumb which can be used in the field to determine the volume of water required to bring the pH to the 6–8 range is as follows:

Step 1: Determine the size of the spill to be diluted in gallons (e.g., there are 10 gallons on the ground).

Step 2: Determine the pH of the spilled material using pH paper or a pH meter (e.g., the spill has a pH of 3).

Step 3: Determine the pH that you want to dilute the spill to (e.g., you want to go to a pH of 6 so that the spill can be safely flushed into the storm system).

Step 4: Determine the number of dilution steps between the starting pH and the ending pH. In our example, we started with a pH of 3 and want to end up with a pH of 6. That is three steps.

Step 5: Add three zeros to the beginning gallonage. In our example, we started with 10 gallons, so we add three zeros, which will give us 10,000. This is the number of gallons of water that must be added to the spilled 10 gallons of acid in order to bring the pH up to the desired level of a pH of 6. This rule can be applied to the entire logarithmic scale no matter where you enter it (e.g., you started with a pH of 4 and want to go to 6, that would be two steps, etc).

Diversion—This is a physical method of confinement by which barriers are constructed on ground or placed in a waterway to intentionally control the movement of a hazardous material into an area where it will pose less harm to the community and the environment.

A flowing, land-based spill can quickly be diverted by placing a barrier (e.g., dirt) ahead of the spill. As when fighting a fast-moving brush fire, the barrier should be placed well ahead of the actual spill. This may require sacrificing some intermediate territory to the hazmat in order to establish complete control at the final diversion site.

Booms can also be placed across streams and waterways to divert the hazardous substance into an area where it can be absorbed or picked up, such as with vacuum trucks.

In constructing a diversion barrier, consider the angle and speed of the oncoming spill. The greater the speed of the flow, the greater the length and angle of the barrier required to slow and divert the flow. For fast-moving spills, barriers constructed at a 45° perpendicular angle will be ineffective; a barrier angle of 60° or more should be used.

Constructing a diversion barrier requires teamwork. When a team with the right equipment works quickly, a large area can be rapidly controlled. A typical four-person crew can build a 20- yard by 8 inch diversion wall in about 10 minutes if the proper materials are available.

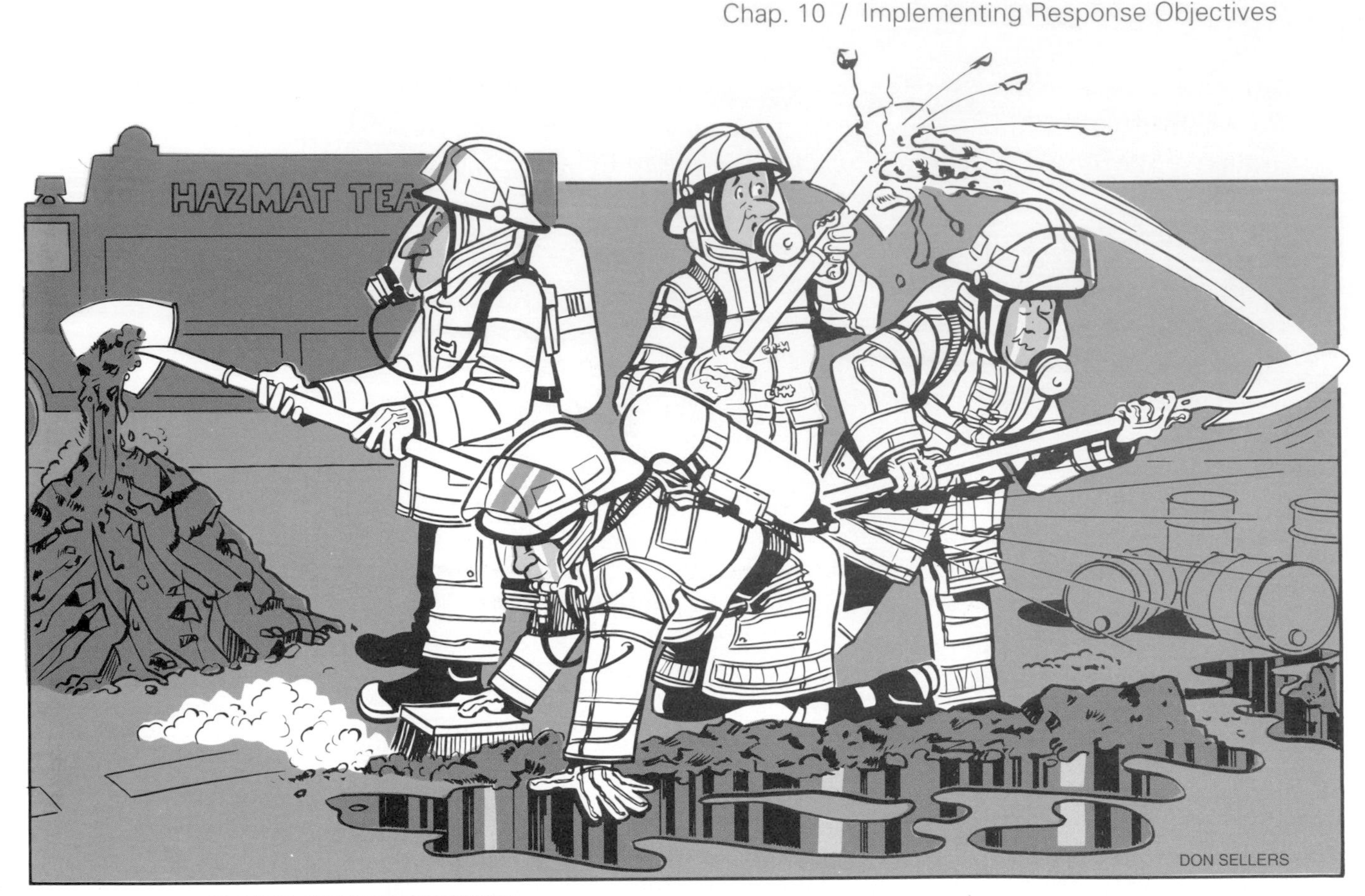

FIGURE 10-4: SPILL CONTROL REQUIRES TEAMWORK

Dispersion—This is a chemical method of confinement by which certain chemical and biological agents are used to disperse or break up the material involved in liquid spills on water. The use of dispersants may result in spreading the hazmat over a larger area.

Dispersants are often applied to hydrocarbon spills, resulting in oil-in-water emulsions and diluting the hazmat to acceptable levels. They do not neutralize or make flammable materials become nonflammable. Experience also shows that some dispersants will "separate" over time. Use of dispersants may require prior approval of the appropriate environmental agencies.

Retention—This is a physical method of confinement by which a liquid is temporarily contained in an area where it can be absorbed, neutralized, or picked up for proper disposal. Retention tactics are intended to be more permanent and may require resources such as portable basins or bladder bags constructed of chemically resistant materials.

Retention tactics can sometimes be implemented independently and act as a back-up to diversion or diking tactics. For example, storm sewer systems can be protected by placing salvage covers or plastic over drains and covering them with dirt. The same procedure can be used for sewer system manways.

When the hazmat is primarily a liquid or slurry, has a specific gravity less than 1.0, and is not water reactive, it may be possible to flood the retention area with water from an engine or hydrant. The hazmat would

NOTES

then float on the water, and any subsequent leakage into the storm system would only be water.

Vapor Dispersion—This is a physical method of confinement by which water spray or fans is used to disperse or move vapors away from certain areas or materials. It is particularly effective on water-soluble materials (e.g., anhydrous ammonia), although the subsequent runoff may involve environmental trade-offs. Fans and positive pressure ventilators may also be used if they are rated for the hazardous atmosphere.

When dealing with flammable materials, such as LP gases, the turbulence created by the water spray may reduce the gas concentration and bring the atmosphere into the flammable range.

Vapor Suppression—This is a physical method of confinement to reduce or eliminate the vapors emanating from a spilled or released material. Operationally, it is an offensive technique used to mitigate the evolution of flammable, corrosive, or toxic vapors and reduce the surface area exposed to the atmosphere. Common examples include the use of firefighting foams and chemical vapor suppressants.

While vapor suppression does not change the nature of a hazardous material, it can greatly reduce the immediate hazard associated with uncontrolled vapors. In addition, it can buy additional time to undertake further measures to control the problem.

LEAK CONTROL/CONTAINMENT OPERATIONS

Leak control strategies and containment tactics are the actions taken to *contain* or keep a material within its container. Typically regarded as offensive operations, containment tactics require personnel to enter the Hot Zone to control the release at its source and should be considered high risk operations. Examples include uprighting leaking containers, closing and tightening container caps and valves, plugging and patching container shells, and depressurizing vessels by isolating valves or shutting down pumping systems.

Containment tactics are often implemented when defensive options have not produced acceptable results or when citizens or employees are at great risk from potential chemical exposures. These tactics should only be approved after conducting a thorough hazard and risk evaluation. No emergency situation is worth taking unreasonable risks. Rapid withdrawal from the Hot Zone is always an option; aggressive/offensive does not mean quick and stupid.

Before initiating containment operations, the IC should consider:

1) What hazardous material(s) are involved?
2) What are its hazards?
3) What are the risks to both ERP and civilians?
4) What are the training levels and physical abilities of the entry team that will perform the operation?
5) Are special tools and equipment needed for the leak control operation? Are they available?

NOTES

6) Are the ERP prepared for emergency care and decontamination if an accident happens?

If these questions cannot be positively answered, leak control operations should be delayed until sufficient information or resources are obtained and the IC feels the operation can be safely conducted.

FIGURE 10-5: CONTAINMENT TACTICS ARE PRIMARILY OFFENSIVE OPERATIONS

Although containment operations may pose higher risks, they may be necessary to:

- **Minimize environmental damage**—This is particularly true for hazardous liquids which threaten storm systems or water supplies.
- **Reduce operating response time**—Leaks confined to the area immediately around the container usually limit the spread of the material and minimize the need for evacuation, particularly when faced with a gas or toxic chemical.
- **Reduce clean-up costs**—Contaminants are usually limited to smaller areas or have not entered impacted ground or surface waters.

Situations well suited for aggressive leak control offensive strategies include the following:

1) The hazmat is in a gaseous form and threatens to migrate away from its container.
2) The hazmat is in a solid, powder form and weather conditions threaten to carry it from its original site.
3) Defensive options have been attempted but have not produced the desired results.

NOTES

4) The situation is getting worse and increasing in risk as time progresses.

Successful offensive operations should be preceded by a thorough reconnaissance. Recon may be as simple as having a trustworthy individual relay his or her observations or as complex as a Recon Team surveying the entire work site with a video camera.

This isn't brain surgery. Always remember the KISS Principle (Keep it Simple, Stupid)—most leak control tactics are pretty simple. Consider the following:

- Isolate the leak by checking the position of upstream valves.
- Check the integrity of container openings—tighten caps, bungs, lids, etc.
- Standing up a leaking liquid container may be sufficient to stop the leak.
- Moving the container to place the hole above the liquid or solid level reduces the hazmat release.
- When dealing with liquefied gases (e.g., chlorine, anhydrous ammonia, LPG), rotate the container to deal with a vapor release rather than a liquid release. If liquid escapes, it will expand the problem and hazard area (e.g., liquid/vapor expansion ratio for chlorine is 460 to 1 and 850 to 1 for ammonia).

When these common-sense techniques aren't effective, try to:

- Plug or patch the opening (i.e., control it at the source).
- Reduce the container pressure (i.e., limit its magnitude).
- Use vapor suppression agents such as foam (i.e., limit its vaporization).
- Neutralize using another chemical (after consulting with a product specialist).
- Control the leak and dispose in-place (e.g., controlled burning, flaring).

CONTAINMENT TECHNIQUES

A number of tactical options are available to achieve the leak control strategic goal. These include both physical and chemical methods. A summary of the various tactical options is given below.

Neutralization—This is a chemical method of containment by which a hazmat is neutralized by applying a second material to the original spill which will chemically react with it to form a less harmful substance. The most common example is the application of a base to an acid spill to form a neutral salt.

The major advantage of neutralization is the significant reduction of harmful vapors being given off. In some cases, the hazmat can be rendered harmless and disposed of at much less cost and effort. However, during the initial phases of combining an acid and a base a tremendous amount of energy may be generated, as well as toxic and flammable vapors.

NOTES

Operationally, many responders recommend that HMRT neutralization operations should be limited to spills less than 55 gallons. It is quite easy to use too much neutralizing agent and end up with a large caustic spill instead of the original acid spill.

Before initiating any neutralization techniques, the following conditions must be satisfied: (1) the hazmat has been positively identified; (2) its physical and chemical properties have been properly researched; and (3) the spill has been controlled and confined to prevent runoff after application of the neutralizing agent. Sufficient neutralizing agent should be on hand to complete the process once it is begun.

To determine the amount of base necessary for an acid spill, consider the following example: A glass bottle containing 1 gallon of 70% nitric acid falls and breaks open. Responders have an ample supply of neutralizing agent (soda ash) available. How much would be required to bring the pH somewhere close to 7 (neutral)?

1) Determine the weight of 1 gallon of the acid in pounds, using an MSDS or the following formula to determine the information:

Quantity of Acid Spilled	X	Specific Gravity	X	8.33 lbs./gal. (weight of water)	X	Percent Acid	=	1 Gallon Weight in Pounds
Example:								
(1 gallon)	X	(1.5)	X	(8.33)	X	(.70)	=	8.75 lbs.

2) Select the appropriate conversion factor:

Neutralizing Agent	Sulfuric Acid	Nitric Acid	Hydrochloric Acid	Phosphoric Acid
Sodium Carbonate (soda ash)	1.082	0.841	1.452	1.622
Calcium Hydroxide (slaked lime)	0.755	0.587	1.014	1.133
Sodium Bicarbonate (baking soda)	1.673	1.302	2.247	2.541

3) Multiply the weight of the spilled acid times the conversion factor for the appropriate neutralizing agent to determine the amount needed to neutralize the acid spill.

8.75 lbs./gallon x 0.841 = 7.4 lbs. of soda ash would be required

Responders can develop a chart by which they can estimate the amount of neutralizing agent available for specific size spills. Using 70% nitric acid as an example:

1-gallon spill	=	7.4 lbs. of soda ash
10-gallon spill	=	74.0 lbs. of soda ash
100-gallon spill	=	740 lbs. of soda ash
1,000-gallon spill	=	7,400 lbs. of soda ash

FIGURE 10-6: NEUTRALIZATION BEGINS AT THE OUTER EDGES OF THE SPILL.

When a decision has been made to neutralize a spill, some consideration should be given to the type of neutralizing agent which will be used. Some materials are more environmentally friendly than others. The key concern is biodegradability. The most widely favored neutralizing agents from an environmental perspective are sodium sesquicarbonate (for acid spills) and acidic acid (for alkali spills). Sodium and calcium hydroxide will not produce a biodegradable end product. If the spill is in an environmentally sensitive area, a specialist should be consulted for advice on which neutralizing agent to use.

Corrosive spills should be covered by shoveling the neutralizing agent from the outermost edge inward, thereby protecting the workers first. Avoid walking through spills, even when wearing proper protective clothing.

Some caustic spills have been neutralized using various types of diluted acids, but this technique should never be attempted without seeking the advice of a product specialist.

NOTES

Neutralizing agents should be purchased in bulk quantities and stored at key locations. Commercial neutralizing kits are also available for handling small spills; these are normally packaged for smaller laboratory or workshop-type spills of one to five gallons.

Overpacking—This is a physical method of containment by which a leaking drum, container, or vessel is placed inside a larger overpack container. Although commonly used for liquid containers, overpacking can also be used for some compressed gas cylinders like chlorine.

Liquid overpacks are constructed of both steel and polyethylene; common sizes include 8, 15, 30, 55, and 85 gallons. Where possible, the leak should be temporarily repaired before the container is placed inside of the overpack. Depending upon container size and weight, mechanical equipment (e.g., forklift or hoist) may be required to raise and lower the leaking container into the overpack. A 55-gallon drum of sulfuric acid can weigh over 800 pounds. In addition, the container may have been weakened as a result of the leak. The overpack container must then be labeled in accordance with DOT hazmat regulations if it will be transported from the scene.

Cylinder overpack devices are used by some compressed gas companies and industrial HMRT's. Once a leaking cylinder is overpacked, it can be transported back to the gas manufacturer or to an approved site for disposal and destruction. Cylinder overpacks do have several drawbacks, including the fact that they aren't readily available, the mobilization process can be very time consuming, and they are extremely expensive to manufacture or purchase ($20,000+).

Patching and Plugging—This is a physical method of containment which uses chemically compatible patches and plugs to reduce or temporarily stop the flow of materials from small container holes, rips, tears, or gashes. Although commonly used on atmospheric pressure liquid and solid containers, some tactics can also be used on pressurized containers.

When dealing with liquid containers, recognize that plastics and metals behave differently when they are breached and that the methods of repair will vary accordingly.

- **Plugging**—Involves putting something into a hole to reduce both the size of the hole and the amount of product flow. The plug must be compatible with both the chemical and the container. For example, a small hole in an aluminum MC-306/DOT-406 cargo tank truck can sometimes be plugged by driving a wooden wedge into the opening with a rubber mallet. However, a soft pine plug would not be compatible with a strong acid leak.

  Plugs can be fabricated on the scene, but you can save a great deal of time by manufacturing a variety of devices before the fact and carrying them on response vehicles. Plugs can be constructed of various materials, including wood, rubber, and metal. Plugs constructed of softwoods, such as yellow pine or Douglas Fir, are quite effective for holes whose area is less than 3 square inches. Preplan potential container leaks and network with other ERP to determine what works.

NOTES

Plugging techniques are usually used in conjunction with synthetic rubber gaskets, lightweight cloth, or special chemical-resistant putty to ensure a good seal by filling the cracks around the plug. Small holes (less than 1/2 inch diameter) which are not under pressure can be filled with putty or epoxy resin compounds. The longevity of these compounds is limited due to material compatibility, the hole size, and the head pressure of the container. These should be viewed as only temporary first aid techniques.

FIGURE 10-7: PLUGGING AND PATCHING OPERATIONS MUST BE DONE USING THE BUDDY SYSTEM

- **Patching**—Involves placing a material or device over a hole to keep the hazmat inside of the container. Patches can include both commercial and home-made devices and are used to repair leaks on container shells, piping systems, and valves. Patches must be compatible with the chemicals involved.

Like plugs, patches can also be fabricated on the scene, but you can save a great deal of time by manufacturing a variety of devices before the fact and carrying them on response vehicles. Responders are only limited by their ingenuity. Common examples include toggle bolt compression patches, gasket patches, glued patches, and epoxy putties.

Container pressure is a critical factor in determining the appropriate patch. Leak bandages and leak sealing kits are effective tools for dealing with liquids with a low head pressure or low pressure gases (100 to 200 psi). Some inflatable air bag patch kits, like the Vetter Bag™, are effective at pressures less than 25 psi. The Chlorine A and B Kits have a side patch kit for container shell leaks on 100- and 150 pound, and 1-ton chlorine cylinders.

NOTES

To properly organize a patching operation, consider the following:

1) Select a patch device half a size larger than the estimated opening. Smaller devices may be drawn inside the container as attachments such as nuts, t-bolts, and hooks are tightened.
2) Ensure that the patch is compatible with the hazmat.
3) Plan the work with SCBA operating times in mind. Several trips may be required. Overlap work crews so that one crew is always working, one crew is ready to step in (back up), and the third is re-servicing equipment and re-habilitating.
4) Brief personnel on the tools required to compete the job. The more complicated the job, the more extensive the briefing. It's easy to forget tools when crews are under pressure. Forgetting one tool may add an hour to the operation since decontamination is required for the entry team member who makes the second trip.

Pressure Isolation and Reduction—This is a physical or chemical method of containment by which the internal pressure of a closed container is reduced. The tactical objective is to sufficiently reduce the internal pressure in order to minimize the potential of a container failure. Pressure reduction tactics are high risk operations which require responders to work in close proximity to the container. Examples include flaring and vent and burn.

Many hazmat containers are designed to store their contents under pressure. Cylinders, process vessels, MC-331 cargo tank trucks, rail cars, and pipelines are examples. It is also possible for nonpressurized containers to become pressurized because of internal chemical reactions, thermal stress, or accidentally diverted pressure.

Pressurized containers are dangerous because:

- They can rupture under stress and travel great distances as fragments or in one piece. This happens quickly and allows no reaction time.
- High pressure kills quickly. ERP cannot usually determine the operating pressure of a given container without close inspection. High pressure can propel valve caps, breach protective clothing, or sever SCBA air lines. Ultra high pressures (5,000 to 15,000 psi) can penetrate the skin and cause an air embolism, which will be followed by death within minutes.
- Cryogenics are stored in pressurized containers at temperatures below minus 150°F. Cryogenics can freeze tissue and damage protective clothing.

Examples of pressure reduction tactics include the following:

- **Isolating Valves**—Pressurized containers often leak in and around valves and fittings. Most valves can be closed by turning the valve wheel clockwise, unless it is damaged. If the leak continues after

NOTES

the valve is closed, try tightening the packing nut on the valve.

However, there are exceptions to these procedures. According to the *Handbook of Compressed Gases*, nearly 200 types of compressed gases were commonly shipped in 1990. More than 12 different cylinder specifications with more than 64 different valve outlets were used for these 200 gases.

In-service containers may also have piping leaks. These leaks usually stop when the supply valve is isolated and blocked in. Depending upon the situation, it may be necessary to isolate the valve both upstream and downstream.

FIGURE 10-8: OPERATORS CAN HELP ISOLATE PRODUCT LINES IN PETROCHEMICAL FACILITIES

- **Isolating Pumps and Pressure/Energy Sources**—Some pressurized vessels are charged by an independent source such as a pumping system or compressor. The magnitude of the leak can be reduced significantly when the pump pressure is lowered or shut down entirely. Specialists familiar with the system must be consulted before any shutdown operations are initiated, as a shutdown may overpressurize other vessels or create related upstream and downstream problems. Lack of pressure may also produce unstable chemical reactions.

  Product specialists and process engineers should also be contacted whenever ERPs face large, complex process units and related pressure vessels. Shutting down electrical power to pressure systems may ruin chemical batch processing equipment, cause dangerous pressure build-ups in other locations, and disable critical safety devices. In some cases, safety systems can kick in and dump hazardous materials to neutralizing scrubbers, flares, or exhaust systems.

NOTES

- **Venting**—This is the controlled release of a material to reduce and contain the pressure and decrease the probability of an explosion. The method of venting will depend upon the nature of the hazardous material and process involved. For example, nontoxic materials may be vented directly to air (e.g., steam), while toxic or corrosive materials may be vented to a scrubber or back-up storage vessel.
- **Flaring**—This is the controlled burning of a liquid or gas material to reduce or control pressure and/or to dispose of a product. Flaring is commonly used at fixed facilities, such as refineries and petrochemical facilities, as an integral part of the fixed process safety system.

  Portable vapor flares may also be used for controlled burns at transportation accidents involving damaged flammable gas cylinders, tank trucks, and rail cars. Liquid flares may be used for the controlled release and burning of liquefied gases and flammable liquids in conjunction, with a pit used to contain any liquid product not completely burned.
- **Hot Tapping**—This technique is used to gain access to a bulk flammable liquid or gas tank, pipeline, or container for the purpose of product removal. It involves the welding of a threaded nozzle to the exterior of a tank or pipeline. A valve is then attached to the threaded nozzle and a hole is drilled through the container shell with a specially designed machine. The drilling machine is equipped with seals that prevent the loss of product during the drilling operation. Hoses are then attached to the valve outlet and the contents are removed.

  Hot tapping is commonly used within the petroleum and petrochemical industries. However, it should only be attempted during an emergency by trained hot tapping specialists.
- **Vent and Burn**—This is a process by which shaped explosive charges are used to cut a hole (or holes) in a pressurized tank car, allowing the contents to flow into a pit for burn-off. This is a highly sophisticated technique that should only be attempted by trained specialists under very specific situations.

  Vent and burn is normally an option of last resort and may be used under the following conditions:

  - The tank car has been exposed to fire, resulting in elevated internal pressures and possible tank damage.
  - Conditions do not allow for the safe transfer, venting, or flaring of the tank car.
  - Site conditions prevent re-railing the damaged tank car.
  - Damage to leaking valves and fittings cannot be repaired.
  - The tank car has been damaged to the extent that it cannot be safely re-railed and moved to an unloading point.

  Vent and burn tactics were used successfully at the Livingston, LA train derailment.

NOTES

Solidification—This is a chemical method of containment whereby a liquid substance is chemically treated so that a solid material results. The primary advantage of this process is that a small spill can be confined relatively quickly and immediately treated.

Solidification is often used for both corrosive and hydrocarbon spills. Commercial formulations are available which can be applied to a liquid acid or caustic spill, neutralize the hazard, and form a neutral salt. Commercially available adsorbents can also be used to solidify nonsoluble oily wastes. The spilled hydrocarbon is adsorbed to granules to form a solid, nonflowing mixture. This resulting mixture is actually safer than the original spilled material and can be easily transported and disposed of at a waste treatment facility.

Vacuuming—This is a physical method of containment by which a hazardous material is placed in a containment system by simply vacuuming it up. The method of vacuuming will depend upon the hazmats involved. Vacuuming is commonly used for containing releases of certain hydrocarbon liquids, solid particulates, asbestos fibers, and liquid mercury.

The primary advantage of vacuuming is that there is no increase in the volume of waste materials. In selecting this tactic, care must be taken to ensure that the vacuum and related equipment is compatible with the hazmats involved and that vacuum exhaust vapors are controlled. The use of larger scale vacuum trucks is discussed later in this chapter in relation to product transfer methods.

FIRE CONTROL OPERATIONS

Considering the target audience for this text and its experience in the "Wide World of Fires," we feel that we do not need to get into the basics or hands-on aspects of firefighting. There are many good texts, videos, and training programs which do a good job of discussing the fundamentals. We have listed many of these at the end of this chapter. In the following Fire Control sections, we will concentrate primarily on the selection and implementation of strategy and tactics. Our primary focus will be on fire situations which usually require hazmat response team involvement for managing flammable liquids, flammable gases, and reactive chemical fires.

The following general factors can be applied uniformly to hazmat fire problems. They should be considered early in the incident as part of the hazard and risk evaluation process.

1) **What hazardous material(s) are involved?** Specifically, are we dealing with flammable liquids, gases, an exotic reactive material, or some combination of both flammable and nonflammable yet extremely hazardous substances?
2) **What are its hazards?** The physical and chemical properties of a material significantly influence the selection of tactics and fire extinguishing agents. What works well on one type of fire won't work on another. Critical questions include determining the material's

NOTES

(1) chemical family (e.g., hydrocarbon or polar solvent); (2) water solubility; (3) specific gravity; (4) water reactivity; and (5) control and extinguishing methods.

3) **What type of container is involved?** This can include storage tanks, pressure vessels, reactors, and pipelines. Responders must also evaluate container features, such as relief devices, valves, fixed foam systems, etc. Many reactive materials are shipped in specialized containers and may have unusual features.
4) **What are the risks to responders, employees, and the community?** What is the likelihood of the incident growing and involving other containers? Are there any significant environmental impacts?
5) **Are specialized resources required? What is their availability?** This could include personnel, supplies, equipment, extinguishing agents, and/or related appliances.
6) **What will happen if I do nothing?** Remember—this is the baseline for hazmat decision-making and should be the element against which all strategies and tactics are compared.

FLAMMABLE LIQUID EMERGENCIES

The majority of flammable liquid emergencies are relatively small, involve 55 gallons of liquid or less, and are successfully handled by fire units. In light of this fact, we would like to concentrate on the tactical problems associated with managing larger flammable liquid fires, such as dealing with bulk liquid storage facilities, transportation containers, and storage tank fires.

HAZARD AND RISK EVALUATION

Flammable liquids are a "good news"–"bad news" type of event. The bad news is that there will be a lot of fire, tremendous amounts of radiant heat, and exposure problems. The good news is that unlike most other hazmat problems, you can usually see where your problem is, and the fire will allow enough time to develop a well-thought-out plan of action. Responders will normally have enough time to gather the necessary resources before mounting an aggressive, offensive-oriented fire attack. This doesn't mean that defensive tactics and exposure protection won't be required (they will!), but only that the time factors are sometimes a little different from those encountered at structural fires or other types of major incidents where life safety is a big issue.

Keeping track of times and key events at a major flammable liquids fire is often difficult. The Incident Commander should acquire the following information during the size-up process:

- Time which the incident started. This may not necessarily be the same time the incident was reported!
- Time at which responders arrived on scene.
- Probability that the fire will be confined to its present size.
- Fuel involved (flammable or combustible liquid), including the quantity, surface area involved, and the depth of the spill.

NOTES

- Hazards involved, including flash point, reactivity, solubility (e.g., hydrocarbon or polar solvent), and specific gravity.
- Estimated pre-burn time. This will help the IC determine factors such as how "hot" the fuel is, identify and prioritize exposures, consider transfer and pump-off options, determine if a heat wave is developing for crude oils, etc.
- Layout of the incident, including the following specific points:
  - Type of storage tank(s) involved. Common aboveground liquid storage tanks are cone roof, open floating roof, covered floating roof, and dome roof tanks. See Figure 6-12 for additional information.
  - Size of the diked area(s) involved.
  - Valves and piping systems stressed or destroyed by the fire.
  - All surrounding exposures, including tanks, buildings, process units, utilities, etc. This should include identifying and prioritizing exposures (e.g., flame impingement, radiant heat exposure). Process unit personnel can help with these decisions.

TACTICAL OBJECTIVES

Hazard and risk evaluation is the cornerstone of decision-making. Based upon the type and nature of risks involved, the Incident Commander will implement the appropriate strategic goals and tactical objectives. Tactical options for flammable liquid emergencies include the following:

- ❑ **Nonintervention**—This is a "no win" situation in which responders assume a passive position (i.e., get out the beach chairs, umbrellas, and suntan lotion, and watch the fire burn itself out). This option is sometimes implemented when there are insufficient water supplies, very little product remaining which can be saved, or no exposures in the immediate area.
- ❑ **Defensive Tactics**—These tactics involve protecting exposures and allowing the fire to burn. In many cases, defensive tactics are a temporary measure until sufficient resources can be assembled to pursue an aggressive, offensive attack.

  The primary concerns during defensive operations are direct flame impingement and radiant heat exposures. Flame impingements must be cooled immediately, while radiant heat exposures should be handled as soon as possible. Exposures should be prioritized in the following manner:

  - **Primary Exposures**—Pressure vessels, closed containers, piping systems, or critical support structures exposed to direct flame impingement. Failure of exposed vessels, tanks, and piping systems is likely unless cooling water is quickly applied.

    If a storage tank is involved, direct flame contact on the tank shell can cause the upper portion of the shell, as well as any associated foam systems, to lose their integrity and fold inward.

NOTES

Streams should cool all surfaces above the liquid level. Remember, cooling water is a valuable resource; don't waste it. See Figure 10-9.

- **Secondary Exposures**—Pressure vessels, closed containers, piping systems, or critical support structures exposed to radiant heat. Failure of structural components is possible if cooling water is not applied.

  Be careful of applying water onto open floating roofs—sinking the roof with water lines can be somewhat embarrassing as well as hazardous! Also, remember that radiant heat will pass through structures with clear glass and windows. In addition to applying exterior cooling lines, firefighters should be sent inside to check for any fire extension.

- **Tertiary Exposures**—Noncritical exposures without life safety concerns.

❑ **Offensive Tactics**—These tactics are implemented when sufficient water and firefighting foam supplies and related resources are available for a continuous, uninterrupted fire attack. Although the primary focus is on fire extinguishment, it may also be necessary to maintain exposure lines.

Time becomes a critical factor for offensive operations. The IC must determine the duration of any fire attack. For example, *NFPA 11—Technical Standard on Low Expansion Foam and Combination Agents*, recommends an application time of 15 minutes for flammable liquid spill fires and 65 minutes for flammable liquid storage tank fires. The IC should document the time foam operations start, the time at which the fire is controlled, and the time at which the fire is extinguished.

A final note about exposure protection. Flammable liquid spill fires confined to diked areas may accumulate a large quantity of water from both fire attack and exposure streams. If these flows and associated runoff are not closely monitored, dikes may overflow and carry the burning flammable liquid outside of the area. In addition, loss of electrical power to the facility may cause the sewer system pumps to lose power and create additional runoff problems. **As a general guideline, water streams applied to exposures inside of a diked area (e.g., adjoining storage tank) should be temporarily shut down when they no longer produce steam at the point where water contacts the hot steel surface.**

EXTINGUISHING AGENTS

Firefighting foam remains the workhorse of flammable liquid firefighting. While other agents, such as dry chemicals, are used to extinguish small fires or deliver the knockout punch for three-dimensional fires (e.g., a flange fire), foam is still the "Big Gun" for large-scale flammable liquid problems. When dealing with three-dimensional fires, be sure that the fuel source can be shut off when the fire is extinguished. Aqueous Film Forming Foam (AFFF) may be used to secure the fuel surface area.

NOTES

FIGURE 10-9: FLAMMABLE LIQUID STORAGE TANK FIRES REQUIRE HEAVY STREAMS

The availability of water and foam concentrate is a critical factor in evaluating the risks involved in a flammable liquid emergency. The IC must determine (1) if an adequate water and foam supply is available, and (2) if there is sufficient capability to deliver the required foam solution to control and extinguish the fire. If an adequate water and foam supply is not available for both protecting exposures and controlling the fire, the IC should consider implementing defensive or nonintervention tactics until sufficient resources are available.

Firefighting foam concentrates should be selected based upon the type of fuel and nature of the fire (e.g., spill vs. storage tank). The most common foam concentrates used today are:

- Fluoroprotein Foam.
- Aqueous Film Forming Foam (AFFF).
- Alcohol Resistant AFFF (ARC)—3% hydrocarbon/6% polar solvent or 3% hydrocarbon/3% polar solvent options.
- Aqueous Film Forming Fluoroprotein Foam (AFFFP).

A few rules should be observed regarding firefighting foam compatibility: (1) different foam concentrates by type or manufacturer are not considered to be compatible; (2) don't mix different kinds of foam concentrates before proportioning; and (3) all finished foams are considered compatible regardless of the type or manufacturer.

DETERMINING FOAM CONCENTRATE REQUIREMENTS

Based upon the recommendations of NFPA 11, hydrocarbon spill fires will require foam application rates ranging from 0.10 to 0.16 gpm/ft^2 depending on the type of foam concentrate used. Polar solvents will

NOTES

require at least 0.20 gpm/ft^2. As a general tactical guideline, do not begin foam application until enough foam concentrate is on site to extinguish 100% of the exposed flammable liquid surface area. However, partial extinguishment of burning surface fires may provide some significant advantages for achieving rescue or exposure protection even if the fire cannot be totally extinguished.

NFPA 11 recommended foam application rates for specific fuels, foams, and applications are listed below.

- 0.1 gpm/ft^2—fixed system application for hydrocarbon fuels (e.g., storage tank with foam chambers).
- 0.1 gpm/ft^2—sub-surface application for hydrocarbon fuels in cone roof tanks.
- 0.10 gpm/ft^2 (AFFF) to 0.16 gpm/ft.^{2} (Fluoroprotein)—portable application for hydrocarbon spills (e.g., 1$^3/_4$-inch handlines with foam nozzles).
- 0.16 gpm/ft^2—portable application for hydrocarbon storage tanks (e.g., portable foam cannons and master stream devices). A foam application rate of 0.18 to 0.20 gpm/ft^2 has been used to successfully extinguish hydrocarbon fires in large diameter tanks using master stream portable nozzles. However, the largest known storage tank extinguished was 150 feet in diameter.
- 0.20 gpm/ft^2—*minimum* recommended rate for polar solvents. Higher flow rates may be required depending on the fuel involved and the foam concentrate used.

How much foam concentrate is enough? Foam concentrate requirements can be determined through the following process:

1) *Determine the type of fuel which is involved—hydrocarbon or polar solvent. This will determine the type of foam to be used.*
2) *Determine the surface area involved.*
   - Storage Tank
     Area = (.785) (Diameter^2) or (.8) (Diameter^2)
   - Dike or Rectangular Area
     Area = Length x Width
3) *Determine the appropriate foam application rate, as noted above.*
4) *Determine the duration of foam application.*
   - Flammable liquid spill = 15 minutes
   - Storage Tank

Flash Point 100–200°F	= 50 minutes
Flash Point < 100°F	= 65 minutes
Crude Oil	= 65 minutes
Polar Solvents	= 65 minutes
Seal Application	= 20 minutes

NOTES

5) Determine the quantity of foam concentrate required. *This figure will be determined by the percentage of foam concentrate used.*

Consider the following example: A 125-ft-diameter open floating roof tank is fully involved in fire. Determine the amount of foam con centrate required to control and extinguish the fire. The fire department is using a 3% x 3% alcohol resistant foam concentrate (ARC).

1) *What is burning?*

Gasoline–hydrocarbon material. The 3% x 3% ARC can be used and will be proportioned at a 3% concentration.

2) *Determine the surface area involved.*

Area = (.8) (Diameter^2)

Area = (.8) (125 ft)^2

Area = 12,500 ft^2

3) *Determine the appropriate foam application rate.*

The fire department is using portable application devices—foam cannons. The foam application rate is 0.16 gpm/ft^2

Foam Application = Area x Recommended Application Rate

Foam Application = (12,500 ft^2) (0.16 gpm/ft^2)

Foam Application = 2,000 gpm

4) *Determine the duration of foam application.*

Gasoline has a flash point of approximately minus 45°F. Therefore, the recommended duration is 65 minutes.

Required Foam Solution = Foam Application x Duration

Required Foam Solution = (2,000 gpm) (65 minutes)

Required Foam Solution = 130,000 gallons of foam solution

5) *Determine the quantity of foam concentrate required.*

The fire department is using a 3% foam concentrate.

Required Amount Foam Concentrate = Required Foam Solution x 3%

Required Amount Foam Concentrate = (130,000 gallons) (0.03)

Required Amount Foam Concentrate = 3,900 gallons of foam concentrate

The required quantities of foam concentrate should be in place before fire extinguishing operations start. Also, ensure that proportioning equipment is available for correctly proportioning the required foam at the prescribed delivery rates.

NOTES

DETERMINING WATER SUPPLY REQUIREMENTS

Flammable liquid storage tank fires can require tremendous quantities of water for a sustained period of time. Before an effective fire attack can be made, it must be determined if the water system is capable of:

- Delivering a water flow rate equal to or greater than that required to control the largest potential fire area.
- Delivering the required flow rates at pressures that can be used effectively by water application devices such as fixed systems, portable monitors, and handlines.

There are several assumptions regarding water supply requirements which are often misunderstood by emergency response personnel. These include:

1) **ASSUMPTION: The fire water system is known to be adequate for fire attack because all of the past fires at the facility have been successfully controlled.**

   REALITY: Most hydrocarbon fires never reach their full potential because they are extinguished in their incipient or small stages. Therefore, they were controlled well before they reached their full-scale potential and never really taxed the water system.

2) **ASSUMPTION: The water flow rate available in the facility is believed to be adequate because the rated flow of the facility's fixed fire pumps exceeds the maximum foreseeable water flow demand.**

   REALITY: Fire pumps are not constant flow and constant pressure devices. Therefore, the sum of their rated flows is not equal to the water flow rate available in the facility. When delivering water, fire pumps must provide sufficient pressure to operate water application devices in the area of the fire and to overcome pressure losses in the piping system. If this pressure is higher than the pumps' rated pressure, they will flow LESS water than the sum of their flows. If the water flow rate demand exceeds the rated flow of the pumps, the pressure available in the fire area will decrease, with possible effects on the operation of water application devices.

   As a result, the water flow rate available in each area of a facility varies depending upon the area's relative elevation and location with respect to the pumps and the pressure requirements of the water application devices to be used. In addition, experience shows that some water systems have been modified or damaged over time. While they may have been properly designed and installed, years of neglect may have rendered them ineffective.

3) **ASSUMPTION: The normal/static system pressure is an adequate indicator of the system's capability (e.g., the system is adequate because it has a pressure of 100 psi when no water is flowing).**

   REALITY: As soon as water begins to flow, the operating pressure will drop. The greatest pressure drop will be in the area where the water is being used. How far the pressure drops will be dependent on the pressure available at the system's water source(s), the distance between the fire and the system's water sources, the size and arrangement of the piping in the fire water system, and the water flow rate. Remember, the greater the flow rate, the greater the pressure drop.

See Figure 10-10 for an example of how these three problems can work against emergency responders.

NOTES

HOW A FIRE WATER SYSTEM REACTS TO THE DEMANDS OF HYDROCARBON FIREFIGHTING OPERATIONS

During the early stages of a fire, several monitors and handlines may be placed in service to protect exposures. As a result, the available water pressure in the immediate fire area drops slightly. See point A in the graph below.

When it is realized that the initial fire attack cannot be handled with monitors and handlines, additional water streams are ordered into service. As these additional hoselines and devices are placed into service, the water pressure continues to drop even further. At this point one of two outcomes is possible: (1) the flow required to control the fire is obtained and the fire is brought under control with the flow and water pressure remaining fairly constant; or (2) the water pressure in the area begins to fall below that required to maintain full firefighting effectiveness. See point B.

At this point, the overall water flow rate cannot be significantly increased because placing additional devices in service will further lower available water pressure, causing the flow out of every line or device already in operation to decrease. As a result, the fire continues to grow, possibly out of control, while the water flow rate remains fairly constant. See point C. This is characteristic of an inadequate water system.

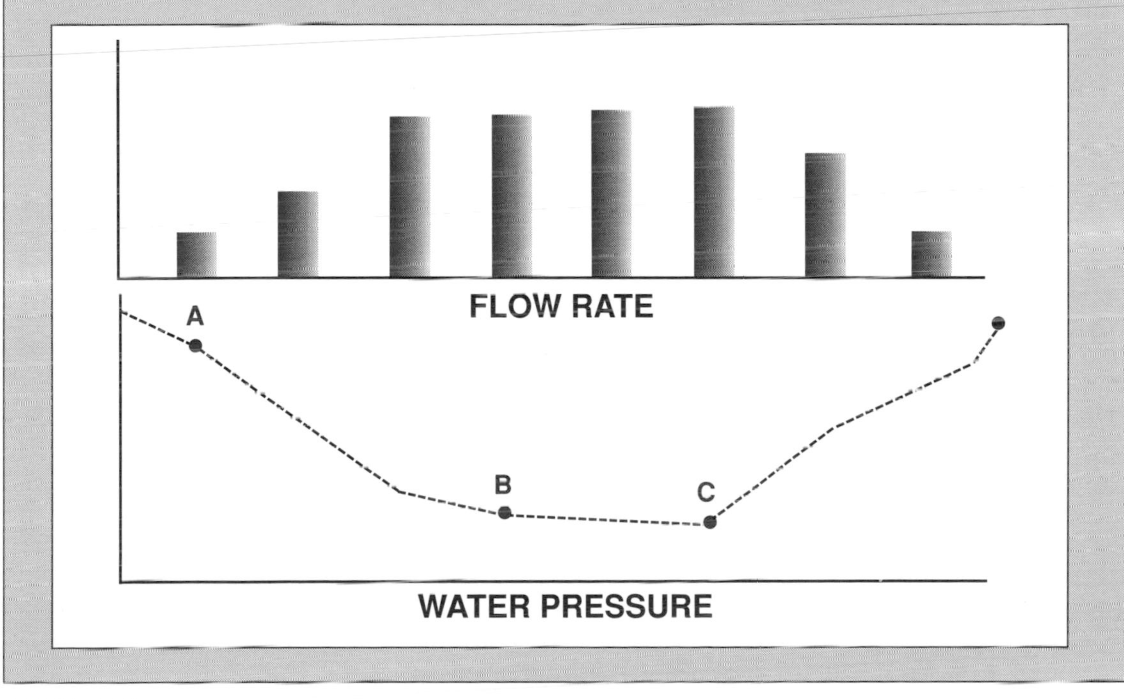

FIGURE 10-10 REPRINTED WITH PERMISSION OF LOSS CONTROL ASSOCIATES, INC.©

NOTES

DETERMINING COOLING WATER REQUIREMENTS FOR EXPOSURES

Water applications for exposure protection usually start before foam application starts. Cooling water for exposure protection may come from both potable or nonpotable sources. As previously noted, exposure lines should be applied when there is direct flame impingement on exposed tanks and/or when radiated heat is sufficient to cause steam at the tank shell when water is applied. Exposure streams should be shut down when steam is no longer produced from the metal surface.

Basic cooling water guidelines for *exposed* tanks and pressure vessels are as follows:

- Atmospheric storage tanks up to 100 ft. diameter require 500 gallons per minute.
- Atmospheric storage tanks from 100 ft. diameter to 150 ft. diameter require 1,000 gallons per minute.
- Atmospheric storage tanks exceeding diameters listed above require 2,000 gallons per minute.
- Pressure vessels should have a *minimum* of 500 gpm applied at the point of fire impingement. This is a widely quoted number which has proven to be a reliable guideline over time. Taking action with less water using offensive and defensive tactics increases risk to personnel significantly. However, lower flow rates may still be effective if they are applied from fixed systems and can be activated without risk to ERP.

FLAMMABLE GAS EMERGENCIES

Flammable gases can involve a wide variety of both fire and hazmat situations. While most responders are very familiar with the potential BLEVE hazards of LPG containers, they often fail to see comparable hazards when dealing with smaller containers with similar hazards. In addition, flammable gases may also possess multiple hazards. Consider the following examples:

- Common Flammable Gases—Propane, butane, natural gas, compressed natural gas (CNG), liquefied natural gas (LNG), liquid hydrogen, methane.
- Flammable and Toxic Gases—Hydrogen sulfide, ethylene oxide, phosphine, arsine.
- Flammable and Corrosive Gases—Methyl mercaptan.
- Flammable and Reactive Gases—Butadiene, acetylene (without acetone stabilizer), silane.
- "Fooler Gases"—Anhydrous ammonia.

HAZARD AND RISK EVALUATION

Keeping track of times and key events is critical. Flammable gas incidents often start out as vapor releases which eventually ignite once they come in contact with an ignition source. It is important that the Incident Commander acquire the following information during the size-up process:

NOTES

- Time which the incident started. Remember, this may not necessarily be the same time the incident was reported!
- Time at which responders arrived on scene.
- Probability that the fire will be confined to its present size.
- Layout of the incident, including the following specific points:

  Size and type of pressure vessel involved. Common pressurized storage tanks include horizontal bullet tanks and spheres. In addition, there are several different railroad tank cars used for transporting flammable gases, as well as the MC-331 high pressure cargo tank truck. See Chapter Six, Problem Identification for additional information.

  Valves and piping systems stressed or destroyed by the fire.

  Presence of any fixed or semi-fixed fire protection systems, such as water spray systems, lab cylinder cabinets, etc.

  All surrounding exposures, including other pressure vessels, tanks, buildings, process units, utilities, etc. This should include identifying and prioritizing exposures (e.g., flame impingement, radiant heat exposure).

Hazard and risk evaluation is essential to the safety of responders and the effectiveness of the response effort. Risk factors will include:

- Nature, quantity, and pressure of the gas involved.
- Design and construction of the container/pressure vessel. Factors should include the container size, type of the container breach, types of safety relief devices present (e.g., safety relief valve, frangible disk, fusible plug), and the ability to isolate the source of the fire/leak.
- Type of stress upon the container or piping system. Although thermal stress is the primary concern in a fire situation, mechanical stress (e.g., overturned tank truck, derailed tank car) may be an equal concern.
- Size and type of area being affected by the fire, as well as the likelihood that the fire will be confined to its present size.
- Identifying and prioritizing exposures. Remember that the highest priority should be given to flame impingement on vessels, piping, and critical support structures.
- Resources available to combat the problem, including the ability to rapidly apply sufficient water to the point of flame impingement.

TACTICAL OBJECTIVES

Based upon the type and nature of risks involved, the Incident Commander will implement the appropriate strategic goals and tactical objectives. Tactical priorities for managing a flammable gas fire are to:

1) Protect primary and secondary exposures to the fire.
2) Isolate the flammable gas source feeding the fire.
3) Reduce the operating pressure of the line feeding the fire.

NOTES

4) Permit the fire to self-extinguish and consume residual flammable gas inside the vessel or piping system.
5) Control and extinguish secondary fires.

Tactical options for flammable gas fires include the following:

- **Nonintervention**—This is a "no win" situation in which responders cannot positively change or influence the sequence of events. The best example is an imminent BLEVE situation. BLEVE means different things to different people, including Blast Levels Everything Very Effectively. See Figure 10-11.

JIM YVORRA

FIGURE 10-11: MEMBERS OF THE PRINCE GEORGE'S COUNTY, MARYLAND HAZMAT TEAM WITHDRAW FROM AN LPG TANK TRUCK ABOUT TO BLEVE

- **Defensive Tactics**—These tactics involve protecting exposures and allowing the fire to burn. This may include implementing a controlled burning of any remaining gas or flaring off the product.

  As with flammable liquids, both direct flame impingement and radiant heat exposures are critical concerns. Critical exposures must be identified and prioritized in a rapid manner. Flame impingements must be cooled immediately, while radiant heat exposures should be handled as soon as possible. Use fixed systems and unmanned portable monitors whenever possible; they are a "Force Multiplier" which frees up personnel for other critical tasks. They also reduce risk to ERP.

- **Offensive Tactics**—These tactics are implemented when sufficient cooling water and related resources are available for a continuous, uninterrupted fire attack. Although the primary focus is toward isolating the fire and the source of the gas, it will be necessary to maintain exposure lines.

TACTICAL CONSIDERATIONS AND LESSONS LEARNED

NOTES

- The nature of pressure vessels and their associated containment systems makes them vulnerable to external heating. A pressure vessel may be thermally stressed and activate one of its safety attachments (e.g., safety relief valve, frangible disk, etc.). As the heat is transmitted internally, the container pressure will increase proportionately. Eventually the gas will escape to the outside atmosphere through one of its attachments, or it will fail violently.
- Pressure-fed flammable gas fires may produce direct flame impingement on nearby vessels and cause catastrophic tank failure within 5 to 20 minutes of exposure. BLEVE's of bulk containers and process vessels can produce severe fire and fragmentation risks within 3,000 feet of the failed container. The IC must evaluate BLEVE situations early in the fire and take immediate action to activate fixed water spray systems, monitor nozzles, provide cooling streams, and evacuate the area.
- The IC must determine (1) if an *adequate water supply* is available to deliver the volume of water required for cooling exposures; and (2) if there is *adequate pumping capacity* to provide the required water pressure at the fire scene. If an adequate water supply is not available for cooling primary exposures, the IC should consider immediate withdrawal of all personnel from the hazard area.
- Effectively placed hose streams can lower the pressure in most small containers (e.g., cylinders); however, the risks associated with advancing hoselines or monitors may exceed the possible benefits.
- Any decision to approach a closed container showing direct flame impingement on its vapor space must be made on a case-by-case basis after evaluating the hazards and risks. Large LPG containers (8,000 to 30,000+ gallons) have violently ruptured and traveled over 3,600 feet, <u>even with functioning relief valves</u>. The November 19, 1984 San Juanico, Mexico LPG Plant fire and explosion killed over 500 people, including the majority of the firefighting team. One LPG bullet tank which BLEVE'd was propelled 3,937 feet into a two-story home. LPG tanks can fail within 10 to 20 minutes of direct flame impingement or even days after the container was initially stressed (e.g., during wreck-clearing operations such as at the Waverly, Tennessee incident).
- Exposures to high velocity jet flame (i.e., pressure-fed fire) will require 500 gallons per minute of water at the point of impingement. Exposures to radiant heat with no direct flame contact will require 0.1 to 0.25 gallons per minute per square foot to maintain the integrity of the exposure.
- Never extinguish a pressure-fed flammable gas fire. Isolate the source of the gas and permit the fire to self-extinguish, thereby consuming any residual gas inside the vessel or piping system. Unignited flammable gases escaping under pressure will rapidly form an unconfined vapor cloud which will usually be re-ignited by ignition sources in

NOTES

the area. Explosions of unconfined vapor clouds can cause major structural damage and quickly escalate the size of the emergency beyond responder capabilities.

FLAMMABLE GAS EMERGENCIES AT PETROCHEMICAL FACILITIES

Incidents at facilities which manufacture, process, store, or use large quantities of flammable gases can create specialized problems for responders. The IC must evaluate the overall fire and hazmat problem, recognizing that products other than flammable gases may be involved.

Specifically, the Incident Commander should determine the following:

- Were there any abnormal operating conditions immediately before the emergency?
- Were there any equipment problems or changes immediately before the emergency (e.g., maintenance operations, changing over pumps, blinding off lines)?
- Are exposures protected with fixed water spray systems or monitors? Are systems operating?
- Are fixed fire protection and chemical mitigation systems available (e.g., scrubbers and neutralizers)? Have they been activated?
- What is the status of the fire pumps? What is the fire water system pressure?
- Is the process isolated?
- What is the structural stability and potential failure of the unit? Is fireproofing in place? Experience in the petrochemical industry shows that in a major fire:
  - Instrumentation lines can begin to fail within 5 minutes.
  - Pressure vessels and other closed containers can begin to fail within 10 minutes.
  - Structural steel will begin to fail within 15 minutes.
- What is the status of the process unit? Has the process been isolated? Is the process stable (e.g., temperature, pressure, and reactions)?
- What types of safety systems are in place? Have they been activated? These would include emergency shutdown systems, pressure relief devices, flares, scrubbers, etc.
- What is the status of the utility systems, including electrical, instrument air, steam, fuel gas, and so forth? **Isolating utilities without coordinating with facility process personnel may create upstream and downstream secondary and tertiary problems greater than the initial event.**

REACTIVE CHEMICALS—FIRES AND REACTIONS

Fires involving reactive chemicals, as well as chemical reactions which have occurred during emergency operations, have historically resulted in numerous responder injuries and deaths. Examples include Norwich, CT, Roseburg, OR, Kansas City, MO, and, most recently, Newton, MA.

NOTES

In Chapter Seven we noted that when a hazmat container breaches, there are only two things jumping out at you—energy and matter. With reactive chemical families, tremendous amounts of energy can be released. Reactive chemical families include oxidizers, organic peroxides, certain flammable solids, pyrophoric materials, and water reactive substances.

Chemical reactivity can cover many hazards and properties. These may include the following:

- **Stability**—The resistance of a chemical to decomposition or spontaneous change. Unstable materials would include those which can rapidly and/or vigorously decompose, polymerize, condense, or become self-reactive.
- **Incompatibility**—The chemical reactions and the products of the reactions as a result of the incompatibility will vary with the nature of the chemicals involved. Typical hazards may include the release of toxic materials, the release of flammable materials, the generation of heat, and the destruction of materials.
- **Decomposition Products**—May be produced in dangerous quantities as a result of a chemical reaction or thermal decomposition. These would include toxic products of combustion (e.g., carbon monoxide, hydrogen cyanide, hydrogen chloride) as well as off-gases created by a chemical reaction.
- **Polymerization**—When polymerization occurs spontaneously or without controls, it can give off tremendous levels of both heat and pressure. Unplanned polymerization can occur as a result of environmental conditions (excessive heat), the depletion of inhibitors, or the inadvertent introduction of a catalyst and can lead to detonation.

HAZARD AND RISK EVALUATION

The hazard and risk evaluation process for reactive chemicals should focus upon the following factors:

- Hazardous nature of the material involved. Key factors would include the nature of its reactivity (e.g., water, air, heat, other materials, etc.).
- Quantity of the material involved. Although risks are often greater when dealing with bulk quantities, small quantities of highly reactive chemicals can pose significant risks. Remember—there is a very fine dividing line between explosives, oxidizers, and organic peroxides. All are capable of releasing tremendous amounts of energy! See Figure 10-12.
- Design and construction of the container. The type of container will vary depending upon whether the chemical is a raw material, an intermediate material being used to form another chemical or product, or the finished product.
- Fixed or engineered safety systems. Reactive chemical processes and facilities may have a number of engineered safety systems in

NOTES

U.S. EPA

FIGURE 10-12: REACTIVE CHEMICALS CAN RELEASE TREMENDOUS AMOUNTS OF ENERGY. CONTROLLED PHOSPHORUS EXPLOSION AT FORT A.P. HILL.

place, including explosion suppression systems, explosion venting systems (e.g., blowout panels), holding tanks, flares, and scrubbers. Individual containers may also have safety relief devices based upon the nature of the hazard.

- Type of stress applied to the hazmat and/or its container. A number of chemical families will react to the presence of heat and elevated temperatures (not necessarily a fire). Likewise, other chemical families (e.g., oxidizers) can become contaminated with water or dirt, develop heat within their container, and overpressurize the container.
- Size and type of area being affected by the fire or reaction, as well as the likelihood that the problem will be confined to its present size.
- Identifying and prioritizing exposures. This should include both the proximity of exposures and the rate of release.
- Level of available resources. Incidents involving reactive chemicals will typically require the expertise of technical information and product specialists who are inherently familiar with the materials involved.

TACTICAL OBJECTIVES

NOTES

Based upon the type and nature of risks involved, the Incident Commander will implement the appropriate strategic goals and tactical objectives. Unlike flammable liquids and gases, however, tactical options will often be limited unless the fire or reaction can be handled in its initial stages.

Tactical options include the following:

- **Nonintervention**—This is a "no win" situation in which responders cannot positively change or influence the sequence of events. As a result, responders withdraw to a safe distance and allow the incident to run its natural course. A classic example of nonintervention would be a vehicle or structure in a remote area containing oxidizers and organic peroxides and which is heavily involved in fire. Given the limited benefits, this should be a "no brainer" type of decision.
- **Defensive Tactics**—These tactics involve protecting exposures and allowing the fire to burn or the reaction to run its course. In some instances, responders must play a "wait and see" game. Defensive tactics may include implementing a controlled burn, remotely transferring the product to another container, remotely injecting a stabilizer into the reaction, or disposing the decomposition products to a flare or scrubber.

  If a building containing reactive chemicals is heavily involved in fire, the only option responders may have is to protect exposures. Do not underestimate the rapid speed at which these fires can move. High temperature accelerants, such as flammable metals and pyrotechnics, have recently been used in a number of arson fires in the United States and Canada. A full-scale test fire in a 30,000 ft^2 vacant shopping center in Puyallup, Washington resulted in the flashover of the entire structure within 2 minutes!
- **Offensive Tactics**—These tactics are implemented when sufficient resources are available to control and extinguish the fire. However, unless the fire or chemical reaction is observed in its initial stages, there are often limited offensive tactics which can be implemented to change the sequence of events. In addition, offensive tactics will expose responders to a significantly higher level of risk.

TACTICAL CONSIDERATIONS AND LESSONS LEARNED

- Don't wait to ask for help. Unless you work with these chemicals every day—and even if you do—quickly seek out information and assistance from product specialists. The costs of screwing up a reactive chemicals incident will often be measured in lives.
- Some reactive chemical incidents may require that both thermal and chemical protective clothing be used simultaneously. For example, a fire or spill of molten sulfur will also result in high levels of sulfur dioxide in the area. When combined with skin moisture and moisture within the respiratory tract, it will also form a mild acidic solution.

NOTES

- ❑ Water reactive materials can react explosively with no warning when they come into contact with even small quantities of water and moisture. When water is applied, results can include steam explosions, burning metal being thrown in all directions, and the production of flammable hydrogen gas, hydroxide compounds, and related toxic gases. Specialized extinguishing agents (i.e., dry powders) will be required for these situations. When dealing with smaller quantities, the best course of action may be to isolate the material outdoors and allow for a controlled burn.
- ❑ The only difference between some oxidizers and explosives is the speed of the reaction. When dealing with large quantities of strong oxidizers and organic peroxides, responders should consider treating the incident like an explosives fire (remember ammonium perchlorate at Henderson, NV?). Use distance to your advantage and keep the problem to the neighborhood of origin.
- ❑ Several major fires have occurred involving pool chemicals, such as calcium hypochlorite and chlorinated isocyanurates. Contamination of these materials can lead to the generation of toxic vapors, heat, and oxygen, which can lead to a fire. If large quantities are involved in fire, manufacturers recommend that large, copious amounts of water be used for fire extinguishment. Of course, this may involve a trade-off between air pollution (decomposition products are toxic and will travel a large distance) versus water pollution (runoff is toxic but may not travel as far and can be controlled).
- ❑ Controlled burning may be an appropriate tactical option if extinguishing a fire will result in large uncontained volumes of contaminated runoff, further threaten the safety of both responders and the public, or lead to more extensive clean-up problems. This option is sometimes used at fires involving agricultural chemical facilities or industrial facilities, where runoff may have significant environmental impacts upon both surface water and groundwater supplies.
- ❑ Chemical process operations at fixed facilities will often have a series of safety features in place in the event of a chemical reaction or decomposition problem. These may include the injection of certain chemicals to reduce the rate of polymerization or "kill" a chemical reaction, the use of emergency tanks in which product can be diverted, gas scrubbers to neutralize toxic or corrosive vapors before being released, and flares to burn off flammable vent gases.

SPECIAL TACTICAL PROBLEMS

CLANDESTINE DRUG LABS

Clandestine drug laboratories are manufacturing facilities set up by criminals for the sole purpose of making illegal drugs. They present new and unusual challenges for fire service personnel and hazmat responders. Drug labs can range from crude makeshift operations to highly sophisticated

NOTES

facilities, some of which are mobile. They can be set up anywhere and can be found in private residences, motel and hotel rooms, apartments, rental trucks, house trailers, houseboats, self-storage warehouses, and commercial occupancies. In many instances, labs are discovered by emergency personnel responding to what they believe is a routine building fire or medical emergency. The initial incident may also be reported as an explosion since many of the feed-stock chemicals used to make the drugs are flammable.

U.S. Drug Enforcement Agency (DEA) statistics show that the majority of clandestine laboratories produce three illegal drugs: methamphetamine (82%), amphetamine (10%), and phencyclidine or PCP (2.5%). As new "designer drugs" are developed in an attempt to circumvent controlled substance laws, other drugs will be produced.

IDENTIFICATION AND RECOGNITION

Federal statistics show that approximately 20% of drug labs are discovered as a result of a fire or explosion. Among the clues for the recognition and identification of a suspected drug lab are:

- Heavily secured doors or bars on the windows, painted or covered windows, and taped door jambs and windows. (Many labs operate 24 hours per day and the operators do not want to draw attention to the fact that the lights are always on.)
- Use of air conditioning, even in the winter months, in order to dissipate odors.
- Delivery of chemicals ranging from 5-gallon containers to 55-gallon drums, in unusual locations, such as residential areas.
- Discovery of similar containers, both empty or full, in areas not normally associated with chemical storage and usage.
- Destruction of plant life around a structure or area, as a result of chemical dumping or emissions.
- Discovery of large quantities of broken glass in the trash, as flasks must typically be discarded after three or four batches.
- Presence of chemical hardware within a structure, including Bunsen burners, heating mantles, beakers, and electric crock pots.

SPECIAL HAZARDS

A clandestine drug lab incident can simultaneously present emergency responders with a fire, a hazmat incident, and a crime scene. Four primary hazards can be present:

1) **Exposure to Fire and Chemical Hazards**—It is difficult to determine which chemicals would be used in any drug lab, as it would depend upon which specific drug(s) were being processed. Hazards present can include flammable and reactive chemicals, acids and caustics, strong oxidizers, and poisons.

   Chemicals are often categorized according to the function they serve in the manufacturing process:

NOTES

- *Precursors* are the raw materials, often a controlled chemical with serious health hazards. Examples include sodium and potassium cyanide.
- *Reagents* react chemically with the precursor but do not become part of the finished product. These often include chemicals that are flammable solids and water reactive, including lithium aluminum hydrate (LAH), magnesium metal turnings, sodium, potassium, and lithium metal.
- *Solvents* are used to dissolve, dilute, separate, and purify other chemicals. Solvents do not react chemically with a precursor or reagent and do not become part of the finished product. The most common solvents are toluene, methanol, ethyl ether, and acetonitrile.

2) **Exposure to Street Drugs and Precursor Chemicals**—Some of these drugs have already been implicated in long-term disabilities (e.g., Parkinson's Disease) among both users and law enforcement officers. There is some evidence, as in the case of PCP and its related precursors, of chemical and neurological disorders in children born to women who were exposed to PCP even before conception. This would include both users and responders.
3) **Bad Guys with Guns and Knives**—The people that work in illegal drug labs are the Bad Guys. They are usually heavily armed and won't think twice about killing the Good Guys just because they are unarmed and arrive at the scene in a fire engine with flashing lights on it. Always call for police assistance if you suspect a drug lab. Avoid the occupants and do not rely on them as a reliable information source.
4) **Weapons and Booby Traps**—Anticipate the presence of booby traps at a suspected drug lab site. As a deterrent to the competition, drug manufacturers sometimes booby trap their lab. Various types of bobby traps may be encountered, including:
   - *Chemical,* such as Armstrong's mixture (potassium chlorate and red phosphorus in an alcohol solution mixed in aluminum foil; when the alcohol evaporates, the compound becomes highly sensitive to shock and heat) and hydrochloric acid/cyanide mixtures.
   - *Electrical,* such as electrified fences, door mats, or door knobs and light bulbs filled with gasoline or smokeless powder inside refrigerators. When the refrigerator is opened, the light bulb filament energizes and initiates an explosion in a confined area.
   - *Mechanical,* such as trip wires anchored to an explosive device, spring-loaded boards with protruding nails, beat traps, etc.
   - *Physical,* including missing stairs and floors and razor wire.

A well-planned booby trap is usually concealed within another object which will not generate suspicion (e.g., a flash light or an AM/FM radio). Activation of the booby trap requires someone to make contact with or handle the device. Stay clear of drug labs until they have been thoroughly evaluated by specialists. **NEVER TOUCH OR HANDLE ANYTHING** in a drug lab until the area has been evaluated and cleared by Bomb Squad personnel who have the proper training in identifying booby traps.

EMERGENCY RESPONSE GUIDELINES

Response guidelines should follow DEA recommendations. Emergency response involvement at a clandestine operation will usually take two forms: (1) fire suppression operations at an actual fire, or (2) stand-by operations to support law enforcement personnel who are dismantling an active laboratory.

FIRE OPERATIONS

Special tactical considerations are needed at these incidents when fire is involved. Command should select an action plan that will minimize ERP exposures and contamination. Tactical considerations should include the following:

- Treat the fire as a hazmat incident. Position all units upwind, if possible.
- Have all personnel wear full protective clothing, including SCBA. Limit the number of personnel and exposure times during rescue and firefighting operations. Avoid standing in or breathing smoke and avoid contact with all spilled chemicals.
- Avoid interior attacks. In most instances, the risk will outweigh the benefits of fire extinguishment. Extinguishment should not be attempted unless personnel safety can be ensured.
- Use a minimum amount of water to avoid environmental damage and glassware breakage. Diking or other spill control measures may be required to minimize runoff and contamination of the area.
- Avoid overhaul activities until the hazards and risks are fully assessed. Make an attempt to preserve the crime scene for investigators. In some situations, chemical protective clothing may be required during this phase of the operation.
- Do not shut off utilities (electric, gas, or water) to equipment that could alter the cooking or cooling processes, since this action may cause an explosion. Don't touch or handle any items on the scene.
- Consider all fire suppression personnel who entered the hot zone and their clothing as being contaminated. Decontaminate all personnel.
- Decontaminate all injured personnel prior to transportation and notify the receiving medical facility of the specifics of the incident. Conduct a medical evaluation for all personnel actively involved in the incident.

Figure 10-13 provides some additional guidelines on lab entry operations.

GENERAL OPERATIONAL GUIDELINES FOR FIRE DEPARTMENT SUPPORT OF CLANDESTINE DRUG LAB ENTRY TEAM OPERATIONS

Hazmat personnel may either provide stand-by services to law enforcement personnel or be an integral element of a lab raid, closure, and dismantling operation. The following checklist was developed by Philip H. McArdle of the New York City Fire Department (FDNY) Hazardous Materials Unit 1 for operations at planned drug lab raids:

NIC, INC., ORIGINAL ART-MAX CRACE©

- ❑ Planning and Notification
  - Identify the possible chemicals in the suspected lab.
  - Determine the proper type and level of personal protective clothing and equipment.
  - Establish staging areas for emergency responders, including fire, EMS, hazmat, law enforcement, and other personnel, as necessary.
- ❑ Initial Entry Operations
  - Law enforcement responsibility only.
  - Protect exposures.
  - Isolate areas of egress into the lab.
  - Perform forcible entry, as necessary.
  - Apprehend the occupants.
  - Secure the perimeter and the area.
- ❑ Hazard and Risk Evaluation
  - Assess the chemical hazards with the lab, including the deactivation of chemical booby traps. This is typically both a law enforcement chemist and HMRT responsibility.
  - Deactivate other booby traps, including electrical, mechanical, and physical traps.
  - Preserve evidence.
  - Perform recon operations and continuously monitor the area.
- ❑ Mitigation and Processing
  - Take samples for chemical analysis.
  - Gather evidence.
  - Photograph and record the scene.
  - Separate and overpack related chemicals and hazardous materials.
  - Dismantle the drug lab operation.

- ❑ Termination Team
  - Review operational procedures and correct any deficiencies.
  - Provide security for the site.
  - Complete paperwork and documentation.
  - Arrange for the removal of chain-of-custody weapons, cash, drugs, and chemicals.

FOR MORE INFORMATION WE SUGGEST READING THE FOLLOWING:

Dyer, Bernard D. et. al., "Clandestine Drug Laboratories: Unusual Problems for the Fire Service." National Fire Academy Executive Fire Officer Program—Research Paper (April, 1988).

Emergency Resource, Inc., SURVIVING THE HAZARDOUS MATERIALS INCIDENT—SERIES II, Fort Collins, CO: Emergency Resource, Inc. (1991).

Federal Emergency Management Agency—National Fire Academy, Hazardous Materials Operating Site Practices Course—Student Manual, Emmitsburg, MD: FEMA.

Gonzales, Joe, DEATHTRAP!: IMPROVISED BOOBY-TRAP DEVICES, Boulder, CO: Paladin Press (1989).

Howard, Hank A., "Clandestine Drug Labs." FIRE ENGINEERING (August, 1986), pages 17–21.

International Association of Fire Fighters, Training Course for Hazardous Materials Team Members—Student Text, Washington, DC: IAFF (1991).

Joint Federal Task Force of The Drug Enforcement Agency, U.S. Environmental Protection Agency, and U.S. Coast Guard, GUIDELINES FOR THE CLEAN-UP OF CLANDESTINE DRUG LABORATORIES (May, 1989).

Hermann, Stephen L., "Drug Labs Are Just Another Hazardous Materials Scene." AMERICAN FIRE JOURNAL (February, 1990), pages 12–13.

James, Randolph D., "Hazards of Clandestine Drug Laboratories." FBI Law Enforcement Bulletin (April, 1989).

Karchmer, Clifford, "Illegal Drug Labs." FIREHOUSE (June, 1989), pages 77–88.

Krebs, Dennis, R., Kenneth C. Henry, and Mark B. Garbriele, WHEN VIOLENCE ERUPTS, St. Louis, MO: C. V. Mosby Company (1990).

Lazarus, Bruce D., "Emergency Response to Illegal Drug Labs." Workshop presented at the 1993 California Hazardous Materials Response Conference, Sacramento, CA (September 10, 1993).

Marquardt, Kathy, Pharm D. and Judith A. Alsop, Pharm. D. "Health, Heat and Water Hazards Associated with Illegal Drug Manufacturing." Report prepared by the University of California Davis—Regional Poison Control Center (March, 1991).

National Fire Protection Association, HAZARDOUS MATERIALS RESPONSE HANDBOOK (2nd edition), Boston, MA: National Fire Protection Association (1993).

Smith, Arthur, "Firefighting Today." FIREHOUSE (July, 1991), pages 48–50.

Washington State Interagency Steering Committee on Illegal Methamphetamine Drug Labs. MODEL LOCAL FIRE DEPARTMENT AND HAZMAT TEAM RESPONSE TO ILLEGAL METHAMPHETAMINE DRUG LABS (January, 1990).

FIGURE 10-13: NYFD CHECKLIST AND DEA SPECIAL WEAPONS AND TACTICS TEAM

NOTES

EXPLOSIVE DEVICE INCIDENTS

Explosive Device Incidents (EDI) usually involve the placement or detonation of an explosive for the sole purpose of hurting someone or creating media attention. Unlike accidents involving military explosives or munitions, most explosive devices are improvised and custom-made by the bomber. Consequently, they require highly trained specialists to evaluate and disarm/dispose of the device.

The emergency response environment is becoming increasingly violent. As we learned the hard way at the World Trade Center, a powerful explosive device is a relatively simple thing to make if the person has the motivation and is willing to take the risk. The explosive device may be made from chemicals and equipment which were legally purchased on the open market, or the device may be made using stolen military or conventional explosives.

Recent experience shows that explosive devices are being used more frequently by criminals as:

- Weapons of terror by drug dealers seeking revenge against rival drug dealers who are infringing on their marketing territory. The consequences of getting caught with the ingredients to make a bomb are sometimes less severe in some jurisdictions than the consequences of carrying a firearm.
- Terrorists who are politically motivated and seek to inflict the most damage which will bring the biggest headline to their cause.
- Weapons of revenge by rival gangs or Hate Groups.
- Threats against hostages taken in a barricade situation. There may or may not actually be an explosive device involved; however, since hostages are involved, the threat is considered to be credible.

Most public safety organizations handle explosive incidents using Special Operations Teams such as Bomb Squads, Explosive Devices Teams, or Explosive Ordinance Disposal (EOD) Teams. While most fire departments or industrial emergency response teams do not get directly involved in the "hands-on" aspects of explosive devices, there is an increasing role for Hazardous Materials Response Teams to support EDI operations.

SPECIAL PROBLEMS

Explosive Devices Incidents present numerous special problems for the Incident Commander which require that they be managed as a hazardous materials incident. Among the more significant problems are:

1) **Risk to the Public**—An explosive device can create significant property damage and casualties. The shock or "blast wave" from the explosion and fragmentation can injure people at distances up to 4,000 feet depending on the type of explosive used and where it was placed. Figure 10-14 provides some estimated effects of blast overpressures on structural elements and people. Remember that the risk of injury is not just from fragmentation; the blast creates a

shock wave that can do serious structural damage at fairly low pressures (e.g., pressures of 0.5 to 1.0 psi can blow out glass which can fall down onto people standing below). See Figure 10-14.

NOTES

Estimated Incident Pressure in p.s.i.	Effects on Structural Elements	Affects on People
0.5–1.0	GLASS WINDOWS SHATTER	PEOPLE KNOCKED OVER
1.0–2.0	CORRUGATED ASBESTOS SIDING SHATTERS	
1.0–2.0	CONNECTIONS FAIL ON CORRUGATED ALUMINUM SIDING	
1.0–2.0	CONNECTIONS FAIL ON WOODEN SIDING	
2.0–3.0	CONCRETE CINDER BLOCK WALL SHATTERS (NONREINFORCED 8 TO 12 INCHES)	EARDRUM RUPTURE
3.0–4.0	SELF-FRAMED STEEL PANEL BUILDING COLLAPSES	
3.0–4.0	CHEMICAL STORAGE TANKS RUPTURE	
4.4–5.1		LUNG DAMAGE
5.0–6.0	WOODEN UTILITY POLES SNAP	
6.5–7.0	LOADED RAIL CARS OVER TURNED	
7.0–8.0	BRICK WALL PANEL SHEARS AND FLEXES (NONREINFORCED 8 TO 12 INCHES)	
11–15		LETHAL

STEVEN J. TUNKEL, P.E., CHIEF TECHNICAL ENGINEER, HAZARDS RESEARCH CORPORATION

FIGURE 10-14: EFFECTS OF BLAST WAVE OVERPRESSURES ON STRUCTURAL ELEMENTS AND PEOPLE

NOTES

2) **Risk to Emergency Responders**—The risk to emergency responders is significant at an explosive device incident. Beyond the obvious risk to Bomb Squad members, firefighters, medics, and police officers engaged in support operations can be exposed to bomb and debris fragmentation, falling shards of glass from buildings, etc. Strict Hazard Control Zones and personnel accountability must be maintained.

3) **Political and Media Attention**—The consequences of an explosive device incident can be serious in terms of public risk. The incident can be disruptive to the community, create fear, or motivate "Copy Cat" incidents. Explosive device incidents are often seen as sensational—good kindling for a fiery story on the 6 o'clock news.

   Explosives incidents require a strong PIO presence and good media relations. The public wants to see that emergency services are professional and in control of the situation. A good command structure can help project that image through the media (we are in control and we know what we are doing).

4) **Coordination Among Agencies**—While there are some exceptions, most explosive device incidents cannot be managed by a single agency. At a minimum, there must be close coordination between law enforcement, the fire department, and EMS agencies. Figure 10-15 provides some simple operational guidelines for coordinating explosive device incidents.

FIGURE 10-15

GENERAL OPERATIONAL GUIDELINES FOR FIRE DEPARTMENT SUPPORT OF BOMB SQUAD OPERATIONS

PRINCE GEORGE'S COUNTY FIRE DEPARTMENT

The following guidelines are excerpted from the Prince George's County, Maryland Fire Department Hazardous Devices Unit (Bomb Squad) Incident Procedures.

FIRST ARRIVING FIRE DEPARTMENT UNITS

- ❑ First arriving emergency response units should report to a location at least 200 yards away from the reported incident location and establish a Staging Area for fire suppression and EMS units. Under no circumstances should personnel approach the area of the suspected package or device without the specific approval of the responding Bomb Squad Commander.
- ❑ Ensure that all personnel remain on their apparatus and wait for the arrival of the Fire Department Bomb Squad.
- ❑ Provide a preplan drawing and area map of the incident location to arriving Bomb Squad personnel.

INCIDENT COMMANDER

- ❑ Assumes Incident Command and establishes a Command Post.
- ❑ Based on available information, the IC will classify the severity of the incident as life threatening or non-life threatening.
- ❑ Identifies the location of the initial evacuation perimeter. Consideration must be given to the nature of the threat; reported location of the item; and the explosive consequences to the surrounding area if there was an explosion, detonation, or a chemical release. Regardless of the area or location of the suspected package or device, evacuation of the area surrounding the device shall be a minimum of 200 yards. Larger areas may be justified. Total evacuation of multi-story occupancies shall be considered on a case-by-case basis.
- ❑ Utilize on-scene Command Officers to ensure coordination of all operations in the "Outer Perimeter."
- ❑ Ensure coordination with all Police and Governmental agencies involved.
- ❑ Designate Sector Officers to include:
    - Entry Sector
    - Evacuation Sector
    - Investigations Sector
    - Operations (Inner Command—Bomb Squad) Sector
    - Emergency Medical Services Sector
    - Fire Suppression Sector
    - Media (Public Information) Sector
- ❑ Ensure that Sector Officers are provided with a sector responsibilities checklist.
- ❑ Provide regular updates to the Bureau of Fire & Rescue Communications which shall include the nature of the incident and its anticipated duration; a list of all streets which will be closed throughout the incident; a list of all business occupancies evacuated or affected by the incident.
- ❑ Notify the Bureau of Fire & Rescue Communications prior to any "disrupt" operations conducted by Bomb Squad personnel.
- ❑ Make a general announcement to all personnel operating on the incident scene, restricting the use of communications equipment, prior to any Bomb Squad personnel entering the Hot Zone.
- ❑ Ensure that the appropriate Fire Department Command Staff personnel are provided with periodic updates.

FIGURE 10-15

FOR MORE INFORMATION WE SUGGEST READING THE FOLLOWING:

Some people don't like the fact that you can buy books on how to make bombs that kill innocent people. We happen to agree with this opinion. Nevertheless, we strongly believe that the Good Guys should be as well trained as the Bad Guys. Bomb Squads and Special Weapons and Tactics Teams have recommended these references to us as background information for HMRT's. They help identify problems to avoid. If this is an area of interest to you, we can recommend the following publications. They are available through NIC, Inc. Law Enforcement Supply, 500 Flournoy Lucas Road, Building # 3, P.O. Box 5950, Shreveport, Louisiana 71135-5950, telephone (318) 688-1365 or fax (318) 688-1367. Some publications have a restricted distribution and may require special ordering procedures (e.g., proper identification or an order placed on department letterhead).

Benson, Ragnar, EOD IMPROVISED EXPLOSIVES MANUAL, Boulder, CO: Paladin Press (1990).

Benson, Ragnar, HOMEMADE C-4: A RECIPE FOR SURVIVAL, Boulder, CO: Paladin Press (1990).

Benson, Ragnar, RAGNAR'S GUIDE TO HOME AND RECREATIONAL USE OF HIGH EXPLOSIVES, Boulder, CO: Paladin Press (1988).

Desert Publications, IMPROVISED MUNITIONS BLACK BOOK—Volume 1, El Dorade, AZ: Desert Publications (1978). (NOTE: Text was originally published by the Frankford Arsenal).

Dmitrieff, George, EXPEDIENT HAND GRENADES, El Dorado, AZ: Desert Publications (1984).

Gonzales, Joe, DEATHTRAP!: IMPROVISED BOOBY-TRAP DEVICES, Boulder, CO: Paladin Press (1989).

Lecker, Seymour, DEADLY BREW: ADVANCED IMPROVISED EXPLOSIVES, Boulder, CO: Paladin Press (1987).

Lecker, Seymour, IMPROVISED EXPLOSIVES: HOW TO MAKE YOUR OWN, Boulder, CO: Paladin Press (1985).

Myers, Lawrence, W., COUNTER BOMB: PROTECTING YOURSELF AGAINST CAR, MAIL AND AREA ENPLACED BOMBS, Boulder, CO: Paladin Press (1991).

Powell, William, THE ANARCHIST COOKBOOK, Barricade Publications (1971).

Saxon, Kurt, POOR MANS JAMES BOND—Volume I, Alpena, AZ: Atlan Formularies (1991).

Uncle Fester, HOME WORKSHOP EXPLOSIVES, Port Townsend, WA: Loompanics Unlimited (1990).

We also recommend THE BLASTERS HANDBOOK, by the E. I. DuPont Company, Wilmington, DE. This is an excellent handbook on the basic principles of explosives. It is well illustrated and is updated on a regular basis.

FIGURE 10-15

TIRE FIRES

Each year, an estimated 240 million vehicle tires are disposed in the United States. Of this number, 72% are stockpiled, while 28% are recycled into other products and uses. The management of scrap tires has become a major economic and environmental issue. New technology is helping to solve the problem by recycling tires; however, there are millions of tires stored throughout the country in dumps, vacant lots, and abandoned buildings. Unfortunately, the solution to many tire disposal problems has been arson.

Tire fires present special problems which often tax the capabilities of local resources. Even relatively small tire fires can require a large resource commitment to resolve the problem. Tires themselves are a not hazardous

NOTES

waste; however, when they burn they produce hazardous byproducts and should be treated as a hazmat incident.

Tire pile fires can be divided into three distinct stages of combustion:

- **Ignition and Propagation Phase**—Tires give off flammable vapors at approximately 1,000°F (538°C). However, once a flame front has developed and elevated temperatures are applied to a large area with a constant radiant heat flow, tires can decompose as low as 410°F (210°C). Once ignited, a tire pile initially burns at a rate of 2 square feet every 5 minutes in the windward direction.
- **Compression Stage**—After the first several minutes of the fire, the top layers of tires will begin to collapse into strips. During this stage, heat and smoke levels increase dramatically. In large, high tire piles, the piles will start to collapse in on themselves within 30 to 60 minutes. Compression causes open flaming to slow down as the internal areas of the tire pile receive less air. The pile continues to collapse, building downward pressure, and forms a semi-solid mass of rubber, tire cords, and steel. Equilibrium also starts to occur.
- **Equilibrium and Pyrolysis Stage**—At this point, the fire is a deep-seated internal fire with low open flames on the surface. The fire is in equilibrium when the level of fuel conversion is approximately equal to the amount of heat, fuel, and oxygen available. Internal temperatures are up to 2,000°F (1,100°C). Tire fires in this stage consume the fuel much more slowly and completely. It is also during this phase where the downward pressure starts to push oil and water runoff into the ground, water, and other areas depending upon the location of the fire.

HAZARDS

The following special hazards can be expected at tire fires:

1) **Health and Environmental Hazards**—The composition of tires varies from one manufacturer to another but usually contains natural and synthetic rubber polymers, oil, fillers, sulfur and sulfur compounds, phenolic resin, wire, clay, petroleum waxes, and pigments such as zinc oxide, titanium dioxide, etc. Studies of tire fires have identified a wide variety of emissions of significant quantities of hazardous byproducts of combustion. These can include polynuclear aromatic hydrocarbons (PNA's), xylene, and benzene. Heavy metals, such as cadmium, lead, and chromium, can also be found when metals are scrapped with tires.

   During the early stages of a fire, tires burn freely, generating large quantities of smoke. Toxic products can be carried downwind, creating public exposure problems which may require protection-in-place or evacuation. However, as combustion becomes less complete, the amount of organics emitted usually increases. This can create special safety problems as firefighters perceive that the risk of exposure has passed.

NOTES

Scrap tire piles can also pose a public health hazard. Tires are a breeding ground for spiders, rodents, and snakes. As the fire grows in intensity, the rats leave the sinking ship.

2) **Fire Hazards**—The average tire has a heating value of 15,000 BTU's per pound (double the output of Class A combustibles) and can generate intense heat during the early stages of the fire. Consider the fact that a tire dump with 5,000 tires has the potential to produce 75 million BTU's of heat. The average passenger car tire also produces about $2^1/_2$-gallons of oil. As a result, tire fires often result in a substantial flammable liquids fire and/or environmental problem.

3) **Physical Hazards**—Tire dumps can contain thousands of tires piled up to 40–60 feet high. The interior spaces of the tires trap oxygen which will continue to support the combustion process. As the burning tires pass through the free-burning stage, the pile begins to compress as individual tires polymerize. Cooling water applied to the outer layer of the burning pile tends to cool the surface while several feet below can reach temperatures in excess of 1,000°F. Walking on or near burning tire piles is extremely dangerous, as high temperature caverns can form below the surface and suddenly collapse.

EMERGENCY RESPONSE GUIDELINES

The following lessons have been learned at major tire fires:

- In a major fire, it is unlikely that initial resources will be sufficient to completely control the fire. Radiant heat will enhance the fire spread and complicate exposure control while limiting the ability of firefighters to approach with handlines.
- High volume fire flows may extinguish the surface fire but don't penetrate to the seat of the fire. In addition, these streams may push the fire into uninvolved areas and increase oil/water runoff problems. Variable gallonage, constant pressure fog nozzles are more effective than solid stream nozzles.
- The use of wetting agents and Class A and B firefighting foams are often of little value in fighting large tire fires.
- Once the fire gains a substantial headway into a tire pile, the best means of control and extinguishment are to use heavy equipment to either bury the pile or break the pile into smaller sections (i.e., divide and conquer).

Large tire fires may require that fire control and spill control strategies be simultaneously initiated. Specific tactical issues include the following:

- ❑ **Fire Control Tactics**—Small tire fires are not too difficult to extinguish if firefighting operations can begin in the very early free-burning stage of the fire (10 to 15 minutes after ignition). Foam has been used effectively for extinguishment early in the fire. However, once the tire pile becomes well involved and enters the advanced free- burning stage, the fire becomes extremely difficult to attack and extinguish. Many

NOTES

Incident Commanders are surprised how ineffective mass applications of water are after the fire has gained a foothold on the pile.

Water supply requirements for fire extinguishment create special problems because tire dumps are often located in remote areas with restricted access. Fire hydrants are rare, and long water supply lines are usually required. As a rule of thumb, an uninterrupted water supply of 1,000 gallons per minute will be required for every 50,000 cubic feet of tires in the pile. That would be approximately 60,000 gallons per hour, or 360,000 gallons of water over a 6-hour period.

- **Spill Control Tactics**—Tire fires will generate huge quantities of oil/water runoff as tires polymerize. The more tires involved, the larger the potential runoff problem. The 1983 Winchester, Virginia Tire Fire is ranked as one of the largest inland oil spills in U.S. history. Approximately 9 million tires produced over 500,000 gallons of runoff; over 250,000 gallons of runoff were recovered in the first two weeks. On the peak day during the initial stages of the fire, the average runoff flow rate at the bottom of the pile was approximately 90 gpm. Be prepared for runoff problems after tires begin to polymerize deep inside the pile.

  The runoff problem can be further complicated as water penetrates deep inside the burning pile and carries waste oil into the soil or out the bottom of the pile. This is one reason why water should not continue to be applied to large burning piles once it has been determined that the fire cannot be successfully extinguished. Dikes and retention ponds should be constructed early in the fire so that positions are not flooded and overrun.

- **Divide and Conquer Tactics**—If you are not willing to wait for the tire pile to burn itself out, then burying and separation become the primary response tactics. A major tactical objective should be to separate the unburned tires from the burning pile. Fire breaks at least 60 feet wide will be required, but even larger fire breaks may be necessary depending on the height of the burning pile and wind conditions.

  Smaller piles already on fire should be left alone to burn out. Larger, burning piles have been successfully separated using heavy equipment; however, this effort is usually unsuccessful without the right type of equipment. Tires are reinforced with steel wire. As the tire burns away, coils of wire remain and make working in and around the pile hazardous. Coiled wire can wrap around truck axles and drive shafts, damaging vehicles. Specialized heavy equipment, such as bulldozers and tracked excavators like the CAT-22 with a long reaching boom, have been used successfully to separate the burning piles.

  Burying the tire pile with dirt or other fill materials is no assurance that the fire will be extinguished. Tire fires can continue to smolder deep in the base of the pile for weeks (Winchester, VA burned for 6 months) and will require continuous observation and environmental monitoring. See Figure 10-16.

CASE STUDY

TIRE DUMP FIRE, WINCHESTER, VIRGINIA
October 31, 1983

At 00:30 hours on Monday, October 31, 1983, an arsonist set fire to a seven-acre tire dump located off Route 608 approximately five miles west of Winchester, Virginia. The fire quickly spread across the surface and eventually involved 9 million tires, which were piled 80 feet high on a 21% grade. At 15:00 hours on Monday, October 31, local officials requested assistance from the Virginia Office of Energy and Emergency Services (OEES).

OEES arrived on the scene at 17:30 hours and assumed command. After assessing the situation, the OEES Incident Commander requested the assistance from the National Response Center, which subsequently mobilized the U.S. Coast Guard Atlantic Coast Strike Team (USCG) and the EPA Region III Response Team (EPA).

At 18:30, on Tuesday, November 2, the EPA Region III Coordinator arrived on the scene and formed a Unified Command, which incorporated the Virginia OEES and Frederick County, Virginia agencies into the command structure. A site operations meeting was held at 19:00 hours to discuss fire-fighting operations and establish a protocol for managing the incident. Approximately 35 different federal, state, and local government agencies, contractors, and volunteer groups were involved in this initial meeting.

The major problems associated with this incident included:

1. There were 9 million tires on fire which could not be quickly extinguished. This created a major air pollution problem downwind of the incident which affected several communities.
2. The burning tires were generating huge quantities of oil, which was flowing out the bottom of the pile and entering a stream which flowed into the Potomac River. The cities of Hagerstown, Maryland and Washington, D.C. obtain their drinking water from the Potomac. Over 250,000 gallons of oil were recovered in the first two weeks of the incident. On the peak runoff day, 80,000 gallons were recovered.

In the first eight days of the incident, approximately $330,000 had been spent on operations. The average cost per day for the first two weeks of the incident was $45,000. Eventually, about 1.2 million dollars were spent simply to contain the fire and control the oil spill. The actual cost of the incident was much higher. The fire was eventually extinguished and the incident declared over in June 1984.

The following major lessons were learned from this incident:

- Pre-established written agreements which outline chain of command for internal and external local, state, and federal operations are necessary for a well-coordinated incident.
- Communications from Section and Branch level leaders are required on a frequent basis to coordinate resources and ensure that the response teams are working toward the same objectives (e.g., Is our objective fighting the fire or controlling the runoff?).
- Complete analysis of the problem from each organization's perspective is important before response objectives are selected and implemented.
- Re-supply and personnel rotation must be anticipated beyond a 24-hour operating period. Running a Unified Command with multiple agencies involved over a prolonged period of time without the proper rest creates stress among the players, which ultimately leads to bad tempers and a breakdown of the Unified Command concept.

Authors' Note: We spent about one week at the scene of this fire assisting with firefighting and oil spill control. During this period we had about four hours sleep per day. The first day was really exciting, the second day was interesting, the third day was a challenge, and by the fourth day the ordeal was starting to get to us mentally and physically. By the fifth day we had all the fun we could stand at one fire. If you work a major oil spill, you better pack a big lunch!

FIGURE 10-16

CONFINED SPACES INCIDENTS

A special study conducted by NIOSH in 1986 revealed that 60% of all fatalities in confined spaces involved rescuers. This is a staggering figure and says something about how ultra-hazardous confined spaces incidents are for emergency responders. A similar study of Selected Occupational Fatalities conducted by the Occupational Safety and Health Administration in July 1985 indicated that between 1974 and 1982 there were 173 fatalities involving workers in confined spaces. Of the 122 cases selected for a more detailed analysis, OSHA found that 56 people were killed as a result of toxic atmospheres in the confined space, 43 from asphyxiating atmospheres and 23 from other physical hazards inside the confined space. **Clearly, confined spaces incidents are extremely dangerous and should be managed using special operating procedures.**

CONFINED SPACES DEFINED

What is a confined space? From an emergency response perspective, it is important to understand what a confined space is and what makes it so dangerous. Recognizing that an area is "confined" is the first step.

NOTES

OSHA (29 CFR 1910.146) defines a confined space as "any area that has limited or restricted means for entry or exit; is large enough and so configured that an employee can bodily enter and perform assigned work; and is not designed for continuous employee occupancy." Examples of confined spaces and related rescue situations include:

FIXED STORAGE TANKS—These could include horizontal or bulk storage tanks or large-diameter open or covered floating roof tanks. Typical rescue scenarios involving these types of spaces include workers who have entered the space for tank cleaning or to perform maintenance and repair work. They may also involve inspectors who have entered the area for tank gauging or to inspect the tank for corrosion.

MOBILE TANK TRUCKS OR RAIL CARS—These could include mobile tank trucks or rail cars which have been entered by personnel to clean and inspect the inside of the tank before switch-loading to another product.

PROCESS VESSELS—These include ASME-rated pressure vessels used in petroleum refining or chemical processing. They are typically entered during plant turnarounds for cleaning and inspection. Access to process vessels may be at grade or in elevated locations, adding the additional high angle rescue problem to the scenario.

PITS—These areas include below-grade pits or ponds designed to contain hazardous waste runoff or drippings used in manufacturing processes (e.g., paint sludge, hydraulic fluid, etc). They also include pits designed for maintenance functions such as those found in automobile and truck repair shops. These spaces are periodically entered to remove heavy hazardous waste which has accumulated at the bottom of the pit.

SHIP AND BARGE COMPARTMENTS—These include below-deck areas, such as dry bulk cargo holds, liquid storage tanks, grain compartments, and bilges. On larger cargo ships, these areas may be many stories below deck and below the water line. They may be entered periodically by maritime inspectors and maintenance personnel.

VATS—These include large containers used in batch food processing, distilleries, and breweries. Vats may also have potentially energized internal equipment such as stirrers and paddles.

EXCAVATIONS—These could include small construction site trenches or very large below-grade excavations (e.g., garage areas for high rise buildings). Confined space rescue situations involving excavations typically have the added hazard of a cave-in.

REACTION VESSELS—These are vessels designed to contain the high or low temperatures and pressures involved in chemical processing. In addition to the obvious flammability, toxicity, and oxygen deficiency problems, they may also have the additional

NOTES

hazards of an exotic catalyst (e.g., platinum). They are normally entered during refinery or chemical plant turnarounds to replace or "recharge" the catalyst bed.

BOILERS—These include the full range of oil-, coal-, and gas-fired commercial and industrial boilers which are shut down periodically for inspection and maintenance.

VENTILATION DUCTS—These can include above- or below-grade metal or masonry structures designed to move large volumes of air. They may be in horizontal, vertical, or diagonal positions.

SEWERS—These include below-grade sanitary, storm, and combination sewers. Rescue situations may involve sanitation workers performing inspection and maintenance functions but may also include children who have entered these areas on "an adventure trip." (Naturally, we never did this type of thing when we were kids.)

TUNNELS—These could include small access tunnels carrying utilities or large transportation tunnels like subways and under-water or below grade highways.

PIPELINES—These include large-diameter bulk liquid and gas transmission pipelines.

SPECIAL HAZARDS AND RISKS

Confined spaces have several characteristics which make them ultra-hazardous. These include:

Hazardous Atmospheres —If you carefully consider the examples of confined spaces listed above, it is hard to imagine ever responding to a confined spaces incident where there won't be the potential for a hazardous atmosphere. Always approach a confined space rescue problem with the attitude that the atmosphere inside the space is flammable, toxic, and oxygen deficient or enriched. Even if monitoring instruments indicate that the atmosphere is entirely clean or within acceptable limits, handle the incident as if the atmosphere were contaminated.

Conditions can change rapidly and there is little or no chance that an ERP will be able to: (1) detect that conditions have become hazardous, and (2) don respiratory protection in time to prevent disorientation and death. Never enter a confined space without positive pressure SCBA or an approved Supplied Air Respirator with an emergency escape bottle, and never remove your facepiece to "revive the victim." If you remove your facepiece in a confined space in a contaminated atmosphere, you are going to die.

Limited Egress —Most confined spaces have only one way in and one way out. This makes an alternate route of escape impossible without forcible entry. If the primary route becomes blocked, you are trapped. Egress is further complicated by restricted access points, such as manways which may be poorly designed and placed at unusual angles. Small manways create special problems for entry teams using SCBA and

NOTES

specialized protective clothing. In some cases, the only way to enter the space is to use a Supplied Air Respirator. Removing the SCBA bottle and harness from the wearer's back while leaving the facepiece in place is a poor option. This is a very difficult task and is dangerous. Many inspectors and maintenance personnel have been overcome by a hazardous atmosphere when they temporarily removed their facepiece to crawl through a manway. Also, accidentally dropping the SCBA harness while crawling through the hole can yank the facepiece off of the wearer's head.

Extended Travel Distances—Many confined spaces require extended travel distances to enter and access the confined area to conduct search and rescue operations. The longer the travel distance, the greater the rate of air consumption. This ultimately translates to a short-duration stay inside the confined area. Much like a diver who must calculate air consumption time to allow for a safe ascent to the surface, a confined spaces rescuer must use special care not to overextend the stay and risk running out of air. Even 45-minute SCBA units will limit the actual on-site working time to five to ten minutes. The use of an SAR can make entry into the space easier and extend the search and rescue time, but these units are limited to 300 feet of air hose. This may not be very practical for confined spaces such as pipelines and sewers.

Unusual Physical Hazards—These can include the hazards of uneven or splippery walking and climbing surfaces, being struck by falling objects, or becoming trapped in inwardly converging areas which slope to a tapered cross section (e.g., a grain elevator chute).

Darkness—Almost every confined space is totally dark—you can't see your hand in front of your face. Search and rescue operations must be carried out using hand lights or portable lights which are brought into the space. The potential for a flammable atmosphere inside most confined areas means that all lighting must be intrinsically safe. In other words, the equipment must be suitable for NEC Class 1, Division 2 locations. How many hand lights does your organization have that meet these requirements? Remember that the typical flash light or hand lantern is an ignition source.

Poor Communications—Confined spaces are notorious for poor radio communications. Below-grade spaces and steel construction do not promote very good radio reception. Add the protective clothing and SCBA facepiece problem and good radio transmissions become virtually impossible. (One ERP we met said that it is like using two old peach cans and a mile of string.) Like hand lights, radio equipment used in confined spaces must meet NEC Class 1, Division 2 atmosphere requirements.

EVALUATING HAZARDOUS ATMOSPHERES

The purpose of evaluating a confined space for a hazardous atmosphere is to determine (1) if the area can be safely entered by emergency responders, and (2) the likelihood of conducting a successful rescue.

1. **Can the confined space be safely entered by emergency responders?**

   Confined spaces may be flammable, toxic, oxygen deficient, or some combination of these.

NOTES

Flammable Atmospheres—Most national standards consider a flammable atmosphere containing 10% or less concentration of flammable vapors as acceptable for conducting rescue operations. Concentrations between 10 and 20% are considered hazardous and should never be entered by rescue teams unless they have the proper PPE and respiratory protection and all electric equipment is rated for Class 1, Division 2 atmospheres. The explosive range of many hydrocarbon vapors range from a 1 to 10% vapor-to-air mixture; however, the explosive range for oxygenated materials like alcohols and glycols is wider. Any mixture of vapor and air between the UEL and LEL will ignite when exposed to an ignition source and should be considered as too dangerous for entry.

Toxic Atmospheres—An atmosphere which is above the OSHA Permissible Exposure Limit (PEL) or the ACGIH Threshold Limit Value (TLV) does not necessarily prohibit entry into a confined space to perform rescue operations. However, realize that the risk to the rescue team increases significantly and there is no margin for error. A rescuer who experiences damaged protective clothing or an air supply problem inside of a confined space with a toxic atmosphere faces almost certain injury or death. Making an entry under these conditions is a judgment call which must be made on a case-by-case basis by the person taking the risk.

Oxygen-Deficient or -Enriched Atmospheres—An oxygen-deficient confined space is considered by most standards to be one in which the oxygen content in air is less than 19.5%. Likewise, most standards recognize an oxygen-enriched atmosphere as being 23.5% or greater. The risk of entering an oxygen deficient atmosphere is similar to entering a toxic atmosphere. Obviously, the less oxygen content, the greater the risk to the rescue team if there is an air supply problem. Oxygen-enriched atmospheres present rescuers with a significant risk because an increase in oxygen content increases the risk of fire. If a flammable atmosphere is present, the explosive range will become wider.

All potentially hazardous atmospheres in confined spaces should be confirmed by monitoring instruments.

2. **Is the purpose of the hazardous entry search and rescue or body recovery?**

   When the human body is deprived of oxygen death occurs in 3 to 5 minutes. The presence of a toxic atmosphere accelerates the timeline. If a flammable hydrocarbon atmosphere is present, the PEL and the TLV-TWA will almost always be exceeded before 10% of the lower flammable atmosphere is reached. Therefore, the atmosphere will almost always be toxic before reaching the flammable concentration.

   As a general guideline, whenever the victim has been subjected to an oxygen deficient atmosphere of less than 19.5% or a flammable atmosphere of 10% or greater or a toxic atmosphere above the PEL/TLV for periods longer than 5 to 15 minutes, the Incident

NOTES

Commander should consider the possibility that there is no real chance for a successful rescue. The lower the oxygen content and the higher the toxic and flammable atmospheres inside the confined space, the less likely that the victim will survive for periods of exposure exceeding five minutes. As is the case with every medical emergency, the condition of the victim, age, pre-existing health conditions, etc., affect the chance for survival.

In addition to these basic medical parameters, the IC must consider the amount of time that it will take to set up and safely conduct search and rescue operations. Darkness, limited access, extended travel distances, and difficult working conditions cause delays which work against the victim's chances for survival.

EMERGENCY RESPONSE GUIDELINES

Conducting an effective confined space rescue requires bringing several emergency response specialties together at the right time, with the right training and equipment to get the job done within an almost unreasonable period of time. These specialties normally include:

- **Rescue Personnel** who are skilled in ropes, slings, harnesses, and confined space rescue techniques which will be required to find, package, and extricate the victim. Lifting a 170 pound person vertically 25 feet through a 30-inch manway wearing protective clothing is next to impossible if you: (1) don't know exactly what you are doing, and (2) don't have the special equipment which can provide the mechanical advantage necessary to remove heavy victims.
- **Hazmat Personnel** who are skilled in monitoring and evaluating hazardous atmospheres and decontamination of the victim and rescue team. If an HMRT is available to support rescue operations, it can be used to conduct a hazard and risk assessment of the confined space while the rescue team is setting up and preparing for search and rescue operations. If the IC gives the "green light" for entry, valuable time can be saved by splitting the duty between the Confined Space Rescue Team and the HMRT.
- **Medical Personnel** who are trained and qualified to provide the advanced medical care that a confined space victim will need. In addition to the obvious respiratory and cardiac problems, the victim may have associated trauma from falling and may require special antidotes and drug therapy.

Before approving a confined space rescue operation, the IC must assure that the following special precautions have been taken:

- ❑ **Confirm that there are victims inside of the confined space.** Are there multiple victims, where are they located, and what condition are they in? How long have they been inside the confined space and what have they been exposed to?
- ❑ **Appoint a Safety Officer and assure that all personnel are qualified for entry and are using the required PPE.** Each member of the rescue team should wear a safety harness. If a vertical rescue will be

required, the rescue harness should be connected to a fall-arresting system. In complex areas, like sewer systems, the entry team's movements should be tracked at the command post and marked on the command and control chart. See Figure 10-17.

NOTES

DELRAY BEACH FIRE RESCUE

Confined Space Entry Permit/Safety Plan Checklist

This checklist to be implemented as part of the entry preparation phase.

Location & Description:________

Purpose of Entry: ________

Date: ________

Incident Commander: ________

Safety Officer: ________

Entry Personnel: ________

Backup Personnel: ________

DETRICK LAWRENCE CORP.

Special Requirements	Yes	No		Yes	No
Lock-out/Tag out			Escape Harness		
Lines Broken–capped or blanked			Tripod w/ winch		
Purge – Flush & Vent			Lifelines		
Ventilation			Lighting		
Secure Area			Protective clothing		
SCBA			Monitoring		

Test to be Taken	P.E.L.	Time / Level	Time / Level	Time / Level	Time / Level	Time / Level
% Oxygen	19.5%–21%					
% L.E.L.	Any over 10%					
Carbon Monoxide	50 PPM					
Aromatic Hydrocarbon	10PPM					
Hydrocyanic Acid	10PPM					
Hydrogen Sulfide	10PPM					
Sulfur Dioxide	5 PPM					
Ammonia	25PPM					

Instruments Used: Name	Type	ID NO.

Responsible Party:

FIGURE 10-17: CONFINED SPACES ENTRY PERMIT/SAFETY PLAN CHECKLIST FROM DELRAY BEACH, FL FIRE DEPARTMENT

NOTES

- ❑ Make sure that the confined space has been de-energized. This means assuring that all valves are closed and blocked in so that hazardous materials do not continue to flow into the confined space. Make sure that rotating machinery is isolated (e.g. paddle wheels in mixers). If key control switches and valves are not locked out and tagged, post personnel at each location to ensure that these systems are not energized while rescue teams are inside the confined space.
- ❑ Make sure that flammability, toxicity, and oxygen deficiency or enrichment readings have been confirmed and are within acceptable limits based on Standard Operating Procedures. Take monitoring readings at different levels outside of the confined space openings if this is possible (e.g., high and low to determine if concentrations have accumulated at high or low points). Atmospheric conditions can change over time. Simply walking around inside the confined space can disturb the atmosphere and increase the flammability or toxicity level. For example, scale or sludge inside a petroleum storage tank can release flammable vapors when the tank is disturbed.
- ❑ If atmospheres are not within acceptable limits, begin ventilation of the confined space to bring conditions within an acceptable range. However, recognize that ventilating a space which is in the upper flammable range will bring the atmosphere into and possibly through the flammable range, thus increasing the risk of an explosion. Stop rescue operations when the atmosphere reaches an unsafe condition. Remember that all forced mechanical ventilators must be rated for Class 1, Division 2 atmospheres. See Figure 7-5 for a detailed description of the various NEC classifications. This disqualifies almost every smoke ejector on every truck company in America.

  Do not rely on water fog to reduce flammability inside of the confined space. Like forced mechanical ventilation, water fog can bring an atmosphere which is in a UEL condition into the flammable range by injecting air entrained in the fog stream into the space. The possibility also exists that you can drown the victim as water accumulates.

  When the IC determines that a confined space rescue is not possible and there is absolutely no doubt that the victim is dead, the operation should be formally stopped. A planning meeting with sector officers should be held by the IC to determine the safest method of conducting body recovery operations. Emphasis should be shifted to maximum safety. Remember that the same basic hazardous conditions still exist inside the confined space.

  Body recovery operations often provide safer options for the entry team which were not previously available because the victims safety would be jeopardized (e.g., inerting a space to assure that the atmosphere will not be flammable, forcing entry in the vicinity of the body, or floating the body out of the space by flooding the area with water, etc.). A meeting with personnel can also help rescuers make the mental shift required to slow down operations to a safer, more controlled pace.

NOTES

TRANSFER AND RECOVERY OPERATIONS

ROLE OF EMERGENCY RESPONDERS

When it is obvious that the incident scene will remain operational after hazmat control operations are completed, the IC should take additional steps to ensure that the transition from the Emergency Response Phase to the Restoration and Recovery Phase is made smoothly and safely.

Public safety emergency response teams have a fairly limited scope of responsibility for clean-up and site restoration. However, more than one Incident Commander has regretted having to return to the same site for "another accident." Sometimes it is prudent to leave a stand-by crew on the scene until clean-up has been completed by contractors.

If the hazmat incident is on public property (e.g., highway), police and fire agencies have a responsibility to assure that public safety is safeguarded during clean-up and recovery operations. For example, it may be necessary to verify that clean-up personnel have the proper HAZWOPER training and documentation.

At an industrial facility such as a chemical plant or refinery, the key players on the Emergency Response Team usually wear two hats. When the emergency is terminated, they may assume the role of plant safety or environmental coordinator and maintain an on-scene presence as other employees and contractors complete clean-up tasks. If public safety agencies have responded to the facility, they may conduct a joint briefing and turn the site over to facility safety representatives, who will supervise the clean-up from that point on.

The extent of ERP involvement in making the transition to the restoration and recovery phase will vary depending on the type of hazards present, location and extent of the incident, jurisdiction, etc. Regardless of the situation encountered, the following general activities should be considered during the restoration and recovery phase:

- ❑ **PPE Requirements**—Personal protective clothing and equipment requirements for restoration should be determined by the Incident Commander in consultation with the Safety Officer and health, safety, or environmental specialists. Requirements should be based upon the results of air monitoring, the potential for re-ignition, the specific tasks to be accomplished, and other related factors. PPE requirements should take into account the overall safety of restoration personnel, including mobility, comfort, and heat stress.
- ❑ **Contractor Safety**—The individual responsible for site safety should verify that any spill contractors used for clean-up and recovery operations have received training, as required by *29 CFR 1910.120, Hazardous Waste Operations and Emergency Response*. Many states now require that hazardous waste contractors be licensed. There are many excellent clean-up contractors in the field today; there are also a few bad ones. Don't contribute to an accident by allowing unqualified workers to handle your clean-up.

NOTES

- **Sampling**—Residual contamination of heavy equipment, soil, pavement, or groundwater may require further assessment to determine if they present a threat to public safety. Depending upon the nature of the incident, sampling may be used to determine whether the surfaces require decontamination. It may also be necessary to sample soil, surface water, sediments, or groundwater to assess environmental impact.

  Sampling procedures should follow regulatory guidelines, including careful sample technique documentation, chain of custody, and quality assurance/quality control (QA/QC) procedures. If contamination is found which will require specialized clean-up, a written remedial action plan should be developed.

- **Disposal Concerns**—It may be necessary to dispose of contaminated protective clothing, decontamination solutions, runoff water, or other materials that may be considered as hazardous waste following an emergency. An environmental specialist should be consulted for waste characterization and disposal, as appropriate.

SITE SAFETY AND CONTROL ISSUES

Product removal operations cannot commence until after the incident site is stabilized. Again, stabilization means that any fires have been extinguished, ignition sources controlled, and all spills and leaks have been controlled, as necessary. Specific site safety considerations which should be addressed during this phase of the incident include:

- When flammable or combustible liquids are involved, ensure that back-up crews with a minimum of $1^1/_2$ or $1^3/_4$-inch foam handlines and at least one 20- to 30-pound dry chemical fire extinguisher are in place to protect all personnel involved in the off-loading and uprighting operation.
- Always have an escape signal understood by everyone and a path for personnel working in the immediate hazard area.
- Continuously monitor the hazard area, as necessary.
- Ensure that all personnel remain alert. Both public safety and industry response personnel may sometimes become sloppy, less attentive, and may attempt shortcuts as the emergency extends over several hours. Frequent relief of personnel can usually minimize this problem.

PRODUCT REMOVAL AND TRANSFER CONSIDERATIONS

Specific procedures for product removal and transfer will vary based upon the hazmat involved, container design and construction, container stress and actual/potential breach, and the position and location of the container. The following are general guidelines and should be used as applicable.

NOTES

SURVEYING THE CONTAINER

The container should first be surveyed to determine the safest method of off-loading. This is particularly true when dealing with bulk transportation containers, such as cargo tank trucks and railroad tank cars. It should be noted that off-loading reduces stress on a container and further reduces the chance that the container will fail mechanically (e.g., during lifting). Factors to evaluate include:

- The pitch and position of the container.
- The position and location of the holes or attachments which will be used for product off-loading.
- If the container is a tank truck (e.g., MC-306/DOT-406), the position of the baffle holes.
- The product being off-loaded (e.g., flammable vs. corrosive vs. poisonous).
- The level of training, resources, and equipment available for product transfer and container uprighting operations.

If a tank truck or rail car are involved, the pitch and position of the container—front to back and left to right— are particularly important. It's possible that the container may move as product is pumped off and the product load shifts. Even where the unit appears stable, consideration must be given to bracing. Bracing materials may include timber, jacks, or air bags.

BONDING AND GROUNDING CONSIDERATIONS

The generation and accumulation of static electrical charges during flammable liquid transfer operations must always be considered. Static electricity can also be an issue when dealing with dusts or powders. To minimize the potential of a flash fire or explosion, this static build-up must be controlled through bonding and grounding.

Bonding is the process of connecting two or more conductive objects together by means of a conductor; it is done to minimize potential differences between conductive objects, thereby minimizing or eliminating the chance of static sparking.

Grounding is the process of connecting one or more conductive objects to the ground through an earthing electrode (i.e., grounding rod); it is done to minimize potential differences between objects and the ground. An **ohm meter** is used to measure the electrical resistance and ensure the electrical continuity of bonding and grounding operations.

In order for static electricity to act as an ignition source, four conditions must be fulfilled:

1) There must be an effective means of static generation. This can occur when a flammable or combustible liquid is moved from one place to another through pipes, filtering, or by pouring.
2) There must be a means of accumulating the static charge build-up.
3) There must be a spark discharge of adequate energy to serve as an ignition source (i.e., incendive spark).
4) The spark must occur in a flammable mixture.

NOTES

Bonding and grounding must be in place before product removal and transfer operations begin. Consider the following operational guidelines:

- ❏ The pump-off vehicle and all pump-off appliances, such as hose couplings, downspouts, and recovery pans and tubes, should all be bonded by connecting a bonding cable from the overturned tank to the appliance. In all appliance bonding operations, the first connection must always start at the damaged unit.
- ❏ Bonding cables must be placed on a clean, grease-free, paint-free surface. Cables with C-clamps are preferable to cables with "alligator clips" because they make better connections.
- ❏ Rubber hoses with a built-in wire will not necessarily provide bonding protection, as the wire within the hose may become broken or the wire may not be properly tied into the coupling.
- ❏ Plastic buckets can pick up static charges and should not be allowed for use as retention basins in an emergency situation.
- ❏ Grounding cables should initially be connected to the damaged container, then moved outward and away from the overturned vehicles. The final connection can be made to a guard rail post, telephone, or electrical pole support rod, if it's deep enough to carry away the charge. In addition, auger-type t-handle grounding rods can be very effective. See Figure 10-18 for an illustration of the bonding and grounding sequence.
- ❏ Periodically monitor all bonding and grounding cable connections to ensure that they remain in place and connected.

BONDING AND GROUNDING SEQUENCE

NOTE: THIS ILLUSTRATION DOES NOT DEPICT THE PROXIMITY OR EXACT SPATIAL LAYOUT OF THE BONDING AND GROUND SYSTEM.

- MAKE CONNECTIONS IN SEQUENCE SHOWN TO AVOID SPARKS IN POTENTIALLY FLAMMABLE AREAS.
- CONNECTIONS A AND B CAN BE MADE ANYTIME PRIOR TO PUMP-OFF.

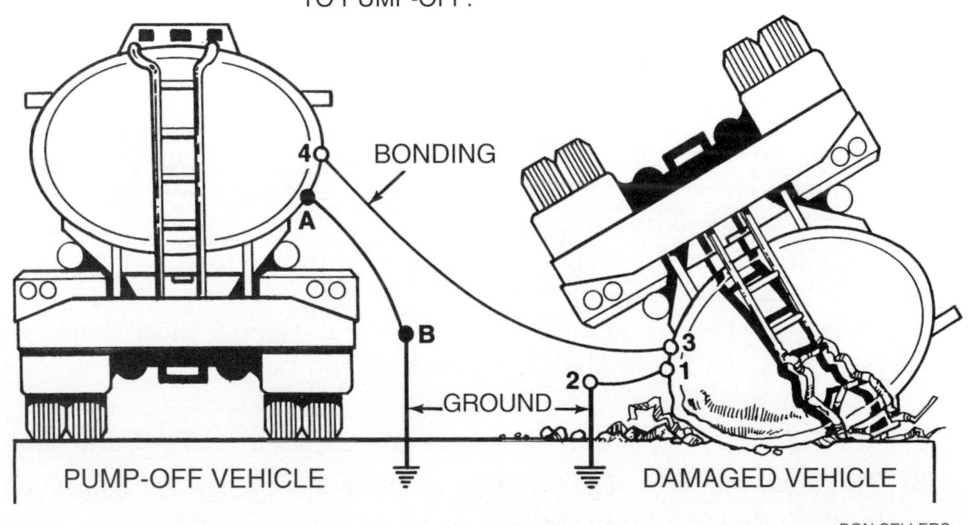

DON SELLERS

FIGURE 10-18: SUGGESTED BONDING PROCEDURE

PRODUCT TRANSFER METHODS

NOTES

Product transfer and removal are normally performed by industrial responders and environmental contractors. However, public safety responders will often continue to be responsible for site safety and will oversee the implementation of all product transfer and removal. The following information is directed toward these oversight operations.

There are three primary methods of product transfer:

1) **Portable Pumps**—Transfer operations using portable pumps may be either "open system" or "closed system." Transfer pumps are categorized by their energy source and include gasoline, diesel, power-take-off (PTO), electrical, water, and air-driven pumps. Key issues include:
   - Energy source and sparking potential of the pump. Gasoline, diesel, PTO, and electrical pumps can act as an ignition source in a flammable environment. Consider the hazardous classification ratings of any electrical equipment used around flammable liquids or gases—Class I, Division 2 should be the *minimum* accepted rating.
   - Hazards of the material being transferred (flammable, corrosive, etc.).
   - Materials of construction. The chemical compatibility of the pump, receiving tank, and all associated hoses and piping is a critical factor. Responders should determine previous products handled by the equipment to ensure that there are no residual chemical contamination or reactivity hazards. In addition, product contamination may also be a concern, particularly when dealing with industrial solvents and other high quality chemicals.
   - Power rating and pressure capacity of the pump, including lift and flow capacities.

2) **Pressure Differential**—Vapors and gases may be transferred through the use of vapor compressors and pressure differential. Materials will seek the path of least resistance and naturally flow from high pressure to low pressure areas. In some cases, a vapor compressor may be used in conjunction with a liquid pump to accelerate the rate of transfer by withdrawing vapors from the receiving container, compressing them, and forcing them into the damaged container. However, the use of the vapor compressor will cause an increase in pressure in the damaged container and should only be used when the pressure increase can be safely accepted.

   Inert gases, such as nitrogen and CO_2, which are compatible with the product can also be used to move the contents of a damaged container into a receiving container. The inert pressure creates a positive pressure differential in the damaged container that pushes the liquid product into the receiving tank. Vapors from the receiving tank may have to be vented to the atmosphere or scrubbed.

3) **Vacuum Trucks**—These are frequently used to remove liquid hazardous materials and hazardous waste from an emergency scene.

NOTES

Depending on their rating and design features, they can handle flammable and combustible liquids, corrosives, and some poisons.

A sewer or septic tank pumper *is not* a vacuum truck rated for hazardous waste. Vacuum trucks are designed for specific operating pressures up to 25 pounds per square inch gage. A vacuum truck generally loads and unloads by reversing its vacuum pump through a four-way valve and manifold, which provides vacuum for loading and pressure for unloading. Some trucks are also equipped with a gear rotary pump for transfer operations.

Vacuum trucks must work close to the damaged hazmat container in order to reach pump-out connections or containment areas. Internal explosions within vacuum truck cargo tanks are rare. However, incidents which have occurred are usually due to pumping incompatible materials inside the tank (e.g., pumping a corrosive into a tank which is not rated for corrosives) or pumping two incompatible hazardous chemicals into the same tank.

Vacuum trucks can be an ignition source and must be operated at the emergency scene with special precautions. Most vacuum truck fires and explosions are due to either operating the vehicle too close to the spill, pick-up, or discharge point, or failing to vent the vacuum pump discharge to a hazard-free area.

The following safety precautions should be observed when operating vacuum trucks within Hazard Control Zones:

- **Flammable Atmosphere Test**—Vacuum trucks should not be permitted to enter the hazardous area until flammable vapor monitoring has been conducted. While a reading of zero is preferred, it may not be realistic when open-air spills are being recovered. As a general guideline, concentrations above 20% of the LEL are considered hazardous. Applying foam to a flammable liquid spill before beginning vacuum operations can help maintain the atmosphere within acceptable limits. Monitoring should also continue during the course of the transfer operation.
- **Grounding**—API Publication 2219, *Safe Operation of Vacuum Trucks in Petroleum Service*, indicates that static electricity does not present an ignition problem with either conductive or nonconductive vacuum truck hoses. However, with nonconductive hoses, any exposed metal, such as a hose flange, can accumulate static electricity and act as an ignition source if the metal touches or comes close to the ground. Since it is often difficult to distinguish between nonconductive and conductive hoses in the field, API recommends that all exposed metal be grounded when any hose is used in other than a closed system with tight connections at both ends of the hose.
- **Venting**—When flammable or toxic liquids are loaded into a vacuum truck, the vacuum pump exhaust should be vented downwind of the truck by attaching a length of hose sufficient to reach an area that is free from hazards and personnel.

NOTES

- **Personnel Safety**—All unnecessary personnel should leave the area during loading. The vacuum truck driver should leave the truck cab and be in proper PPE. Strict control of ignition sources should be maintained within 100 feet of the truck, the discharge of the vacuum pump, or any other vapor source.

SUMMARY

Implementing Response Objectives is the phase in a hazmat emergency when the Incident Commander implements the best available strategic goals and tactical objectives which will produce the most favorable outcome. The operational strategy for an incident is developed based upon the IC's evaluation of the current conditions and forecast of future conditions. The effectiveness of this phase of the incident is directly related to how well the hazards were identified and the risks evaluated.

Primary hazmat strategic goals include rescue, public protective actions, spill control (confinement), leak control (containment), fire control, and transfer and recovery.

REFERENCES AND SUGGESTED READINGS

American Petroleum Institute, FIRE PROTECTION CONSIDERATIONS FOR THE DESIGN AND OPERATION OF LIQUEFIED PETROLEUM GAS (LPG) STORAGE FACILITIES, (1st edition), API Publication 2510-A, Washington, DC: American Petroleum Institute (1989).

American Petroleum Institute, GUIDE FOR FIGHTING FIRES IN AND AROUND FLAMMABLE AND COMBUSTIBLE ATMOSPHERIC PETROLEUM STORAGE TANKS, (3rd edition), API Publication 2021, Washington, DC: American Petroleum Institute (1991).

American Petroleum Institute, GUIDELINES FOR WORK IN CONFINED SPACES IN THE PETROLEUM INDUSTRY, (1st edition), API Publication 2217, Washington, DC (1984).

American Petroleum Institute, GUIDELINES FOR WORK IN INERT CONFINED SPACES IN THE PETROLEUM INDUSTRY, (1st edition), API Publication 2217A, Washington, DC (November 1987).

American Petroleum Institute, GUIDELINES FOR SAFE DESCENT ONTO FLOATING ROOFS IN PETROLEUM SERVICE, (1st edition), API Publication 2026, Washington, DC (1988).

American Petroleum Institute, PROTECTION AGAINST IGNITIONS ARISING OUT OF STATIC, LIGHTNING, AND STRAY CURRENTS ,(5th edition), API Publication 2003, Washington, DC: American Petroleum Institute (1991).

American Petroleum Institute, SAFE OPERATION OF VACUUM TRUCKS IN PETROLEUM SERVICE, (5th edition), API Publication 2219, Washington, DC: American Petroleum Institute (1986).

American Society for Testing and Materials, FIRE AND EXPLOSION HAZARDS OF PEROXY COMPOUNDS, SPECIAL PUBLICATION NO. 394, September 1965.

Andrews, Jr., Robert C., "The Environmental Impact of Firefighting Foam." INDUSTRIAL FIRE SAFETY (November/December, 1992), pages 26–31.

Arnold, David, "Water-Sodium Mix Set off Newton Blast," THE BOSTON GLOBE, Thursday, October 28, 1993, page 22.

Benner, Ludwig, Jr., "D.E.C.I.D.E. In Hazardous Materials Emergencies." FIRE JOURNAL (July, 1975), pages 13–18.

Benner, Ludwig, Jr., HAZARDOUS MATERIALS EMERGENCIES (2nd Edition), Oakton, VA: Lufred Industries, Inc. (1978).

Bradish, Jay, "The Fatal Explosion," A Special Report of the NFPA Investigations and Applied Research Division on the December 27, 1983 Propane Explosion in Buffalo, New York. FIRE COMMAND (March 1984) pages 28–33.

Chemical Manufacturers Association, "Flammable Liquids: Proper and Safe Handling" (videotape). Washington, DC: Chemical Manufacturers Association (1993).

Chemical Manufacturers Association, "Flammable Solids: Identification, and Safe Handling" (videotape). Washington, DC: Chemical Manufacturers Association (1992).

Chemical Manufacturers Association, "Oxidizers: Identification, Properties and Safe Handling" (videotape). Washington, DC: Chemical Manufacturers Association (1990).

Chemical Manufacturers Association, "Poisons: Identification, Toxicity and Safe Handling" (videotape), Washington, DC: Chemical Manufacturers Association (1991).

Coastal Video Communication Corp., "Solving the Mystery: Static Electricity" (videotape and booklet), Virginia Beach, VA: Coastal Video Communication Corp. (1990).

Compressed Gas Association, Inc., HANDBOOK OF COMPRESSED GASES (3rd Edition), New York, NY: Van Nostrand Reinhold Inc. (1990).

CONTAINING LEAKS IN PRESSURIZED CYLINDERS (videotape), by Action Video, Portland, Oregon (1988).

Devonshire, Jim, "A Quick Guide to Foam/Water Sprinkler Systems." INDUSTRIAL FIRE CHIEF (March/April, 1993), pages 37–39.

DIKING, DIVERTING, AND RETAINING SPILLS (videotape), by Action Video, Portland, Oregon (1989).

Dunn, Vincent, "BLEVE: The Propane Cylinder." FIRE ENGINEERING (August, 1988), pages 63–70.

Emergency Film Group, Foam (videotape), Plymouth, MA: Emergency Film Group (1993).

Emergency Film Group, Oil Spill Response (four-videotape series), Plymouth, MA: Emergency Film Group (1994).

Emergency Resource, Inc., Surviving the Hazardous Materials Incident—Series II, Fort Collins, CO: Emergency Resource, Inc. (1991).

Epperson, Jimmy C., Jr., "Preparing for Cylinder Emergencies." INDUSTRIAL FIRE SAFETY (November/December, 1992), pages 32–40.

Federal Emergency Management Agency—National Fire Academy, Hazardous Materials Operating Site Practices Course—Student Manual, Emmitsburg, MD: FEMA.

Federal Emergency Management Agency—U.S. Fire Administration, Technical Report on High Temperature Accelerant Arson Fires, Emmitsburg, MD: FEMA (1991).

Fire, Frank L., THE COMMON SENSE APPROACH TO HAZARDOUS MATERIALS. Saddle Brook, NJ: Fire Engineering (1986).

Hawthorne, Ed, PETROLEUM LIQUIDS—FIRE AND EMERGENCY CONTROL. Englewood Cliffs, NJ: Prentice Hall (1987).

Hildebrand, Michael, S., "The American Petroleum Institute Response to the Winchester, Virginia Oil Spill and Tire Dump Fire." A technical paper presented to the API Operating Practices Committee, OPC Paper 9.2-20 (D-120) (1984).

Hildebrand, Michael, S., HAZMAT RESPONSE TEAM LEAK AND SPILL CONTROL GUIDE, Stillwater, OK: Fire Protection Publications, Oklahoma State University (1992).

Hildebrand, Michael, S. and Gregory G. Noll, GASOLINE TANK TRUCK EMERGENCIES: GUIDELINES AND PROCEDURES, Stillwater, OK: Fire Protection Publications, Oklahoma State University (1992).

International Association of Fire Chiefs et al., GUIDELINES FOR THE PREVENTION AND MANAGEMENT OF SCRAP TIRE FIRES. Washington, DC: International Association of Fire Chiefs (1993).

International Association of Fire Fighters, Training Course for Hazardous Materials Team Members—Student Text, Washington, DC: IAFF (1991).

Isman, Warren, E. and Gene, P. Carlson, HAZARDOUS MATERIALS, Encino, CA: Glencoe Publishing Company, Inc. (1980).

Lesak, David M., "GEDAPER for Haz Mats—The Final Steps." FIRE ENGINEERING (November, 1989), pages 56–60.

Meal, Larie, "Static Electricity." FIRE ENGINEERING (May, 1989), pages 61–64.

Meidl, James H., EXPLOSIVE AND TOXIC HAZARDOUS MATERIALS. Encino, CA: Glencoe Publishing Company, Inc. (1979).

Meidl, James H., FLAMMABLE HAZARDOUS MATERIALS. Encino, CA: Glencoe Publishing Company, Inc. (1979).

Meyer, Eugene, CHEMISTRY OF HAZARDOUS MATERIALS. Englewood Cliffs, NJ: Prentice Hall (1977).

Meyer, Eugene, "Volatile Chemical Drama Was Ended By Secret Convoy." THE WASHINGTON POST, Thursday, October 25, 1979, page C-6.

Monsanto Chemical Company, Olin Chemicals and PPG Industries, GUIDELINES FOR THE SAFE HANDLING AND STORAGE OF CALCIUM HYPOCHLORITE AND CHLORINATED ISOCYANURATE POOL CHEMICALS, St. Louis, MO: Monsanto Chemical Company et al. (1989).

National Fire Protection Association, FLAMMABLE AND COMBUSTIBLE LIQUIDS CODE HANDBOOK (4th edition), Boston, MA: National Fire Protection Association (1991).

National Fire Protection Association, HAZARDOUS MATERIALS RESPONSE HANDBOOK (2nd edition), Boston, MA: National Fire Protection Association (1992).

National Fire Protection Association, LIQUEFIED PETROLEUM GASES HANDBOOK (3rd edition), Boston, MA: National Fire Protection Association (1992).

Calvin and Hobbes
by Bill Watterson
HERE WE FIND A THRIVING CITY: BRAND NEW BUILDINGS, A BUSTLING ECONOMY.
A SCENIC THOROUGHFARE WINDS THROUGH THIS HAPPY MUNICIPALITY. HERE, A FARMER DRIVES HIS LIVESTOCK TO MARKET.
© 1985 Universal Press Syndicate
WATTERSON
11-27
TRAGICALLY, THIS SERENE METROPOLIS LIES DIRECTLY BENEATH THE HOOVER DAM...

DECONTAMINATION

OBJECTIVES

1) Define the following terms:
    - Contaminant
    - Contamination
    - Decontamination (Decon)
    - Decontamination Corridor
    - Decontamination Officer
    - Technical Decontamination
    - Emergency Decontamination
2) Define surface and permeation contamination and their significance in decontamination operations.
3) Describe the difference between direct contamination and cross contamination and their significance in site safety operations.
4) Define the following terms and describe their potential harmful effects to the human body relating to contamination:
    - Highly acute toxicity contaminants
    - Moderate to highly chronic toxicity contaminants
    - Embryotoxic contaminants
    - Allergenic contaminants
    - Flammable contaminants
    - Highly reactive or explosive contaminants

"Fear is an emotion indispensable for survival."

Hannah Arendt,
German/American
Political Theorist

NOTES

- Etiologic contaminants
- Radioactive contaminants

5) Define the following terms and describe their application as general decontamination methods:
   - Physical methods
   - Chemical methods
6) Identify the advantages and limitations, and describe the application of each of the following physical methods of decontamination:
   - Brushing or scraping
   - Dilution
   - Absorption and adsorption
   - Heating and freezing
   - Blowing and vacuuming
   - Isolation and disposal
7) Identify the advantages and limitations, and describe the application of each of the following chemical methods of decontamination:
   - Chemical degradation
   - Neutralization
   - Solidification
   - Disinfection and sterilization
8) Describe three methods for evaluating the effectiveness of decontamination.
9) Describe the duties and responsibilities of the Decontamination Officer.
10) Describe the basic criteria for selecting the decontamination site.
11) Describe the general conditions which require an Emergency Decontamination.
12) Describe the nine general stations in the decontamination sequence for conducting field decontamination.
13) Define the following terms:
    - Gross Decontamination
    - Secondary Decontamination
14) List three types of fixed or engineered safety systems which may be used to assist ERP in decontamination within special hazmat facilities.
15) List four general emergency medical concerns when handling a contaminated patient.
16) Define the term "Infection Control" as it applies to decontamination.

NOTES

17) Define the following terms as they relate to infection control:
    - Body fluids
    - Contaminated
    - Disinfection
    - Exposure
    - Fluid-Resistant Clothing
    - Leakproof bags
    - Medical Gloves
    - Medical Waste
    - Mucous Membrane
    - Sharps Container
    - Splash-Resistant Eyewear
    - Sterilization
    - Universal Precautions
18) Define the term "Clean-up" as it applies to decontamination.
19) Describe four general clean-up concerns when decontaminanting equipment.

INTRODUCTION

This chapter will describe the seventh step in the **Eight Step Process©—Decontamination (Decon)**. Proper decon is essential to ensure the safety of emergency responders and the public. Although decon is typically implemented after entry operations, the determination of decon methods and procedures must be considered early in the incident as part of the Hazard and Risk evaluation process. It does not make much sense to conduct entry operations if you do not understand the contaminant and have no way to perform decontamination.

The basic concepts of decontamination are relatively simple. In fact, the simplicity of decontamination has been compared to changing a diaper; remove it from others, keep it off yourself, and don't spread it around! In other words, don't get contaminated and you won't have a decontamination problem.

Over the last ten years, many organizations and individuals have added to the "body of knowledge" which makes up what we now routinely call decontamination. As techniques have improved, new terminology has emerged to describe the different phases and stages required to make decon effective and simple for field application.

As is the case with most operating procedures, each organization has its own way of doing things. Likewise, regional preferences for approaching the same problem can vary from one place to another. For example, the field decontamination problems in February for Detroit, Michigan are going to be different from Phoenix, Arizona. Who wants to take a shower outside when it's ten below zero?

In developing this chapter, we found many inconsistencies in decontamination terminology. Minor differences in terminology like "phases,"

NOTES

"stages," or "steps" should not become an obstacle for adopting a good technique. Decon is still decon!

For the purposes of this text, we have adopted the decontamination terminology referenced in NFPA-472 and 473. Where standard terms or technical guidance did not exist in these standards, we tried to adopt the more popular terms and organizational structure found in the references at the end of this chapter.

The bottom line is that the safety and health hazards of the contaminants at any incident define how complex decon operations will be. A minimal hazard, such as a lightly contaminated oil, can be partially decontaminated by simply wiping or flushing it from protective clothing. In contrast, a major hazard, such as a highly poisonous material, will require implementing a detailed procedure which includes several intermediate cleaning steps. Exactly how many steps and where they are performed is best determined locally. It's OK to be different from one organization to the next as long as you use consistent terminology and follow safe operating procedures.

As you review the subjects in this chapter, you will note that we also address decon concerns involving biological agents and infectious diseases. While these topics are often viewed as strictly an EMS concern, hazmat responders may encounter these materials and associated decon problems at medical facilities and research labs.

To understand the materials in this chapter, you must be familiar with some basic terminology.

DEFINITIONS AND TERMINOLOGY

Contaminant—A hazardous material that physically remains on or in people, animals, the environment, or equipment, thereby creating a continuing risk of direct injury or a risk of exposure outside of the hot zone.

Contamination—An uncontained substance or process that poses a threat to life, health, or the environment.

Decontamination— (Decon). The physical and/or chemical process of reducing and preventing the spread of contamination from persons and equipment used at a hazardous materials incident. (Also referred to as "contamination reduction.") Note: OSHA defines decontamination as the removal of hazardous substances from employees and their equipment to the extent necessary to preclude foreseeable health effects.

Decontamination Corridor—A distinct area within the "Warm Zone" that functions as a protective buffer and bridge between the "Hot Zone" and the "Cold Zone," where decontamination stations and personnel are located to conduct decontamination procedures.

NOTES

Decontamination Officer—A position within the Hazmat Branch which has responsibility for identifying the location of the decontamination corridor, assigning stations, managing all decontamination procedures, and identifying the types of decontamination necessary.

Decontamination Team—(Decon Team). A group of hazmat trained personnel and resources operating within a decontamination corridor.

Emergency Decontamination—The physical process of immediately reducing contamination of individuals in potentially life-threatening situations without the formal establishment of a decontamination corridor.

Exposure—The subjection of a person to a toxic substance or harmful physical agent through any route of entry (e.g., inhalation, ingestion, skin absorption, or direct contact).

Technical Decontamination —The process of reducing contamination on personal protective clothing and equipment of emergency response personnel to a level which is As Low As Reasonably Achievable (ALARA).

UNDERSTANDING THE BASICS OF CONTAMINATION

Before ERP can understand the basics of decontamination, it is necessary to establish a foundation by learning some important points about how contamination occurs. The four most important basic concepts of contamination are:

1. How to prevent contamination.
2. Surface vs. Permeation Contamination.
3. Direct vs. Cross Contamination.
4. Types of Contaminants.

PREVENTING CONTAMINATION

ERP sometimes get the terms "contamination" and "exposure" confused. They actually have very different meanings. Knowing the difference between them can help a great deal in understanding the importance of preventing contamination.

Contamination is any form of hazardous material (solid, liquid, or gas) that physically remains on people, animals, or objects. In the emergency response business, "contamination" generally means any contaminant which is on the outside of PPE or ERP equipment.

Exposure means that a person has been subjected to a toxic substance or harmful physical agent through any route of entry into the body (e.g., inhalation, ingestion, skin absorption, or by direct contact).

NOTES

A person who has been "contaminated" has not necessarily been "exposed." For example, an entry team can be contaminated with pesticide dust on the outside of their PPE without having their respiratory system or skin exposed to the contaminant. The exposure could occur if the PPE were accidentally breached *or if they were not properly decontaminated.*

If contact with the contaminant can be controlled, the risk of exposure is reduced, and the need for decontamination can be minimized. Consider the following points to prevent contamination:

- Stress work practices that minimize contact with hazardous substances. Don't walk through areas of obvious contamination, and don't touch potentially hazardous substances as reasonably practical.
- If contact is made with a contaminant in the Hot Zone, move contaminated ERP to a Zone of Refuge until they can be decontaminated. Remove the contaminant as soon as possible.
- Wear a disposable outer garment over other chemical protective clothing to minimize exposures and contamination.
- Wear limited use or disposable protective clothing and equipment, where appropriate.
- Protect monitoring instruments by covering or bagging them within a clear plastic bag. However, detecting elements, intake paths, and exit ports must be kept uncovered. Also, keep in mind that some instruments can overheat when covered, and plastic can make it difficult to read meters and to adjust settings.

SURFACE VS. PERMEATION CONTAMINATION

Contaminants can present problems in any physical state (i.e., solid, liquid, or gas). There are two general types of contamination—surface and permeation.

Surface Contaminants. Contaminants which are on the *outer* surface of a material and which have not been absorbed *into* the material. Surface contaminants are normally easy to detect and remove to a reasonably achievable and safe level. Examples include dusts, powders, fibers, etc.

Permeation Contaminants. Contaminants which are absorbed into a material at the molecular level. Permeated contaminants are often difficult or impossible to detect and remove. If the contaminants are not removed, they may continue to permeate through the material. If the material is protective clothing, it could cause an "exposure" on the inside of the suit. If the material is a tool or piece of equipment, it could lead to the failure of the item (e.g., an airline hose). See Figure 8-1 in Chapter Eight for a more detailed explanation of this concept. Remember that permeation can occur with any porous material, not just PPE.

Factors which influence permeation include:

- **Contact Time**—The longer a contaminant is in contact with an object, the greater the probability and extent of permeation.
- **Concentration**—Molecules of the contaminant will flow from areas of high concentration to areas of low concentration.

NOTES

All things being equal, the greater the concentration of the contaminant, the greater the potential for permeation to occur.

- **Temperature**—Increased temperatures generally increase the rate of permeation. Conversely, lower temperatures will generally slow down the rate of permeation.
- **Physical State**—As a rule, gases, vapors, and low viscosity liquids tend to permeate more readily than high viscosity liquids or solids.

A single contaminant can present both a surface and permeation threat. This is especially the case when liquids are involved.

DIRECT VS. CROSS CONTAMINATION

Exposures such as people, animals, property, and the environment may be contaminated either directly or indirectly (cross contamination). A brief summary of each follows.

Direct Contamination results from direct contact with the contaminate. This may occur during entry into the Hot Zone or by coming into direct contact with the contaminant in the Decon Area. Examples of common reasons for direct contamination include:

- Poor Site Management practices (see Chapter Five).
- Selection of incompatible protective clothing (see Chapter Eight).
- Failure to use formal Checklists to identify potential safety problems.
- Running out of breathing air in the decontamination area.
- Failure to decontaminate.

Cross Contamination results whenever a "clean" or uncontaminated element (person, animal, or object) comes into direct contact with a "dirty" or contaminated element. This is also sometimes known as secondary contamination. Several examples of how cross contamination may occur include:

- A contaminated patient comes into physical contact with a first responder or ERP.
- A bystander or site worker crosses into the Hot Zone or comes into contact with a contaminated object which was in the Hot Zone.
- A decontaminated ERP re-enters the decontamination area and comes into contact with a contaminated ERP or object.

The variety of ways contamination can occur makes a strong case for establishing and enforcing clearly marked and communicated Hazard Control Zones. The more people directly or indirectly exposed to the contaminate, the bigger the decontamination problem will become.

TYPES OF CONTAMINANTS

Some decontamination operations turn into long drawn-out events that are later judged by others to be "overkill." Unfortunately, a lot of this criticism is valid. The primary cause of the problem is usually failure to

NOTES

understand the contaminant. You really can't blame ERP for going to extremes when they did not understand the hazards and risks of the contaminant in the first place. The more you know about the contaminant, the faster and more focused your decontamination operation can be.

The types of contaminants can be divided into eight different categories. These include:

1. Highly Acute Toxicity Contaminants
2. Moderate to Highly Chronic Toxicity Contaminants
3. Embryotoxic Contaminants
4. Allergenic Contaminants
5. Flammable Contaminants
6. Highly Reactive or Explosive Contaminants
7. Etiologic Contaminants
8. Radioactive Contaminants

In the following sections, we will examine each of these eight categories individually; however, as we have discussed in Chapter Seven, Hazard and Risk Assessment, a single hazmat may have more than one hazardous characteristic (e.g., a flammable liquid may also be poisonous). **Beware that a contaminant may also have more than one hazardous characteristic.**

Highly Acute Toxicity Contaminants

An acutely toxic substance causes damage to the human body as a result of a single or short-duration exposure. These can be found in solid, liquid, or gas forms and will present risks to ERP from any route of exposure. Examples include carbon disulfide, carbon tetrachloride, and potassium cyanide.

Moderate to High Chronic Toxicity Contaminants

Contaminants of moderate to high chronic toxicity include, but are not limited to, certain heavy metals, their derivatives, and potent carcinogens. They can be found in solid, liquid, or gas forms and will present risks to ERP from any route of exposure. Examples include ethylene dibromide (EDB) and benzene.

Embryotoxic Contaminants

Embryotoxins are substances that can act during pregnancy to cause adverse effects on the fetus, including death, malformation, retarded growth, and post-natal functional deficits. These substances may also be called "teratogens." The period of greatest susceptibility to embryotoxins is the first eight to twelve weeks of pregnancy. This includes a period when a woman may not know she is pregnant; therefore, special precautions must be taken at all times. However, as an additional precaution, pregnant members of an HMRT should avoid working in contaminated areas involving known embryotoxins. Examples of these substances include lead compounds and formamide.

NOTES

Allergenic Contaminants

Allergens are substances which produce skin and respiratory hypersensitivity. ERP are at risk from allergens from both inhalation and direct skin contact. Two people exposed to the same allergen at the same level of exposure may react differently. A sudden outbreak of a "mystery rash" among one ERP who was at the scene of a hazmat incident, while none of the other members of the group show effects, is not cause to dismiss the individual's signs or symptoms. Examples include diazomethane, chromium, dichromates, formaldehyde, and isocyanates.

Flammable Contaminants

Flammable and combustible substances are those which catch fire readily and burn in air. They may be found as a liquid or a gas. While flammable and combustible contaminants demand respect because of their obvious fire hazard, they also include a wide range of secondary and tertiary hazards. For example, ERP contaminated with toluene may experience respiratory distress if they remove their SCBA facepiece before adequately decontaminating their protective clothing. This may lead to pulmonary edema and eventually pneumonia. Always expect flammables and combustibles to present more than one contamination problem. Examples of flammable or combustible contaminants include acetone, benzene, and ethanol (liquids), and butane and hydrogen (gases).

Highly Reactive or Explosive Contaminants

Highly reactive or explosive contaminants include peroxides, peroxide-forming compounds, and explosive materials. Some of these materials can react with the oxygen present in the atmosphere. Others are heat, shock, and friction sensitive.

Among the most dangerous are the peroxide-forming compounds. Classes of hazardous materials that can form peroxides include aldehydes, ethers, most alkenes, and vinyl and vinyldene compounds.

Specific examples of highly reactive or explosive contaminants include cyclohexane, P-dioxane, diethyl ether, diisopropyl ether, tetrahydrofuran (THF), and tetralin.

The concentration of these contaminants plays an important role in determining the risk involved in decontamination. For example, hydrogen peroxide below 30% concentration presents no serious fire or explosion hazard. However, as the concentration increases, particularly above 52%, the hazard increases. It is also important to recognize that the evaporation and distillation processes of these materials can create high risk situations.

Metal tools such as spatulas and shovels should not be used to clean up peroxides contaminants because metal contamination can lead to explosive decomposition. Obviously, ERP should avoid friction, grinding, and other forms of impact.

Etiologic Contaminants

Etiologic (biological) contaminants are microorganisms such as viruses, fungi, and bacteria or their toxins that can cause illness, disease, or death.

NOTES

Etiologics can enter the body by ingestion, direct contact, or through the respiratory system. They are not always labeled or clearly identifiable (e.g., in research facilities, clinical labs, etc.). Without proper identification labels and packaging it is difficult for ERP to determine whether etiologic contaminants are present. Specific examples of these materials include bacillus anthracis (anthrax), clostridium botulinum (botulism), and hepatitis A, B, C, D, and E.

The fact that an area is contaminated with an etiologic does not necessarily mean a person has been exposed or, even if exposed, will be susceptible to its effects. Taking some basic precautions like using PPE with positive pressure SCBA can significantly reduce the risk to ERP. Good pre-entry planning using safe work practices can reduce risk even further. For example, if an etiologic has been spilled, there will be an aerosol hazard. Waiting 30 to 60 minutes for the aerosol to settle before entry can significantly reduce the risk of exposure.

There are four basic factors that influence an etiologic's ability to invade and alter the human body. These include:

Virulence—This is the etiologic's ability to cause disease. It is important to realize that there are a wide variety of different etiologic organisms. Their ability to survive, reproduce, and cause harm will depend on the type of organism. If the organism is not strong enough to survive in its environment long enough to enter the human body, or if it is too weak or lacking in its ability to cause illness, then the potential harm to an exposed individual is significantly reduced. Some etiologics cannot cause disease unless the body's defense system has been compromised. For example, if a person's skin is intact and the respiratory system has not been compromised, many organisms cannot penetrate the body. However, cuts, abraded or chapped skin, or exposed mucous membranes increase the opportunity for the organism to enter the body.

Dose—This means the number of organisms which have been ingested, absorbed, or inhaled during an exposure period. The size, composition, and population of an etiologic will determine its ability to effect an exposed person. If an organism is not compatible with the host, or if there are not enough of the organisms to alter the natural balance within the human body, then the etiologic cannot survive, regardless of the dose.

Physical Environment—The physical environment can also determine the ability of the etiologic to enter the body. Factors such as heat, cold, skin and membrane acidity, and alkalinity can affect the invading organism's potential to survive and cause harm.

ERP's Health Status—If the exposed person is in good health, with a fully functioning immune system, then the etiologic will have more difficulty surviving.

Fortunately, most etiologic contaminants are relatively easy to kill using a wide variety of commercially available decontamination solutions. These are divided into disinfectants and antiseptics and are discussed in more detail later in this chapter.

NOTES

Radioactive Contaminants

Radioactive contaminants may be in the form of alpha, beta, or gamma radiation. Specific examples of these materials include cesium 137, cobalt 60, radon 222, uranium hexafluoride, and plutonium.

If alpha-emitting contaminants are airborne (e.g., carried in dust or smoke), they can become surface contamination on protective clothing or tools carried into the Hot Zone. These solid contaminants can enter the body through any one of the routes of exposure described above, although inhalation is the most common.

Beta emitters present both the possibility of becoming airborne and creating high radiation levels near the surface of the contaminated area. As is the case of alpha emitters, beta contaminants may enter the body through any route of exposure.

If the contaminant is a gamma emitter, there is an additional hazard of gamma rays coming off the tiny bits of radioactive materials which may be spread all around the area. Of course, the possibility exists that all three types of emitters may be present.

The single most important factor in eliminating radioactive material contamination problems is to avoid contact with the material. It is very easy to spread radioactive contaminants over a wide area by cross contamination. If the radioactive contaminant has a short half-life, it is sometimes simpler to isolate contaminated equipment until the radioactivity has died off by the natural process of passing through its half-lives. For example, if a reading of 100 milliroentgens (mR) were taken from a gamma emitter with a 24 hour half-life, in 24 hours the reading would be down to 50 mR; in 48 hours it would be down to 25 mR, and in 72 hours it would be down to 12.5 mR. Waiting several days to cleanup equipment significantly reduces the risk to ERP.

A good rule of thumb for making an on-scene decision to cleanup contaminated equipment now or wait out the half-life is the fact that the passage of 7 half-lives will bring a radiation level down to 1% of what it is at the time you take the first reading. In 10 half-lives (10 days), the level will be down to 0.1%. See Figure 11-1.

Remember the basics. The more time, distance, and shielding between you and the radioactive material, the lower the risk will be.

DECONTAMINATION METHODS

Decontamination is a widely recognized term; however, "Contamination Reduction" more accurately describes what we are trying to accomplish. **Total decontamination** is our ultimate goal, but the complications involved in achieving this under actual field conditions make this impractical. A more realistic goal is to attempt to reduce the contamination to an acceptable level (e.g., below the contaminant's TLV, PEL, or its ceiling value).

A wide range of decontamination techniques are available for emergency response work. Most of the techniques described in this chapter have been revised and improved through trial and error. The more

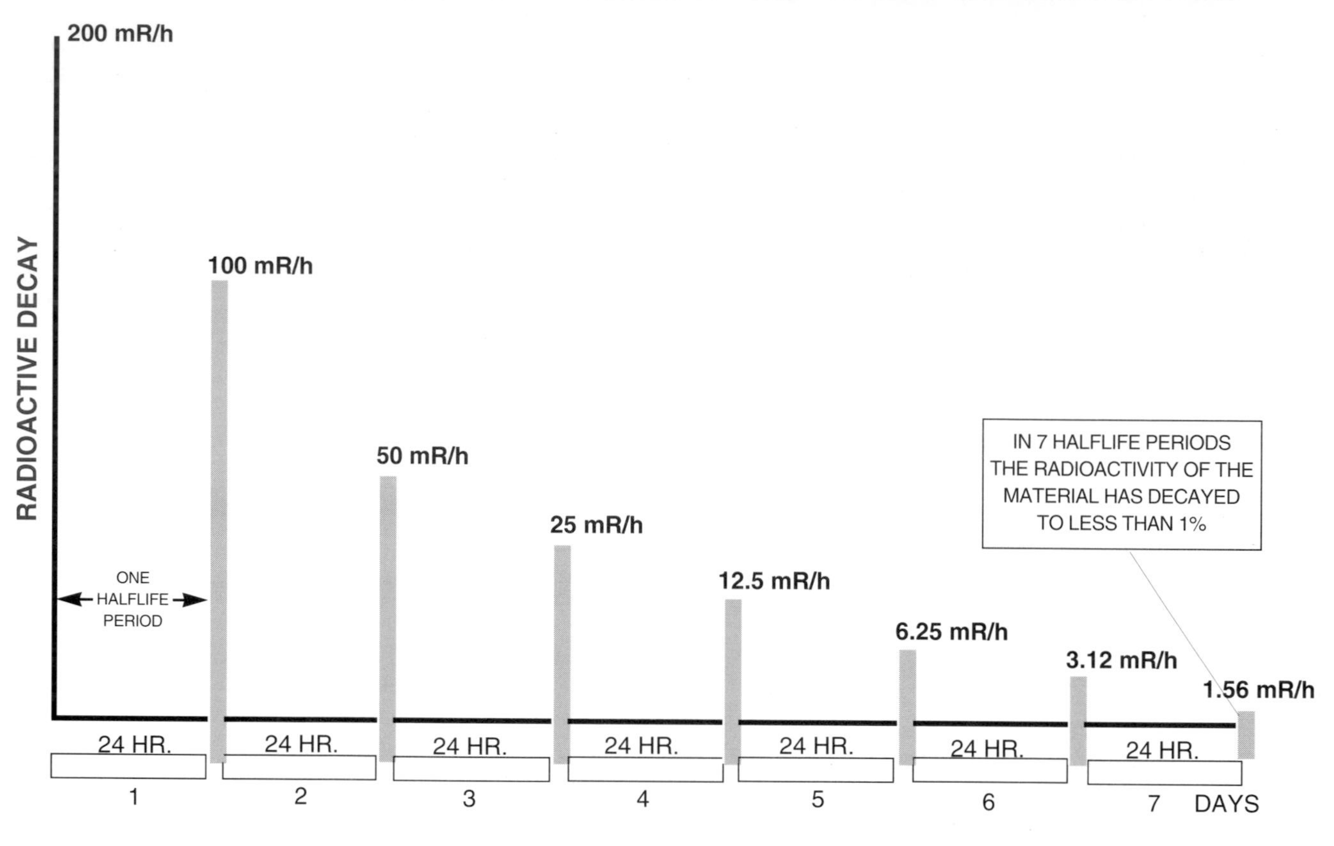

FIGURE 11-1

focused you are on a particular class of hazardous material, the better and more reliable your technique can become. For example, a chemical plant that manufactures specific types of hazardous materials should have perfected its decontamination techniques; they know what gets results and what doesn't. This is one reason why product specialists are often the best source of decontamination information for public safety organizations. On the other hand, unlike fixed industrial facilities, a public safety HMRT must be prepared to handle a wide range of hazardous materials decon problems. The following discussion has been written with the special needs of these ERP in mind.

Decontamination methods can be divided into two basic categories: physical and chemical.

Physical Methods generally involve "physically" removing the contaminate from the contaminated person or object. While these methods may dilute the contaminant's concentration (reducing its harmful effects), it generally remains chemically unchanged. Examples of physical methods include:

- Dilution
- Brushing and Scraping
- Absorption and Adsorption
- Heating and Freezing
- Blowing and Vacuuming
- Isolation and Disposal

NOTES

Chemical Methods generally involve removing the contaminant through some type of chemical process. In other words, the contaminant is undergoing some type of chemical change which renders it less harmful. In the case of etiologic contaminants, chemical methods are actually biologically "killing" the organism. Examples of chemical methods include:

- Chemical Degradation
- Neutralization
- Solidification
- Disinfection or Sterilization

The following sections briefly review various types of physical and chemical decon methods.

PHYSICAL METHODS

Dilution

Dilution is simply the use of water or soap and water solutions to flush the hazmat from protective clothing and equipment. It is the most commonly used method since large amounts of water are almost always available. Obviously one of the best resources for dilution is the tank water carried on any engine company or from hydrants. Dilution is especially effective for use on water-soluble hazmats like anhydrous ammonia and chlorine. Dilution can also be effective in removing some solids. However, the application of water only reduces the contaminant's concentration; it does not change it chemically. The more water you use, the more hazardous waste you generate and must dispose.

While dilution may be an effective method of removing non-water-soluble dusts and fibers, care must be used to prevent the spread of these contaminants to larger areas (e.g., asbestos fibers suspended in water). Before using water, consideration should be given to whether the contaminant will react with water, be soluble in water, or spread the contaminants to a larger area. Simple "dry decon" may be a better alternative.

Brushing or Scraping

Brushing or scraping basically involves using elbow grease to remove the contaminant. The object is to remove as much of the big stuff as possible before moving on to another technique. For example, contaminated mud stuck on the soles of boots should be scraped off before stepping into showers or rinse basins.

Absorption and Adsorption

Absorption is the process of "soaking up" a liquid hazardous material to prevent enlargement of the contaminated area. It is primarily used in decontamination for wiping down equipment and property. Beyond wiping off PPE with towels or rags, it has limited application for decontaminating personnel.

Like dilution, contaminants in absorbants remain chemically unchanged. In other words, a gallon of spilled PCB-contaminated transformer oil

NOTES

still has the same properties once it has been soaked up into the absorbent.

With some exceptions, use of absorbents is limited to flat surfaces (e.g., soaking up liquids on the ground). The most readily available absorbents are soil, diatomaceous earth, and vermiculite. Other acceptable materials include anhydrous fillers, sand, and commercially available products. Absorbent materials should be inert or have no active properties.

Adsorption is the process of adhering the contaminant to a surface. The adhesion takes place in an extremely thin layer of molecules between the contaminant and the adsorbant.

An easy way to remember the difference between these two methods is that an **AB***sorbant* works by "soaking up" the contaminant while an **AD***sorbant* adds itself or "sticks to" the contaminant.

Heating and Freezing

Heating usually involves the use of high temperature steaming in conjunction with high pressure water jets to heat up and blast away the contaminant. When detergent or solvents are added, this technique can be very effective on petroleum-based materials like used motor oil. Heating may also be used to simply evaporate the contaminant.

Freezing has been used in some special research facilities to literally "freeze" the contaminant to an object. One version of the process uses cryogenic liquid nitrogen to turn sticky liquids into a solid so that they can be scraped or flaked off and handled for a limited time as a solid.

Blowing and Vacuuming

Pressurized air hoses can be used to blow dusts and liquids out of hard-to-get at places like cracks and crevices. The down side is that the contaminant is blown into the surrounding atmosphere, where it will contaminate other people and objects.

Vacuuming involves the use of a special High Efficiency Particulate Air (HEPA) vacuum cleaner. This equipment has practical application when the contaminant is a hazardous dust, powder, or fiber. HEPA vacuums can normally filter particles of at least 0.3 microns. HEPA filters physically capture the contaminant by allowing air to pass through the filter while capturing the larger particulate floating in the air. To be effective, filters must be replaced periodically.

Isolation and Disposal

Isolation and disposal is a two-step process. First the contaminated article is removed and isolated in a designated area. When enough contaminated items are collected (e.g., disposable coveralls), they are bagged and tagged. The final step involves packaging the bagged contaminated material in a container suitable for transportation to an approved hazardous waste facility. The bagged material is either incinerated or buried in a landfill approved for hazardous waste.

Contaminated water can be disposed of in a variety of ways, including filtration, evaporation, and deep well injection. (The contaminated water is actually legally injected back into the earth, thousands of feet below the surface.)

NOTES

CHEMICAL METHODS

Chemical Degradation

Chemical degradation is the process of altering the chemical structure of the hazardous material. Commonly used agents include sodium hypochlorite (household bleach), sodium hydroxide as a saturated solution (household drain cleaner), sodium carbonate slurry (washing soda), calcium oxide slurry (lime), liquid household detergents, and isopropyl alcohol. Figure 11-2 lists some of the more widely used chemical degradation solutions.

DECONTAMINATION SOLUTIONS

FOR UNKNOWN PRODUCTS

Solution A: Five percent (5%) sodium carbonate and five percent (5%) trisodium phosphate. Mix four (4) pounds of commercial-grade trisodium phosphate with each ten (10) gallons of water.

Solution B: Solution containing ten percent (10%) calcium hypochlorite. Mix eight (8) pounds with ten (10) gallons of water.

Rinse Solution: To be used for both solutions. Five percent (5%) solution of trisodium phosphate with each ten (10) gallons of water.

FOR KNOWN PRODUCTS WITHIN THE 10 HAZARD CLASSES

Solution A: A solution containing five percent (5%) sodium carbonate and five percent (5%) trisodium phosphate.

Solution B: A solution containing ten percent (10%) calcium hypochlorite.

Solution C: A solution containing five percent (5%) trisodium phosphate which can be used as a general-purpose rinse.

Solution D: A dilute solution of hydrochloric acid (HCl). Mix one (1) pint of concentrated HCl into ten (10) gallons of water (acid to water only). Stir with wood or plastic stirrer.

GUIDELINE FOR SELECTING DEGRADATION CHEMICALS FOR SPECIFIC TYPES OF HAZARDS

Hazard	Solution
1. Inorganic acids, metal processing wastes	Solution A
2. Heavy metals: mercury, lead, cadmium, etc.	Solution B
3. Pesticides, chlorinated phenols, dioxins, PCP's	Solution B
4. Cyanides, ammonia, and other nonacidic inorganic wastes	Solution B
5. Solvents and organic compounds such as trichloroethylene, chloroform, and toluene	Solution C (or A)
6. PBB's and PCB's	Solution C (or A)
7. Oily, greasy, unspecified wastes not suspected to be contaminated with pesticides	Solution C
8. Inorganic bases, alkali, and caustic wastes	Solution D

U.S. Environmental Protection Agency

FIGURE 11-2: CHEMICAL DEGRADATION SOLUTIONS

NOTES

Technical advice for chemical degradation procedures should be obtained from the manufacturer to ensure the solution used is not reactive with the contaminant. Degradation chemicals should never be applied directly to the skin. See Figure 11-3.

FIGURE 11-3: NEVER APPLY DEGRADATION CHEMICALS TO THE SKIN

The physical and chemical compatibility of the decon solutions must be determined before they are used. Any decon method that permeates, degrades, damages, or otherwise impairs the safe function of PPE should not be used unless there are plans to isolate and dispose of the PPE.

Neutralization

Neutralization techniques involve applying another material (usually an acid or caustic) to a corrosive liquid spill. The neutralizing agent reacts chemically with the corrosive to form a less harmful corrosive substance. The objective is to bring the spilled material's pH towards neutral (pH of 7).

Solidification

Solidification is a process that chemically bonds the contaminant to another object or encapsulates it. For example, if the contaminated item is a fairly large piece of equipment, it may be treated with a cementitious material so that the contaminate is permanently and chemically fixed to the object. The solidified material is normally buried in a hazardous waste landfill.

NOTES

Large contaminated objects may also be mixed with cement and entombed. (This procedure was used at Chernobyl.)

Disinfection and Sterilization

Disinfection is the process used to inactivate (kill) virtually all recognized pathogenic microorganisms. Proper disinfection results in a reduction in the number of viable organisms to some acceptable level. *It does not produce a 100% kill of the microorganism you are trying to remove.* Consequently, it is important that ERP obtain expert advice on whether disinfection is the right technique to use. Likewise, some disinfectants work better on certain etiologics than others. Commercial disinfectants usually include a detailed brochure inside the box which describes what its limitations and capabilities are. If you respond to research labs, hospitals, and universities in your area, you should do your homework and familiarize yourself with the specific types of biological hazards present and the best disinfectant for the type of hazard you may encounter.

There are two major categories of disinfectants: chemical and antiseptic.

Chemical Disinfectants are the most practical for field use. The most common types are commercially available chemical disinfectants. These include: phenolic compounds, quaternary ammonium compounds, chlorine compounds, iodine and iodophors, alcohols, glutaraldehydes, and formaldehyde-alcohol compounds.

Antiseptic Disinfectants are designed primarily for direct application to the skin. These include alcohol, iodine, hexachlorophene, and quaternary ammonium compounds. Some of these compounds are also classified as disinfectants, but alterations in concentration allow them to be classified as antiseptics.

The terms "disinfection" and "sterilization" are sometimes used interchangeably by ERP. However, it is important to recognize that **sterilization is not the same as disinfection.** A decontamination recommendation from an etiologic specialist to sterilize a piece of equipment must not be misunderstood to mean disinfect it.

Sterilization is the process of destroying **all microorganisms** in or on an object. The most common method of sterilization is by using steam, concentrated chemical agents, or ultraviolet light radiation. Obviously, sterilization has no field application and cannot be used to decontaminate ERP, but it does play an important role in decontaminating medical equipment. Contaminated medical equipment is sometimes disinfected at the site, then transported as contaminated equipment to a special facility, where it is then sterilized or disposed of.

EVALUATING THE EFFECTIVENESS OF DECON OPERATIONS

Decon methods vary in their effectiveness for removing different substances. The effectiveness of any decon method should be assessed at the beginning of the decon operation and periodically throughout the operation. If contaminated materials are not being removed or are permeating through protective clothing, the decon operation must be revised.

NOTES

Some simple criteria which can be used for evaluating decon effectiveness during field operations include:

1. No personnel are exposed to concentrations above the TLV during decon operations.
2. Personnel are not exposed to skin contact with materials presenting a skin hazard during decon operations.
3. Contamination levels are reduced as personnel move through the Decon Corridor.
4. Contamination is confined to the Hot Zone and Decon Corridor.
5. Contamination is reduced to a level which is as low as reasonably achievable (ALARA).

Methods which may be useful in assessing the effectiveness of decontamination include:

Visual Observation—stains, discolorations, corrosive effects, etc.

Monitoring Devices— Devices such as photo-ionization detectors (PID's), detector tubes, geiger counters, and survey meters can show that contamination levels are at least below the device's detection limit.

Wipe Sampling—Provides after-the-fact information on the effectiveness of decon. Once a wipe swab is taken, it is analyzed in a laboratory. Both protective clothing, equipment, and skin may be tested using wipe samples.

It should be noted that for etiologic hazards, there is no practical way to determine the effectiveness of decontamination in the field.

SITE SELECTION AND MANAGEMENT

The success of decontamination is directly tied to how well the Incident Commander and the Decon Officer control on-scene personnel and their operations. Before initiating decontamination, the IC and the Decon Officer must decide:

1. How much and what type of decon is necessary.
2. To what extent it will be accomplished in the field.

These decisions should be based on the answers to the following questions:

- Can decontamination be conducted safely? Dilution, for example, may be impractical due to cold weather or because it presents an unacceptable risk to emergency personnel.
- Are existing resources immediately available to decon personnel and equipment? If not, where can they be obtained, and how long will it take to get them?
- Can the equipment used be decontaminated? The toxicity of some materials may render certain equipment unsafe for further use. In these cases, disposal may be the only safe alternative.

NOTES

DECONTAMINATION GROUP

At working hazardous materials incidents, a Decontamination Group should be established to manage and coordinate all decon operations. An ERP or HMRT member should be designated as the Decon Officer.

The Decon Officer should assume the responsibility for establishing a decontamination site and ensuring that all personnel, clothing, and equipment are cleaned prior to their being returned to service. The Decon Officer is responsible for the following activities:

- Determine the appropriate level of decontamination to be provided.
- Ensure that proper decon procedures are used by the Decon Team, including decon area set-up, decon methods and procedures, staffing, and protective clothing requirements.
- Coordinate decon operations with the Entry Officer and other personnel within the Hazmat Branch.
- Coordinate the transfer of decontaminated patients requiring medical treatment and transportation with the Hazmat Medical Group.
- Ensure that the Decon Area is established before any entry personnel are allowed to enter the Hot Zone.
- Monitor the effectiveness of decon operations.
- Control all personnel entering and operating within the decon area.

The Decon Officer should use a formal checklist to assure that important items are not overlooked. Figure 11-4 provides an example.

DECON SITE SELECTION

When the incident is outdoors, the decon site should be accessible from a hard-surfaced road. Water supply, access to safety showers, runoff potential, and proximity to any environmentally sensitive areas, such as streams or ponds, should be considered. If the decon will be conducted indoors, consideration should be given to quick access such as hallways, type and slope of floor, drains, and ventilation airflows in the area.

The ideal outdoor decon site is upwind and uphill from the incident and remote from drains, manholes, and waterways, but close enough to the scene to limit the spread of contaminants. Unfortunately, it is often not possible to actually choose such an ideal site. Shifting winds and dispersing gases further complicate your choice for a good location. Such real-life problems may force the movement of the decon area once it has been in operation if the site planning has been hasty.

Complete decon may be impractical to achieve at one location, so a combination of on-site and off-site contingencies may be necessary. Any time decon is conducted off-site (at fire stations, for example), the entire operation will become more complicated due to the logistics of moving people off-site. If you intend to use an off-site location for decontamination, special preparations will need to be made to prevent the spread of contaminants.

FIGURE 11-4

REFINERY TERMINAL FIRE COMPANY
HAZARDOUS MATERIALS RESPONSE PROCEDURES

DECONTAMINATION OFFICER WORKSHEET
"DECON"

I. GENERAL INFORMATION

The Decontamination Officer Worksheet is designed to be used by those designated individuals who will function as the Decontamination Officer. The objective of the worksheet is to ensure the safe and effective implementation of decontamination operations at emergencies involving spills and releases of hazardous materials.

The Decontamination Officer shall assume responsibility for the development and implementation of the decontamination operation and ensuring that all personnel, clothing, and equipment are cleaned prior to their being returned to service. The Decon Officer reports to and must coordinate with the Hazmat Sector Officer from the time the decon area is being set up until the incident is terminated.

The radio designation for the Decontamination Officer will be "DECON."

II. TASKS AND DUTIES

❏ The DECON Officer shall ensure that the decontamination area is set up and that sufficient materials are available for the decon operations.

❏ The selection of DECON operations, procedures, and protective clothing shall be coordinated with the Haz Mat Safety Officer and the Haz Mat Sector Officer, as appropriate.

❏ The DECON Officer shall follow and complete the Hazardous Materials Decontamination Worksheet.

III. DECON INFORMATION

❏ HAZARDOUS MATERIAL(S) INVOLVED

____

____

____

❏ IMMEDIATE HEALTH EFFECTS OF EXPOSURE

____

____

____

❏ EFFECTS OF HAZARDOUS MATERIALS ON CLOTHING AND EQUIPMENT

____

____

❏ REACTIVITY OF HAZARDOUS MATERIALS WITH WATER?

YES __________ NO __________

- ❑ REACTIVITY OF HAZARDOUS MATERIALS WITH DECON SOLUTIONS?

  YES __________ NO __________

- ❑ SELECTION OF DECON METHOD/PROCEDURE COORDINATED WITH HAZ MAT SAFETY AND FACILITY TECHNICAL SUPPORT PERSONNEL?
- ❑ DECON PROCEDURES DETERMINED

  PPE ______________________________

  EQUIPMENT ______________________________

  SKIN/BODY ______________________________

  EMERGENCY ______________________________

NOTES:

DECON SITE SELECTION

- ❑ DECON AREA LOCATED IN THE WARM ZONE AT EXIT FROM HOT ZONE
- ❑ DECON AREA POSITIONED BASED UPON GROUND/FLOOR CONTOUR AND WIND DIRECTION/AIR FLOWS? (i.e., uphill, location of drains, wind direction, air flows, etc.)
- ❑ DECON AREA LEVEL OR SLOPED TOWARD ENTRANCE

DECON RESOURCE REQUIREMENTS

- ❑ WATER SUPPLY ESTABLISHED, IF NECESSARY
- ❑ SUFFICIENT AMOUNT OF DECON SOLUTIONS AND SUPPORTING EQUIPMENT AVAILABLE
- ❑ DECON TEAM PROTECTIVE CLOTHING COORDINATED WITH SAFETY

  PROTECTIVE CLOTHING ______________________________

  GLOVES ______________________________

  BOOTS ______________________________

  RESPIRATORY PROTECTION ______________________________

- ❑ SUFFICIENT RESPIRATORY PROTECTION DEVICES AVAILABLE (i.e., extra SCBA, air bottles, filters, etc.)

❏ PERSONAL SHOWERING REQUIREMENTS AND FACILITIES ESTABLISHED

_____ On Incident Site _____ Supervised Shower Within Facility After Incident

DECON SITE SET-UP

❏ DECON STATION IS WELL MARKED AND DESIGNATED

❏ RUN-OFF CONSIDERATIONS—RECHECK LOCATION OF SEWERS/DRAINS NEAR DECON STATION

MUST BE CONTAINED __________

PERMITTED INTO FACILITY SEWERS __________ (Note Person Granting Approval)

❏ CONTAINMENT BASINS SET UP WITHIN CONTAINMENT AREA

❏ WATER FLOW ESTABLISHED

❏ SUFFICIENT DISPOSAL CONTAINERS AVAILABLE AND IN PLACE FOR CONTAMINATED CLOTHING AND EQUIPMENT DROP-OFF

❏ SPARE RESPIRATORY PROTECTION DEVICES AVAILABLE FOR DECON PERSONNEL AND ENTRY CREWS

❏ DECON SOLUTIONS MIXED

❏ ENTRY AND EXIT POINTS ARE WELL MARKED

❏ EMS PERSONNEL ADVISED AND PREPARED

DECON PRE-ENTRY CHECKLIST

❏ ALL DECON / CLEANING EQUIPMENT IS IN POSITION

❏ DECON TEAM IN PROTECTIVE CLOTHING

❏ ENTRY TEAM BRIEFED ON DECON PROCEDURES

DECON PROCEDURES

❏ PERSONNEL ENTER THE DECON AREA

___ Drop tools on the dirty side ___ Confirm that entry personnel are okay and air supply adequate

❏ REMOVE CONTAMINANTS

___ Step into containment basin

___ Protective clothing examined for cuts and breaches

___ Scrub contaminated entry personnel

❏ REMOVE / REPLACE RESPIRATORY PROTECTION

___ Vapor-tight (Level A) chemical suits = First open suit

___ Disconnect SCBA low pressure hose or mask-mounted regulator while personnel hold breath

___ INSERT new regulator into low pressure hose or facepiece while personnel exhale

- ❑ REMOVE PROTECTIVE CLOTHING
  - ___ Remove duct tape or bands, if used
  - ___ Unzip protective clothing and remove, turning inside out during removal
  - ___ Undress entry crew
  - ___ Place contaminated clothing in disposal containers
- ❑ REMOVE PERSONAL CLOTHING (IF REQUIRED)
- ❑ REMOVE UNDERGARMENTS AND SHOWER (IF REQUIRED)
- ❑ DRY OFF AND RE-DRESS INTO CLEAN CLOTHING
- ❑ EMS EVALUATION

INCIDENT TERMINATION

- ❑ DISPOSABLE MATERIALS ARE ISOLATED, BAGGED, AND PLACED INTO APPROVED CONTAINERS OR PLASTIC BAGS
- ❑ ALL CONTAINERS ARE SEALED, MARKED, AND ISOLATED
- ❑ ALL EQUIPMENT CLEANED AND ACCOUNTED FOR
- ❑ DOES ANY EQUIPMENT REQUIRE ISOLATION FOR FURTHER ANALYSIS OR DECONTAMINATION?

  YES __________ NO __________

  SPECIFY:
- ❑ HAS ALL CONTRACTOR EQUIPMENT BEEN DECONTAMINATED?

  YES __________ NO __________

  SPECIFY:
- ❑ DECON PERSONNEL CLEANED?
- ❑ ALL ENTRY PERSONNEL CLEANED?
- ❑ DECON SOLUTIONS CONTAINED AND DISPOSED OF PROPERLY
- ❑ REPLENISH DECON SUPPLIES
- ❑ TERMINATE DECON OPERATIONS

NOTES:

FIGURE 11-4: EXAMPLE OF A DECONTAMINATION OFFICER'S CHECKLIST

NOTES

Figure 11-5 provides an example of where decon should be placed within the Hazard Control Zone.

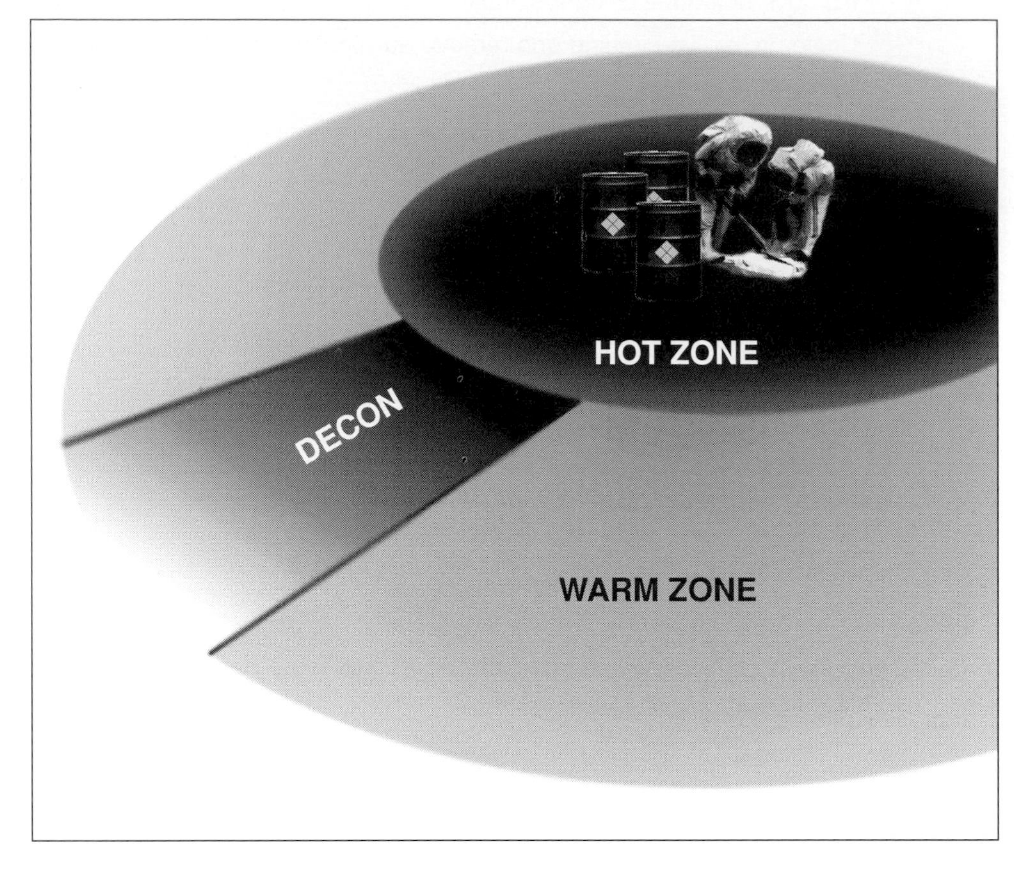

FIGURE 11-5: DECON WORKS BEST WHEN SET UP IN THE WARM ZONE

Once a decon site has been selected, a decontamination corridor should be marked. The decon corridor is simply a pathway from the Hot Zone into the decon area. The decon corridor and the boundaries of the decon area should be clearly identified. Examples include using fence post stakes and colored fire line tape, traffic cones, etc. Signs may also be used to indicate the entry and exit points to the decon area. Pre-sized tarps with various stations marked on them can help organize the decon area. (Imagine a cattle chute going into a corral.)

When the IC is faced with decon locations more than 100 yards from the incident, team transportation is an additional requirement. Vehicles will be necessary to carry the team itself and their SCBA, personal protective clothing, and equipment to the scene. Experience has shown that a pick-up truck provides the easiest access to the Hot Zone for team members and their equipment. Beware of driving too close when flammable atmospheres are present. The driver must also wear the appropriate level of personal protection, including SCBA. This can be accomplished by placing the SCBA beside the driver on the seat. Placing a plastic tarp in the pick-up truck bed can minimize the spread of contaminants. (Nobody said this was going to be easy.)

NOTES

PERSONAL PROTECTION OF THE DECON TEAM

When setting up the decon area, some consideration must be given to how the decon operation will be staffed and the safety of the Decon Team. The team must be properly protected and provided with an uninterrupted air supply. Airline breathing units (discussed in Chapter Eight) can significantly reduce the time required for the Decon Team to reservice their own units. Extended air supply and increased mobility are two advantages.

In some situations, decon personnel should wear the same level of PPE as ERP who are operating in the Hot Zone. However, in most cases, decon personnel may be sufficiently protected by wearing PPE one level below that of the entry crews.

When using a multi-station decon system, decon personnel who initially come in contact with personnel and equipment leaving the hot zone will often require more protection than those decon workers assigned to the final station in the decon corridor. All decon personnel must also be decontaminated before leaving the decon area. The extent of their decontamination process will be determined by the types of contaminants involved in the emergency and the type of work performed within the decon operation.

FIELD DECONTAMINATION PROCEDURES

Decontamination can be carried out safely and smoothly if ERP follow a standard operating procedure which is practical and suitable for use in the field. As is the case with any emergency response procedure, implementing it successfully requires regular training and practice.

Many good "model" procedures are available to serve as a basis for developing and improving your own system. While there are a variety of different approaches to decon, the better standard operating procedures are built around some basic principles:

- Contaminated people and equipment generally flow from the dirty end (area of highest contamination) to the clean end (area of least contamination). The best analogy we have come across to explain this concept is an automated car wash. You drive your dirty car in one end and it comes out clean at the other end.
- Just like the car wash, decontamination requires a multiple-step process to reduce contaminants to an acceptable level. There are two main reasons for adopting a multi-step cleaning process:
  1) Conducting all of the cleaning process at one station concentrates all of the contamination in one area. The more things you clean in the same spot, the greater the contamination level becomes. Would you really want to step out of your protective clothing into a highly contaminated puddle of water?
  2) Multiple decontamination stations make you cleaner. The further along the decon line you progress, the cleaner you should become. (Think about it, does a dirty shirt get cleaner the more times you launder it?)

NOTES

- Supplies and equipment needed in the decontamination area should flow from the uncontaminated (clean) side to the contaminated (dirty) side.

As long as your local Emergency Response Team has incorporated these basic concepts into the procedure, it really doesn't matter too much whether you have two or eight different decon stations or what you call them (e.g., stations, steps, phases, etc.).

The decision to implement all or part of a decon procedure should be based upon a field analysis of the hazards and risks involved. This generally consists of checking technical reference sources to determine the general hazards, such as flammability and toxicity, and then evaluating the relative risks. As discussed in Chapter Seven, Hazard and Risk Assessment, there is no one best resource for field decon. The usual reference manuals will not spell out how to decon personnel or equipment. However, some on-line computer data bases will include this type of data.

Technical data and advice from product specialists may be difficult to obtain early in the incident when you need it the most. Response personnel need to plan before the incident. Learn who the best sources for reliable information are, the right questions to ask them, and how to reconcile conflicting information.

EMERGENCY DECONTAMINATION

Situations can occur when ERP become unknowingly or accidentally contaminated. This problem is normally associated with routine fire or accident responses to buildings or locations which are not normally viewed as hazardous materials occupancies (e.g., an apartment building).

If hazmat identification clues are not present (e.g., placards, labels, etc.), ERP can become contaminated as they operate in or near the area. Some typical problems include:

- Structural fires in residential and commercial occupancies where hazardous materials such as cleaning supplies, garden fertilizers, and pesticides may be stored. Firefighters engaged in overhaul can become contaminated when they break open damaged hazmat containers.
- Fires in storage lockers, garages, and sheds where a variety of small hazmat containers may be stored. Containers can explode or rupture, contaminating firefighters.
- Fires in dumpsters, trash compactors, and garbage cans where discarded small hazmat containers may be hidden.
- Fires in Clandestine Drug Labs. See Chapter Ten for more information on these illegal facilities.
- Motor vehicle accidents and fires where batteries can explode spraying sulfuric acid onto EMS personnel and rescue personnel.

When ERP's have been accidentally contaminated under similar conditions, emergency decontamination must be performed in an expedient manner. Remember, this is not a controlled HMRT situation. No chemical protective clothing is in place, and there is usually no specialized equipment or expertise on scene to assist.

NOTES

Some key points to remember in these situations are:

1. Establish an Area of Refuge as soon as possible. (Corral them).
2. Establish a gross decon area using a hoseline and an improvised basin constructed from salvage covers, plastic sheeting, and ladders, etc. (Clean them.)

Emergency decon operations are graded on speed, not neatness. The sooner you decontaminate the better.

MULTIPLE-STATION DECONTAMINATION

The typical hazmat incident does not require an extensive or elaborate decon procedure. The following example is designed to convey the full range of decontamination steps that may be required for a worst-case scenario. The procedure shown in the photographs uses a chemical vapor protective suit (EPA Level A) with an inside-the-suit positive pressure SCBA. The HMRT member is wearing an ice vest, radio with head set, and a PASS device. A triple glove combination is used to demonstrate the proper method of removing gloves. The tape shown around the boots is specially designed by 3-M for use with PPE. Begin the sequence by studying the set-up shown in Figure 11-6.

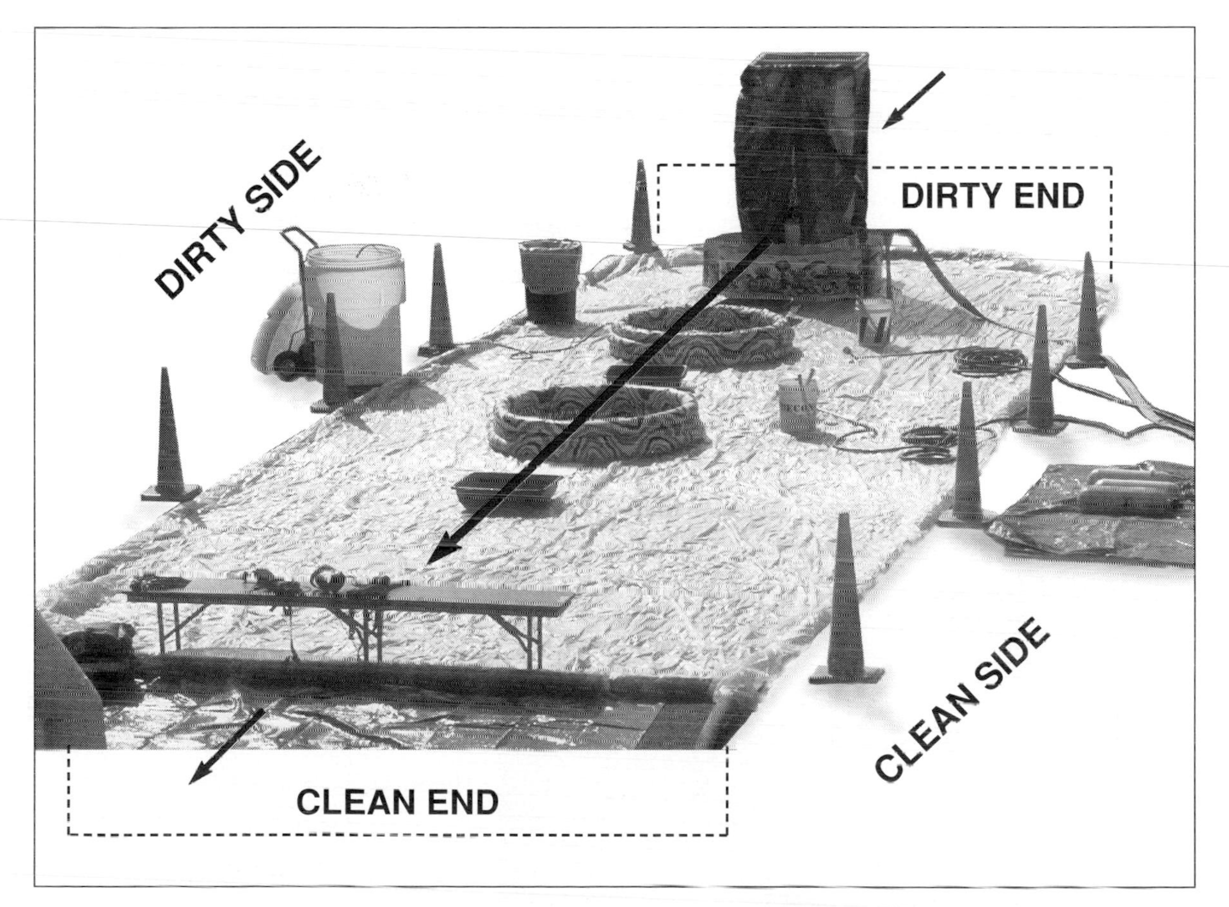

FIGURE 11-6: DECON IS A MULTI-STEP PROCESS

Station 1: Establish an Entry Point and Handle Emergency Decontamination as Necessary.

An "entry point" should be established and clearly marked in order to guide contaminated personnel into the decon area. Make sure the entrance can be easily seen and quickly accessed. If the operation will occur at night, the area should be well lighted on both sides to avoid shadows, which will make cleaning more difficult for the Decon Team.

As ERP approach the decon area, the Decon Team should establish radio or visual contact and determine if there are any problems (e.g., air supply or illness). Look for the prearranged distress signal. Most HMRT's adopt a standard hand signal. This is conveyed to the entry team as part of their pre-entry briefing.

Obvious medical emergencies should be dealt with immediately. If time permits, conduct an on-the-spot gross rinse of the PPE; then open the suit to gain access to the distressed ERP's SCBA facepiece. If the ERP is in respiratory distress or out of breathing air, immediately remove the facepiece and establish an airway. Every attempt should be made to minimize interior contamination, especially if the contaminant is acutely toxic. If other medical problems exist, follow the general emergency medical guidelines discussed later in this chapter. See Figures 11-7 and 11-8.

FIGURE 11-7: EMERGENCY DECON PROCEDURE

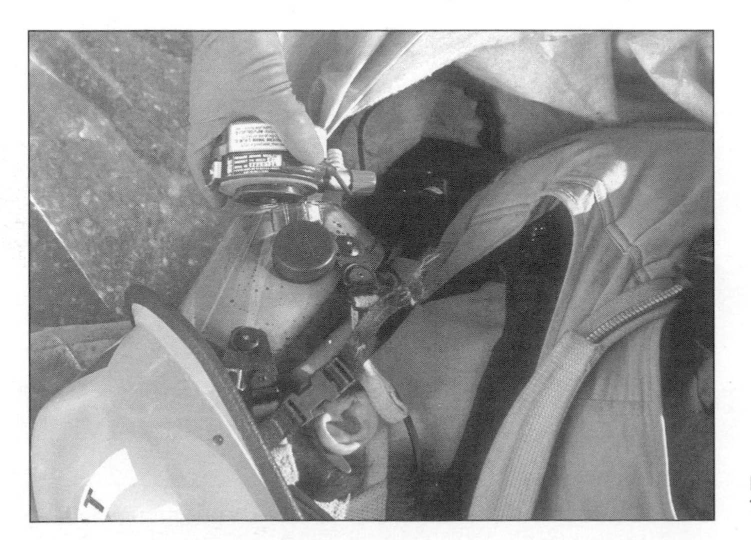

FIGURE 11-8: GAIN ACCESS TO THE FACEPIECE

Any tools or instruments which may be needed by other team members in the Hot Zone should be left here. Tools may be picked up again by the next entry team. If a team rotation schedule is being used (see Chapter Ten), tools should be left near the work site. See Figure 11-9.

FIGURE 11-9: PLACE TOOLS ON A PLASTIC SHEET

Station 2: Technical Decon

With PPE still in place, the ERP should step down the decon line for gross decon. If the situation calls for a "wet decon," the ERP steps into a portable shower, where high pressure and low volume water flow is used to rinse off and dilute contaminants. See Figure 11-10.

FIGURE 11-10: GROSS DECON SHOWERS CAN BLAST AWAY AND DILUTE CONTAMINANTS

For tough contaminants, several intermediate cleaning steps may be necessary. In these cases, one to three different wash and rinse stations may be used. This is sometimes referred to as secondary or technical decontamination, although there is no widely accepted terminology.

FIGURE 11-11: A SECONDARY DECON INVOLVES SEVERAL WASH AND RINSE STATIONS

Normally, secondary wash and rinse stations are supervised by Decon Team members who scrub the outside of the PPE from top to bottom. Care should be used to minimize overspray and splashing by using low water pressure. Specially designed long-neck shower heads can direct water exactly where it is needed and minimize the possibility of cross contamination. See Figure 11-11.

If the entry team has walked through the contaminants, an additional station may be needed to brush and scrape them away. Areas needing special attention are the bottom of boots, gloves, and the kneecap areas. Experience shows that decontamination to this level of detail requires two Decon Team members. One person stabilizes the ERP as the right and left foot are raised for scrubbing. The ERP being cleaned maintains balance by grabbing onto the handle of a scrub brush which is held out by the Decon Team member. See Figure 11-12.

FIGURE 11-12: STABILIZE THE ERP BEING CLEANED WITH A SCRUB BRUSH OR BROOM STICK

The Decon Team can also provide stability while scrubbing the boots by using two scrub brushes simultaneously. One brush is placed under the boot while the other brush cleans the sole. A "Walker" commonly used by the disabled can provide the same type of balance and stability. See Figure 11-13.

FIGURE 11-13: A DOUBLE SCRUB BRUSH COMBINATION CAN PROVIDE BALANCE AND STABILITY FOR THE ERP BEING SCRUBBED

Station 3: SCBA Removal

If the SCBA is worn inside of the PPE as shown in our example, there should be no contamination inside the suit. This is especially the case when primary and secondary decontamination has been conducted correctly. The Decon Team unzips the PPE and peels back the suit, exposing the SCBA. The PPE should be folded back so that the dirty side is facing inward. If the procedure is done properly, there should be no ERP contact with the dirty side. With the assistance of the Decon Team, the ERP simply removes the SCBA facepiece and moves on down the line to complete the removal of the PPE, SCBA, and other support equipment. See Figures 11-14 and 11-15.

FIGURE 11-14: UNZIP THE PPE AND PEEL IT BACK, EXPOSING THE SCBA

FIGURE 11-15

Anything you do in the way of decon from this point on is often overkill. However, as stated above, for purposes of this section, we have assumed that a worst-case situation exists. In the example provided, we will assume that contamination may have passed to the inside of the PPE and that we are dealing with a serious problem.

High risk exposures that present serious inhalation hazards may require a complete SCBA change to a clean, serviced unit. An alternative to this is plugging into a supplied airline respirator (SAR). The idea is that if the inside of the suit is believed to have been contaminated, supplying the respiratory system with clean air as long as possible will buy time to further assess the level of contamination and conduct more detailed decontamination.

FIGURE 11-16: SCBA CHANGEOVER SEQUENCE

An SCBA changeover can be accomplished by having one member of the Decon Team hold the serviced SCBA while the second Decon Team member disconnects the regulator from the ERP's facepiece. The regulator from the serviced SCBA is then connected to the ERP's facepiece. See Figures 11-16, 11-17, and 11-18.

FIGURE 11-17

FIGURE 11-18

FIGURE 11-19: PLACE SCBA ON A TARP FOR CLEAN-UP LATER

Changing the entire SCBA, rather than just the cylinder, ensures a higher safety level for the entry person. SCBA units which are removed should be stacked on plastic tarps on the dirty side of the decon area or placed in individual plastic bags. See Figure 11-19.

From a support point of view, sufficient fire/rescue apparatus or breathing air units should be on hand to completely replace or resupply SCBA. When an extended Hot Zone operation requiring several entries is anticipated, additional air supply units should be requested.

When it is necessary to run air compressors in the field, it should be done in an area remote from the incident to guarantee that contaminated air is not drawn into the compressor pump and passed on to refilled cylinders. Personnel engaged in decontamination should not participate in reservicing any SCBA's.

Station 4: Removal and Isolation of PPE.

Regardless of the decon situation, a systematic approach should be used to minimize the spread of contaminants. If you practice this technique every time you take off your PPE, you will minimize the possibility of direct and cross contamination when you run into a serious problem.

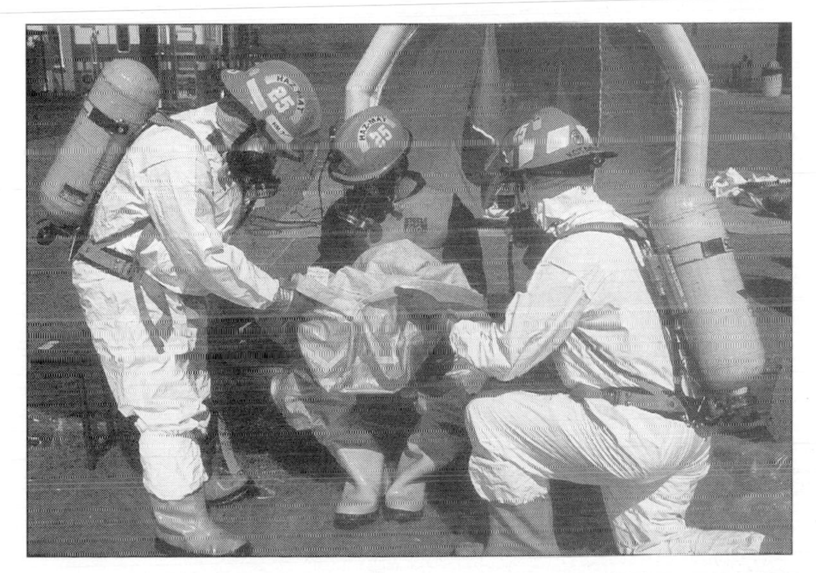
FIGURE 11-20: PPE IS EASIER TO REMOVE WHEN ERP ARE SEATED

The general approach to removing protective clothing is to turn as much of the PPE inside out as possible. This folds any remaining trace contaminants back inside the suit or glove.

The order of removing individual garments, boots, gloves, etc. will be different from one ensemble to the next. Removing PPE is a lot easier if the ERP is seated and doesn't help too much. See Figure 11-20.

The helmet, boots, and other support equipment like ice vest and radio should be removed along with the SCBA unit. The facepiece is usually left on and taken off last as an additional safeguard against trace contaminants. Equipment should be placed on the dirty side of the decon area until clean-up begins. See Figures 11-21 and 11-22.

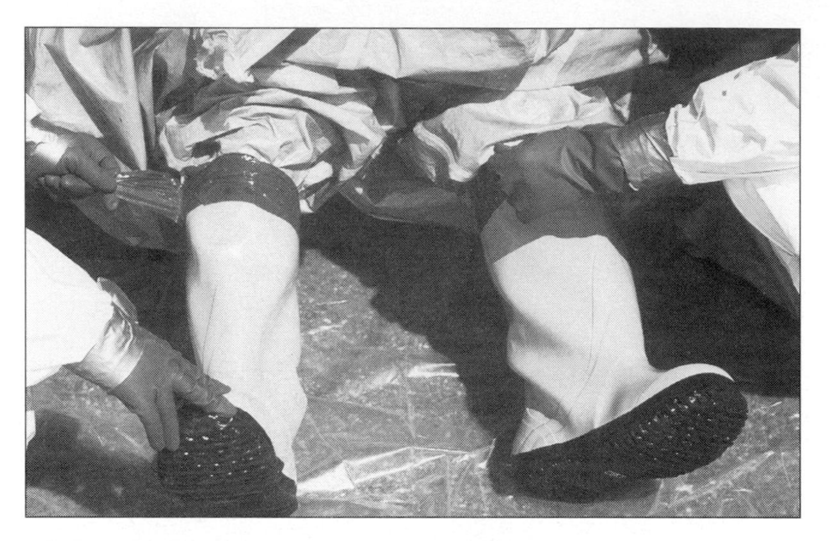

FIGURE 11-21: REMOVAL OF BOOTS AND HELMET

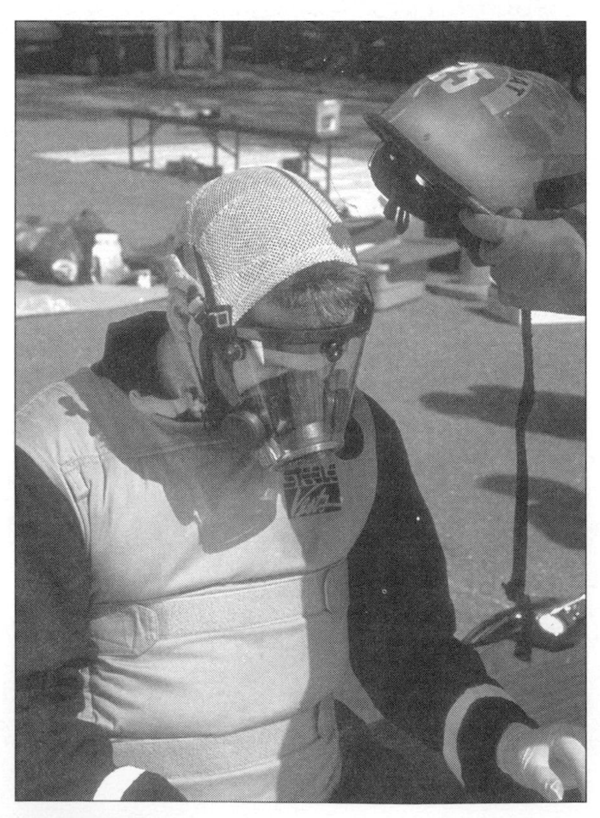

FIGURE 11-22

Gloves should be removed in sequence and turned inside out. Outer gloves may need to be removed earlier, along with protective clothing, depending on the ensemble and level of contamination. Gloves should be sorted into individual containers for clean-up or disposal. See Figures 11-23 through 11-30.

REMOVE GLOVES IN SEQUENCE AND TURN THEM INSIDE OUT

FIGURE 11-23

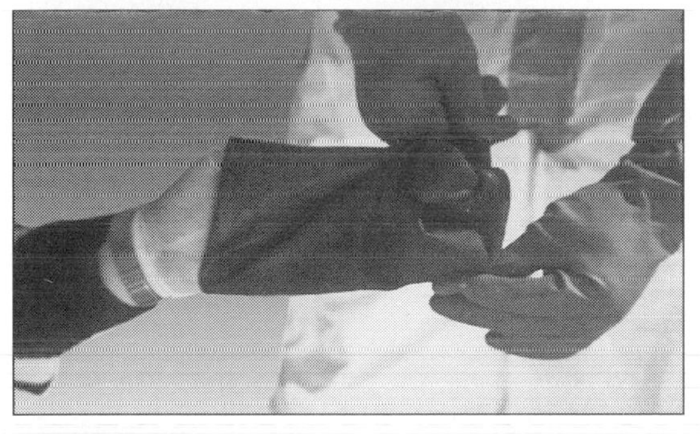

FIGURE 11-24

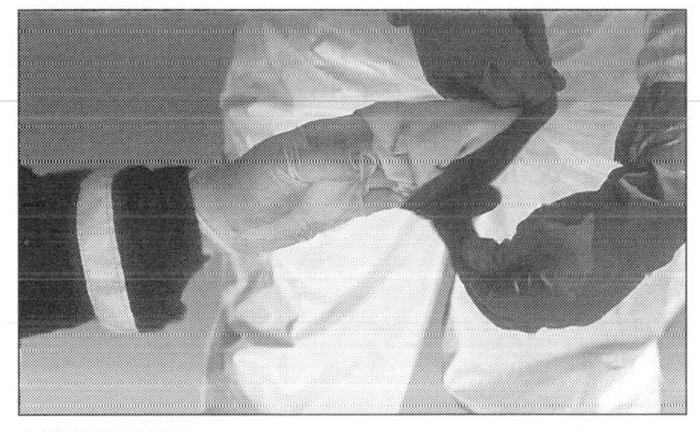

FIGURE 11-25

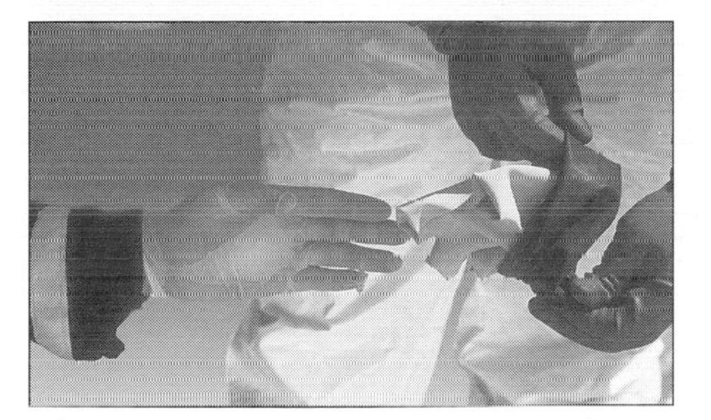

FIGURE 11-26

FIGURE 11-27

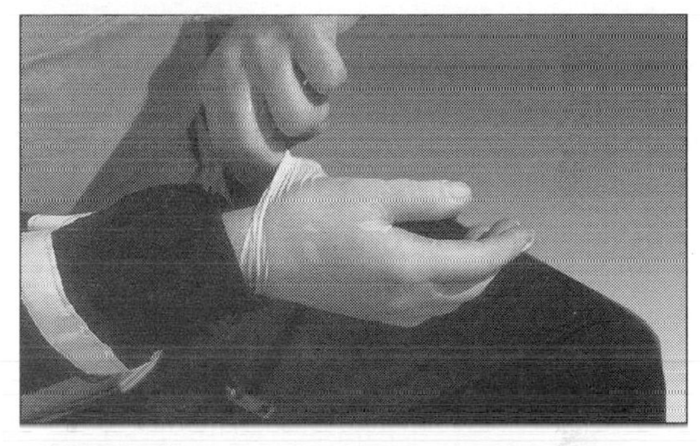

FIGURE 11-28

FIGURE 11-29

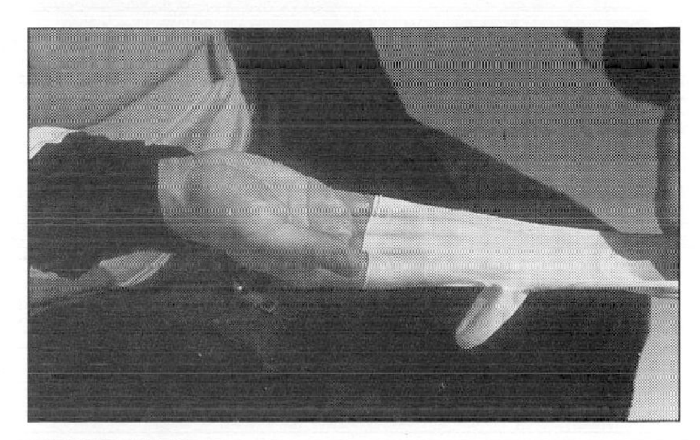

FIGURE 11-30

Station 5: Removal of Personal Clothing.

If personal clothing has been contaminated, it should be removed as soon as possible. Provisions should be made to bag clothing individually in the event uncontaminated clothing can be salvaged.

Prior to entering contaminated areas, all personal items such as rings, watches, wallets, and jewelry should be removed to prevent their possible contamination. It is a good idea for the Entry Team and Decon Team to automatically remove all such items whenever work in the Hot Zone and decon area are required. Make sure that their security and identification have been provided for.

If personal clothing or work uniforms have been confirmed as contaminated, ERP should progress to a showering station. This may be performed on-site, inside of a portable enclosure. See Figure 11-31.

If the contaminant is not acutely toxic and presents no immediate safety problems, ERP's can be driven to a showering facility such as a fire station or plant change-house. This is done primarily as a precaution and for personal hygiene reasons.

FIGURE 11-31: REMOVE CONTAMINATED CLOTHING IMMEDIATLEY

Station 6: Wash the Body.

There are a number of very good inflatable and portable showering facilities available for use in the field. These can be set up by two or three ERP by using a portable hand pump or an SCBA. The interior of the temporary structure usually has three chambers. The first is used for undressing, the second for showering, and the third for dressing in clean clothing. The better models have complete containment systems and drainage ports which can be connected to a hand pump for transferring liquid waste to a 55-gallon holding drum. Showering tents are usually placed in proximity to the end of the decon line. See Figure 11-32.

FIGURE 11-32: PORTABLE SHOWERING TENTS SHOULD BE PLACED NEAR THE END OF THE DECON LINE

Trailer mounted mobile decontamination units are an alternative to inflatable and portable showering facilities. While they are significantly more expensive, mobile units offer the ability to process large numbers of people over an extended period of time in the field. They also may include creature comforts like heated water, electric lights, heating and air conditioning, and better showering and dressing areas.

An overhead shower produces much better results than a hose line. Ample soap should be applied to all areas of the body, especially the head and groin. (This is where teamwork ends. You are on your own here.)

Liquid surgical soaps placed in plastic squeeze bottles get the best results. Small brushes and sponges should be used for body surface scrubbing. Pre-packaged cleaning kits are available through local hospitals. All cleaning items used should be bagged and marked for disposal. Likewise, the runoff should be controlled whenever possible.

If an emergency shower was required, contact lens users should be re-evaluated. If any discomfort or irritation is noted, remove the lens and flush the eyes to remove any contaminants collected under the lens. (Any emergency eye treatment would have been initiated immediately at the first station using emergency decon procedures.)

Station 7: Dry off the Body and Put on Clean Clothing.

Towels or sheets should be used to dry off. Each towel is used once and placed in bags on the contaminated side for laundering or disposal. Plastic garbage cans lined with plastic bags are good temporary storage containers.

Clean clothes are next. Cotton coveralls, disposable paper coveralls, and hospital surgical gowns are examples of items commonly used. These garments can be pre-packaged with inexpensive footwear like sneakers and labeled according to size. Pre-packaged units are stored in kits for easy access and transportation. See Figure 11-33.

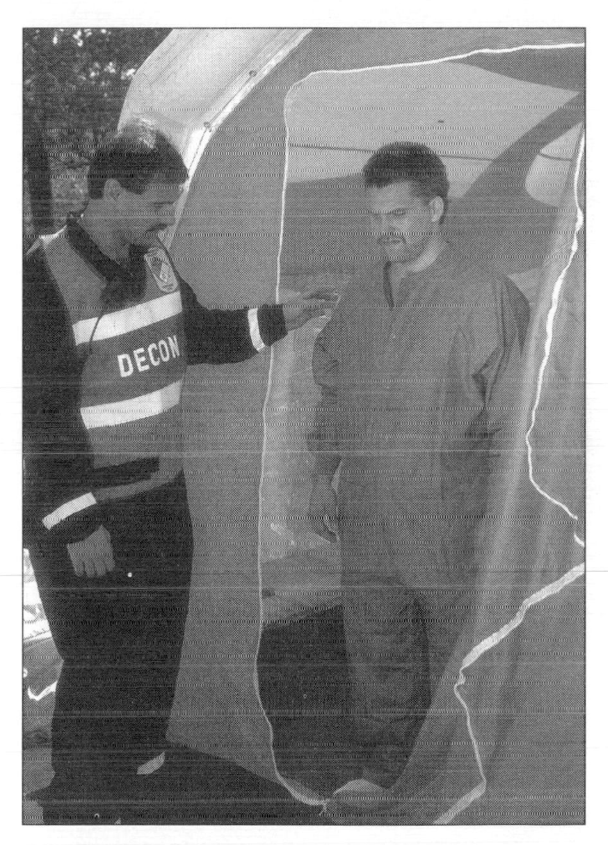

FIGURE 11-33: SHOWER AND CHANGE CLOTHES

Station 8: On-Site Medical Evaluation.

Once personnel are thoroughly decontaminated, they should proceed to an EMS station for evaluation by the Hazmat Medical Group. Those staffing the medical area should be familiar with correct procedures for treating chemical injuries and have a communications link with other advanced care specialists, including a medical command facility and Poison Control Center. Medical staff should have access to HMRT baseline physical data.

Vital signs should be noted for each person leaving decon and compared to the initial Entry Team's baseline screening data. They should be regularly monitored and recorded throughout the incident for later evaluation.

Any open wounds or breaks in skin surface should be immediately reported to the medical control. Unless advised otherwise by the medical command physician, all open wounds should be cleaned on the scene. The EMS Sector should be alerted immediately of possible contaminant hazards as soon as they confirmed. See Figure 11-34.

Station 9: Advanced Medical Evaluation.

If ERP have received a chemical exposure or have been injured by a hazardous material, they should be transported to the appropriate hospital for further evaluation and monitoring.

Each person who passed through the decon line should be documented by the Decon Officer along with the decontamination method used. Documentation should be included in the termination activity report. See Figure 11-35.

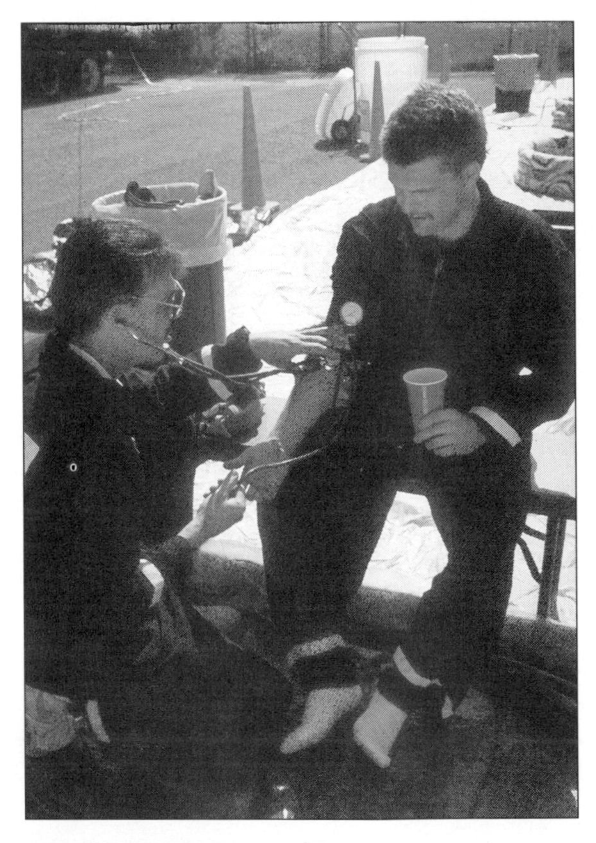

FIGURE 11-34: CONDUCT ON-SITE MEDICAL EVALUATION AFTER LEAVING THE DECON AREA

FIGURE 11-35: THE DECON OFFICER SHOULD DOCUMENT THE EXPOSURE AND METHOD OF DECONTAMINATION USED

NOTES

MEDICAL EMERGENCIES REQUIRING DECON

While this text has not been written to address the wide range of hazmat-related medical emergencies, we thought that it was important to review some of the basic emergency medical procedures which apply to hazardous materials decontamination of a patient or ERP.

There are several emergency medical texts which should be consulted for more specialized information. We recommend the following:

- EMERGENCY CARE FOR HAZARDOUS MATERIALS EXPOSURE, by Dr. Alvin C. Bronstein and Philip L. Currance.
- HAZARDOUS MATERIALS EXPOSURE: Emergency Response and Patient Care, Dr. Jonathan Borak, M.D., and Michael Callan, and William Abott.
- HAZARDOUS MATERIALS INJURIES: A Handbook for Pre-Hospital Care, by Dr. Douglas R. Stutz and Stanley J. Janusz.
- PARAMEDIC EMERGENCY CARE, by Bryan E. Bledsoe, D.O., Robert S. Porter and Bruce R. Shade.

The following general emergency medical guidelines have been referenced from these excellent publications.

EMERGENCY MEDICAL DECONTAMINATION

Medical emergencies involving hazardous materials are best handled in the field. As much decontamination as possible should be performed before the patient is transported from the scene of the emergency. The primary objective is to minimize the spread of contaminants downstream from the patient to emergency medical personnel and eventually to the hospital staff.

Basic points that should be considered when assessing emergency decontamination and patient care are as follows:

- ❑ The injured person should be removed from the contaminated area by ERP using appropriate PPE. *Rescue should not be attempted by untrained or improperly equipped personnel.*
- ❑ Minimize the amount of emergency care performed in the contaminated area. Make an attempt to open the patient's airway and immobilize the cervical spine if there is any possibility of injury to the head or neck.
- ❑ Following removal of the patient to the decon area, basic care and decontamination can begin. Maintain the airway. All ERP and emergency medical personnel involved in the decon operation must be properly protected. For EMS personnel, this would include disposable or limited-use chemical coveralls with a hood, a face mask and/or eye protection, and double gloves.
- ❑ Carefully remove and isolate contaminated clothing, eye glasses, jewelry, and shoes. Save all contaminated articles that are removed from the patient, place them in separate plastic bags, and mark them with the individual's name.

NOTES

- ❑ Brush any solid or particle contaminants off the skin as gently and completely as possible before washing to reduce the chance of reaction with water. Brush contaminants away from wounds. Blot heavy liquid contaminants from the body before washing to reduce the chance of dilution or increased absorption. Exercise caution not to cause any skin damage.
- ❑ Once any visible contaminant is removed, rinse and wash the person. Soaps which are used for patient decon should be mild and nonabrasive. Tincture of green soap is desirable because it approximates the skin's pH level. Its alcohol base also helps to remove hydrocarbons and solvents from the skin. If green soap is not available, any mild liquid soap such as dishwashing detergent will work. Never use decon solutions on the skin, as they may cause burns and further injury.
- ❑ Begin decontamination at the head and face to allow for proper airway control and respiratory support. Be careful not to flush contaminants into the eyes or ears. Some teams protect the patient by placing goggles over the eyes. Then clean areas of gross contamination and soft tissue damage next (e.g., burns, bruises, lacerations, etc.). Care must be taken not to flush contaminants into wounds. Carefully wash and rinse wound areas from the center outward. After decon, cover areas of soft tissue damage with a water-occlusive dressing or a plastic wrap to prevent secondary contamination.
- ❑ If the eyes require emergency irrigation due to contamination, continue this process enroute to the hospital.
- ❑ Once all wound areas are clean, the remainder of the body can then be decontaminated. Pay special attention to ear and nose cavities, hair, nail beds, and skin folds. Soft brushes and sponges may be used. Be careful not to abrade the skin. Use extra caution over bruised or broken skin areas.
- ❑ Rinse the patient with large quantities of water. Use low water pressure and a gentle spray to avoid aggravating any soft tissue damage. Try to contain all runoff, but do not delay treatment in life-threatening situations if containment is not practical. Use warm water to provide for patient comfort and reduce the potential for hypothermia. If warm water is not available, cold water can be used, but it will increase the chance of hypothermia. **NEVER USE HOT WATER!**
- ❑ Under ideal circumstances, the patient should be fully decontaminated prior to transportation to a medical facility. In most cases, this will eliminate the chance of secondary contamination of both emergency medical personnel and hospital staff. However, when dealing with emergencies involving multiple injuries or secondary problems, total commitment may not be focused solely on patient decon. Advise both the Emergency Medical Services (EMS) unit and the receiving hospital when handling a chemically contaminated patient.
- ❑ The patient should be packaged to prevent the spread of any remaining contaminants. Some EMS personnel have used body bags for this

NOTES

purpose; however, experience shows that they can quickly overheat the patient. (Body bags probably don't do much for the patients' morale either.) Consider using disposable fabric bags (eg-Tyvek™).

- ❑ Ambulances and Medic Units should be adequately prepared before transporting a patient who is still contaminated. Plastic sheeting or specially manufactured cargo bags can be hung inside the EMS unit to prevent the spread of contaminants to the truck cab or to on-board equipment. Never transport a patient who is grossly contaminated. Toxic, flammable, or corrosive contaminants can accumulate in poorly ventilated vehicles and further expose the patient as well as the EMS team. See Figure 11-36.

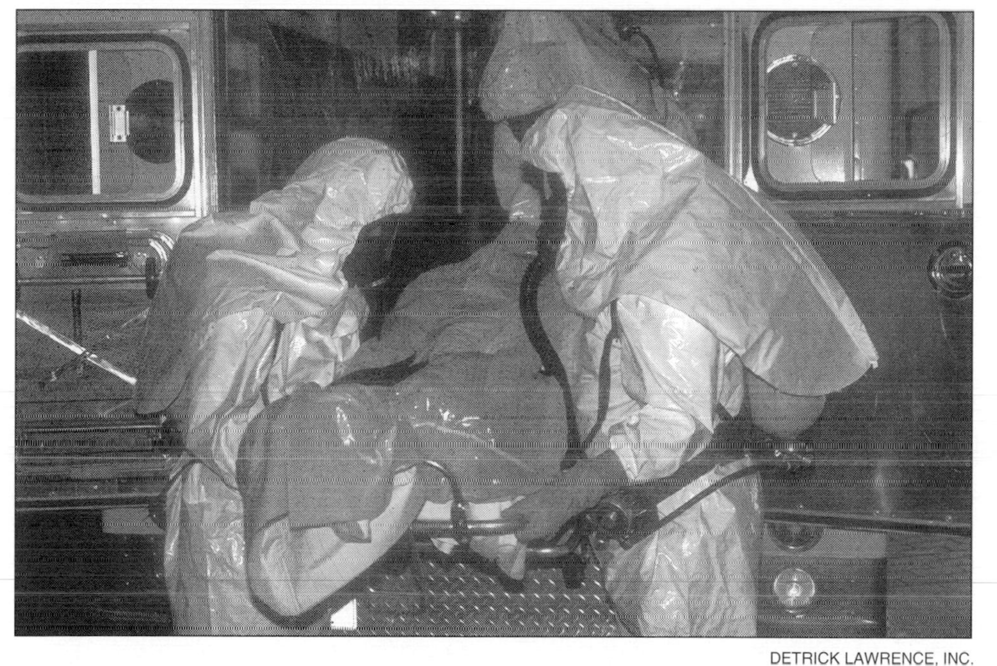

DETRICK LAWRENCE, INC.

FIGURE 11-36: AMBULANCES AND EMS PERSONNEL MUST BE PROTECTED FROM CONTAMINANTS

- ❑ Fatalities should be handled with the same decontamination precautions that are recommended for patients. Contaminated corpses or severed body parts should be properly decontaminated at the scene of the incident before transporting the body. Body identification tags should be used to indicate the type of contaminant.

DECON AND INFECTION CONTROL

Emergency Response Teams can be called to assist with a wide range of problems at facilities which may involve bloodborne pathogens. Examples include hospitals, clinics, medical research labs, mortuaries, medical waste disposal facilities, blood banks, etc. Personnel incur the risk of infection and subsequent illness each time they are exposed to blood or other potentially infectious materials. Any emergency response to a fixed facility where bloodborne pathogens may be involved in an accident or spill should be handled with special precautions.

NOTES

In an emergency medical situation like a rescue situation, the infectious disease status of patients is usually unknown by ERP's; therefore, all patients must be considered infectious. Infectious diseases include bloodborne pathogens, such as the Hepatitis B Virus (HBV) and the Human Immunodeficiency Virus (HIV).

ERP are at no greater risk of contracting bloodborne diseases than members of the general population. Neither HBV nor HIV is transmitted by casual contact in the workplace. Understanding some basic terminology and applying Universal Precautions can help ERP deal with this special hazard.

INFECTION CONTROL TERMINOLOGY

The following terms and definitions are used throughout the Infection Control Guidelines:

BODY FLUIDS—Fluids that the body makes including, but not limited to, blood, saliva, semen, mucus, feces, urine, vaginal secretions, breast milk, amniotic (fetus) fluids, cerebrospinal (brain) fluid, synovial (joint) fluid, pericardial (heart) fluid, and fluids that might contain concentrated HIV or HBV viruses.

CONTAMINATED—Having come in contact with body fluids.

DISINFECTION—The process used to inactivate virtually all recognized pathogenic microorganisms but not necessarily all microbial forms, such as bacterial endospores. Disinfection is not the same as sterilization.

EXPOSURE—Contact with an infectious agent, such as body fluids, through inhalation, accidental needle stick, or contact with an open wound, nonintact skin, or mucous membrane.

FLUID-RESISTANT CLOTHING—Clothing that provides a barrier against splashing or spraying of body fluids or other potentially infectious material.

LEAKPROOF BAGS—Bags that are sufficiently sturdy to prevent leaking or breaking and can be sealed securely to prevent leakage. Such bags are red in color or display the universal biohazard symbol.

MEDICAL GLOVES—Gloves that are designed to provide a barrier against body fluids meeting the requirements of ASTM D 3578, Standard Specification for Rubber Examination Gloves (nonsterile).

MEDICAL WASTE—Human waste, human tissue, blood or body fluids or items contaminated by these materials for which special handling precautions are necessary.

MUCOUS MEMBRANE—A moist layer of tissue that lines the mouth, eyes, nostrils, vagina, anus, or urethra.

SHARPS CONTAINER—Containers that are closable, puncture-resistant, disposable, leakproof on the sides and bottom, that are red in color or display the universal biohazard symbol, and that are designed to store sharp objects like needles and syringes after use.

SPLASH-RESISTANT EYEWEAR—Safety glasses, prescription eyewear, goggles, or chin-length face shields that, when properly worn, provide limited protection against splashes, spray, spatter, droplets, or aerosols of body fluids or other potentially infectious material.

STERILIZATION—The destruction of all microorganisms in or about an object, as by steam, chemical agents, or ultraviolet light radiation.

NOTES

UNIVERSAL PRECAUTIONS

Universal Precautions refers to a system of infectious disease control which assumes that every contact with body fluids is infectious and requires every employee exposed to direct contact with body fluids to be protected as though body fluids were HBV or HIV infected. Universal precautions are intended to protect ERP from needle puncture, mucous membrane, and injured skin exposures to bloodborne pathogens.

EMERGENCY MEDICAL PERSONNEL PROTECTION

- Prior to any contact with patients, ERP should cover all areas of abraded, lacerated, chapped, irritated, or otherwise damaged skin with adhesive dressings.
- Any ERP who has skin contact with body fluids must thoroughly wash the exposed area immediately using water or saline on mucosal surfaces and soap and running water on skin surfaces. If soap and running water are not available, alcohol or other skin cleaning agents that do not require running water should be used until soap and running water can be obtained.

INFECTION CONTROL GARMENTS AND EQUIPMENT

- Medical gloves must be single-use, disposable, and meet the requirements of ASTM D 3578.
- ERP engaging in any emergency patient care must don medical gloves prior to initiating emergency care due to the variety of diseases, modes of transmission, and unpredictable nature of the work environment. Medical gloves must be a standard component of emergency response equipment. See Figure 11-37.
- Medical gloves must be removed as soon as possible after the termination of patient care, taking care to avoid skin contact with the glove exterior. Gloves must be disposed of and hands should be washed as specified in the Cleaning, Disinfecting, and Disposal Section.
- Medical personnel involved in rescue operations should wear structural firefighting gloves over medical gloves since extrication frequently requires removal of work gloves to touch the patient.

NOTES

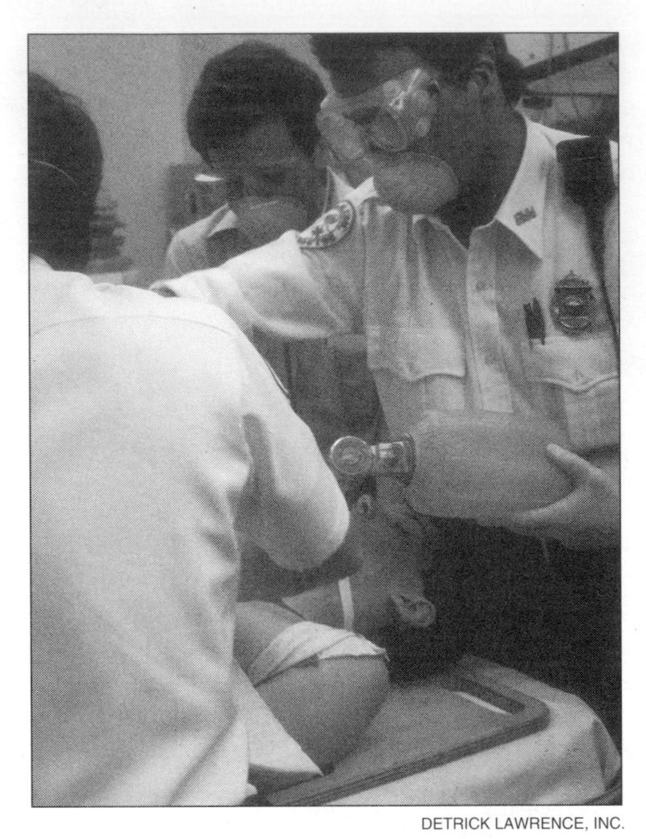

DETRICK LAWRENCE, INC.

FIGURE 11-37: INFECTION CONTROL PROCEDURES REQUIRE PROPER PROTECTIVE CLOTHING BEFORE HANDLING A PATIENT

- Cleaning gloves must be worn during the cleaning or disinfecting of clothing or equipment involved in emergency medical operations. Cleaning gloves must be re-usable, heavy-duty, mid-forearm-length gloves. These gloves should be designed to provide limited protection from abrasions, cuts, snags, and punctures and provide a barrier against body fluids and disinfectants.
- ERP should not eat, drink, smoke, apply cosmetics or lip balm, or handle contact lenses while wearing gloves.
- Masks, splash-resistant eyewear, and fluid-resistant clothing should be donned by ERP prior to any patient care situations during which contamination of the eyes, mouth, or nose with body fluids or large splashes of body fluids can occur. Examples would include spurting blood and childbirth.
- Resuscitation equipment, including pocket masks and bag-valve masks, must be used when providing airway management.

HANDLING SHARP OBJECTS

- When paramedics are administering IV's and drugs during the stabilization and treatment of the patient, all ERP should take precautions to prevent injuries caused by needles and other sharp instruments or devices.

NOTES

- All used sharp objects, such as needles, scalpels, catheter stylets, and other contaminated sharp objects, should be considered infectious and must be handled with extraordinary care.
- Needles must not be re-capped, bent or broken by hand, removed from disposable syringes, or otherwise manipulated by hand. Safe enclosure devices, such as a Sharps Container, should be used.

CLEANING, DISINFECTING, AND DISPOSAL

SKIN WASHING

- Skin surfaces not covered by work uniforms, protective clothing, or infection control garments should be washed after providing emergency patient care.
- Hands should be washed after each emergency medical incident, after cleaning and disinfecting emergency medical equipment, after cleaning protective clothing or equipment, before and after using the bathroom, before and after handling food or cooking utensils, and before and after handling cleaned and disinfected emergency medical equipment.
- Hands and contaminated skin surfaces must be washed with soap and water by lathering the skin and vigorously rubbing together all lathered surfaces for at least 10 seconds, followed by thorough rinsing under running water.

EMERGENCY MEDICAL EQUIPMENT

- All cleaning must be conducted in an area which is physically separated from areas used for food preparation, cleaning of food and cooking utensils, personal hygiene, and sleeping or living areas.
- Infection control garments and equipment for cleaning and disinfecting must be used whenever there is a potential for exposure to body fluids or infectious material during cleaning.
- Prior to cleaning, dirty or contaminated emergency medical equipment should be stored separately from other cleaned and disinfected emergency medical equipment.
- Re-usable emergency medical equipment that comes in contact with mucous membranes will require cleaning and a high level of disinfection or sterilization. The medical equipment manufacturer's instructions should be followed.
- One of the following disinfectant/solutions can be used for the clean-up of blood and/or body fluids from equipment, vehicles, and facilities:
    1) Chemical germicides approved for use as hospital disinfectants and which are tuberculocidal (good enough to kill tuberculosis bacteria) when used at recommended dilutions and exposure times.
    2) A solution of 5.25% sodium hypochlorite (household bleach) diluted between 1:10 and 1:100 with water and prepared daily.

NOTES

Commercially available chemical germicides may be more compatible with medical equipment that is susceptible to corrosion. This is particularly true when more concentrated 1:10 mixtures are used.

3) Look for products registered by EPA as being effective against HIV with an acceptable "HIV (AIDS Virus)" label for clean-ups in which HIV is the only infectious matter of concern (e.g., fluids found in HIV experimental laboratories).

CLOTHING

- Small stains from body fluids can be spot cleaned and then disinfected. The stain must be initially cleaned with a mild detergent and water. The affected area should then be disinfected only with disinfectants that are chemically compatible with the clothing.
- Clothing that is contaminated with large amounts of body fluids must be placed in leakproof bags, sealed, and transported for proper cleaning or disposal.

DISPOSAL OF MATERIALS

- Contaminated disposable medical supplies and equipment, contaminated disposable infection control garments, and contaminated wastes must be placed in leakproof bags, sealed, and disposed of as medical waste. If the exterior of the leakproof bag is contaminated, it should be placed inside another clean bag.
- Noncontaminated disposable medical supplies, infection control garments, and wastes should be collected in closeable waste containers and disposed of properly.
- If it is determined that normally re-usable items cannot be disinfected, they should be placed in leakproof bags, sealed, and disposed of as medical waste.

SPECIAL FACILITIES AND EQUIPMENT

Special research and manufacturing facilities are often engineered and constructed to handle emergencies involving contamination of laboratory and chemical process areas. Some examples of the features that ERP may be able to take advantage of include:

- **Positive and Negative Pressure Atmospheres**—These systems usually maintain a positive air pressure on the outside of the hazardous work area and a negative pressure inside. If an incident occurs, the flow of air is from the outside toward the inside. This prevents contaminants (dust, vapors, and gases) from spreading beyond the containment room. ERP must not shut down the power supply which maintains these special systems.
- **Safety Showers**—Safety showers are usually located throughout these facilities and are restricted for emergency use in the event of

NOTES

accidental chemical contamination. If a hazmat incident has occurred indoors, consider conducting gross decon using the facility fixed safety showers. Safety showers have special "deluge" heads that deliver 30 to 50 gallons per minute, far more than a bathroom shower head. This large flow is essential in the initial stage to sweep away, rather than just to dilute, the strong contaminant. Safety showers are good gross decon options if they have provisions to confine runoff and have drains.

- **Emergency Eyewash Fountains**—Emergency eyewash fountains or hoses are located throughout most industrial facilities. Their location should be identified for ERP during the pre-entry briefing.
- **Fixed Ventilation Systems**—Most facilities engaged in hazardous materials research or manufacturing have specially engineered ventilation systems to reduce employee exposure to hazardous materials. (Removal of air contaminants at the source is the most effective method of preventing employee chemical exposure.)

  These systems are usually designed based on the physical, chemical, and toxicological characteristics of the materials being handled. ERP should consider the use of these specialized systems as a method of reducing the potential contaminants within the Hot Zone. Activation of fixed ventilation systems may significantly reduce the contamination level before entry.

Always consult with the facility building engineer and safety personnel before using fixed ventilation options. Activation of a ventilation system not designed and rated for the hazards present under emergency conditions could make the situation worse (e.g., cause an explosion or spread the contaminants to a larger area).

If special ventilation systems are available, they are usually monitored for toxicity, flammability, and oxygen content. This information is typically available from a central control room.

CLEAN-UP

WHAT IS CLEAN-UP?

The term "clean-up" means different things to different emergency response personnel. For our purposes, clean-up activities consist of any work performed at the emergency scene, by emergency response personnel, which is directed toward removing contaminated protective clothing, tools, dirt, water, etc. Clean-up may also involve ERP decontamination of some debris, damaged containers, etc.

From a decontamination perspective, not all work related to restoring the contaminated site to its previous (nonpolluted) state is considered "clean-up." Chapter Ten provides a detailed review of the short- and long-term site restoration and recovery activities associated with hazmat emergency response.

NOTES

GENERAL CLEAN-UP OPTIONS

Generally speaking, the IC has two clean-up options:

1. Conduct a limited-scale clean-up of key emergency response equipment such as fire apparatus. The objective of this option is to place essential equipment back in service as soon as it is safely possible. ERP may also get involved in the more technical aspects of clean-up by working directly under the supervision of an outside agency or contractor.
2. Conduct clean-up using a qualified and authorized outside contractor. This option is usually exercised when large pieces of heavy equipment have been contaminated and are not part of the emergency response organization's fleet. Examples of such equipment are bulldozers and end loaders used by ERP to construct dikes inside the Hot Zone.

Before jumping into clean-up activities, make sure that you have a well-thought-out and coordinated plan. Once ERP are decontaminated, the rules of the game shift from one of being an Emergency Responder to Hazardous Waste Generator. Make sure that your clean-up activities are within regulatory guidelines.

Agencies which should be consulted in the development of equipment decon plans include:

- **Water/Sewage Treatment Facilities**—Prior arrangements will be necessary before large quantities of waste water can be flushed into storm/sewer systems via drains connected to the street or facilities such as fire stations, gyms, etc. As a general rule, all waste should be contained until permission is received for disposal.
- **Pollution Control (Environmental Protection Agency or U.S. Coast Guard)**—The Incident Commander's authority to create a "runoff" situation during an emergency involving life-threatening materials is well established. Adding to this runoff with equipment decontamination is questionable in most cases. Failure to isolate the runoff could easily result in a citation from regulatory agencies, result in bad publicity, or increase your civil liability.
- **Product Specialists**—May be able to provide clean-up recommendations based on their practical knowledge. Some chemical manufacturers are trained and experienced in equipment decon and can be invaluable when their information is correct and very damaging when wrong. Product specialists should be interviewed to establish credibility as discussed in Chapter Seven. They do not have legal authority and cannot assume the responsibility for approving on- or off-site disposal.

EQUIPMENT CLEAN-UP

General Guidelines

Equipment and apparatus decon can be difficult and very expensive. Liquids can soak into wood and flow into metal cracks and seams or under bolts. Begin decon by consulting the hazmat manufacturer for clean-up recommendations. An ERP with authority should supervise this phase to ensure that proper planning and coordination with local officials takes place.

While decontaminating, avoid direct contact with contaminated equipment. Brooms and sponge mops can be used to apply cleaning agents to equipment. Protective clothing and respiratory protection must be worn unless proven to be unnecessary by technical specialists who have conducted an appropriate analysis of the contaminants.

NOTES

Small and Portable Equipment Clean-Up

All small- to medium-sized equipment such as monitoring instruments, hand tools, etc. should be decontaminated before leaving the site. The following issues and concerns should be considered:

Hand Tools

- Hand tools may be cleaned for re-use or disposed. Cleaning methods include hand cleaning or, more commonly, pressure washing or steam cleaning. One must weigh the cost of the item against the cost of decontamination and the probability that it can be completely cleaned. Wooden and plastic handles on tools should be evaluated to determine if they can be completely decontaminated. The scientific literature referenced at the end of this chapter provides overwhelming evidence that many pesticide-contaminated wooden parts, like shovel handles, *cannot be adequately decontaminated.* Always consult with product specialists for the best advice.

Monitoring Instruments

- Always follow any manufacturer recommendations with respect to decon. If the instrument becomes damaged or disabled during the emergency, most instrument manufacturers will not accept the device for repair unless it has been properly decontaminated.
- If instruments were covered with protective plastic, remove and discard the plastic covering and tape properly.
- Decontaminate the instrument. If possible and where recommended by the manufacture, disassemble the instrument to assure thorough cleaning. Avoid introducing water or decon solutions into sensitive portions of the instrument.

FIRE HOSE

- Fire hose should be cleaned following the manufacturer's recommendations. For most materials, detergents will perform adequately. However, strong detergents and cleaning agents may damage the fire hose fibers. The fire hose should be thoroughly rinsed to prevent any fiber weakening. The hose should then be marked and pressure tested before being placed back in service.

NOTES

Motor Vehicle and Heavy Equipment Clean-Up

If a large number of vehicles need to be decontaminated, consider implementing the following recommendations:

- Establish a decon pad as a primary wash station. The pad may be a concrete slab or a pool liner covered with gravel. Each of these should be bermed or diked with a sump or some form of water recovery system to collect the resulting rinse. Get some engineering help and do it right the first time.
- Completely wash and rinse vehicles several times with detergent. Pay particular attention to wheel wells and the chassis. Depending upon the nature of the contaminant, it may be necessary to collect all runoff water from the initial gross rinse, particularly if there is contaminated mud and dirt on the underside of the chassis.
- Engines exposed to toxic dusts or vapors should have their air filters replaced. Mechanics sometimes blow dust out of air filters during routine maintenance, exposing themselves unnecessarily to this hazardous dust. Contaminated air filters should be properly disposed of.

If vehicles have been exposed to minimal contaminants such as smoke and vapors, they may be decontaminated on site and then driven to an off-site car wash for a second, more thorough washing. Car washes may be suitable if the drainage area is fully contained and all runoff drains into a holding tank. Car washes are not recommended if they drain into the sanitary sewer. Car washes used for decon should be inspected and approved as acceptable in advance.

When a vehicle is exposed to corrosive atmospheres, it should be inspected by a mechanic for possible motor damage. Equipment sprayed with acids should be flushed or washed as soon as possible with a neutralizing agent such as baking soda, and then flushed again with rinse water.

POST-INCIDENT DECON CONCERNS

GENERAL CONCERNS

Debriefing

A debriefing should be held for those involved in decontamination and clean-up as soon as practical. (See Chapter Twelve for more information on debriefings.) Exposed ERP or contractors should be provided with as much information as possible about the delayed health effects of the hazmats.

If necessary, follow-up examinations should be scheduled with medical personnel and exposure records maintained for future reference by the individual's personal physician.

NOTES

Site Security and Custody

If contaminated materials must remain at the incident scene until they can be removed for off-site cleaning or disposal, special precautions should be taken.

- Take appropriate security measures. Additional lighting may be necessary when the materials remain overnight. Always secure hazardous waste and ensure that a chain of custody is maintained.
- Make sure appropriate warning signs are posted and labels are attached to containers.
- Make sure containers are properly sealed.

HAZARDOUS WASTE HANDLING AND DISPOSAL

Regulatory Compliance

ERP are required by local, state, and federal regulations to dispose of hazardous wastes in a specific manner. All personnel involved in the disposal of hazardous waste must be trained in the provisions of the Federal Resource Conservation and Recovery Act (RCRA), any related state or local regulations, and in the procedures for waste disposal.

Hazardous Waste Containers

All containers that contain materials designated as hazardous waste should be visibly identified with the proper markings. Containers should not be handled or utilized unless the contents are properly identified on a label. Containers used for the accumulation of waste should be labeled so that anyone working in the area will be aware of the contents. Any containers stored outdoors should be labeled in a manner that will withstand the elements.

Only approved chemically compatible containers of sufficient strength should be used for hazardous waste. The containers should be kept covered at all times and arranged so that easy access exists. Care should be taken during all handling to maintain the integrity of the container. Any container stored outdoors must be waterproof.

SUMMARY

Decontamination is the process of making people, equipment, and the environment safe from hazardous materials contaminants. The more you know about how contamination occurs and spreads, the more effective decontamination will be.

The basic concepts of decontamination are relatively simple. If contact with the contaminant can be controlled and minimized, the need for decontamination can be reduced.

The safety and health hazards of the contaminants at any incident will define how complex decon operations will be. The best field decontamination procedures emphasize the need to confine contaminants to a limited area. Establishing a designated decontamination corridor and decontamination area are the first steps in limiting the spread of contaminants.

Regardless of the number of decontamination steps required, decontamination is most effective when it is carried out by a trained Decontamination Team using multiple cleaning stations.

REFERENCES AND SUGGESTED READINGS

Action Video, DECONTAMINATION PROCEDURES, (videotape), Portland, OR: (1989).

Action Video, HANDLING CONTAMINATED VICTIMS (videotape), Portland, OR: (1990).

Action Video, WHY DECONTAMINATE?, (videotape), Portland, OR: (1989).

Agency for Toxic Substances and Disease Registry, MANAGING HAZARDOUS MATERIALS INCIDENTS—EMERGENCY MEDICAL SERVICES, Atlanta, GA: ATSDR (1992).

Agency for Toxic Substances and Disease Registry, MANAGING HAZARDOUS MATERIALS INCIDENTS—HOSPITAL EMERGENCY DEPARTMENTS, Atlanta, GA: ATSDR (1992).

Emergency Film Group. AIDS, HEPATITIS, AND THE EMERGENCY RESPONDER (videotape), Plymouth, MA: Emergency Fim Group (1989).

Ballinger, Walter, F., Jacquelyn C. Treybal and Ann, B. Vose, ALEXANDERS CARE OF THE PATIENT IN SURGERY (5th edition), St. Louis, MO: C. V. Mosby Company (1972).

Black, R. H., "Protecting and Cleaning Hands Contaminated by Synthetic Fallout Under Field Conditions.", INDUSTRIAL HYGIENE JOURNAL, (April, 1960), pages 162–168.

Bledsoe, Bryan E., D.O., Robert S. Porter and Bruce R. Shade, PARAMEDIC EMERGENCY CARE, (2nd edition), Englewood Cliffs, NJ: Prentice Hall, Inc. (1993).

Borak, Jonathan, M.D., Michael Callan, and William Abbott, HAZARDOUS MATERIALS EXPOSURE: EMERGENCY RESPONSE AND PATIENT CARE, Englewood Cliffs, NJ: Prentice Hall, Inc. (1991).

Brannigan, Francis, L., "Living with Radiation: Fundamentals No. 1," U.S. Atomic Energy Commission, Washington, DC (Undated).

Bronstein, Alvin C. and Philip L. Currance, EMERGENCY CARE FOR HAZARDOUS MATERIALS EXPOSURE, St. Louis, MO: C. V. Mosby Company (1988).

Callan, Michael, J., "Building a Decon Program." FIRE JOURNAL, (November/December, 1993), page 16.

"Care and Cleaning of Your Gear," Northeastern Pennsylvania Volunteer Firemen's Federation, (March, 1993), pages 21–24.

Carroll, Todd, R., "Contamination and Decontamination of Turnout Clothing," Washington, DC: Federal Emergency Management Agency (April 1993).

"Cleaning Pesticide Contaminated Clothing: A Special Report on Safety, Pest Control Technology Magazine,"(November, 1984) pages 42–44.

Coleman, Ronald, J., "Decontamination: Keeping Personnel Safe." FIRE CHIEF MAGAZINE, (June, 1993) pages 43–45

Dawson, Gaynor, W., and B. W. Mercer, HAZARDOUS WASTE MANAGEMENT, New York, NY: John Wiley and Sons (1986).

Deater, John, William, Medic, USN, "Considerations for Etiological Hazardous Material Exposure," special report prepared to fulfill the requirements of FS-205, Montgomery College, Rockville, MD (1985).

Docimo, Frank, "Decontamination...Or How Dirty Are We", THE FIREFIGHTER NEWS, (June–July 1991).

DOL/HHS Joint Advisory Notice, "Protection Against Occupational Exposure to HBV and HIV," U.S. Department of Labor (DOL) and U.S. Department of Health and Human Services (HHS) Publications (October, 1987).

Easley, J. M., R. E. Laughlin, and K. Schmidt, "Detergents and Water Temperature as Factors in Methyl Parathion Removal from Denim Fabrics." BULLETIN OF ENVIRONMENTAL CONTAMINATION AND TOXICOLOGY (1982), pages 241–244.

Emergency Film Group. DECON TEAM (video tape), Plymouth, MA: Emergency Fim Group (1992).

Federal Emergency Management Agency, Hazardous Materials Workshop for Hospital Staff, Washington, DC, (July, 1992).

Finley, E. L., G. I. Metcalfe, and F. G. McDermott, "Efficacy of Home Laundering in Removal of DDT, Methyl Parathion and Toxaphene Residues From Contaminated Fabrics." BULLETIN OF ENVIRONMENTAL CONTAMINATION AND TOXICOLOGY (1974), pages 268–274.

Finley, E. L., and R. B. Rogillio, "DDT and Methyl Parathion Residues Found in Cotton and Cotton-Polyester Fabrics Worn in Cotton Fields." BULLETIN OF ENVIRONMENTAL CONTAMINATION AND TOXICOLOGY (1962), pages 343–351.

Friedman, William, J., "Decontamination of Synthetic Radioactive Fallout from Intact Human Skin," INDUSTRIAL HYGIENE JOURNAL (February, 1958).

Ganelin, Robert, M.D., Gene Allen Mail, and L. Cueto, Jr., "Hazards of Equipment Contaminated with Parathion." ARCHIVES OF INDUSTRIAL HEALTH (June, 1961). Pages 326–328.

Grant, Harvey D., Robert H. Murray, Jr., and David Bergeron. EMERGENCY CARE (6th edition), Englewood Cliffs, NJ: Prentice Hall, Inc. (1993).

Gold, Avram, William A. Burgess, and Edward, V. Clough, "Exposure of Firefighters to Toxic Air Contaminants."AMERICAN INDUSTRIAL HYGIENE ASSOCIATION JOURNAL (July, 1978).

Hildebrand, Michael, S., "Complete Decontamination Procedures for Hazardous Materials: The Nine Step Process, Part-1." FIRE COMMAND, (January, 1985), pages 18–21.

Hildebrand, Michael, S., "Complete Decontamination Procedures for Hazardous Materials: The Nine Step Process, Part-2." FIRE COMMAND (February, 1985), pages 38–41.

NOTES

Hildebrand, Michael, S., HAZMAT RESPONSE TEAM LEAK AND SPILL CONTROL GUIDE, Oklahoma State University, Stillwater, OK (1984).

Hughes, Stephen M., David, W. Berry, and Edward D. Hartin, "What Does A Car Wash and a Baby Have to Do with Hazardous Materials Decon?" An independent technical paper prepared by HazMat-TISI, Columbia, MD, (1992).

Isman, Warren, E. and Gene P. Carlson, HAZARDOUS MATERIALS, Encino, CA: Glencoe Publishing Company (1980).

Laughlin, J.M., C.B. Easly, R. E. Gold, and D. R. Tupy, "Methyl Parathion Transfer from Contaminated Fabrics to Subsequent Laundry and to Laundry Equipment." BULLETIN OF ENVIRONMENTAL CONTAMINATION AND TOXICOLOGY, (1981), pages 518–523.

Leahy, John, R. Jr. and Roger A. McGary, "Be Prepared for Decontamination at Hazardous Materials Incidents." FIRE ENGINEERING, (July, 1982). Pages 12–17.

LeMaster, Frank, "Why Protective Clothing Must Be Cleaned." THE VOICE, (August/September 1993), pages 19–20.

Lillie, T. H., R. E. Hampson, Y. A. Nishioka, and M. A. Hamlin, "Effectiveness of Detergent and Detergent Plus Bleach for Decontaminating Pesticide Applicator Clothing." BULLETIN OF ENVIRONMENTAL CONTAMINATION AND TOXICOLOGY, (1982), pages 89–94.

McGary, Roger, A., "Disinfection of SCBA." THE VOICE (October, 1993), pages 26–28.

Action Training Systems, MEDICAL OPERATIONS AT HAZMAT INCIDENTS, videotape produced by the Emergency Film Group, Plymouth, MA (1993).

National Fire Protection Association, NFPA 1581—FIRE DEPARTMENT INFECTION CONTROL PROGRAM, Boston, MA: National Fire Protection Association (1991).

Olson, Kent R., M.D., POISONING AND DRUG OVERDOSE (San Francisco Bay Area Regional Poison Control Center), Norwalk, CT and San Mateo, CA: Appleton and Lange (1990).

OSHA Instruction CPL 2–2.44B, "Enforcement Procedures for Occupational Exposure to HBV and HIV," U.S. Department of Labor (DOL) and U.S. Department of Health and Human Services (HHS) Publications (February 27, 1990).

Perkins, John, J., PRINCIPLES AND METHODS OF STERILIZATION IN HEALTH SCIENCES (2nd edition), Chicago, IL: Charles Thomas Publishing Co. (1980).

Ronk, Richard, and Mary Kay White, "Hydrogen Sulfide and the Probabilities of Inhalation Through a Tympanic Membrane Defect." JOURNAL OF OCCUPATIONAL MEDICINE, (May, 1985), pages 337–340.

Stutz, Douglas R. and Stanley J. Janusz, HAZARDOUS MATERIALS INJURIES: A HANDBOOK FOR PRE-HOSPITAL CARE (2nd edition), Beltsville, MD: Bradford Communications Corp. (1988).

Teller, Robert, "Developing Proper Decontamination Procedures for Emergency Response." ECON MAGAZINE (March 1993), pages 32–33.

U.S. Centers for Disease Control (CDC) Publications, A CURRICULUM GUIDE FOR PUBLIC SAFETY EMERGENCY RESPONSE WORKERS: PREVENTION OF HUMAN IMMUNODEFICIENCY VIRUS AND HEPATITIS B VIRUS (February, 1989).

U. S. Centers for Disease Control (CDC) Publications, GUIDELINE FOR HANDWASHING AND HOSPITAL ENVIRONMENTAL CONTROL (1985).

U. S. Centers for Disease Control (CDC) Publications, GUIDELINE FOR PREVENTION OF HIV AND HBV EXPOSURE TO HEALTH CARE AND PUBLIC SAFETY WORKERS (February, 1989).

U. S. Environmental Protection Agency, GUIDE FOR INFECTIOUS WASTE MANAGEMENT (1986).

NOTES

TERMINATING THE INCIDENT

OBJECTIVES

1. Identify the need to conduct effective incident termination activities.
2. List and describe the three basic phases of incident termination.
3. Describe the procedures for conducting an on-scene incident debriefing and its significance in terminating a hazmat incident.
4. Describe the Post-Incident Analysis as a method of documenting incident activities.
5. Identify the regulatory reporting requirements of federal, state, and local agencies.
6. Describe the importance of documentation for a hazardous materials incident including training records, exposure records, incident reports, and critique reports.
7. Describe the procedure for conducting a critique of a hazardous materials incident.

"Experience is the name everyone gives to their mistakes."

Oscar Wilde,
Irish Poet

INTRODUCTION

Termination is the final step in the **Eight Step Incident Management Process©**. It is the transition phase between the end of the emergency and the initiation of restoration and recovery operations.

NOTES

It is important to clarify that there is a distinct difference between the restoration and recovery phases and the termination phase of an emergency. Restoration and recovery activities are more operational, while termination activities are primarily administrative. For the purposes of this chapter, we have limited our discussion to termination. The more detailed operational aspects of recovery and restoration are discussed in *Chapter Ten, Implementing Response Objectives*.

While it may sound a little silly to "officially end" an emergency, it is an important part of an emergency response. Unlike fire emergencies where it is usually obvious that the fire is out (e.g., there is a large smoking hole in the ground where the building used to be), hazardous materials incidents do not always have a clear ending.

Consider the following examples:

- Liquid hazmats can soak into the ground and continue to present a health and environmental hazard even though it is not obvious.
- Containers involved in the incident may not show visible signs of stress and may breach while being moved or off-loaded.
- Flammable vapors can re-accumulate in confined areas and enter the flammable range.

Emergency responders and support personnel sometimes get mixed signals concerning whether there are actually hazards present at this phase of the operation. If spills have been controlled and personnel are standing around waiting for product transfer operations to begin, boredom and complacency set in. Control of the perimeter becomes relaxed, protective clothing comes off, and the incident scene becomes dangerous again.

TERMINATION ACTIVITIES

Termination activities should concentrate on funnelling accurate information to the people who need it the most. Initially, this group is a small number of on-scene emergency responders who may be briefed on the signs and symptoms of a particular substance or on special recovery procedures. On larger incidents, the number of people with a "need to know" expands and may even include the accident investigation team or representatives from contractors or other agencies.

Release of inaccurate information may have many long-reaching effects. Incorrect hazard data could result in illness to those exposed, improper clean-up techniques, and unsafe disposal procedures. Failure to properly manage termination activities may also result in undesirable opinions of your organization from the public, your peers, and the news media.

The termination process is divided into three phases: debriefing, post-incident analysis, and critique. See Figure 12-1.

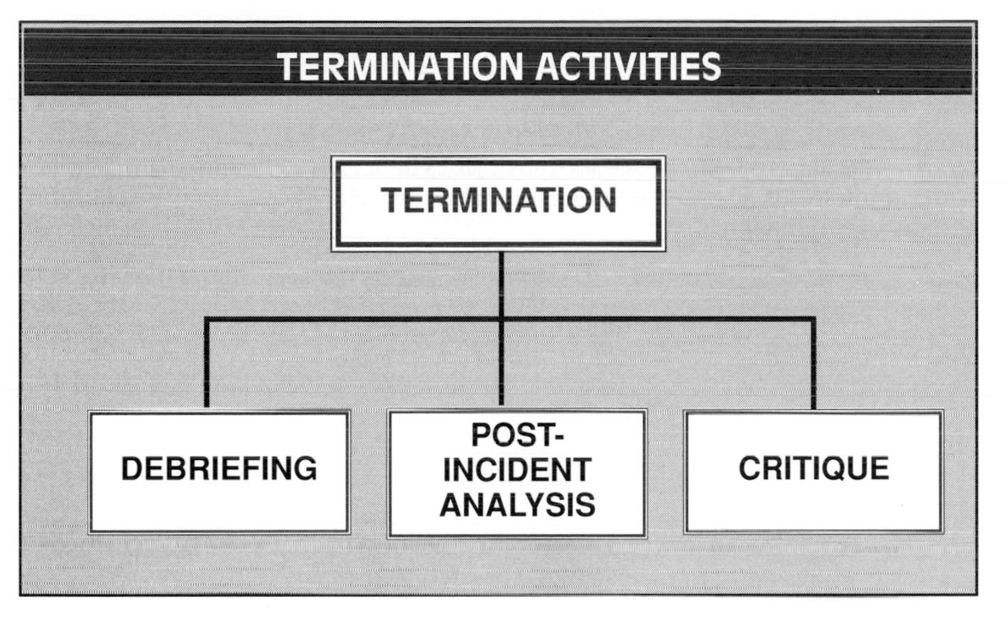

FIGURE 12-1: TERMINATION ACTIVITIES INVOLVE THREE PRINCIPAL PHASES

NOTES

DEBRIEFING

In Chapter Three we discussed how sectoring provides the best way to manage a hazmat incident. However, this process also tends to separate personnel from information that may be important at a later date. An Entry Team may not be concerned about laundering procedures for uniforms during leak control operations, but they do need to know this information before they are released from the scene.

An effective debriefing should:

- Inform responders exactly what hazmats they were (potentially) exposed to and their signs and symptoms.
- Identify damaged equipment requiring servicing, replacement, or repair.
- Identify equipment or expended supplies that will require specialized decontamination or disposal.
- Identify unsafe site conditions which will impact the clean-up and recovery phase. Owners and contractors should be formally briefed on these problems before responsibility for the site is turned over to them.
- Assign information gathering responsibilities for a post-incident analysis and critique.
- Assess the need for a Critical Incident Stress Debriefing.

Debriefings should begin as soon as the "emergency" phase of the operation is completed. Ideally, this should be before any responders leave the scene, and it should include the hazmat response team, sector officers, and other key players such as public information officers and agency representatives who the Incident Commander determines have a need to know.

NOTES

On larger incidents, these representatives will return to their personnel and pass on the essential information, including, where applicable, who to contact for more information.

Debriefings should be conducted in areas which are free from distractions. In poor environmental conditions, such as extremely cold or hot weather or loud ambient noise, the debriefing should be conducted in a nearby building or vehicle.

The debriefing should be conducted by one person acting as the leader. The Incident Commander may not be the best facilitator for the debriefing because special knowledge of the hazards may be required. Also, the IC may not be available for the entire meeting. The IC should at least be present to summarize feelings about the entire incident and to reinforce its positive aspects.

The debriefing should be limited to a maximum of 15 minutes. Its intent is to briefly review the incident and send everyone home, not analyze every action of every player. If more interaction is needed on a specific subject or operation, continue it after the debriefing.

Debriefings should cover certain subjects, in the following order:

1. **Health information**

   Exactly what each ERP has been (possibly) exposed to and signs and symptoms of oncoming illness. Some substances may not reveal signs and symptoms of exposure for 24 to 48 hours. When appropriate, cover responsibilities for follow-up evaluations and complete health exposure forms.

2. **Equipment and apparatus exposure review**

   Ensure that equipment and apparatus that is unfit for service is clearly marked and plans are made for special cleaning or equipment disposal. It is common to have firefighting gear and personal clothing laundered upon return to the station. Someone must be specifically delegated the responsibility for assuring that contaminated garments are properly laundered or disposed of.

3. **Provide a follow-up contact person**

   Ensure that anyone involved after the release of ERP from the scene, such as clean-up contractors and investigators, has access to a single information source who can share the needed data. The contact person is also responsible for collecting and maintaining all incident documents until they are delivered to the appropriate investigator or critique leader.

4. **Identify problems requiring immediate action**

   Equipment failures, safety, major personnel problems, or potential legal issues should be quickly reviewed on-scene. If it is not crucial, save it for the critique.

5. **Say "thank you"**

   Most hazmat incidents are hard work and often test personal endurance and everyone's sense of humor. Reinforcement of the things that went right, a commitment to work on the problems uncovered, and a thank you from the boss go a long way. Never end on a sour note.

POST-INCIDENT ANALYSIS

NOTES

The term **Post-Incident Analysis (PIA)** is used in this section as it relates to improving emergency response. It may have different meanings to various organizations responsible for fact-finding missions.

Post-Incident Analysis is the reconstruction of the incident to establish a clear picture of the events that took place during the emergency. It is conducted to:

- Assure that the incident has been properly documented and reported.
- Determine the level of financial responsibility (Who pays?).
- Establish a clear picture of the emergency response for further study.
- Provide a foundation for the development of formal investigations, which are usually conducted to establish the probable cause of the accident for administrative, civil, or criminal proceedings.

There are many agencies and individuals who have a legitimate need for information about any significant hazmat incident. They may include manufacturing and shipping company representatives, insurance companies, government agencies, and even citizens groups. A formal PIA is one method for coordinating the release of factual information to those who have a need to know.

The PIA begins with the designation of one person (or office) to collect information about the response. This person is usually appointed during the on-scene debriefing. The PIA coordinator should have the authority to determine who will have access to information. This method guarantees that sensitive or unverified information (e.g., injured personnel) is not released to the wrong organization or in an untimely manner.

The PIA should focus on five key topics:

- ❑ **Command and Control.** Was the Incident Management System established and was the emergency response organized according to the existing Emergency Response Plan and/or SOP's? Did information pass from Section personnel to the Incident Commander or through appropriate channels? Were response objectives communicated to field personnel who were expected to implement them?
- ❑ **Tactical Operations.** Were tactical operations ordered by the On-Scene Incident Commander and implemented by the emergency teams effective? What worked? What did not? Were tactical operations conducted in a timely and coordinated fashion? Do revisions need to be made to tactical procedures or worksheets?
- ❑ **Resources.** Were resources adequate to conduct the response effort? Are improvements needed to equipment or facilities? Were mutual aid agreements implemented effectively?
- ❑ **Support Services.** Were support services adequate and provided in a timely manner? What is needed to increase the provision of support to the necessary level?

NOTES

- ❑ **Plans and Procedures.** Were the Emergency Response Plan and associated Tactical Procedures current? Did they adequately cover notification, assessment, response, recovery, and termination? Were roles and assignments clearly defined? How will plans and procedures be upgraded to reflect the "lessons learned"?
- ❑ **Training.** Did this event highlight the need for additional basic or advanced training? Multi-agency training? Were personnel trained adequately for their assignments?

The PIA should attempt to gather factual information concerning the response as soon as possible. The longer the delay in gathering information, the less likely it will be accurate and available. Some suggested sources of information include:

- ❑ Verification of shipping documents or Material Safety Data Sheets
- ❑ Owner/operator information
- ❑ Chemical hazard information from checklists, computer printouts, etc.
- ❑ Cassette tapes from the Command Post
- ❑ Notes by IMS Officers and Command Staff
- ❑ Photographs, videos, and sketches. If photographs or videotapes are taken, copies should also be obtained for the incident file. If future litigation is a concern, the following information should be recorded on photos and videotapes:
  - Time, date, location, direction, and weather conditions
  - Description or identification of subject and relevance of photographs
  - Sequential number of photos and film roll number(s)
  - Camera type and serial number
  - Name, telephone number, and social security number of the photographer
- ❑ Records on levels of contamination or documentation of exposures from Decon and EMS Sections
- ❑ Incident reports
- ❑ Incident command charts
- ❑ Business cards or notes from agency, organization, or company representatives
- ❑ Tape recordings of the communications center(s) involved
- ❑ Videotape recordings made by the media
- ❑ Photographs, film, and videotape taken by ERP or bystanders
- ❑ Interviews of witnesses conducted by investigators which may help establish where ERP were located at the incident scene
- ❑ Interviews of ERP

NOTES

As soon as practical, construct a brief chronological review of who did what, when, and where during the incident. A simple timeline placing the key players at specific locations at different times is a good start.

Cooperation between the PIA coordinator and other official investigators will save time and combine resources to reconstruct the incident completely.

Once all available data has been assembled and a rough draft report developed, the entire package should be reviewed by the key responders to verify that the available facts are arranged properly and actually took place.

Recording and Reporting of Incidents

Each emergency response organization has its own unique requirements for recording and reporting hazmat incidents. These requirements may be self-imposed as administrative and management controls or be mandatory under federal or state laws.

For private shippers, carriers, and manufacturers, the regulatory reporting requirements for leaks, spills and other releases of specified chemicals into the environment are significant. These include the following:

- Section 304 of the Superfund Amendments and Reauthorization Act (SARA, Title III)
- Section 103 of the Comprehensive Environmental Response, Compensation and Liability Act (CERCLA)
- 40 CFR Part 110—Discharge of Oil
- 40 CFR Part 112—Oil Pollution Prevention
- Any additional local, state, or regional reporting requirements

Under CERCLA, the responsible party must report to the National Response Center (NRC) any spill or release of a specified hazardous substance in an amount equal to or greater than the reportable quantity (RQ) specified by EPA.

In addition, SARA, Title III requires that releases be reported immediately to the National Response Center and the Local Emergency Planning Committee (LEPC).

Many industrial organizations have developed initial incident reporting forms as a way to ensure that key corporate and regulatory reporting requirements are correctly documented. See Figures 12-2 and 12-3 for examples. Chapter One provides an overview of some of the more important federal laws which have reporting provisions.

Most major public fire departments participate in the National Fire Incident Reporting System. NFIRS is a product of the National Fire Information Council and is sponsored by the U.S. Fire Administration. The system includes a special category for recording hazardous materials incidents.

INITIAL INCIDENT ASSESSMENT AND NOTIFICATION SHEET

ACTIONS

Who have you notified? ____________________

External ____________________

What actions have you taken so far? ____________________

CASUALTIES

How many injured employees? __________ How serious? __________

How many injured civilians? __________ How serious? __________

RELEASE

Has there been a release? ____________________

What was released? ____________________

Estimated size? ____________________

Worst-case scenario? ____________________

____

Has the release entered drains or waterways? ____________________

Describe the flow ____________________

OTHER CONCERNS

Has there been a fire? ____________________

Confined? ____________________

Potential for further damage ____________________

____

Additional facilities at risk ____________________

Impact on facility operations ____________________

____

Has the incident impacted outside the fence? ____________________

Describe ____________________

News coverage thus far? RADIO / TV / PRINT? ____________________

From where? ____________________

Expected duration of the incident?

Hours ____________________

Days ____________________

Weeks ____________________

FIGURE 12-2: INITIAL INCIDENT ASSESSMENT AND NOTIFICATION SHEET

1. Date:	**2.** Time ❑ AM ❑ PM	**3.** Duration of Release:

4. Location:

5. Identity of released chemical (or its components):

6. Quantity of material released:

7. Did any released material leave the company property? ❑ Yes ❑ No
If yes, answer a) and b) below.

a. How?

b. Where?

8. Medium or media into which release occurred: ❑ Air ❑ Water ❑ Land
Description:

9. Any known or anticipated acute or chronic risks associated with the release and, where appropriate, advice regarding medical attention necessary for exposed individuals?

10. Any precautions to take as a result of the release, including evacuation:

11. Name(s) and telephone number(s) of the person(s) to be contacted for information:

a.	d.
b.	e.
c.	f.

12. Steps being taken to clean up release/spill:

13. Conclusion: Incident ❑ Does ❑ Does Not* require Section 304 reporting.
(*—Accumulate record to Section 313 file)

14. Notification to Authorities if 304 Release	Contact Name	Time
1) National Response Center (800-424-8802)		
2) County LEPC		
3) EPA		
4) Local Police		
5) County/State Emergency Preparedness		

FIGURE 12-3: SARA/CERCLA REVIEW FOR EVALUATING HAZARDOUS MATERIALS RELEASE INTO THE ENVIRONMENT

NOTES

CRITIQUE

Many injuries and fatalities have been prevented as a result of lessons learned through the critique process. An effective critique program must be supported by top management and is the single most important way for an organization to "self-improve" over time. OSHA requires that a critique be conducted of every hazardous materials emergency response.

The primary purpose of a critique is to develop recommendations for improving the emergency response system rather than to find fault with the performance of individuals. A good critique promotes:

- System-dependent operations rather than people-dependent organizations.
- A willingness to cooperate through teamwork.
- Improvement of safe operating procedures.
- Sharing information among emergency response organizations.

The critique leader is the crucial player in making the critique session a positive learning experience. A critique leader can be anyone who is comfortable and effective working in front of a group.

The critique leader need not necessarily be part of the emergency response team. For example, an organization may select one or two respected and credible individuals to act as neutral parties to critique the larger, more sensitive incidents. Examples of outsiders who may be effective critique leaders include Fire Science or Law Enforcement faculty from a local community college, Regional Fire Training Coordinators, or professionals in your community who are experienced with "people skills" such as coaches, counselors, and arbitrators.

Although every organization has a tendency to develop its own critique style, never use a critique to assign blame (public meetings are the worst time to discipline personnel). Do use it as a valuable learning experience (everyone came to the incident with good intentions).

The critique leader should:

- ❑ Control the critique. Introduce the players and procedures, keep it moving, and end it on schedule. Experience shows that critiques lasting longer than 60 to 90 minutes quickly lose their effectiveness and "quality" of discussion.
- ❑ Ensure that direct questions receive direct answers.
- ❑ Ensure that all participants play by the critique rules.
- ❑ Ensure that individual observations are shared with the group.

The following is a recommended critique format for large-scale emergency responses:

Participant-Level Critique. After explaining the rules for the critique, the leader calls on each key player to make an individual statement relevant to his or her on-scene activities and what he or she feels are the major issues. Depending upon time, more detail may be added. There should be no interruptions during this phase. For obvious time reasons, the leader should limit this phase of the process to two or three minutes per person.

DON SELLERS

FIGURE 12-4: THE CRITIQUE SHOULD NEVER BECOME A FREE-FOR-ALL

NOTES

Operations-Level Critique. After determining a feel for the group, the leader moves on to a structured review of emergency operations. Through a spokesperson, each section/sector presents an activity summary of challenges encountered, unanticipated events, and lessons learned. Each presentation should not exceed five minutes.

Group-Level Critique. At the end of the operations level critique, the leader moves the meeting into a wider and more open forum. The facilitator encourages discussion, reinforces constructive comments, and records important points.

As the critique draws to a close, the leader should summarize the more important observations and conclusions revealed by the participants. For large groups, the critique leader should have one or two assistants who act as recording secretaries throughout the session. These notes become the beginning point for writing a post-critique report.

Critique reports should be short and to the point. Simply describe what happened in one or two pages and move on to the lessons learned. If recommendations for improvement are appropriate, they should be listed at the end of the report.

When larger incidents are involved or injuries have occurred, formal critique reports should be circulated so that everyone in the response system can share the lessons learned. Other forums which may be appropriate to share lessons learned include trade magazines and technical conferences.

It is important that lessons learned which have been identified through the critique process be converted into changes and improvements to the

NOTES

emergency response system. Set aside a semi-annual review date to make sure that action items have been addressed.

LIABILITY ISSUES

Many managers express concerns that the critique process can expose weaknesses which can be exploited to build a liability case against emergency responders. There is no question that:

- Civil suits and regulatory citations against government and industry for emergency response operations are dramatically increasing.
- Your chances of losing the suit or citation are increasing.
- Fines and award sizes have increased more than five-fold.

Nevertheless, organizations must balance the potential negatives against the benefits that are gained through the critique process. Remember—the reason for doing the critique in the first place is to improve your operations.

There are five primary reasons for liability problems in emergency response work. *They are worth considering as a case for actually building a strong critique program.*

1. **Problems with Planning.** Plans and procedures are poorly written, out-of-date, and unrealistic. In addition, they are not used in the field.
2. **Problems with Training.** No training, unsafe training, and undocumented training.
3. **Problems with Identification of Hazards.** Hazards were not identified, were not prioritized, or were ignored.
4. **Problems with Duty to Warn.** Warnings were not given or were improper.
5. **Problems with Negligent Operations.** Equipment was not employed properly, plans and procedures were not followed, and equipment was not maintained.

A little common sense goes a long way when developing a critique policy. If the critique and follow-up report are properly written in the first place, there will be little useful information that attorneys may use against emergency responders. While the critique report is a critical item, recognize that the legal process of discovery can also reveal organizational shortcomings in other ways. For example, official investigations conducted by OSHA, the fire marshal, or an insurance company can be used to build a case against you or your organization.

Don't let attorneys make management decisions for your organization. A quality emergency response system is your best liability defense.

NOTES

SUMMARY

It is important that every hazmat incident be formally terminated by a specific, written procedure. This process documents safety procedures, site operations, hazards faced, and lessons learned. It also provides a record of resources and events which may affect the public health, financial resources, and political well-being of a community. Lastly, it provides the data which may be required to comply with local, state, and federal laws.

Termination activities are divided into three phases: a debriefing, a post-incident analysis of what actually happened, and a formal critique designed to emphasize successful, as well as unsuccessful, operations.

REFERENCES AND SUGGESTED READINGS

Benner, Ludwig, Jr., and Hildebrand, Michael, S., HAZARDOUS MATERIALS MANAGEMENT SYSTEMS: THE M.A.P.S METHOD, Prentice-Hall, (1981).

Emergency Management Institute, LIABILITY ISSUES IN EMERGENCY MANAGEMENT, Federal Emergency Management Agency, Washington, D.C. (April, 1992).

Energy Research and Development Administration, ACCIDENT/INCIDENT INVESTIGATION MANUAL: The Management and Risk Oversight and Risk Tree, ERDA-76-20, Washington, D.C. (July 1975).

Hildebrand, Michael, S., "An Effective Critique Program." FIRE CHIEF, Volume 27, No.4 (April, 1983).

International Fire Service Training Association, PHOTOGRAPHY FOR THE FIRE SERVICE, IFSTA-204, Fire Protection Publications, Oklahoma State University (1976).

Kutner, Kenneth, C., "CISD: Critical To Health and Safety." FIRE ENGINEERING MAGAZINE (April, 1992).

Lorenzo, D. E., "A Manager's Guide to Reducing Human Errors." Washington, D.C. , Chemical Manufacturers Association (July, 1980).

Murley, Thomas, E., "Developing a Safety Culture." A technical paper presented at the Nuclear Regulatory Commission, Regulatory Information Conference, Washington, D.C. (April, 1989).

GLOSSARY OF TERMS

A

Absorbent Material. A material designed to pick up and hold liquid hazardous material to prevent contamination spread. Materials include sawdust, clays, charcoal and polyolefin-type fibers.

Absorption. (1) The process of absorbing or "picking up" a liquid hazardous material to prevent enlargement of the contaminated area. Common physical method for spill control. (2) Movement of a toxicant into the circulatory system by oral, dermal, or inhalation exposure.

ACGIH. (See American Conference of Governmental Industrial Hygienists.)

Activity. The number of radioactive atoms that will decay and emit radiation in 1 second of time. Measured in curies (1 curie = 37 billion disintegrations per second), although it is usually expressed in either millicuries or microcuries. Activity indicates how much radioactivity is present and not how much material is present.

Acute Effects. Results from a single dose or exposure to a material. Signs and symptoms may be immediate or may not be evident for 24 to 72 hours after the exposure.

Acute Exposures. An immediate exposure, such as a single dose that might occur during an emergency response.

Administration/Finance Section. Responsible for all costs and financial actions of the incident. Includes the Time Unit, Procurement Unit, Compensation/Claims Unit, and the Cost Unit.

Adsorption. Process of adhering to a surface. Common method of spill control.

Aerosols. Liquid droplets, or solid particles dispersed in air, that are of fine enough particle size (0.01 to 100 microns) to remain dispersed for a period of time.

Agency for Toxic Substances and Disease Registry (ATSDR). An organization within the Center for Disease Control, it is the lead federal public health agency for hazmat incidents and operates a 24-hour emergency number for providing advice on health issues involving hazmat releases.

Air Monitoring. To measure, record, and/or detect contaminants in ambient air.

Air Purifying Respirators (APR). Respirators or filtration devices which remove particulate matter, gases, or vapors from the atmosphere. These devices range from full-facepiece, dual-cartridge masks with eye protection, to half-mask, facepiece-mounted cartridges. They are intended to be used only in atmospheres where the chemical hazards and concentrations are known.

American Conference of Governmental Industrial Hygienists (ACGIH). A professional society of individuals responsible for full-time industrial hygiene programs who are employed by official governmental units. Its primary function is to encourage the interchange of experience among governmental industrial hygienists and to collect and make information available of value to them. ACGIH promotes standards and techniques in industrial hygiene and coordinates governmental activities with community agencies.

American National Standards Institute (ANSI). Serves as a clearinghouse for nationally coordinated voluntary safety, engineering, and industrial consensus standards developed by trade associations, industrial firms, technical societies, consumer organizations, and government agencies.

American Petroleum Institute (API). Professional trade association of the United States petroleum industry. Publishes technical standards and information for all areas of the industry, including exploration, production, refining, marketing, transportation, and fire and safety.

Anhydrous. Free from water, dry. For example, anhydrous ammonia, anhydrous hydrogen chloride.

API Uniform Marking System. American Petroleum Institute marking system used at many petroleum storage and marketing facilities to identify hydrocarbon pipelines and transfer points. Classified hydrocarbon fuels and blends into leaded and unleaded gasoline (regular, premium, super). gasoline additives (methyl tertiary butyl ether) and distillates, and fuel oils.

Area of Refuge. Area within the hot zone where exposed or contaminated personnel are protected from further contact and/or exposure. This is a "holding area" where personnel are controlled until they can be safely decontaminated or treated.

Aromatic Hydrocarbons. A hydrocarbon which contains the benzene "ring" which is formed by six carbon atoms and contains resonant bonds. Examples include benzene (C^6H^6) and toluene (C^7H^8).

Asphyxiation Harm Events. Those events related to oxygen deprivation and/or asphyxiation within the body. Asphyxiants can be classified as simple or chemical.

Association of American Railroads (AAR). Professional trade association which coordinates technical information and research within the United States railroad industry. Publisher of emergency response guidebooks.

Atmosphere-supplying Devices. Respiratory protection devices coupled to an air source. The two types are self-contained breathing apparatus (SCBA) and supplied air respirators (SAR).

B

B-End. The end of a railroad car where the handbrake is located. Is typically used as the initial reference point when communicating railroad car damage.

Boiling Liquid Expanding Vapor Explosion (BLEVE). A container failure with a release of energy, often rapidly and violently, which is accompanied by a release of gas to the atmosphere and propulsion of the container or container pieces due to an overpressure rupture.

Boiling Point. The temperature at which a liquid changes its phase to a vapor or gas. The temperature where the vapor pressure of the liquid equals atmospheric pressure. Significant property for evaluating the flammability of a liquid, as flash point and boiling point are directly related. A liquid with a low flash point will also have a low boiling point, which translates into a large amount of vapors being given off.

Bonding. A method of controlling ignition hazards from static electricity. The process of connecting two or more conductive objects together by means of a conductor; it is done to minimize potential differences between conductive objects.

Boom. A floating physical barrier serving as a continuous obstruction to the spread of a contaminant.

Branch. That organizational level within the Incident Management System having functional/geographic responsibility for major segments of incident operations (e.g., Hazmat Branch). The Branch level is organizationally between Section and Division/Sector/Group.

Breach Event. The event causing a hazmat container to open up or "breach." It occurs when a container is stressed beyond its limits of recovery (ability to hold contents). Different containers breach in different ways—disintegration, runaway cracking, failure of container attachments, container punctures, and container splits or tears.

Breakthrough Time. The elapsed time between initial contact of the hazardous chemical with the outside surface of a barrier, such as protective clothing material, and the time at which the chemical can be detected at the inside surface of the material.

Buddy System. A system of organizing employees into work groups in such a manner that each employee of the work group is designated to be observed by at least one other employee in the work group (per OSHA 1910.120 (a)(3)).

Bulk Packaging. Bulk packaging has an internal volume greater than 119 gallons (450 liters) for liquids, a capacity greater than 882 pounds (400 kg) for solids, or a water capacity greater than 1,000 pounds (453.6 kg) for gases. It can be an integral part of a transport vehicle (e.g., cargo tank truck, railroad tank car, and barges), packaging placed on, or in a transport vehicle (e.g., portable tanks, intermodal portable tanks, ton containers), or fixed on processing containers.

Bung. A plug used to close a barrel or drum bung hole.

C

CAA. (See the Clean Air Act.)

Canadian Transport Emergency Center (CANUTEC). Operated by Transport Canada, it is a 24-hour, government sponsored hot line for chemical emergencies. (The Canadian version of CHEMTREC.)

Cancer. A process in which cells undergo some change that renders them abnormal. They begin a phase of uncontrolled growth and spread.

Carboy. Glass or plastic bottles used for the transportation of liquids. Range in capacity to over 20 gallons. May be encased in an outer packaging, such as polystyrene boxes, wooden crates, or plywood drums. Often used for the shipment of corrosives.

Carcinogen. A material that can cause cancer in an organism. May also be referred to as "cancer suspect" or "known carcinogens."

Cargo Tanks. Tanks permanently mounted on a tank truck or tank trailer which is used for the transportation of liquefied and compressed gases, liquids, and molten materials. Examples include MC-306, DOT-406, MC-307/DOT-407, MC-312/DOT-412, MC-331, and MC-338. May also be any bulk liquid or compressed gas packaging not permanently attached to a motor vehicle which (because of its size, construction, or attachment to the vehicle, can be loaded or unloaded without being removed from the vehicle.

CAS Number. The Chemical Abstract Service number. Often used by state and local Right-to-Know regulations for tracking chemicals in the workplace and the community. Sometimes referred to as a chemical's "social security number." Sequentially assigned CAS numbers identify specific chemicals and have no chemical significance.

Catalyst. Used to control the rate of a chemical reaction by either speeding it up or slowing it down. If used improperly, catalysts can speed up a reaction and cause a container failure due to pressure or heat build-up.

Center for Disease Control (CDC). The federally funded research organization tasked with disease control and research.

Chemical Degradation. The altering of the chemical structure of a hazmat during the process of decontamination. Commonly used agents include sodium hypochlorite (household bleach) and sodium hydroxide.

Chemical Interactions. Reaction caused by mixing two or more chemicals together. Chemical interaction of materials within a container may result in a build-up of heat and pressure, leading to container failure. In other situations, the combined material may be more corrosive than the container was originally designed to withstand and cause the container to breach.

Chemical Manufacturers Association. Professional trade association of the United States chemical industry. The parent organization that operates CHEMTREC™.

Chemical Protective Clothing (CPC). Single or multi-piece garment constructed of chemical protective clothing materials designed and configured to protect the wearer's torso, head, arms, legs, hands, and feet. Can be constructed as a single or multi-piece garment. The garment may completely enclose the wearer either by itself or in combination with the wearer's respiratory protection, attached or detachable hood, gloves, and boots.

Chemical Protective Clothing Material. Any material or combination of materials used in an item of clothing for the purpose of isolating parts of the wearer's body from contact with a hazardous chemical.

Chemical Resistance. The ability to resist chemical attack. The attack is dependent on the method of test and its severity is measured by determining the changes in physical properties. Time, temperature, stress, and reagent may all be factors that affect the chemical resistance of a material.

Chemical Resistant Materials. Materials that are specifically designed to inhibit or resist the passage of chemicals into and through the material by the processes of penetration, permeation, or degradation.

Chemical Stress. The result of a chemical reaction of two or more materials. Examples include corrosive materials attacking a metal, the pressure or heat generated by the decomposition or polymerization of a substance, or any variety of corrosive actions.

Chemical Transportation Emergency Center (CHEMTREC™). The Chemical Transportation Center, operated by the Chemical Manufacturers Association (CMA), can provide information and technical assistance to emergency responders. (Phone number: 1-800-424-9300)

Chemical Vapor Protective Clothing. The garment portion of a chemical protective clothing ensemble that is designed and configured to protect the wearer against chemical vapors or gases. Vapor chemical protective clothing must meet the requirements of NFPA 1991. This type of protective clothing is a component of EPA Level A chemical protection.

Chlorine Emergency Plan (CHLOREP). Chlorine industry emergency response system operated by the Chlorine Institute and activated through CHEMTREC.

Chlorine Kits. Standardized leak control kits used for the control of leaks in chlorine cylinders (Chlorine A kit), 1 ton cylinders (Chlorine B kit), and tank cars, tank trucks, and barges (Chlorine C kit). These kits are commercially available and are built to specifications developed by the Chlorine Institute.

Chronic Effects. Result from a single exposure or from repeated doses or exposures over a relatively long period of time.

Chronic Exposures. Low exposures repeated over time.

Clandestine Laboratory. An operation consisting of a sufficient combination of apparatus and chemicals that either have been or could be used in the illegal manufacture/synthesis of controlled substances.

Classes. As used in NFPA 70, *The National Electric Code*, used to describe the type of flammable materials that produce the hazardous atmosphere. There are three classes:

Class I Locations—Flammable gases or vapors may be present in quantities sufficient to produce explosive or ignitible mixtures.

Class II Locations—Concentrations of combustible dusts may be present (e.g., coal or grain dust).

Class III Locations—Areas concerned with the presence of easily ignitible fibers or flyings (e.g., cotton milling).

Clean Air Act (CAA). Federal legislation which resulted in EPA regulations and standards governing airborne emissions, ambient air quality, and risk management programs.

Clean Water Act (CWA). Federal legislation which resulted in EPA and state regulations and standards governing drinking water quality, pollution control, and enforcement. The Oil Pollution Act (OPA) amended the CWA and authorized regulations pertaining to oil spill preparedness, planning, response, and clean-up.

Cleanup. Incident scene activities directed toward removing hazardous materials, contamination, debris, damaged containers, tools, dirt, water, and road surfaces in accordance with proper and legal standards and returning the site to as near a normal state as existed prior to the incident.

Code of Federal Regulations (CFR). A collection of regulations established by federal law. Contact with the agency that issues the regulation is recommended for both details and interpretation.

COFC. (See container-on-flat-car.)

Cold Zone. The control zone of a hazmat incident that contains the command post and other support functions as are deemed necessary to control the incident. This zone may also be referred to as the clean zone or the support zone.

Coliwasa (Composite Liquid Waste Sampler). A glass or plastic waste sampling kit commonly used for collecting samples from drums and other containerized wastes.

Colorimetric Tubes. Glass tubes containing a chemically treated substrate that reacts with specific airborne chemicals to produce a distinctive color. The tubes are calibrated to indicate approximate concentrations in air.

Combination Package. Packaging consisting of one or more inner packagings and a non-bulk outer packaging. There are many different types of combination packagings.

Combined Liquid Waste Sampler (Coliwasa). A tool designed to provide stratified sampling of a liquid container.

Combined Sewers. Carries domestic wastewater as well as stormwater and industrial wastewater. It is quite common in older cities to have an extensive amount of these systems. Combined sewers may also have regulators or diversion structures that allow overflow directly to rivers or streams during major storm events.

Command. The act of directing, ordering, and/or controlling resources by virtue of explicit legal, agency, or delegated authority.

Command Post. The location from which all incident operations are directed and planning functions are performed. The communications center is often incorporated into the command post.

Command Staff. The command staff consists of the Public information Officer, the Safety Officer, and the Liaison Officer, who report directly to the Incident Commander.

Community Awareness and Emergency Response (CAER). A program developed by the Chemical Manufacturers Association (CMA) to provide guidance for chemical plant managers to assist them in developing integrated hazmat emergency response plans between the plant and the community.

Compatibility. The matching of protective chemical clothing to the hazardous material involved to provide the best protection for the worker.

Compatibility Charts. Permeation and penetration data supplied by manufacturers of chemical protective clothing to indicate chemical resistance and breakthrough time of various garment materials as tested against a battery of chemicals. These test data should be in accordance with ASTM and NFPA standards.

Composite Packaging. Packaging consisting of an inner receptacle, usually made of glass, ceramic, or plastic, and an outer protection (e.g., sheet metal, fiberboard, etc.) so constructed that the receptacle and the outer package form an integral packaging for transport purposes. Once assembled, it remains an integral single unit.

Compound. Chemical combination of two or more elements, either the same elements or different ones, that is electrically neutral. Compounds have a tendency to break down into their component parts, sometimes explosively.

Comprehensive Environmental Response, Compensation, and Liability Act (CERCLA). Known as CERCLA or SUPERFUND, it addresses hazardous substance releases into the environment and the clean-up of inactive hazardous waste sites. It also requires those who release hazardous substances, as defined by the Environmental Protection Agency (EPA), above certain levels (known as "reportable quantities") to notify the National Response Center.

Compressed Gas. Any material or mixture having an absolute pressure exceeding 40 psi in the container at 70°F, having an absolute pressure exceeding 104 psi at 130°F, or any liquid flammable material having a vapor pressure exceeding 40 psi at 100°F as determined by testing. Also includes cryogenic liquids with boiling points lower than 130°F at 1 atmosphere.

Computer Aided Management of Emergency Operations (CAMEO). A computer data base storage-retrieval system of pre-planning and emergency data for on-scene use at hazardous materials incidents.

Computerized Telephone Notification System (CT/NS). A computerized autodial telephone system which can be used for notifying a potentially large number of people in a short period of time. CT/NS systems are often used around high hazard facilities to ensure the timely notification of nearby citizens. Systems are capable of making call-backs to unanswered phones, keeping track of both who is notified and the time of notification, and providing pre-recorded messages and instructions to residents.

Concentration. The percentage of an acid or base dissolved in water. Concentration is *not* the same as strength.

Confined Space. A space that (1) is large enough and so configured that an employee can bodily enter and perform assigned work; (2) has limited or restricted means for entry or exit (e.g., tanks, vessels, silos, storage bins, hoopers, vaults, and pits are spaces that may have limited means of entry); and (3) is not desinged for continuous employee occupancy.

Confined Space (Permit Required). Has one or more of the following characteristics:

1) Contains or has the potential to contain a hazardous atmosphere. A hazardous atmosphere would be created by any of the following, including:
   a) Vapors exceed 10% of the lower explosive limit (LEL).
   b) Airborne combustible dust exceeds its LEL.
   c) Atmospheric oxygen concentrations below 19.5% or above 23.5%.
   d) Atmospheric concentration of any substance for which a dose or PEL is published and which could result in employee exposure in excess of these values.
   e) Any other atmospheric condition which is immediately dangerous to life or health (IDLH).
2) Contains a material that has the potential for engulfing an entrant.
3) Has an internal configuration such that a person could be trapped or asphyxiated by inwardly converging walls or by a floor which slopes downward and tapers to a smaller cross section; or
4) Contains any other recognized serious safety or health hazard.

Confinement. Procedures taken to keep a material in a defined or localized area once released.

Consignee. Person or company to which a material is being shipped.

Consist. A railroad shipping document that lists the order of cars in a train.

Contact. Being exposed to an undesirable or unknown substance that may pose a threat to health and safety.

Container. Any vessel or receptacle that holds a material, including storage vessels, pipelines, and packaging. Includes both bulk and nonbulk packaging, and fixed containers.

Container-on-Flat-Car (COFC). Intermodal containers which are shipped on a railroad flat cars.

Containment. Actions necessary to keep a material in its container (e.g., stop a release of the material or reduce the amount being released).

Contaminant. A hazardous material that physically remains on or in people, animals, the environment, or equipment, thereby creating a continuing risk of direct injury or a risk of exposure outside of the hot zone.

Contamination. An uncontained substance or process that poses a threat to life, health, or the environment.

Control. The offensive or defensive procedures, techniques, and methods used in the mitigation of a hazardous materials incident, including containment, extinguishment, and confinement.

Control Zones. The designation of areas at a hazardous materials incident based upon safety and the degree of hazard. Many terms are used to describe these control zones; however, for the purposes of this text, these zones are defined as the hot, warm, and cold zones.

Controlled Burn. Defensive or nonintervention tactical objective by which a fire is allowed to burn with no effort to extinguish the fire. In some situations, extinguishing a fire will result in large volumes of contaminated runoff or threaten the safety of emergency responders. Consult with the appropriate environmental agencies when using this method.

Corrosive. A material that causes visible destruction of, or irreversible alterations to, living tissue by chemical action at the point of contact.

Corrosivity Harm Events. Those events related to severe chemical burns and/or tissue damage from corrosive exposures.

Crack. Narrow split or break in the container metal which may penetrate through the container metal (may also be caused by fatigue). It is a major mechanism which could cause catastrophic failure.

Crisis. An unplanned event that can exceed the level of resources, has the potential to significantly impact an organization's operability and credibility, or poses a significant environmental, economic, or legal liability.

Critical Temperature and Pressure. Critical temperature is the minimum temperature at which a gas can be liquefied no matter how much pressure is applied. Critical pressure is the pressure that must be applied to bring a gas to its liquid state. Both terms relate to the process of liquefying gases. A gas cannot be liquefied above its critical temperature. The lower the critical temperature, the less pressure required to bring a gas to its liquid state.

Critique. An element of incident termination which examines the overall effectiveness of the emergency response effort and develops recommendations for improving the organization's emergency response system.

Cryogenic Liquids. A gas with a boiling point of minus 150°F or lower. Cryogenic liquid spills will vaporize rapidly when exposed to the higher ambient temperatures outside of the container. Expansion ratios for common cryogenics range from 694 (nitrogen) to 1,445 (neon) to 1.

D

Dam. A defensive confinement procedure consisting of constructing a dike or embankment to totally immobilize a flowing waterway contaminated with a liquid or solid hazardous substance.

Damage Assessment. The process of gathering and evaluating container damage as a result of a hazmat incident.

Dangerous Cargo Manifest. A list of the hazardous materials carried as cargo on board a vessel. Includes the location of the hazmat on the vessel.

Dangerous Goods. In Canadian transportation, hazardous materials are referred to as "dangerous goods."

Debriefing. An element of incident termination which focuses on the following factors:

1) Informing responders exactly what hazmats they were (possibly) exposed to and the signs and symptoms of exposure.
2) Identifying damaged equipment requiring replacement or repair.
3) Identifying equipment or supplies requiring specialized decontamination or disposal.
4) Identifying unsafe work conditions.
5) Asking information gathering responsibilities for a post-incident analysis.

Decon. Popular abbreviation referring to the process of decontamination.

Decontamination. The physical and/or chemical process of reducing and preventing the spread of contamination from persons and equipment used at a hazardous materials incident. (Also referred to as "contamination reduction.")

Decontamination Corridor. A distinct area within the "Warm Zone" that functions as a protective buffer and bridge between the "Hot Zone" and the "Cold Zone," where decontamination stations and personnel are located to conduct decontamination procedures.

Decontamination Officer. A position within the Hazmat Branch which has responsibility for identifying the location of the decontamination corridor, assigning stations, managing all decontamination procedures, and identifying the types of decontamination necessary.

Decontamination Team (Decon-Team). A group of personnel and resources operating within a decontamination corridor.

Defensive Tactics. These are less aggressive spill and fire control tactics where certain areas may be "conceded" to the emergency, with response efforts directed toward limiting the overall size or spread of the problem.

Degradation. The physical destruction or decomposition of a clothing material due to exposure to chemicals, use, or ambient conditions (i.e., storage in sunlight). Degradation is noted by visible signs such as charring, shrinking, swelling, color change or dissolving, or by testing the clothing material for weight changes, loss of fabric tensile strength, etc.

Degree of Solubility. An indication of the solubility and/or miscibility of the material.

Negligible—less than 0.1%

Slight—0.1 to 1.0%

Moderate—1 to 10%

Appreciable—greater than 10%

Complete—soluble at all proportions

Dent. Deformation of the tank head or shell. It is caused from impact with a relatively blunt object (e.g., railroad coupler, vehicle). If the dent has a sharp radius, there is the possibility of cracking.

Dermatotoxins. Toxins of the skin which may act as irritants, ulcers, chloracne or cause skin pigmentation disorders (e.g., halogenated hydrocarbons, coal tar compounds).

Detonation. An explosive chemical reaction with a release rate less than 1/100th of a second. This gives responders *no* time to react. Examples include military munitions, dynamite, and organic peroxides.

Dike. A defensive confinement procedure consisting of an embankment or ridge on ground used to control the movement of liquids, sludges, solids, or other materials. Barrier which prevents passage of a hazmat to an area where it will produce more harm.

Dike—Overflow. A dike constructed in a manner that allows uncontaminated water to flow unobstructed over the dike while keeping the contaminant behind the dike.

Dike—Underflow. A dike constructed in a manner that allows uncontaminated water to flow unobstructed under the dike while keeping the contaminant behind the dike.

Dilution. Application of water to water-miscible hazmats to reduce to safe levels the hazard they represent. It can increase the total volume of liquid which will have to be disposed of. In decon applications, it is the use of water to flush a hazmat from protective clothing and equipment, and it is the most common method of decon.

Direct Contact. Direct skin contact with some chemicals, such as corrosives, will immediately damage skin or body tissue upon contact.

Direct-Reading Instruments. Provide information at the time of sampling. They are used to detect and monitor flammable or explosive atmospheres, oxygen deficiency, certain gases and vapors, and ionizing radiation.

Dispersants. The use of certain chemical agents to disperse or break down liquid hazmat spills. The use of dispersants may result in spreading the hazmat over a larger area. Dispersants are often applied to hydrocarbon spills, resulting in oil-in-water emulsions and diluting the hazmat to acceptable levels. Use of dispersants may require prior approval of the appropriate environmental agencies.

Dispersion. To spread, scatter, or diffuse through air, soil, surface, or groundwater.

Diversion. A defensive confinement procedure to intentionally control the movement of a hazardous material into an area where it will pose less harm to the community and the environment.

Divisions. As used in NFPA 70, *The National Electric Code*, describe the types of location that may generate or release a flammable material. There are two divisions:

> **Division I**—Location where the vapors, dusts, or fibers are continuously generated and released. The only element necessary for a hazardous situation is a source of ignition.
>
> **Division II**—Location where the vapors, dusts, or fibers are generated and released as a result of an emergency or a failure in the containment system.

Dome. Circular fixture on the top of a pressurized railroad tank car containing valves, pressure relief valve, and gauging devices.

Dose. The amount of a substance ingested, absorbed, and/or inhaled during an exposure period.

Dose–Response Relationship. Basic principle of toxicology. The intensity of a response elicited by a chemical within a biologic mechanism is a function of the administered dose.

Doublegloving. Involves the use of latex surgical gloves under a work glove. It permits the wearing of the work glove without compromising exposure protection and also provides an additional barrier for hand protection. Doublegloving also reduces the potential for hand contamination when removing protective clothing during decon procedures.

Drums. Cylindrical packagings used for liquids and solids. Constructed of plastic, metal, fiberboard, plywood, or other suitable materials. Typical drum capacities range up to 55 gallons.

E

Element. Pure substance that cannot be broken down into simpler substances by chemical means.

Elevated Temperature Materials. Materials which, when offered for transportation in a bulk container, are (1) liquids at or above 212°F (100°C); (2) Liquids with a flash point at or above 100°F (37.8°C) that are intentionally heated and are transported at or above their flash point, and (3) solids at a temperature at or above 464°F (240°C).

Emergency Breathing Apparatus (EBA). Short duration (e.g., 5–10 minutes) respiratory protection devices developed for use by the general public. Typically consist of a small breathing air cylinder and a clear plastic hood assembly which is placed over the head of the wearer to provide a fresh breathing air supply.

Emergency Broadcast System (EBS). The national emergency notification system that uses commercial AM and FM radio stations for emergency broadcasts. The EBS is usually initiated and controlled by Emergency Management agencies.

Emergency Contact. The telephone number for the shipper or shipper's representative that may be accessed 24 hours a day, 7 days a week in the event of an accident.

Emergency Decontamination. The physical process of immediately reducing contamination of individuals in potentially life-threatening situations without the formal establishment of a decontamination (or contamination reduction) corridor.

Emergency Medical Services (EMS). Functions as required to provide emergency medical care for ill or injured persons by trained providers.

Emergency Operations Center (EOC). The secured site where government or facility officials exercise centralized direction and control in an emergency. The EOC serves as a resource center and coordination point for additional field assistance. It also provides executive directives to and liaison for government and other external representatives, and it considers and mandates protective actions.

Emergency Response. Response to any occurrence which has or could result in a release of a hazardous substance.

Emergency Response Organization. An organization that utilizes personnel trained in emergency response. This would include fire, law enforcement, EMS, and industrial emergency response teams.

Emergency Response Personnel. Personnel assigned to organizations that have the responsibility for responding to different types of emergency situations.

Emergency Response Plan. A plan that establishes guidelines for handling hazmat incidents as required by regulations such as SARA, Title III and HAZWOPER (29 CFR 1910.120).

Emergency Response Planning Guidelines (ERPG-2). The maximum airborne concentration below which it is believed that nearly all individuals could be exposed for up to one hour without experiencing or developing irreversible or serious health effects or symptoms which could impair an individual's ability to take protective action.

Emergency Response Team (ERT). Crews of specially trained personnel used within industrial facilities for the control and mitigation of emergency situations. May consist of both shift personnel with ERT responsibilities as part of their job assignment (e.g., plant operators) or volunteer members. ERT's may be responsible for fire, hazmat, medical, and technical rescue emergencies depending upon the size and operation of the facility.

Emergency Traffic. A priority radio message to be immediately broadcast throughout the emergency scene.

Endothermic. A process or chemical reaction which is accompanied by the absorption of heat.

Engulfing Event. Once the hazmat and/or energy is released, it is free to travel or disperse, engulfing an area. The farther the contents move outward from their source, the greater the level of problems. How quickly they move and how large an area they engulf will depend upon the type of release, the nature of the hazmat, the physical and chemical laws of science, and the environment.

Environmental Protection Agency (EPA). The purpose of the EPA is to protect and enhance our environment today and for future generations to the fullest extent possible under the laws enacted by Congress. The Agency's mission is to control and abate pollution in the areas of water, air, solid waste, pesticides, noise, and radiation. EPA's mandate is to mount an integrated, coordinated attack on environmental pollution in cooperation with state and local governments.

EPA. (See Environmental Protection Agency.) *EPA Levels of Protection.* EPA system for classifying levels of chemical protective clothing.

Level A: Chemical vapor protective suit.

Level B: Chemical liquid splash protective suit with SCBA.

Level C: Chemical liquid splash protective suit with air purifying respirator.

EPA Registration Number. Required for all agricultural chemical products marketed within the United States. It is one of three ways to positively identify an ag chemical. The others are by the product name or chemical ingredient statement. The registration number will appear as a two- or three-section number.

Etiological Harm Events. Those harm events created by uncontrolled exposures to living microorganisms. Diseases commonly associated with etiological harm include hepatitis, typhoid, and tuberculosis. It is often difficult to detect when and where the physical exposure to the etiological agent occurred and the route(s) of exposure.

Evacuation. A public protective option which results in the removal of fixed facility personnel and the public from a threatened area to a safer location. It is typically regarded as the controlled relocation of people from an area of known danger or unacceptable risk to a safer area, or one in which the risk is considered to be acceptable.

Expansion Ratio. The amount of gas produced by the evaporation of one volume of liquid at a given temperature. Significant property when evaluating liquid and vapor releases of liquefied gases and cryogenic materials. The greater the expansion ratio, the more gas that is produced and the larger the hazard area.

Explosion-Proof Construction. Encases the electrical equipment in a rigidly built container so that (1) it withstands the internal explosion of a flammable mixture, and (2) prevents propagation to the surrounding flammable atmosphere. Used in Class I, Division 1 atmospheres at fixed installations.

Exposure. The subjection of a person to a toxic substance or harmful physical agent through any route of entry (e.g., inhalation, ingestion, skin absorption, or direct contact).

Exposures. Items which may be impinged upon by a hazmat release. Examples include people (civilians and emergency responders), property (physical and environmental), and systems disruption.

Exothermic. A process or chemical reaction which is accompanied by the evolution of heat.

Extremely Hazardous Substances (EHS). Chemicals determined by the Environmental Protection Agency to be extremely hazardous to a community during an emergency spill or release as a result of their toxicities and physical/chemical properties (U.S. Environmental protection Agency, 40 CFR 355).

F

Failure of Container Attachments. Attachments which open up or break off the container, such as safety relief valves, frangible disks, fusible plugs, discharge valves, or other related appliances.

Fire Entry Suits. Suits which offer complete, effective protection for short-duration entry into a total flame environment. Designed to withstand exposures to radiant heat levels up to 2,000°F. Entry suits consist of a coat, pants, and separate hood assembly. They are constructed of several layers of flame-retardant materials, with the outer layer often aluminized.

First Responder. The first trained person(s) to arrive at the scene of a hazardous materials incident. May be from the public or private sector of emergency services.

First Responder, Awareness Level. Individuals who are likely to witness or discover a hazardous substance release who have been trained to initiate an emergency response sequence by notifying the proper authorities of the release. They would take no further action beyond notifying the authorities of the release.

First Responder, Operations Level. Individuals who respond to releases or potential releases of hazardous substances as part of the initial response to the site for the purpose of protecting nearby persons, property, or the environment from the effects of the release. They are trained to respond in a defensive fashion without actually trying to stop the release. Their function is to contain the release from a safe distance, keep it from spreading, and prevent exposures.

Flammable (Explosive) Range. The range of gas or vapor concentration (percentage by volume in air) that will burn or explode if an ignition source is present. Limiting concentrations are commonly called the "lower flammable (explosive) limit" and the "upper flammable (explosive) limit." Below the lower flammable limit, the mixture is too lean to burn; above the upper flammable limit, the mixture is too rich to burn. If the gas or vapor is released into an oxygen enriched atmosphere, the flammable range will expand. Likewise, if the gas or vapor is released into an oxygen-deficient atmosphere, the flammable range will contract.

Flaring. Controlled burning of a high vapor pressure liquid or compressed gas in order to reduce or control the pressure and/or dispose of the product.

Flash Point. Minimum temperature at which a liquid gives off enough vapors that will ignite and flashover but will not continue to burn without the addition of more heat. Significant in determining the temperature at which the vapors from a flammable liquid are readily available and may ignite.

Form. Refers to the physical form of a material—solid, liquid, or gas. Significant factor in evaluating both the hazards of a material and tactics for controlling a release. In general, gases and vapor releases cause the greatest problems for emergency responders.

Full Protective Clothing. Protective clothing worn primarily by firefighters which includes helmet, fire-retardant hood, coat, pants, boots, gloves, PASS device, and self-contained breathing apparatus designed for structural fire fighting. It does not provide specialized chemical splash or vapor protection.

Fumes. Airborne dispersion consisting of minute solid particles arising from the heating of a solid material (e.g., lead), indistinction to a gas or vapor. This physical change is often accompanied by a chemical reaction, such as oxidation. Odorous gases and vapors should not be referred to as vapors.

G

Gelation. The process of forming a gel. Gelling agents are used on some hazmat spills to produce a gel that is more easily cleaned up.

Gouge. Reduction in the thickness of the tank shell. It is an indentation in the shell made by a sharp, chisel-like object. A gouge is characterized by the cutting and complete removal of the container or weld material along the track of contact.

Gross Decontamination. The initial phase of the decontamination process during which the amount of surface contaminant is significantly reduced. This phase may include mechanical removing and initial rinsing.

Grounding. A method of controlling ignition hazards from static electricity. The process of connecting one or more conductive objects to the ground; it is done to minimize potential differences between objects and the ground.

Groups. As used in NFPA 70, *The National Electric Code*, are products within a Class. Class I is divided into four groups (Groups A–D) on the basis of similar flammability characteristics. Class II is divided into three groups (Groups E–G). There are no groups for Class III materials.

H

Half-Life. The time it takes for the activity of a radioactive material to decrease to one half of its initial value through radioactive decay.

Halogenated Hydrocarbons. A hydrocarbon with halogen atom (e.g., chlorine, fluorine, bromine, etc.) substituted for a hydrogen atom. They are often more toxic than naturally occurring organic chemicals, and they decompose into smaller, more harmful elements when exposed to high temperatures for a sustained period of time.

Harm Event. Pertains to the harm caused by a hazmat release. Harm events include thermal, radiation, asphyxiation, toxicity, corrosivity, etiologic, and mechanical.

Hazard. Refers to a danger or peril. In hazmat operations, usually refers to the physical or chemical properties of a material.

Hazard Analysis. Part of the planning process, it is the analysis of hazmats present in a facility or community. Elements include hazards identification, vulnerability analysis, risk analysis, and evaluation of emergency response resources. Hazards analysis methods used as part of Process Safety Management (PSM) include HAZOP Studies, Fault Tree Analysis, and What If Analysis.

Hazard and Risk Evaluation. Evaluation of hazard information and the assessment of the relative risks of a hazmat incident. Evaluation process leads to the development of Incident Action Plan.

Hazard Class. The hazard class designation for the material as found in the Department of Transportation regulations, 49 CFR. There are currently 9 DOT hazard classes which are divided into 22 divisions.

Hazard Communication (HAZCOM). OSHA regulation (29 CFR 1910.1200) which requires hazmat manufacturers to develop MSDS's on specific types of hazardous chemicals and provide hazmat health information to both employees and emergency responders.

Hazardous Chemicals. Any chemical that would be a risk to employees if exposed in the workplace (U.S. Occupational Safety and Health Administration, 29 CFR 1910).

Hazardous Materials. Any substance or material in any form or quantity capable of posing an unreasonable risk to safety and health and property when transported in commerce (U.S. Department of Transportation, 40 CFR 171).

Hazardous Materials General Behavior Model (GHMBMO). Process for visualizing hazmat behavior. Applies the concept of events analysis which is simply breaking down the overall incident into smaller, more easily understood parts for purposes of analysis.

Hazardous Materials Response Team (HMRT). An organized group of employees, designated by the employer, who are expected to perform work to handle and control actual or potential leaks or spills of hazardous substances requiring possible close approach to the substance. A Hazmat Team may be a separate component of a fire brigade or a fire department or other appropriately trained and equipped units from public or private agencies.

Hazardous Materials Specialists. Individuals who respond and provide support to Hazardous Materials Technicians. While their duties parallel those of the Technician, they require a more detailed or specific knowledge of the various substances they may be called upon to contain. Would also act as a liaison with federal, state, local, and other governmental authorities in regards to site activities.

Hazardous Materials Technicians. Individuals who respond to releases or potential releases of hazardous materials for the purposes of stopping the leak. They generally assume a more aggressive role in that they are able to approach the point of a release in order to plug, patch, or otherwise stop the release of a hazardous substance.

Hazardous Substances. Any substance designed under the Clean Water Act and the Comprehensive Environmental Response, Compensation, and Liability Act (CERCLA) as posing a threat to waterways and the environment when released (U.S. Environmental Protection Agency, 40 CFR 302). Hazardous substances as used within OSHA 1910.120 refer to every chemical regulated by EPA as a hazardous substance and by DOT as a hazardous material.

Hazardous Waste Manifest. Shipping form required by the EPA and DOT for all modes of transportation when transporting hazardous wastes for treatment, storage, or disposal.

Hazardous Wastes. Discarded materials regulated by the Environmental Protection Agency because of public health and safety concerns. Regulatory authority is granted under the Resource Conservation and Recovery Act (RCRA). (U.S. Environmental Protection Agency, 40 CFR 260–281.)

Hazmat. Acronym used for hazardous materials.

Hazmat Branch. Responsible for all hazmat operations which occur at a hazmat incident. Functions include safety, site control, information, entry, decontamination, hazmat medical, and hazmat resources.

Hazmat Branch Director. Officer responsible for the management and coordination of all functional responsibilities assigned to the Hazmat Branch. Must have a high level of technical knowledge and be knowledgeable of both the strategical and tactical aspects of hazmat response. Reports to the Operations Section Chief.

Hazmat Entry Function. Responsible for all entry and back-up operations within the Hot Zone, including reconnaissance, monitoring, sampling, and mitigation.

Hazmat Decontamination Function. Responsible for the research and development of the decon plan, set-up, and operation of an effective decontamination area capable of handling all potential exposures, including entry personnel, contaminated patients, and equipment.

Hazmat Information Function. Responsible for gathering, compiling, coordinating, and disseminating all data and information relative to the incident. This data and information will be used within the Hazmat Branch for assessing hazard and evaluating risks, evaluating public protective options, selecting the PPE, and developing the incident action plan.

Hazmat Medical Function. Responsible for pre- and post-entry medical monitoring and evaluation of all entry personnel, and provides technical medical guidance to the Hazmat Branch as requested.

Hazmat Resource Function. Responsible for control and tracking of all supplies and equipment used by the Hazmat Branch during the course of an emergency, including documenting the use of all expendable supplies and materials. Coordinates, as necessary, with the Logistics Section Chief.

Hazmat Safety Function. Primarily the responsibility of the Incident Safety Officer and the Hazmat Safety Officer. Responsible for ensuring that safe and accepted practices and procedures are followed throughout the course of the incident. Possesses both the authority and responsibility to stop any unsafe actions and correct unsafe practices.

Hazmat Site Control Function. Establish control zones, establish and monitor access routes at the incident site, and ensure that contaminants are not being spread.

HAZWOPER. Acronym used for the OSHA Hazardous Wastes Operations and Emergency Response regulation (29 CFR 1910.120).

Heat Affected Zone. Area in the undisturbed tank metal next to the actual weld material. This area is less ductile than either the weld or the steel plate due to the effect of the heat of the welding process. This zone is most vulnerable to damage, as cracks are likely to start here.

Heat Cramps. A cramp in the extremities or abdomen caused by the depletion of water and salt in the body. Usually occurs after physical exertion in an extremely hot environment or under conditions that cause profuse sweating and depletion of body fluids and electrolytes.

Heat Exhaustion. A mild form of shock caused when the circulatory system begins to fail as a result of the body's inadequate effort to give off excessive heat.

Heat Rash. An inflammation of the skin resulting from prolonged exposure to heat and humid air and often aggravated by chafing clothing. Heat rash is uncomfortable and decreases the ability of the body to tolerate heat.

Heat Stroke. A severe and sometimes fatal condition resulting from the failure of the temperature regulating capacity of the body. It is caused by exposure to the sun or high temperatures. Reduction or cessation of sweating is an early symptom. Body temperature of 105°F or higher, rapid pulse, hot and dry skin, headache, confusion, unconsciousness, and convulsions may occur. Heat stroke is a true medical emergency requiring immediate transport to a medical facility.

Hematotoxins. A toxin of the blood system (e.g., benzene, chlordane, DDT).

Hepatotoxin. A toxin destructive of the liver (e.g., carbon tetrachloride, vinyl chlorise monomer).

High Temperature Protective Clothing. Protective clothing designed to protect the wearer against short-term high temperature exposures. Includes both proximity suits and fire entry suits. This type of clothing is usually of limited use in dealing with chemical exposures.

HMRT. (See Hazardous Materials Response Team.)

Hot Tapping. An offensive technique for welding on and cutting holes through liquid and/or compressed gas vessels and piping for the purposes of relieving the internal pressure and/or removing the product.

Hot Zone. An area immediately surrounding a hazardous materials incident, which extends far enough to prevent adverse effects from hazardous materials releases to personnel outside the zone. This zone is also referred to as the "exclusion zone," the "red zone," and the "restricted zone" in other documents.

Housing. Fixture on the top of a nonpressurized railroad tank car designed to provide protection for valves, pressure relief valve, and/or gauging devices.

Hydrocarbons. Compounds primarily made up of hydrogen and carbon. Examples include LPG, gasoline, and fuel oils.

Hygroscopic. A substance that has the property of absorbing moisture from the air, such as.

Hypergolic. Two chemical substances that spontaneously ignite upon mixing.

I

Ignition (Autoignition) Temperature. Minimum temperature required to ignite gas or vapor without a spark or flame being present. Significant in evaluating the ease at which a flammable material may ignite.

Immediately Dangerous to Life or Health (IDLH). An atmospheric concentration of any toxic, corrosive or asphyxiant substance that poses an immediate threat to life or would cause irreversible or delayed adverse health effects or would interfere with an individual's ability to escape from a dangerous atmosphere.

Impingement Event. As the hazmat and/or its container engulf an area, they will impinge or come in contact with exposures. They may also impinge upon other hazmat containers, producing additional problems.

Incident. A release or potential release of a hazardous material from its container into the environment.

Incident Action Plan. The strategic goals, tactical objectives, and support requirements for the incident. All incidents require an action plan. For simple incidents (Level I) the action plan is not usually in written form. Large or complex incidents (Level II or III) will require that the action plan be documented in writing.

Incident Commander (IC). The person responsible for the management of all incident operations. The IC is in charge of the incident site. May also be referred to as the On-Scene Incident Commander as defined in 29 CFR 1910.120.

Incident Management System (IMS). An organized system of roles, responsibilities, and standard operating procedures used to manage and direct emergency operations. May also be referred to as Incident Command System (ICS).

Inert Gas. A nonreactive gas, such as argon, helium, and neon.

Ingestion. The introduction of a chemical into the body through the mouth. Inhaled chemicals may be trapped in saliva and swallowed. Exposed personnel should be prohibited from smoking, eating, or drinking except in designated rest and rehab areas after being decontaminated.

Ingredient Statement. The statement on all agricultural chemical labels which breaks down the chemical ingredients by their relative percentages or as pounds per gallon of concentrate. "Active" ingredients are the active chemicals within the mixture. They must be listed by chemical name, and their common name may also be shown. "Inert" ingredients have no ag chem/pesticide activity and are usually not broken into specific components, only total percentage.

Inhalation. The introduction of a chemical or toxic products of combustion into the body by way of the respiratory system. Inhalation is the most common exposure route and often the most damaging. Toxins may be absorbed into the bloodstream and carried to other internal organs, or they may affect the upper and/or lower respiratory tract. Resulting respiratory injuries include pulmonary edema and respiratory congestion.

Inhibitor. Added to products to control their chemical reaction with other products. If the inhibitor is not added or escapes during an incident, the material will begin to polymerize, possibly resulting in container failure.

Inorganic Materials. Compounds derived from other than vegetable or animal sources which lack carbon chains but may contain a carbon atom (e.g., sulfur dioxide—SO^2).

Instability. (See Reactivity.)

Intrinsically Safe Construction. Equipment or wiring is incapable of releasing sufficient electrical energy under both normal and abnormal conditions to cause the ignition of a flammable mixture. Commonly used in portable direct-reading instruments for operations in Class I, Division 2 hazardous locations.

Intermodal Tank Containers. Specific class of portable tanks specifically designed for international intermodal use. Most common types are the IM 101, IM 102, and the DOT Spec. 51 portable tanks.

Ionizing Radiation. Characterized by the ability to create charged particles or ions in anything which it strikes. Exposure to low levels of ionizing radiation can produce short-term or long-term cellular changes with potentially harmful effects, such as cancer or leukemia.

Isolating the Scene. The process of preventing persons and equipment from becoming exposed to a actual or potential hazmat release. Includes establishing isolation perimeter and control zones.

Isolation Perimeter. The designated crowd control line surrounding the Hazard Control Zones. The isolation perimeter is always the line between the general public and the Cold Zone.

J

Jacket. Outer metal covering of a railroad tank car that protects the tank's insulation and keeps it in place.

L

Lab Pack. An overpack drum or disposal container which contains multiple, smaller chemical containers with compatible chemical characteristics. Absorbent materials are usually placed within the overpack container to minimize potential for breakage and/or leakage.

Leak. The uncontrolled release of a hazardous material which could pose a threat to health, safety, and/or the environment.

Leak Control Compounds. Substances used for the plugging and patching of leaks in nonpressure containers (e.g., putty, wooden plugs, etc.).

Leak Control Devices. Tools and equipment used for the plugging and patching of leaks in nonpressure and some low pressure containers, pipes, and tanks (e.g., patch kits, Chlorine kits, etc.).

LEPC. (See Local Emergency Planning Committee.)

Lethal Concentration, 50 Percent Kill (LC-50). Concentration of a material, expressed as parts per million (Ppm) per volume, which kills half of the lab animals in a given length of time. Refers to an inhalation exposure, the LC-50 may also be expressed as mg/liter or mg/cubic meter. Significant in evaluating the toxicity of a material; the lower the value, the more toxic the substance.

Lethal Concentration Low (LC_{Low}). The lowest concentration of a substance in air reported to have caused death in humans or animals. The reported concentrations may be entered for periods of exposure that are less than 24 hours (acute) or greater than 24 hours (subacute and chronic).

Lethal Dose, 50 Percent Kill ($LD\text{-}_{50}$). The amount of a dose which, when administered to lab animals, kills 50% of them. Refers to an oral or dermal exposure and is expressed in terms of mg/kg. Significant in evaluating the toxicity of a material; the lower the value, the more toxic the substance.

Lethal Dose Low (LDLow). The lowest amount of a substance introduced by any route, other than inhalation, reported to have caused death to animals or humans.

Level I Staging. Initial arriving emergency response units go directly to the incident scene taking standard positions (e.g., upwind, uphill as appropriate), assume command, and begin site management operations. The remaining units stage at a safe distance away from the scene until ordered into action by the Incident Commander.

Level II Staging. Used for large, complex, or lengthy hazmat operations. Additional units are staged together in a specific location under the command of a Staging Officer. May be referred to as "Base" within the Firescope System.

Liaison Officer. The point of contact for assisting or coordinating agencies. Member of the Command Staff.

Limited-Use Materials. Protective clothing materials which are used and then discarded. Although they may be reused several times (based upon chemical exposures), they are often disposed of after a single use. Examples include Tyvek™ QC, Tyvek™/Saranex™ 23-P, Barricade™, Kappler CPF™ III and CPF™ IV, Chemrel Max™, and the Lifeguard Responder™.

Liquid Chemical Splash Protective Clothing. The garment portion of a chemical protective clothing ensemble that is designed and configured to protect the wearer against chemical liquid splashes but not against chemical vapors or gases. Liquid splash chemical protective clothing must meet the requirements of NFPA 1992. This type of protective clothing is a component of EPA Level B chemical protection.

Local Emergency Planning Committee (LEPC). A committee appointed by a State Emergency Response Commission, as required by SARA Title III, to formulate a comprehensive emergency plan for its region.

Lower Detection Limit (LDL). The lowest concentration to which a monitoring instrument will respond. The lower the LDL, the quicker contaminant concentrations can be evaluated.

M

Manifest. A shipping document that lists the commodities being transported on a vessel.

Markings. The required names, instructions, cautions, specifications, or combinations thereof found on containers of hazardous materials and hazardous wastes.

Material Safety Data Sheet (MSDS). A document which contains information regarding the chemical composition, physical and chemical properties, health and safety hazards, emergency response, and waste disposal of the material as required by 29 CFR 1910.1200.

Mechanical Harm Events. Those harm events resulting from direct contact with fragments scattered because of a container failure, explosion, or shock wave.

Mechanical Stress. The result of a transfer of energy when one object physically contacts or collides with another. Indicators include punctures, gouges, breaks or tears in the container.

Medical Monitoring. An ongoing, systematic evaluation of individuals at risk of suffering adverse effects of exposure to heat, stress, or hazardous materials as a result of working at a hazmat emergency.

Medical Surveillance. Comprehensive medical program for tracking the overall health of its participants (e.g., HMRT personnel, public safety responders, etc.). Medical surveillance programs consist of pre-employment screening, periodic medical examinations, emergency treatment provisions, nonemergency treatment, and recordkeeping and review.

Melting Point. The temperature at which a solid changes its phase to a liquid. This temperature is also the freezing point depending on the direction of the change. For mixtures, a melting point range may be given. Significant property in evaluating the hazards of a material as well as the integrity of a container (e.g., frozen material may cause its container to fail).

Minimum Detectable Permeation Rate (MDPR). The minimum permeation rate that can be detected by the laboratory analytical system being used for the permeation test.

Miscible. Refers to the tendency or ability of two or more liquids to form a uniform blend or to dissolve in each other. Liquids may be totally miscible, partially miscible, or nonmiscible.

Mitigation. Any offensive or defensive action to contain, control, reduce, or eliminate the harmful effects of a hazardous materials release.

Mixture. Substance made up of two or more compounds, physically mixed together. A mixture may also contain elements and compounds mixed together.

Monitoring. The act of systematically checking to determine contaminant levels and atmospheric conditions.

Monitoring Instruments. Monitoring and detection instruments used to detect the presence and/or concentration of contaminants within an environment. They include:

Combustible Gas Indicator (CGI): Measure the concentration of a combustible gas or vapor in air.

Oxygen Monitor: Measures the percentage of oxygen in air.

Colorimetric Indicator Tubes: Measures the concentration of specific gases and vapors in air.

Specific Chemical Monitors: Designed to detect a large group of chemicals or a specific chemical. Most common examples include carbon monoxide and hydrogen sulfide.

Flame Ionization Detector (FID): A device used to determine the presence of organic vapors and gases in air. Operates in two modes—survey mode and gas chromatograph.

Gas Chromatograph: Instruments used for identifying and analyzing specific organics compounds.

Photoionization Detector (PID): A device used to determine the total concentration of many organic and some inorganic gases and vapors in air.

Radiation Monitors: Instruments used to measure accumulated radiation exposure. Include alpha, beta, and gamma survey detectors.

Instruments which measure the amount of radiation to which a person has been exposed.

Corrosivity (pH) Detector: A meter, paper, or strip that indicates the relative acidity or alkalinity of a substance, generally using an international scale of 0 (acid) through 14 (alkali-caustic). (See pH.)

Indicator Papers: Special chemical indicating papers which test for the presence of specific hazards, such as oxidizers, organic peroxides, and hydrogen sulfide. Are usually part of a hazmat identification system.

MSDS. (See Material Safety Data Sheet.)

Multi-Use Materials. Based upon the chemical exposure, multi-use materials are designed and fabricated to allow for decontamination and re-use. Generally thicker and more durable than limited-use garments, they are used for chemical splash and vapor protective suits, gloves, aprons, boots, and thermal protective clothing. The most common materials include butyl rubber, Viton, polyvinyl chloride (PVC), neoprene rubber, and Teflon™.

Mutagen. A material that creates a change in gene structure which is potentially capable of being transmitted to the offspring.

N

National Contingency Plan (NCP). Outlines the policies and procedures of the federal agency members of the National Oil and Hazardous Materials Response Team (also known as the National Response Team or the NRT).

Provides guidance for emergency responses, remedial actions, enforcement, and funding for federal government response to hazmat incidents.

National Fire Protection Association (NFPA). An international voluntary membership organization to promote improved fire protection and prevention, establish safeguards against loss of life and property by fire, and writes and publish national voluntary consensus standards (e.g., NFPA 472, *Professional Competence of Responders to Hazardous Materials Incidents*).

National Institute for Occupational Safety and Health (NIOSH). A Federal agency which, among other activities, tests and certifies respiratory protective devices and air sampling detector tubes and recommends occupational exposure limits for various substances.

National Interagency Incident Management System (NIIMS). A standardized systems approach to incident management that consists of five major sub-divisions collectively providing a total systems approach to all-risk incident management.

National Response Center (NRC). Communications center operated by the U.S. Coast Guard in Washington, DC. It provides information on suggested technical emergency actions and is the federal spill notification point. The NRC must be notified within 24 hours of any spill of a reportable quantity of a hazardous substance by the spiller. Can be contacted at (800) 424-8802.

National Response Team (NRT). The National Oil and Hazardous Materials Response Team consists of fourteen federal government agencies which carry out the provisions of the National Contingency Plan at the federal level. The NRT is chaired by EPA, while the vice-chairperson represents the U.S. Coast Guard.

National Transportation Safety Board (NTSB). Independent federal agency charged with responsibility for investigating serious accidents and emergencies involving the various modes of transportation (e.g., highway, pipeline, air) as well as hazardous materials. Issues investigation reports and nonbinding recommendations for action.

Nephrotoxins. Toxins which attack the kidneys (e.g., mercury, halogenated hydrocarbons).

Neurotoxins. Toxins which attack the central nervous system (e.g., organophosphate pesticides).

Neutralization. The process of neutralizing a hazmat liquid spill by applying another material to the spill which will react chemically with it to form a less harmful substance. Those materials which can be used to neutralize the effects of a corrosive material (e.g., acids and bases).

Nonbulk Packaging. Any packaging having a capacity meeting one of the following criteria:

- Liquid—internal volume of 119 gallons (450 L) or less;
- Solid—capacity of 882 lbs. (400 kg) or less; and
- Compressed Gas—water capacity of 1,001 lbs. (454 kg) or less.

Nonintervention Tactics. Essentially "no action." It is useful at certain fire emergencies where the potential costs of action far exceed any benefits (e.g., BLEVE scenario).

Nonionizing Radiation. Waves of energy, such as radiant heat, radio waves, and visible light. The amount of energy in these waves is small as compared to ionizing radiation. Examples include infrared waves, microwaves, and lasers.

Normalized Breakthrough Time. A calculation, using actual permeation results, to determine the time at which the permeation rate reaches

0.1 μg/cm^2/min. Normalized breakthrough times are useful for comparing the performance of several different protective clothing materials. Note that in Europe, breakthrough times are normalized at 1.0 μg/cm^2/min, a full order of magnitude less sensitive.

Not Otherwise Specified (NOS). A shipping paper notation which indicates that the material meets the DOT definition for a hazardous material but is not listed by a generic name within the DOT Regulations. The technical name of the material must be entered in parenthesis with the basic description. For example, Flammable Liquid, n.o.s. (contains methanol).

O

Occupational Safety and Health Administration (OSHA). Component of the United States Department of Labor; an agency with safety and health regulatory and enforcement authorities for most United States industries, businesses and states.

Odor Threshold (TLV_{Odor}). The lowest concentration of a material's vapor in air that is detectable by odor. If the TLVodor is below the TLV/TWA, odor may provide a warning as to the presence of a material.

Offensive Tactics. Aggressive leak, spill, and fire control tactics designed to quickly control or mitigate the problem. Although increasing risks to emergency responders, offensive tactics may be justified if rescue operations can be quickly achieved, if the spill can be rapidly confined or contained, or if the fire can be quickly extinguished.

Off-Site Specialist Employee A. Those persons who are specially trained to handle incidents involving chemicals and/or containers for chemicals used in their organization's area of specialization. Consistent with the organization's response plan and standard operating procedures, the Off-Site Specialist Employee A shall have the ability to analyze an incident involving chemicals within the organization's area of specialization, plan a response to that incident, implement the planned response within the capabilities of the resources available, and evaluate the progress of the planned response.

Off-Site Specialist Employee B. Those persons who in the course of their regular job duties, work with or are trained in the hazards of specific chemicals and/or containers for chemicals used in their individual area of specialization. Because of their education, training or work experience, the Off-Site Specialist Employee B may be called upon to gather and record information, provide technical advice, and provide technical assistance (including work within the hot zone) at an incident involving chemicals consistent with their organization's emergency response plan and standard operating procedures and the local emergency response plan.

Off-Site Specialist Employee C. The Off-Site Specialist C should be able to provide information on a specific chemical or container and have the organizational contacts needed to acquire additional technical assistance. This individual need not have the skills or training necessary to conduct control operations. This individual is generally found at the command post providing the IC or his or her designee with technical assistance.

Oil Pollution Act (OPA). Amended the Federal Water Pollution Act, OPA's scope covers both facilities and carriers of oil and related liquid products, including deepwater marine terminals, marine vessels, pipelines, and rail cars. Requirements include the development of emergency response plans, training and exercises, and verification of spill resources and contractor capabilities.

On-Scene Coordinator (OSC). The federal official pre-designated by EPA or the USCG to coordinate and direct federal responses and removals under the National Contingency Plan.

On-Scene Incident Commander. (See Incident Commander.)

Operations Section. Responsible for all tactical operations at the incident. The Hazmat Branch falls within the Operations Section.

Organic Materials. Materials which contain two or more carbon atoms. Organic materials are derived from materials that are living or were once living, such as plants or decayed products. Most organic materials are flammable. Examples include methane (CH_4) and propane (C_3H_8).

Organic Peroxide. Strong oxidizers, often chemically unstable, containing the -o-o- chemical structure. May react explosively to temperature and pressure changes as well as contamination.

Other Regulated Materials D (ORM D). A material, such as a consumer commodity, which presents a limited hazard during transportation due to its form, quantity, or packaging.

Overgarments. Protective clothing ensembles which are worn over chemical vapor protective clothing to provide either additional flash protection or low temperature protection.

Overgloving. The wearing of a second glove over the work glove for additional chemical and/or abrasion protection during entry operations.

Overpack. (1) A packaging used to contain one or more packages for convenience of handling and/or protection of the packages; (2) a term used to describe the placement of damaged or leaking packages in an overpack or recovery drum; (3) the outer packaging for radioactive materials.

Overpacking. Use of a specially constructed drum to overpack damaged or leaking containers of hazardous materials for shipment. Overpack containers should be compatible with the hazards of the materials involved.

Oxidation Ability. The ability of a material to (1) either give up its oxygen molecule to stimulate the oxidation of organic materials (e.g., chlorate, permanganate and nitrate compounds), or (2) receive electrons being transferred from the substance undergoing oxidation (e.g., chlorine and fluorine). Result of either activity is the release of energy.

Oxidizer. A chemical, other than a blasting agent or an explosive, that initiates or promotes combustion in other materials. This action may either cause the material to ignite or release oxygen or other gases, which causes the ignition of other surrounding materials.

Oxygen-Deficient Atmosphere. An atmosphere which contains an oxygen content less than 19.5% by volume at sea level.

P

Packaging. Any container that holds a material (hazardous and nonhazardous). Packaging for hazardous materials includes nonbulk and bulk packaging.

Packing Group. Classification of hazardous materials based on the degree of danger represented by the material. There are three groups: Packing Group I indicates great danger, Packing Group II indicates medium danger, and Packing Group III indicates minor danger.

Patching (Plugging). The use of chemically compatible patches and plugs to reduce or temporarily stop the flow of materials from small holes, rips, tears or gashes in containers.

PCB Contaminated. Any equipment, including transformers, that contains 50 to 500 ppm of PCB's.

Penetration. The flow or movement of a hazardous chemical through closures, seams, porous materials, and pinholes or other imperfections in the material. While liquids are most common, solid materials (e.g., asbestos) can also penetrate through protective clothing materials.

Permeation. The process by which a hazardous chemical moves through a given material on the molecular level. Permeation differs from penetration in that permeation occurs through the clothing material itself rather than through the openings in the clothing material.

Permeation Rate. The rate at which a chemical passes through a given chemical protective clothing material. Expressed as micrograms per square centimeter per minute ($\mu gm/cm_2/min$). For reference purposes, 0.9 $\mu gm/cm^2/min$ is equal to approximately 1 drop/hour.

Permissible Exposure Limit (PEL). The maximum time-weighted concentration at which 95% of exposed, healthy adults suffer no adverse effects over a 40-hour work week and are comparable to ACGIH's TLV/TWA. PEL's are used by OSHA and are based on an eight-hour, time-weighted average concentration.

Personal Protective Equipment (PPE). Equipment provided to shield or isolate a person from the chemical, physical, and thermal hazards that may be encountered at a hazardous materials incident. Adequate personal protective equipment should protect the respiratory system, skin, eyes, face, hands, feet, head, body, and hearing. Personal protective equipment includes: personal protective clothing, self-contained positive pressure breathing apparatus, and air purifying respirators.

pH (Power of Hydrogen). Acidic or basic corrosives are measured to one another by their ability to dissociate in solution. Those that form the greatest number of hydrogen ions are the strongest acids, while those that form the hydroxide ion are the strongest bases. The measurement of the hydrogen ion concentration in solution is called the pH (power of hydrogen) of the compound in solution. The pH scale ranges from zero to 14, with strong acids having low pH values and strong bases or alkaline materials having high pH values. A neutral substance would have a value of 7.

Physical State. The physical state or form (solid, liquid, gas) of the material at normal ambient temperatures (68°F to 77°F).

Planning Section. Responsible for the collection, evaluation, dissemination and use of information about the development of the incident and the status of resources. Includes the Situation Status, Resource Status, Documentation, and Demobilization Units as well as Technical Specialists.

Plume. A vapor, liquid, dust, or gaseous cloud formation which has shape and buoyancy.

Pneumatic Hopper Trailer. Covered hopper trailers that are pneumatically unloaded and used for transporting solids. Have a capacity up to 1,500 cubic feet.

Polymerization. A reaction during which a monomer is induced to polymerize by the addition of a catalyst or other unintentional influences, such as excessive heat, friction, contamination, etc. If the reaction is not controlled, it is possible to have an excessive amount of energy released.

Portable Bin. Portable tanks used to transport bulk solids. Are approximately 4 feet square and 6 feet high, with weights up to 7,700 pounds. Normally loaded through the top and unloaded from the side or bottom.

Portable Tank. Any packaging (except a cylinder having 1,000 lbs. or less water capacity) over 110 gallons capacity and designed primarily to be loaded into, on, or temporarily attached to a transport vehicle or ship and equipped with skids, mountings, or accessories to facilitate handling of the tank by mechanical means.

Post-Emergency Response. That portion of an emergency response performed after the immediate threat of a release has been stabilized or eliminated and the clean-up of the site has begun.

Post-Incident Analysis. An element of incident termination that includes completing the required incident reporting forms, determining the level of financial responsibility, and assembling documentation for conducting a critique.

Process Safety Management (PSM). The application of management principles, methods and practices to prevent and control releases of hazardous chemicals or energy. Focus of both OSHA 1910.119, *Process Safety Management of Highly Hazardous Chemicals, Explosives and Blasting Agents* and EPA Part 68, *Risk Management Programs for Chemical Accidental Release Prevention.*

Product Name. Brand or trade name printed on the front panel of a hazmat container. If the product name includes the term "technical," as in Parathion Technical, it generally indicates a highly concentrated pesticide with 70% to 99% active ingredients.

Proper Shipping Name. The DOT designated name for a commodity or material. Will appear on shipping papers and on some containers. May also be referred to as shipping name.

Protection in-Place. Directing fixed facility personnel and the general public to go inside of a building or a structure and remain indoors until the danger from a hazardous materials release has passed. It may also be referred to as in-place protection, sheltering-in-place, sheltering, and taking refuge.

Protective Clothing. Equipment designed to protect the wearer from heat and/or hazardous materials contacting the skin or eyes. Protective clothing is divided into four types:

- Structural firefighting protective clothing
- Liquid splash chemical protective clothing
- Vapor chemical protective clothing
- High temperature protective clothing

Proximity Suits. Designed for exposures of short duration and close proximity to flame and radiant heat, such as in aircraft rescue firefighting (ARFF) operations. The outer shell is a highly reflective, aluminized fabric over an inner shell of a flame-retardant fabric such as Kevlar™ or Kevlar™/PBI™ blends. These ensembles are not designed to offer any substantial chemical protection.

Public Information Officer. The individual responsible for interface with the media or other appropriate agencies requiring information direct from the incident scene. Member of the Command Staff.

Public Protective Actions. The strategy used by the Incident Commander to protect unexposed people from the hazardous materials release by evacuating or protecting-in-place. This strategy is usually implemented after the IC has established an isolation perimeter and defined the Hazard Control Zones for emergency responders.

Purging. Totally enclosed electrical equipment is protected with an inert gas under a slight positive pressure from a reliable source. The inert gas provides positive pressure within the enclosure and minimizes the development of a flammable atmosphere. Used in Class I, Division 1 atmospheres at fixed installations.

Pyrophoric Materials. Materials that ignite spontaneously in air without an ignition source.

R

Radiation Harm Events. Those harm events related to the emission of radioactive energy. There are two types of radiation—ionizing and nonionizing.

Radioactivity. The ability of a material to emit any form of radioactive energy.

Rail Burn. Deformation in the shell of a railroad tank car. It is actually a long dent with a gouge at the bottom of the inward dent. A rail burn can be oriented circumferentially or longitudinally in relation to the tank shell. The longitudinal rail burns are the more serious because they have a tendency to cross a weld. A rail burn is generally caused by the tank car passing over a stationary object, such as a wheel flange or rail.

Reactivity/Instability. The ability of a material to undergo a chemical reaction with the release of energy. It could be initiated by mixing or reacting with other materials, application of heat, physical shock, etc.

Recommended Exposure Levels (REL). The maximum time-weighted concentration at which 95% of exposed, healthy adults suffer no adverse effects over a 40-hour work week and are comparable to ACGIH's TLV/TWA. REL's are used by NIOSH and are based upon a 10-hour, time-weighted average concentration.

Regional Response Team (RRT). Established within each federal region, the RRT follows the policy and program direction established by the NRT to ensure planning and coordination of both emergency preparedness and response activities. Members include EPA, USCG, state government, local government, and Indian tribal governments.

Rehabilitation (Rehab). Process of providing for EMS support, treatment, and monitoring, food and fluid replenishment, mental rest and relief from extreme environmental conditions associated with a hazmat incident. May function as either a sector or group within the Incident Management System.

Release Event. Once a container is breached, the hazmat is free to escape (be released) in the form of energy, matter, or a combination of both. Types of release include detonation, violent rupture, rapid relief, and spills or leaks.

Reportable Quantity (RQ). The designated amount of a hazardous substance that, if spilled or released, requires immediate notification to the National Response Center (NRC).

Reporting Marks and Number. The set of initials and a number stenciled on both sides and both ends of railroad cars. These markings can be used to obtain information on the contents of the car from either the railroad or the shipper.

Residue. The material remaining in a package after its contents have been emptied and before the packaging is refilled, cleaned, or purged of vapor to remove any potential hazard.

Resource Conservation and Recovery Act (RCRA). Law which establishes the regulatory framework for the proper management and disposal of all hazardous wastes, including treatment, storage, and disposal facilities. It also establishes installation, leak prevention, and notification requirements for underground storage tanks.

Respiratory Protection. Equipment designed to protect the wearer from the inhalation of contaminants. Respiratory protection includes positive-pressure self-contained breathing apparatus (SCBA), positive-pressure airline respirators (SAR's), and air purifying respirators.

Respiratory Toxins. Toxins which attack the respiratory system (e.g., asbestos, hydrogen sulfide).

Response. That portion of incident management in which personnel are involved in controlling (offensively or defensively) a hazmat incident. The activities in the response portion of a hazmat incident include analyzing the incident, planning the response, implementing the planned response, and evaluating progress.

Responsible Party (RP). A legally recognized entity (e.g., person, corporation, business or partnership, etc.) that has a legally recognized status of financial accountability and liability for actions necessary to abate and mitigate adverse environmental and human health and safety impacts resulting from a nonpermitted release or discharge of a hazardous material. The person or agency found legally accountable for the clean-up of an incident.

Retention. A defensive spill confinement method. Temporary containment of a hazmat in an area where it can be absorbed, neutralized, or picked up for proper disposal. Retention tactics are intended to be more permanent and may require resources such as portable basins, bladders, or other special material.

Risks. The probability of suffering a harm or loss. Risks are variable and change with every incident.

Risk Analysis. A process to analyze the probability that harm may occur to life, property, and the environment and to note the risks to be taken to identify the incident objectives.

Risk Management Programs. Required under EPA's proposed 40 CFR Part 68, risk management programs consist of three elements: (1) hazard assessment of the facility; (2) prevention program; and (3) emergency response considerations.

Roentgen. A measure of the charge produced in air created by ionizing radiation, usually in reference to gamma radiation.

Roentgen Equivalent Man (REM). The unit of dose equivalent; takes into account the effectiveness of different types of radiation.

Runaway Cracking. Cracking occurring in closed containers under pressure, such as liquid drums or pressure vessels. A small crack in a closed container suddenly develops into a rapidly growing crack which circles the container. As a result, the container will generally break into two or more pieces.

S

Safety Officer. Responsible for monitoring and assessing safety hazards and unsafe conditions and developing measures for ensuring personnel safety.

Member of the Command Staff. The Safety Officer is a required position at a hazmat incident based upon the requirements of OSHA 1910.120.

Sampling. The process of collecting a representative amount of a gas, liquid, or solid for evidence or analytical purposes.

Sampling Kit. Kits assembled for the purpose of providing adequate tools and equipment for taking samples and documenting unknowns to create a "chain of evidence."

Sanitary Sewer. A "closed" sewer system which carries wastewater from individual homes, together with minor quantities of stormwater, surface water, and groundwater that are not admitted intentionally. May also collect wastewater from industrial and commercial businesses. The collection and pumping system will transport the wastewater to a treatment plant, where the wastewater is processed.

SAR. (See Supplied Air Respirator.)

SARA. (See Superfund Amendments & Reauthorization Act.)

Saturated Hydrocarbons. A hydrocarbon possessing only single covalent bonds. All of the carbon atoms are saturated with hydrogen. Examples include methane (CH_4), propane (C_3H_8), and butane (C_4H_{10}).

SCBA. (See Self-Contained Breathing Apparatus.)

Scene. The location impacted or potentially impacted by a hazard.

Score. Reduction in the thickness of the container shell. It is an indentation in the shell made by a relatively blunt object. A score is characterized by the reduction of the container or weld material so that the metal is pushed aside along the track of contact with the blunt object.

Secondary Contamination. The process by which a contaminant is carried out of the hot zone and contaminates people, animals, the environment, or equipment outside of the hot zone.

Section. That organization level within the Incident Management System having functional responsibility for primary segments of incident operations, such as Operations, Planning, Logistics and Administration/Finance. The Section level is organizationally between Branch and the Incident Commander.

Sector. Either a geographic or functional assignment. Sector may take the place of either the Division or Group or both.

Self-Accelerating Decomposition Temperature (SADT). Property of organic peroxides. When this temperature is reached by some portion of the mass of the organic peroxide, irreversible decomposition will begin.

Self-Contained Breathing Apparatus (SCBA). A positive pressure, self-contained breathing apparatus (SCBA) or combination SCBA/supplied air breathing apparatus certified by the National Institute for Occupational Safety and Health (NIOSH) and the Mine Safety and Health Administration (MSHA) or the appropriate approval agency for use in atmospheres that are immediately dangerous to life or health (IDLH).

Sensitizer. A chemical that causes a substantial proportion of exposed people or animals to develop an allergic reaction in normal tissue after repeated exposure to the chemical. Skin sensitization is the most common form, while respiratory sensitization to a few chemicals is also known to occur.

SERC. (See State Emergency Response Commission.)

Shipper. A person, company, or agency offering material for transportation.

Shipping Documents/Papers. Generic term used to refer to documents that must accompany all shipments of goods for transportation. These include Hazardous Waste Manifest, Bill of Lading, and Consists, etc. Shipping documents should provide the following:

- Proper shipping name.
- Hazard classification.
- Four-digit identification number(s), as required.
- Number of packages or containers.

- Type of packages.
- Total quantity by weight, volume, and/or packaging.

Shipping Name. The proper shipping name or other common name for the material; also any synonyms for the material.

Single Trip Container (STC). Container that may not be refilled or reshipped with a DOT-regulated material except under certain conditions.

Site Management and Control. The management and control of the physical site of a hazmat incident. Includes initially establishing command, approach, and positioning, staging, establishing initial perimeter and hazard control zones, and implementing public protective actions.

Size-Up. The rapid yet deliberate consideration of all critical scene factors.

Skilled Support Personnel. Personnel who are skilled in the operation of certain equipment, such as cranes and hoisting equipment, and who are needed temporarily to perform immediate emergency support work that cannot reasonably be performed in a timely fashion by emergency response personnel.

Skin Absorption. The introduction of a chemical or agent into the body through the skin. Skin absorption can occur with no sensation to the skin itself. Do not rely on pain or irritation as a warning sign of absorption. Skin absorption is enhanced by abrasions, cuts, heat, and moisture. The rate of skin absorption can vary depending upon the body part that's exposed.

Slurry. Pourable mixture of a solid and a liquid.

Sludge. Solid, semi-solid, or liquid waste generated from a municipal, commercial, or industrial waste treatment plant or air pollution control facility, exclusive of treated effluent from a wastewater treatment plant.

Solubility. The ability of a solid, liquid, gas, or vapor to dissolve in water or other specified medium. The ability of one material to blend uniformly with another, as in a solid in liquid, liquid in liquid, gas in liquid, or gas in gas. Significant property in evaluating the selection of control and extinguishing agents, including the use of water and firefighting foams.

Solution. Mixture in which all of the ingredients are completely dissolved. Solutions are composed of a solvent (water or another liquid) and a dissolved substance (known as the solute).

Specialist Employee. Employees who, in the course of their regular job duties, work with and are trained in the hazards of specific hazardous substances and who will be called upon to provide technical advice or assistance to the Incident Commander at a hazmat incident.

Specific Gravity. The weight of the material as compared with the weight of an equal volume of water. If the specific gravity is less than one, the material is lighter than water and will float. If the specific gravity is greater than one, the material is heavier than water and will sink. Most insoluble hydrocarbons are lighter than water and will float on the surface. Significant property for determining spill control and clean-up procedures for water-borne releases.

Specification Marking. Found in various locations on railroad tank cars, intermodal portable tanks, and cargo tank trucks; it indicates the standards to which the container was built.

Spill. The release of a liquid, powder, or solid hazardous material in a manner that poses a threat to air, water, ground, and the environment.

Stabilization. The point in an incident at which the adverse behavior of the hazardous materials is controlled.

Staging. The management of committed and uncommitted emergency response resources (personnel and apparatus) to provide orderly deployment. See Level I Staging and Level II Staging.

Staging Area. The safe area established for temporary location of available resources closer to the incident site to reduce response time.

Standard of Care. The minimum accepted level of hazmat service to be provided as may be set forth by law, current regulations, consensus standards, local protocols and practice, and what has been accepted in the past (precedent).

Standard Transportation Commodity Code (STCC). A number which will be found on all shipping documents accompanying rail shipments of hazmats. A seven-digit number assigned to a specific material or group of materials and used in determination of rates. For a hazardous material, the STCC number will begin with the digits "49." Hazardous wastes may also be found with the first two digits being "48." This code will also be found when intermodal containers are changed from rail to highway movement.

State Emergency Response Commission. Formed under SARA, Title III, the SERC is responsible for developing and maintaining the statewide hazmat emergency response plan. This includes ensuring that planning and training are taking place throughout the state as well as providing assistance to local governments and LEPC's, as appropriate.

Statement of Practical Treatment. Located near the signal word on the front panel of an agricultural chemical or poison label, it is also referred to as the "First Aid Statement" or "Note to Physician." It may have precautionary information as well as emergency procedures. Antidote and treatment information may also be added.

Storm Sewer. An "open" system which collects stormwater, surface water, and groundwater from throughout an area but excludes domestic wastewater and industrial wastes. A storm sewer may dump runoff directly into a retention area which is normally dry or into a stream, river, or waterway without treatment.

Strategic Goals. The overall plan that will be used to control an incident. Strategic goals are broad in nature and are achieved by the completion of tactical objectives.

Street Burn. Deformation in the shell of a highway cargo tank. It is actually a long dent that is inherently flat. A street burn is generally caused by a container overturning and sliding some distance along a cement or asphalt road.

Strength. The degree to which a corrosive ionizes in water. Those that form the greatest number of hydrogen ions are the strongest acids (e.g., $pH < 2$), while those that form the hydroxide ion are the strongest bases ($pH > 12$).

Stress Event. An applied force or system of forces that tend to either strain or deform a container (external action) or trigger a change in the condition of the contents (internal action). Types of stress include thermal, mechanical, and chemical.

Structural Firefighting Protective Clothing. Protective clothing normally worn by firefighters during structural fire fighting operations. It includes a helmet, coat, pants, boots, gloves, PASS device, and a hood to cover parts of the head not protected by the helmet. Structural firefighting clothing provides limited protection from heat but may not provide adequate protection from harmful liquids, gases, vapors, or dusts encountered during hazmat incidents. May also be referred to as turnout or bunker clothing.

Sublimation. The ability of a substance to change from the solid to the vapor phase without passing through the liquid phase. An increase in temperature can increase the rate of sublimation. Significant in evaluating the flammability or toxicity of any released materials which sublime. The opposite of sublimation is deposition (changes from vapor to solid).

Subsidiary Hazard Class. Indicates a hazard of a material other than the primary hazard assigned.

Superfund Amendments & Reauthorization Act (SARA). Created for the purpose of establishing federal statutes for right-to-know standards and emergency response to hazardous materials incidents. Re-authorized the federal Superfund program and mandated states to implement equivalent regulations/requirements.

Supplied Air Respirator (SAR). Positive pressure respirator which is supplied by either an airline hose or breathing air cylinders connected to the respirator by a short airline (or pigtail). When used in IDLH atmospheres, require a secondary source of air supply.

Synergistic Effect. The combined effect of two or more chemicals which is greater than the sum of the effect of each agent alone.

System Detection Limit (SDL). The minimum amount of chemical breakthrough that can be detected by the laboratory analytical system being used for the permeation test. Lower SDL's result in lower (or earlier) breakthrough times.

Systemic. Pertaining to the internal organs and structures of the human body.

T

Tactical Objectives. The specific operations that must be accomplished to achieve strategic goals. Tactical objectives must be both specific and measurable. Tactical level officers are Division/Group/Sector.

Technical Information Specialists. Individuals who provide specific expertise to the Incident Commander or the HMRT either in person, by telephone, or through other electronic means. They may represent the shipper, manufacturer or be otherwise familiar with the hazmats or problems involved.

Technical Name. Identifies the recognized chemical name currently used in scientific and technical handbooks, journals, and texts.

Teratogen. A material that affects the offspring when the embryo or fetus is exposed to that material.

Termination. That portion of incident management in which personnel are involved in documenting safety procedures, site operations, hazards faced, and lessons learned from the incident. Termination is divided into three phases: debriefing, post-incident analysis, and critique.

Thermal Harm Events. Those harm events related to exposure to temperature extremes.

Thermal Stress. Hazmat container stress generally indicated by temperature extremes, both hot and cold. Examples include fire, sparks, friction or electricity, and ambient temperature changes. Extreme or intense cold, such as that found with cryogenic materials, may also act as a stressor. clues of thermal stress include the operation of safety relief devices or the bulging of containers.

Threshold. The point at which a physiological or toxicological effect begins to be produced by the smallest degree of stimulation.

Threshold Limit Value/Ceiling (TLV/C). The maximum concentration that should not be exceeded, even instantaneously. The lower the value, the more toxic the substance.

Threshold Limit Value/Short-term Exposure Limit (TLV/STEL). The 15-minute, time-weighted average exposure which should not be exceeded at any time nor repeated more than four times daily with a 60-minute rest period required between each STEL exposure. The lower the value, the more toxic the substance.

Threshold Limit Value/Skin (Skin). Indicates a possible and significant contribution to overall exposure to a material by absorption through the skin, mucous membranes, and eyes by direct or airborne contact.

Threshold Limit Value/Time Weighted Average (TLV/TWA). The airborne concentration of a material to which an average, healthy person may be exposed repeatedly for 8 hours each day, 40 hours per week, without suffering adverse effects. The young, old, ill, and naturally susceptible will have lower tolerances and will need to take additional precautions. TLV's are based upon current available information and are adjusted on an annual basis by organizations such as the American Conference of Governmental Industrial Hygienists (ACGIH). As TLV's are time weighted averages over an 8-hour exposure, they are difficult to correlate to emergency response operations. The lower the value, the more toxic the substance.

Threshold Planning Quantity (TPQ). The quantity designated for each extremely hazardous substance (EHS) that triggers a required notification from a facility to the State Emergency Response Commission (SERC) and the Local Emergency Planning Committee (LEPC) that the facility is subject to reporting under SARA Title III.

TOFC. (See trailer-on-flat-car.)

Toxic Products of Combustion. The toxic byproducts of the combustion process. Depending upon the materials burning, higher levels of personal protective clothing and equipment may be required.

Toxicity. The ability of a substance to cause injury to a biologic tissue. Refers to the ability of a chemical to harm the body once contact has occurred.

Toxicity Harm Events. Those harm events related to exposure to toxins. Examples include neurotoxins, nephrotoxins, and hepatotoxins.

Toxicity Signal Words. The signal word found on product labels of poisons and agricultural chemicals which indicates the relative degree of acute toxicity. Located in the center of the front label panel, it is one of the most important label markings. The three toxicity signal words and categories are DANGER (high), WARNING (medium), and CAUTION (low).

Toxicology. The study of chemical or physical agents that produce adverse responses in the biologic systems with which they interact.

Trailer-on-Flat-Car (TOFC). Truck trailers which are shipped on a railroad flat cars.

Transfer. The process of physically moving a liquid, gas, or some forms of solids either manually, by pump, or by pressure transfer from a leaking or damaged container. The transfer pump, hoses, fittings, and container must be compatible with the hazardous materials involved. When transferring flammable liquids, proper bonding and grounding concerns must be addressed.

Transportation Index (TI). The number found on radioactive labels which indicates the maximum radiation level (measured in milli-roentgens/hour—mR/hr) at 1 meter from an undamaged package. For example, a TI of 3 would indicate that the radiation intensity that can be measured is no more than 3 mR/hr at 1 meter from the labeled package.

U

Unified Command. The process of determining overall incident strategies and tactical objectives by having all agencies, organizations, or individuals who have jurisdictional responsibility, and in some cases those who have functional responsibility at the incident, participate in the decision-making process.

UN/NA Identification Number. The four-digit identification number assigned to a hazardous material by the Department of Transportation; on shipping documents may be found with the prefix "UN" (United Nations) or "NA" (North American). The ID numbers are not unique, and more than one material may have the same ID number.

Unsaturated Hydrocarbons. A hydrocarbon with at least one multiple bond between two carbon atoms somewhere in the molecule. Generally, unsaturated hydrocarbons are more active chemically than saturated hydrocarbons, and are considered more hazardous. May also be referred to as the alkenes and alkynes. Examples include ethylene (C_2H_4), butadiene (C_4H_6), and acetylene (C_2H_2).

V

Vacuuming. Use of vacuums for picking up hazmat releases (e.g., mercury, asbestos). The method of vacuuming will depend upon the nature of the hazmat.

Vapor. An air dispersion of molecules in a substance that is normally a liquid or solid at standard temperature and pressure.

Vapor Density. The weight of a pure vapor or gas compared with the weight of an equal volume of dry air at the same temperature and pressure. The molecular weight of air is 29. If the vapor density of a gas is less than one, the material is lighter than air and may rise. If the vapor density is greater than one, the material is heavier than air and will collect in low or enclosed areas. Significant property for evaluating exposures and where hazmat gas and vapor will travel.

Vapor Dispersion. Use of water spray to disperse or move vapors away from certain areas or materials. Note that reducing the concentration of a material through the use of a water spray may bring the material into its flammable range.

Vapor Pressure. The pressure exerted by the vapor within the container against the sides of a container. This pressure is temperature dependent; as the temperature increases, so does the vapor pressure. Consider the following three points:

1) The vapor pressure of a substance at 100°F is always higher than the vapor pressure at 68°F.
2) Vapor pressures reported in millimeters of mercury (mm Hg) are usually very low pressures. 760 mm Hg is equivalent to 14.7 psi or 1 atmosphere. Materials with vapor pressures greater than 760 mm Hg are usually found as gases.
3) The lower the boiling point of a liquid, the greater vapor pressure at a given temperature.

Vapor Suppression. Offensive control techniques used to mitigate the evolution of flammable, corrosive or toxic vapors and reduce the surface area exposed to the atmosphere. Includes the use of firefighting foams and hazmat vapor suppressants.

Vent and Burn. The use of shaped explosive charges to vent the high pressure at the the top of a pressurized container and then, with additional explosive charges, release and burn the remaining liquid in the container in a controlled fashion. This is a highly sophisticated technique that is only used under very controlled conditions.

Venting. The controlled release of a liquid or compressed gas to reduce the pressure and diminish the probability of an explosion. The method of venting will depend upon the nature of the hazmat.

Violent Rupture. Associated with chemical reactions having a release rate of less than one second (i.e., deflagration). There is no time to react in this scenario. This behavior is commonly associated with runaway cracking and overpressure of closed containers.

Viscosity. Measurement of the thickness of a liquid and its ability to flow. High viscosity liquids, such as heavy oils, must first be heated to increase their fluidity. Low viscosity liquids spread more easily and increase the size of the hazard area.

Volatility. The ease with which a liquid or solid can pass into the vapor state. The higher a material's volatility, the greater its rate of evaporation. Significant property in that volatile materials will readily disperse and increase the hazard area.

W

Warm Zone. The area where personnel and equipment decontamination and hot zone support takes place. It includes control points for the access corridor and thus assists in reducing the spread of contamination. This is also referred to as the "decontamination," "contamination reduction," "yellow zone," "support zone," or "limited access zone" in other documents.

Water Reactivity. Ability of a material to react with water and release a flammable gas or present a health hazard.

Waybill. A railroad shipping document describing the materials being transported. Indicates the shipped, consignee, routing, and weights. Used by the railroad for internal records and control, especially when the shipment is in transit.

Wheel Burn. Reduction in the thickness of a railroad tank shell. It is similar to a score but is caused by prolonged contact with a turning rail-car wheel.

STREET SMARTS

There is an old saying..."It takes about 20 years to get 20 years experience." Some things just can't be learned from books.

A quick scan of this text for terms like *"Remember"*, *"Never"*, *"Always"*, and *"Must"* turned up some observations which reflect our experiences over the years. They are a good review of the important points that are not necessarily tied into the objectives listed at the beginning of each chapter.

CHAPTER 1
THE HAZARDOUS MATERIALS MANAGEMENT SYSTEM

- Remember, there is no single agency (public or private) which can effectively manage a major hazmat emergency alone.
- The purpose of OSHA 1910.120 and NFPA 471 is to outline the minimum *(not the maximum)* requirements that should be considered when dealing with responses to hazmat incidents.
- "Standard of care" can be defined as the minimum accepted level of hazmat service to be provided as may be set forth by law, current regulations, consensus standards, local protocols and practice, and what has been accepted in the past (precedent).

CHAPTER 2
HEALTH AND SAFETY

- Safety must be more than a policy or procedure...it is both an attitude and a responsibility that must be shared by all responders.
- Safety is both an attitude and a behavior. Safety MUST be an inherent part of all operations from the development of SOP's to the selection and purchase of PPE. The operating philosophy of every emergency response organization should be, "If we cannot do this safely, then we will not do it at all.
- Protective clothing is not your first line of defense, but is your last line of defense.
- Final accountability always rest with the Incident Commander.
- Remember the basics—upwind, uphill, and always have an escape route and pre-designated withdrawal signal.
- Remember that a material does not have to be a gas in order to be inhaled—solid materials may generate fumes or dusts in a dry powdered

form, while liquid chemicals will generate vapors, mists or aerosols which can be inhaled.

- Remember—dose makes the poison!
- Remember that there are some chemicals that cannot be excreted and will start to accumulate within the body. These include PCB's (body fat) and hydrogen fluoride (bone).
- REMEMBER—THE LOWER THE REPORTED CONCENTRATION, THE MORE TOXIC THE MATERIAL.
- Exposure values should be regarded only as guidelines, not absolute boundaries between safe and dangerous conditions. ....To be most effective, exposure values must be combined with monitoring instrument readings and interpreted by response personnel who are familiar with their proper application and limitations.
- When dealing with a potential inhalation hazard, response personnel must use positive pressure self-contained breathing apparatus (SCBA) until the Incident Commander determines through the use of air monitoring that a decreased level of respiratory protection will not result in a hazardous exposure.
- There is nothing wrong with taking a risk. However, always remember that there are good risks and bad risks—if there is much to be gained, then perhaps much can be risked. Of course, if there is little to be gained, then little should be risked.

CHAPTER 3
THE INCIDENT MANAGEMENT SYSTEM

- Remember, the ability to mount a safe and effective response builds upon what is accomplished during planning and preparedness activities.
- Remember—the Incident Safety Officer is responsible for the safety of all personnel operating at the emergency, while the Hazmat Safety Officer is responsible for all operations within the Hazmat Branch and within the Hot and Warm Zones. This includes having the authority to stop or prevent unsafe actions and procedures during the course of the incident.
- Command should never be transferred or passed to an officer who is not on the scene.

CHAPTER 4
THE POLITICS OF HAZMAT INCIDENT MANAGEMENT

- Never say never, particularly when dealing with a long-term, campaign operation. History is full of incidents where tactical options which appeared to be totally unrealistic during the first hour, eventually looked real good and were implemented during the twentieth hour.
- Never put the media area at the command post. Think Hollywood—perceptions are reality. Visual shots and pictures of emergency responders in action help to "sell" and "market" the capabilities of emergency response agencies.

- Never lie to the media, speculate or give personal opinions. Avoid "spinning" the story.
- Never provide comments and information to the media which are "off the record."
- Remember the "Twelve Second Rule"—A TV reporter will look for a sound bite or about 12 seconds worth of interview in response. The first statement should be a concise, positive summation of the incident and the facility or community's response to the incident.
- Talk to be understood. Always remember who your audience is. Avoid technical jargon.
- Remember—consult, document, and do not assume!
- Remember, initial observations often underestimate the significance of a problem. Command should always be prepared to implement an alternative action plan if the current plan fails.
- Remember the difference between "street smarts" and "technical smarts"—one doesn't need a degree in chemistry to know when something just doesn't look right.
- Having an alphabet behind one's name does not automatically mean that an individual necessarily understands the world of emergency response and operations in a field setting.

CHAPTER 5
SITE MANAGEMENT AND CONTROL

- An experienced commander only gives up the advantage of a stationary command post when it is absolutely necessary for the IC to personally provide one-on-one direction to ERP operating in forward positions. In either case, the IC must maintain a command presence on the radio.
- The IC must appreciate that the decision to evacuate initiates a process that is expensive and difficult to call off once it is started. Before initiating any evacuation, the IC should consider whether the population at risk can be protected just as well in place, within the confines of the structure where they are presently located.
- Don't look stupid because you didn't have a plan.

CHAPTER 6
IDENTIFY THE PROBLEM

- Be practical and realistic—the routine establishes the foundation upon which the non-routine must build. If you don't "have your act together" managing high probability flammable liquid and gas emergencies, it's unlikely that you'll perform very well trying to manage an incident involving a more exotic hazardous material.
- Responders must recognize that hazmat exposures are no longer limited to industrial or transportation emergencies. Not every emergency is a hazmat emergency; however, responders have the potential of being exposed to hazmats at ANY emergency.

- Senses are not a primary identification tool. In most cases, if you are close enough to smell, feel, or hear the problem, you are probably too close to operate safely.

CHAPTER 7
HAZARD AND RISK EVALUATION

- Risks can't be determined from books or pulled from computerized data bases—they are those intangibles that are different at every hazmat incident and must be evaluated by a knowledgeable Incident Commander.
- The objective of ERP operations is to minimize the level of risk to ERP, the community, and the environment. Hazmat responders must see their role as RISK EVALUATORS, not RISK TAKERS!
- Remember that an oxygen deficient atmosphere will shorten the flammable range, while an oxygen-enriched atmosphere will expand the flammable range.
- Remember that groundwater supplies can become contaminated by concentrations as small as 200 parts per billion (ppb). Preventing spills and releases from entering the soil is a critical element in many areas.
- Remember—your job is to be a risk evaluator, not a risk taker. Bad risk takers get buried; effective risk evaluators come home.

CHAPTER 8
PERSONAL PROTECTIVE CLOTHING AND EQUIPMENT

- Remember that the longer the breakthrough time, the better the level of protection.
- Remember- structural firefighting clothing is not designed for chemical protection.

CHAPTER 9
INFORMATION MANAGEMENT AND RESOURCE COORDINATION

- Remember—bad news doesn't get better with time. If there's a problem, the earlier you know about it, the sooner you can start to fix it!
- Remember—a computer's outstanding attributes of speed, storage capacity, consistency, and the ability to process complex logical instructions are of no value unless they are applied within a good management process.
- Don't allow external resources to "free-lance" or do the "end-run".
- Don't build your system around one individual! The same case can be made for every piece of equipment in your inventory.
- The most simple and reliable method of coordinating information between the various Hazmat Branch functions is to use the checklist system. Checklists don't forget, people do.

- Experience shows that equipment which works well in an office environment often won't hold up in the emergency response field. Junk yards are full of equipment that was rattled to pieces by the smooth ride of a 10-ton hazmat squad. Dust, cold weather, Mongo the Hazmat Tech, and a little spilled coffee on the key board often equals a dead computer.

CHAPTER 10
IMPLEMENTING RESPONSE OBJECTIVES

- Never extinguish a pressure-fed flammable gas fire. Isolate the source of the gas and permit the fire to self-extinguish, thereby consuming any residual gas inside the vessel or piping system.
- NEVER TOUCH OR HANDLE ANYTHING in a drug lab until the area has been evaluated and cleared by Bomb Squad personnel who have the proper training in identifying booby traps.
- Never enter a confined space without positive pressure SCBA or an approved Supplied Air Respirator with an emergency escape bottle, and never remove your facepiece to "revive the victim". If you remove your facepiece in a confined space in a contaminated atmosphere you are going to die.
- Remember—without an oxygen supply to the brain, clinical death occurs in three minutes, and biological death occurs in five minutes. Exposure to hazardous atmospheres accelerates the time line.
- What will happen if I do nothing? Remember—this is the baseline for hazmat decision-making and should be the element against which all strategies and tactics are compared.
- Remember, cooling water is a valuable resource, don't waste it.
- Be careful of applying water onto open floating roofs—sinking the roof with water lines can be somewhat embarrassing, as well as hazardous!
- Identifying and prioritizing exposures. Remember that the highest priority should be given to flame impingement on vessels, piping, and critical support structures.
- Remember—there is very fine dividing line between explosives, oxidizers and organic peroxides. All are capable of releasing tremendous amounts of energy!
- Remember that "outcomes" are measured in terms of fatalities, injuries, property and environmental damage, and systems disruption.
- Remember, all hazmat incidents follow a natural time line. Leaks and spills usually get worse all by themselves before the situation gets better.
- Remember—surprises are nice on your birthday, but not on the emergency scene.
- This isn't brain surgery. Always remember the KISS Principle (Keep it Simple Stupid)—most leak control tactics are pretty simple.

CHAPTER 11
DECONTAMINATION

- Remember that permeation can occur with any porous material, not just PPE.
- Remember the basics. The more time, distance, and shielding between you and the radioactive the material, the lower the risk will be.
- Degradation chemicals should never be applied directly to the skin.
- Never transport a patient who is grossly contaminated. Toxic, flammable, or corrosive contaminants can accumulate in poorly ventilated vehicles and further expose the patient as well as the EMS team.

CHAPTER 12
TERMINATING THE INCIDENT

- Although every organization has a tendency to develop its own critique style, never use a critique to assign blame (public meetings are the worst time to discipline personnel).
- Organizations must balance the potential negatives against the benefits that are gained through the critique process. Remember – the reason for doing the critique in the first place is to improve your operations.

NOTES